ANNUAL REVIEW OF NUCLEAR SCIENCE

ANNUAL REVIEW OF NUCLEAR SCIENCE

EMILIO SEGRÈ, *Editor*
University of California, Berkeley

J. ROBB GROVER, *Associate Editor*
Brookhaven National Laboratory

H. PIERRE NOYES, *Associate Editor*
Stanford University

VOLUME 25

1975

ANNUAL REVIEWS INC.   4139 EL CAMINO WAY   PALO ALTO, CALIFORNIA 94306

ANNUAL REVIEWS INC.
Palo Alto, California, USA

International Standard Book Number: 0-8243-1525-1
Library of Congress Catalog Card Number: 53-995

REPRINTS

The conspicuous number aligned in the margin with the title of each
article in this volume is a key for use in ordering reprints. Available
reprints are priced at the uniform rate of $1 each postpaid. Effective
January 1, 1975 the minimum acceptable reprint order is 10 reprints
and/or $10.00, prepaid. A quantity discount is available.

PRINTED AND BOUND IN THE UNITED STATES OF AMERICA

CONTENTS

ANNUAL REVIEWS INC. is a nonprofit corporation established to promote the advancement of the sciences. Beginning in 1932 with the *Annual Review of Biochemistry,* the Company has pursued as its principal function the publication of high quality, reasonably priced Annual Review volumes. The volumes are organized by Editors and Editorial Committees who invite qualified authors to contribute critical articles reviewing significant developments within each major discipline.

Annual Reviews Inc. is administered by a Board of Directors whose members serve without compensation.

Annual Reviews are published in the following sciences: Anthropology, Astronomy and Astrophysics, Biochemistry, Biophysics and Bioengineering, Earth and Planetary Sciences, Ecology and Systematics, Entomology, Fluid Mechanics, Genetics, Materials Science, Medicine, Microbiology, Nuclear Science, Pharmacology, Physical Chemistry, Physiology, Phytopathology, Plant Physiology, Psychology, and Sociology. The *Annual Review of Energy* will begin publication in 1976. In addition, two special volumes have been published by Annual Reviews Inc.: *History of Entomology* (1973) and *The Excitement and Fascination of Science* (1965).

KNOCK-OUT PROCESSES AND REMOVAL ENERGIES

×5557

A. E. L. Dieperink and T. de Forest Jr.

Institute for Nuclear Physics Research (IKO), Oosterringdijk 18,
Amsterdam, The Netherlands

CONTENTS

1 INTRODUCTION

The independent-particle shell model (IPSM) plays a very important role in the description of nuclei. This simple model that provides the basis of nearly all microscopic nuclear theories has proven to give an excellent first approximation to the structure. The single-particle energies ε_α and wave functions ϕ_α which specify this model are thus of particular interest. The motivation for high-energy single-nucleon knock-out experiments (1–3) such as (p, 2p) and (e, e'p) is that they provide a very direct means of measuring these single-particle aspects of nuclear structure. In particular, these reactions can explore the structure of the deeply bound shells, to which most other processes are insensitive. The shell structure of

1

the inner orbitals has been observed for the first time in the pioneering (p, 2p) experiments by Tyrén et al (4).

The cross section for such processes (2, 3, 5) is basically proportional to a spectral function $P(\mathbf{k}, E)$, which is the combined probability that one can remove a nucleon with momentum $\mathbf{k}$ from the target nucleus, leaving the final nucleus with excitation energy E (with respect to the ground state energy of the target nucleus). These quantities are determined by the momentum and energy balance of the external particles. In the extreme IPSM, for example, $P(\mathbf{k}, E)$ has the simple form

$$P(\mathbf{k}, E) = \sum_{\alpha < F} |\phi_\alpha(\mathbf{k})|^2 \delta(E + \varepsilon_\alpha), \qquad\qquad 1.1$$

which allows one to directly extract the single-particle energies ε_α and momentum distributions $|\phi_\alpha(\mathbf{k})|^2$ for the occupied states ($\alpha < F$).

In more realistic situations the form of the spectral function is more complicated, for the hole state is not in general an eigenstate, and thus the hole strength is fragmented over several states. Also correlations must be included in the wave function of the target-nucleus. This allows knock-out from states above the Fermi surface, and gives rise to a corresponding reduction of strength for the normally occupied states. In this paper we discuss these effects, which represent deviations from the pure IPSM, in some detail. In particular we review the problem of defining removal energies and extracting them from the spectral function, and the interpretation of these quantities in terms of the single-particle energies of specific theories.

We note here that there are many other direct reactions of the single-nucleon removal type [pickup, $(\pi^\pm, N)$, etc] whose cross sections, in the simplest approximation, are also proportional to the spectral function. The advantage of high-energy knock-out processes is twofold. First, one can completely map out the spectral function; in other processes $\mathbf{k}$ and E tend to be correlated or their range is restricted. Second, the main corrections to the basic knock-out process, at least in (e, e′p), tend to have a simple form, mainly an overall reduction of the cross section for each shell, resulting from the absorption of the knocked-out protons, which is rather easy to take into account. In pickup reactions like (d, ^{3}He), by contrast, the absorption inside the nucleus is very strong, so that this region is not explored, and therefore the $\mathbf{k}$ dependence of the spectral function cannot really be obtained. This absorption furthermore prevents one from observing the deeply bound levels [$E \gtrsim 30$ MeV (3)]. In addition the analysis of the latter is hampered by the contributions from two-step processes. On the other hand such reactions do provide valuable complementary information, because the orbital angular momentum of the removed nucleon can be determined more reliably than in knock-out processes (3), and because of the better energy resolution.

The interest in knock-out processes has been renewed by recent $(e, e'p)$ experiments with incident electron energies in the range 500–2500 MeV, performed at Saclay (6, 7), Tokyo (8, 9), DESY (10), and Kharkov (11). In the energy spectra single-particle strength has been observed in several light nuclei up to missing energy $E \sim 80$ MeV. The best energy and momentum resolution has been obtained at Saclay ($\Delta E \sim 1.5$ MeV, $\Delta k \sim 6$ MeV/c).

The use of knock-out processes to study structure information is not restricted to nuclear physics. In fact an analysis very similar to the one described here has been used for $(e, 2e)$ reactions in atomic physics (12) to study electron orbitals and energies in atoms and molecules.

2   CROSS SECTION FOR KNOCK-OUT REACTIONS

In this section we review the derivation of the cross section for knock-out reactions, treating $(e, e'N)$ as an example. For simplicity, spin and isospin labels are suppressed in the following. In the impulse approximation (IA), treating the electrons as plane waves, the cross section for the $(e, e'p)$ process [the same formalism is applicable to $(e, e'n)$] shown in Figure 1 takes a particularly simple, intuitive form (2, 5, 13–15):

$$d^4\sigma/d\varepsilon_2\, d\varepsilon_p\, d\Omega_2\, d\Omega_p = K \sum_f |\hat{T}_{ep}\, M_{fi}|^2$$
$$\times \delta[\varepsilon_1 - \varepsilon_2 - \varepsilon_p - E_R - (E^f_{A-1} - E_A)]. \qquad 2.1$$

Here ε_p (**p**) is the energy (momentum) of the knocked-out proton, ε_1 (**k**$_1$) and ε_2 (**k**$_2$) those of the incident and scattered electrons, $\mathbf{q} = \mathbf{k}_1 - \mathbf{k}_2$

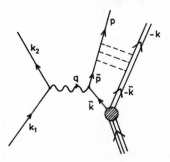

Figure 1   Diagram of the $(e, e'p)$ reaction. The final state interactions which give rise to the "distorted momentum" $\tilde{\mathbf{p}} = \mathbf{q} + \tilde{\mathbf{k}}$ are indicated by the dashed lines.

E_A is the (binding) energy of the target nucleus, E_{A-1}^f that of the residual $A-1$ nucleus in state f, and $E_R = (\mathbf{q}-\mathbf{p})^2/[2(A-1)m]$ is the recoil energy of the residual nucleus; $\hat{T}_{\mathrm{ep}}$ represents the (off-shell) electron-proton scattering amplitude.[1] The nuclear matrix element M_{fi}, which contains the nuclear structure information, is given by

$$M_{fi}(\mathbf{p}, \mathbf{q}) = (A)^{1/2} \langle \psi_{A-1}^f(\mathbf{r}_1' \ldots \mathbf{r}_{A-1}') \chi_{\mathbf{p}'}(\mathbf{r}_A') \exp{(i\mathbf{q} \cdot \mathbf{R}_A)} |$$

$$\times \exp{(i\mathbf{q} \cdot \mathbf{r}_A)} | \psi_A(\mathbf{r}_1' \ldots \mathbf{r}_A') \rangle$$

$$= \int d\mathbf{r}_A' \chi_{\mathbf{p}'}^*(\mathbf{r}_A') \exp{\left(i \frac{A-1}{A} \mathbf{q} \cdot \mathbf{r}_A' \right)} \Phi_{fi}(\mathbf{r}_A'), \qquad 2.2$$

where

$$\Phi_{fi}(\mathbf{r}_A') = (A)^{1/2} \langle \psi_{A-1}^f(\mathbf{r}_1' \ldots \mathbf{r}_{A-1}') | \psi_A(\mathbf{r}_1' \ldots \mathbf{r}_A') \rangle \qquad 2.3$$

represents the so-called overlap integral. The relative coordinates $\mathbf{r}_i' = \mathbf{r}_i - \mathbf{R}_{A-1}$ are defined with respect to the center-of-mass of the residual nucleus, $\mathbf{R}_{A-1} \equiv 1/(A-1) \sum_i^{A-1} \mathbf{r}_i$, and $\mathbf{R}_A \equiv 1/A \sum_i^A \mathbf{r}_i$; $\chi_{\mathbf{p}'}(\mathbf{r}_A')$ represents the wave function of the outgoing proton with asymptotic momentum (with respect to $\mathbf{r}_A'$) $\mathbf{p}' = \mathbf{p} - \mathbf{q}/A$.

In the plane wave impulse approximation (PWIA), that is with the further assumption that the outgoing proton can be represented by a plane wave

$$\chi_{\mathbf{p}'}(\mathbf{r}_A') = (2\pi)^{-3/2} \exp{(i\mathbf{p}' \cdot \mathbf{r}_A')}, \qquad 2.4$$

equation 2.2 reduces to

$$M_{fi}(\mathbf{p}, \mathbf{q}) = (2\pi)^{-3/2} \int d\mathbf{r}_A' \exp{\left[-i(\mathbf{p}-\mathbf{q}) \cdot \mathbf{r}_A' \right]} \Phi_{fi}(\mathbf{r}_A') \equiv \Phi_{fi}(\mathbf{p}-\mathbf{q}), \qquad 2.5$$

i.e. the Fourier transform of the overlap integral. Furthermore in PWIA, $\hat{T}_{\mathrm{ep}}$ is determined by the observed asymptotic momenta and thus can be factorized from M_{fi}. One thus obtains

$$d^4\sigma/d\varepsilon_2 \, d\varepsilon_{\mathrm{p}} \, d\Omega_2 \, d\Omega_{\mathrm{p}} = d\sigma/d\Omega|_{\mathrm{ep}}(\varepsilon_{\mathrm{p}}+m) p P(\mathbf{k}, E), \qquad 2.6$$

where $d\sigma/d\Omega|_{\mathrm{ep}}$ is an off-shell electron-proton cross section. The proton spectral function $P(\mathbf{k}, E)$, which depends only on the net (intrinsic) energy, $E = \varepsilon_1 - \varepsilon_2 - \varepsilon_{\mathrm{p}} - E_R$, and recoil momentum, $-\mathbf{k}(\mathbf{k} = \mathbf{p} - \mathbf{q})$, transferred to the residual $A-1$ nucleus, is defined as (3, 5)

$$P(\mathbf{k}, E) = \sum_f |\Phi_{fi}(\mathbf{k})|^2 \, \delta[E - (E_{A-1}^f - E_A)]$$

$$= \langle \psi_A | a_{\mathbf{k}}^+ \, \delta(E + E_R - H + E_A) a_{\mathbf{k}} | \psi_A \rangle. \qquad 2.7$$

[1] Because the structure of the electron-proton (in contrast e.g. to the proton-proton) interaction is quite well known, so is the off-shell scattering amplitude, $\hat{T}_{\mathrm{ep}}$. The operator notation is used to indicate that $\hat{T}_{\mathrm{ep}}$ depends not only on the electron kinematics but also on the initial momentum of the proton $\tilde{\mathbf{k}}$.

where a_k^+ and a_k represent proton creation and annihilation operators. It contains the desired nuclear structure information and therefore is the central physical quantity that is discussed in the succeeding sections.

While the PWIA results most clearly illustrate the physics of knock-out reactions, they represent an oversimplification of the actual situation. In particular high-energy protons are rather strongly absorbed by the nucleus and thus χ cannot be represented by the plane wave 2.4. Given an appropriate optical potential, however, the distorted wave functions χ can be found. Fortunately, calculations tend to indicate (2) that this correction results mainly in an overall reduction (which depends, however, on the final state f) of the $M_{fi}(\mathbf{k})$ and thus the physics of the PWIA results remain largely intact. There is, however, some ambiguity in the choice of the optical potential due to nonlocality effects (16) and to the fact that in general the residual nucleus is in an excited state (and thus can decay before the knocked-out proton has left the nucleus).

In addition there are corrections to the basic IA: for example, as a result of processes in which the observed proton is not the nucleon that interacted with the electron, or in which, because of multiple-scattering, the proton loses energy in leaving the nucleus. The effect of these physical processes is partially contained in the absorptive part of the optical potential, but this does not describe the "background" caused by the reemission of the proton in other channels (15, 17). This contribution appears to be rather small. Finally there are corrections associated with the scattering of the electron: (a) distortion of the electron wave functions which are generally small for high-energy electrons (18); and (b) radiative corrections (19), which are rather small and in any case readily calculable. The situation for (p, 2p) is similar, but because here the effects of distortions on the projectile must be taken into account, a much more complicated and therefore more uncertain analysis is involved.

We note here that the kinematic invariance of the PWIA-result—that is, that because the spectral function depends only on $\mathbf{k}$ and E the same result can be obtained under a variety of kinematical conditions—offers an experimental method of investigating these corrections which in general do not satisfy this invariance.

In the formalism presented above, translational invariance of the nuclear wave functions has implicitly been assumed (20). Thus it cannot be directly applied to most nuclear theories, which violate this invariance. The corresponding results for the latter can be obtained by formally letting $A \to \infty$: thus $\mathbf{p}_i' \to \mathbf{p}_i$, $\mathbf{r}_i' \to \mathbf{r}_i$, etc. Equation 2.7 is, however, valid for both formalisms. In the following we do not discuss this point further except in section 6 where we consider some explicit effects resulting from center-of-mass corrections.

3   STRUCTURE OF THE SPECTRAL FUNCTION

As shown in the previous section, the basic nuclear structure information
that one is probing in knock-out reactions is contained in the spectral
function. In this section we consider the structure of the spectral function
—that is, what information it contains and how it can be extracted.

To begin with, before considering more realistic situations, we briefly
review the spectral function for the IPSM for three reasons: 1. This
model has provided the basic motivation for doing knock-out experiments. 2. Most knock-out experiments to date have been analyzed in
terms of this model. 3. Most microscopic theories for finite nuclei are
based on this model.

3.1   Independent-Particle Shell Model

Due to the basic assumption of independent particle motion in the IPSM,
the hole state created by single-nucleon knock-out is by definition an
eigenstate. One thus directly finds that the overlap integrals Φ_{fi} (equation
2.3) are equal to the single-particle wave functions for the occupied
states $\phi_\alpha(\mathbf{r}) = R_a(r) Y_\alpha(\hat{\mathbf{r}})$, where we used the notation $\alpha = \{a, m_\alpha\} = \{n_\alpha, l_\alpha, j_\alpha, m_\alpha\}$; therefore the spectral function (equation 2.7) takes on the
form

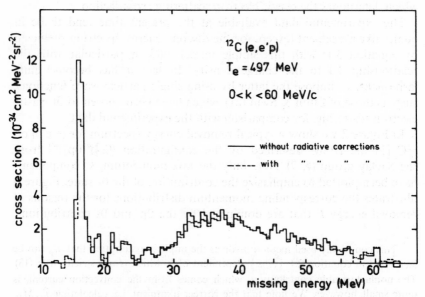

Figure 2   Missing energy spectrum for the (e, e'p) reaction on ^{12}C (6).

$$P(\mathbf{k}, E) = \sum_{\alpha < F} P_\alpha(E) |\phi_\alpha(\mathbf{k})|^2 \qquad\qquad 3.1$$

with $P_\alpha(E) = \delta(E + \varepsilon_a)$. Assuming, as we do from now on, that the target nucleus is a closed shell nucleus, equation 3.1 reduces to

$$P(\mathbf{k}, E) = 1/4\pi \sum_{\alpha < F} N_a |R_a(k)|^2 \delta(E + \varepsilon_a), \qquad\qquad 3.2$$

where

$$R_a(k) = (2/\pi)^{1/2} \int dr\, r^2 j_{l_a}(kr) R_a(r), \qquad\qquad 3.3$$

and $N_a = 2j_a + 1$ is the number of protons in the shell a; $j_l(kr)$ denotes a spherical Bessel function. Within the scope of this model it is thus possible (assuming PWIA) to directly measure the single-particle energies, $\varepsilon_a (\equiv \varepsilon_\alpha)$, and the momentum distributions, $|R_a(k)|^2$.

In order to take into account the effects of the optical potential on the outgoing proton, the distorted momentum distributions, obtained by replacing the ϕ_α in equation 3.1 by M_{fi} (equation 2.2) calculated with the optical model wave functions χ, can be used[2] (13).

The consistency of this model can be checked by computing the charge density distribution

$$\rho(r) = 1/4\pi \sum_{a < F} N_a |R_a(r)|^2, \qquad\qquad 3.4$$

which determines the elastic electron scattering cross section.

The experimental data available at the present time tend to be in qualitative agreement (except for the discrete energy spectrum predicted by equation 3.1) with this simple model, and in particular with the relationship 3.4 to the charge density. In fact it has become quite fashionable to impose the latter by using single-particle wave functions, such as those of Elton & Swift (21), which have been chosen to fit elastic electron scattering, for comparison with the experimental data.

In Figure 2 we show a typical removal energy spectrum for (e, e'p) on ^{12}C [in the IPSM described by the configuration $(0s\tfrac{1}{2})^4 (0p\tfrac{3}{2})^8$] from the Saclay group (6, 7). Here only the low momentum, k, components have been plotted to emphasize the contribution of the 0s state. Figure 3 illustrates the corresponding momentum distributions for the regions of removal energy E that are dominated by the 0p and 0s contributions.

---

[2] This commonly used method involves the implicit assumption that $\hat{T}_{ep}$ can be factorized in equation 2.1, i.e. is independent of $\mathbf{k}$, which is not strictly true (13). The nonfactorizable contribution, which comes from the convection current, is quite small, however. We note that the correct treatment, i.e. calculating $\hat{T}_{ep} M_{fi}$, is no more difficult than calculating M_{fi} itself.

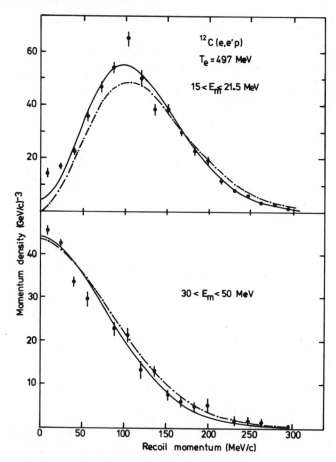

Figure 3 Momentum distributions for the (e, e'p) reaction on ^{12}C in two different regions of removal energy, E_m, corresponding to knock-out from the 0p and 0s shells (6). The calculated results using DWIA (PWIA) are given by the solid (dot-dashed) curves and have been normalized to the experimental data.

The agreement with the theoretical curves obtained with slightly modified Elton-Swift-type wave functions (21) is excellent, considering that no adjustable parameters (except for normalization) have been used.

3.2   Generalized Shell Model

Although the IPSM works quite well it does not, nor is it designed to, explain the widths of the peaks in Figure 2. These are due to the residual interaction that prevents the simple hole state from being an eigenstate, thus leading to a fragmentation or spreading of the hole strength. In addition there are other effects that occur when one looks at the situation more realistically, such as the reduction of strength below the Fermi surface due to ground state correlations. In order to interpret these effects, and especially to relate them to the results of more sophisticated (than the IPSM) theories, we now consider the general form of the spectral function. We do, however, for reasons both of clarity and practicality, follow the spirit of the IPSM picture as closely as possible.

The spectral function can be expanded in a complete set of orthonormal orbitals

$$P(\mathbf{k}, E) = \sum_{\alpha\beta} P_{\alpha\beta}(E)\phi_\alpha^*(\mathbf{k})\phi_\beta(\mathbf{k}), \qquad\qquad 3.5$$

where

$$P_{\alpha\beta}(E) = P_{\alpha\beta}(E)\delta_{l_\alpha l_\beta}\delta_{j_\alpha j_\beta} = \langle\psi_A|a_\alpha^+\delta(E-H+E_A)a_\beta|\psi_A\rangle. \qquad 3.6$$

It is convenient to choose the orbitals (which in principle are arbitrary) so as to diagonalize the $P_{\alpha\beta}(E)$ as much as possible, especially for small E where the IPSM effects are seen. They thus will correspond quite closely to the single-particle wave functions of microscopic theories. One then has a situation (3) similar to that for the IPSM

$$P(\mathbf{k}, E) = \sum_\alpha P_\alpha(E)|\phi_\alpha(\mathbf{k})|^2. \qquad\qquad 3.7$$

The spectral function $P_\alpha(E)$ $[\equiv P_{\alpha\alpha}(E)]$, which describes the distribution of single-particle strength, can be obtained by comparing the momentum distribution to the predictions of "reasonable" single-particle wave functions. Provided one restricts the basis of states (e.g. those below and just above the Fermi surface) one can separate the overlapping contributions which begin to occur as one goes beyond the 0p shell nuclei.

In Figure 4 some examples of the energy distributions $P_\alpha(E)$ analyzed in this way are shown for ^{28}Si [(e, e'p) (7)] and ^{40}Ca [(e, e'p) (7) and (p, 2p)(22)]. Note the large splitting of the 0p strength, and the observation of strength in the normally empty ls state in ^{28}Si. The bump in the 0s state component observed in the (p, 2p) reaction on ^{40}Ca probably contains a large background contribution (22).

It should be kept in mind that the $P_\alpha(E)$ obtained in this way depends on some rather specific assumptions, based on the concept of the IPSM. We note in particular that the set $\{\phi_\alpha \phi_\beta\}$ forms an overcomplete basis for the determination of $P(\mathbf{k}, E)$. It is thus in principle not possible to extract $P_{\alpha\beta}(E)$ from equation 3.5. This problem is alleviated by the assumption that only a restricted number of orbitals contribute for a given E. In practice these difficulties manifest themselves in the fact that as more and more orbitals are included it becomes very difficult (if not impossible)

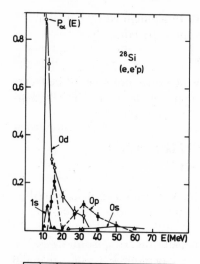

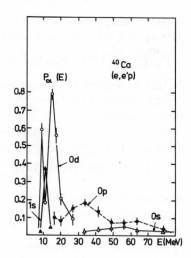

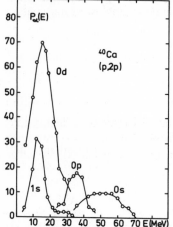

Figure 4 Spectral functions for the reactions ^{28}Si(e, e'p) (*upper left*), ^{40}Ca(e, e'p)(*upper right*) and ^{40}Ca(p, 2p) (*lower left*) obtained by the orbital analysis (equation 3.7).

to distinguish between the various possible combinations of contributions; furthermore the results depend strongly on the assumed form of the single-particle wave functions. For these reasons it is also impossible in general to establish any direct connection between the spectral function and the charge density

$$\rho(\mathbf{r}) = \sum_{\alpha\beta} \int dE \, P_{\alpha\beta}(E)\phi_\alpha^*(\mathbf{r})\phi_\beta(\mathbf{r}). \qquad 3.8$$

The latter, however, does give a constraint on the $P_{\alpha\beta}(E)$ and thus can be used to check the validity of the analysis.

We note in passing that it would be extremely useful to have knowledge of the off-diagonal spectral function $P(\mathbf{k}, \mathbf{k}', E)$ (defined by equation 2.7 with $a_{\mathbf{k}} \rightarrow a_{\mathbf{k}'}$), for then one could obtain the $P_{\alpha\beta}(E)$ directly

$$P_{\alpha\beta}(E) = \int d\mathbf{k} \int d\mathbf{k}' \, P(\mathbf{k}, \mathbf{k}', E)\phi_\alpha(\mathbf{k})\phi_\beta^*(\mathbf{k}'). \qquad 3.9$$

In view of the lack of this information, which does not seem to be forthcoming in the near future, we will confine ourselves to the standard type of analysis.

4 OCCUPATION PROBABILITIES AND MEAN REMOVAL ENERGIES

Assuming that the orbital expansion of the spectral function as described in the previous section is performed, we now consider the question of how one can relate the energy dependence of the spectral function $P_\alpha(E)$ to the results of microscopic nuclear structure theories. To describe the energy spectrum of the $A-1$ nucleus one can exploit the idea that the most relevant physical information is contained in the lowest moments.

4.1 Definitions

The energy integral or zeroth moment of the spectral function defines the correlation matrix

$$\rho_{\alpha\beta} = \int dE \, P_{\alpha\beta}(E) = \langle \psi_A | a_\alpha^+ a_\beta | \psi_A \rangle. \qquad 4.1$$

The diagonal matrix elements $\rho_{\alpha\alpha} = \rho_\alpha$ are usually referred to as occupation probabilities (3). They are related to the spectroscopic strength $N_a = \sum_{m_\alpha} \rho_\alpha = \sum_f S_a^f$ determined in transfer reactions, where the spectroscopic factors S_a^f are given by

$$S_a^f = \sum_{m_\alpha} |\langle \psi_{A-1}^f | a_\alpha | \psi_A \rangle|^2. \qquad 4.2$$

It is worth noting that in principle one can choose a single-particle basis that diagonalizes $\rho_{\alpha\beta}$, the so-called natural orbital basis (23). Because this presumes knowledge of the structure of the true wave function ψ_A it is not a convenient choice for practical purposes.

The first moment of the spectral function defines the mean removal energy (mre) (24–26):

$$\varepsilon_a = 1/\rho_\alpha \int dE(-E)P_\alpha(E) = -1/\rho_\alpha \langle \psi_A | [a_\alpha^+, H]a_\alpha | \psi_A \rangle. \qquad 4.3$$

More explicitly, the mre represents the energy-weighted sum over spectroscopic factors S_a^f

$$\varepsilon_a = 1/N_a \sum_f S_a^f (E_A - E_{A-1}^f), \qquad 4.4$$

which is the same as the centroid energy that is often used in pick-up reactions (3). Because it involves an averaging over final states, the mre depends only on the physics of the A-nucleus and thus is theoretically a simple quantity. It has received much attention (2, 24–27) recently as it seems to provide a useful meeting ground between experiment and theory.

Some typical results for the mre and occupation probabilities obtained in several experiments are given in Table 1. It should be kept in mind that there are basic uncertainties involved in the analysis, for example because of possible contributions from outside the region that is explored experimentally, or because of the problem of resolving contributions from overlapping orbitals. The latter, for example, may be the cause of the relatively large value for $|\varepsilon_{0p}|$, obtained in the Saclay experiment (7) for ^{40}Ca, which in Figure 4 is seen to be due to the strength in the region of the 0s state. In most cases the results for the mre are reasonably consistent with each other; on the other hand, there are rather large differences in the observed occupation probabilities, presumably due to differences in the treatment of the distortion effects.

4.2   Sum Rules

Before we turn to a review of the detailed information contained in the spectral function some sum rules that can be obtained from quite general considerations are discussed. First, one has the well-known model-independent spectroscopic sum rule

$$\int d\mathbf{k} \int dE\, P(\mathbf{k}, E) = \sum_{\alpha} \rho_{\alpha} = Z. \qquad 4.5$$

In knock-out processes a violation of equation 4.5 can be ascribed to the fact that some of the single-particle strength can be located outside the experimentally accessible region of energy and momentum transfer, and to uncertainties in the corrections to PWIA [absorption, $(e, e'2p)$ processes, etc]. Assuming that the former is small (or calculable), an accurate measurement of this sum rule could give a valuable check on the method used to make such corrections. This, however, would require rather accurate absolute cross sections.

More interesting is the energy-weighted sum rule (25). If the nuclear Hamiltonian $H = \hat{T} + \hat{V}$ contains no more than two-body interactions, one can construct the following operator identity (28):

$$-\int d\mathbf{k}[a_{\mathbf{k}}^{+}, H]a_{\mathbf{k}} = \hat{T} + 2\hat{V}. \qquad 4.6$$

By taking the expectation value with respect to ψ_A and using equation 2.7 one obtains the sum rule:

Table 1 Experimental proton occupation numbers N_a and removal energies ε_a (MeV)[a]

Nucleus	Orbit	Saclay[b]		Tokyo[c]		Liverpool[d]
		N_a	$|\varepsilon_a|$	N_a	$|\varepsilon_a|$	$|\varepsilon_a|$
^{12}C	0s	1.1	38.1 ± 0.8	1.74	37.5 ± 0.3	38
	0p	2.6	17.5 ± 0.4	2.84	15.3 ± 0.1	17
^{40}Ca	0s	1.5	56	3.7	58.4 ± 1.1	48.5
	0p	5.7	42	10.2	35.3 ± 0.5	35
	0d5/2 }	7.7	14.9 ± 0.8	4.7	18.4 ± 1.6 }	16
	0d3/2 }			4.2	10.4 ± 1.4 }	
	1s1/2	1.3	11.2 ± 0.3	2.0	13.6 ± 0.4	12

[a] For a discussion of the assigned uncertainties the reader is referred to the cited references.
[b] (e, e'p) at 500 MeV, DWIA (6, 7).
[c] (e, e'p) at 700 MeV, DWIA (9).
[d] (p, 2p) at 385 MeV, PWIA (22).

$$E_A = \tfrac{1}{2} \int d\mathbf{k} \int dE \, P_A(\mathbf{k}, E)(k^2/2m - E_{\text{Lab}}), \qquad\qquad 4.7$$

where P_A is the sum of the proton and neutron spectral functions. In applying equation 4.7 it should be noted that the energy variable E_{Lab} is the energy transfer in the laboratory frame,

$$E_{\text{Lab}} = \varepsilon_1 - \varepsilon_2 - \varepsilon_{\text{p}} = E^f_{A-1} - E_A + E_{\text{R}},$$

where $E_{\text{R}} = k^2/2(A-1)m$, is the recoil energy of the $A-1$ nucleus. In practice it is customary, however, to use as energy parameter the missing mass $E = \varepsilon_1 - \varepsilon_2 - \varepsilon_{\text{p}} - E_{\text{R}} = E^f_{A-1} - E_A$, in terms of which the sum rule reads (6, 20)

$$E_A = \tfrac{1}{2} \int d\mathbf{k} \int dE \, P_A(\mathbf{k}, E)\left(\frac{A-2}{A-1}\frac{k^2}{2m} - E\right). \qquad\qquad 4.8$$

Equivalently, in a shell model basis, equation 4.8 reads

$$E_A = \tfrac{1}{2}\left\{[(A-2)/(A-1)]\langle \psi_A|T|\psi_A\rangle + \sum_\alpha \rho_\alpha \varepsilon_\alpha\right\}. \qquad\qquad 4.9$$

Keeping only the proton spectral function in the right-hand side of equation 4.7, and dividing by the normalization integral $\int d\mathbf{k} \int dE \, P(\mathbf{k}, E)$ to remove the uncertainties in the absolute cross section, the binding energy per proton, E_Z/Z, can be compared with that obtained from nuclear masses (with the appropriate Coulomb corrections).

This sum rule was first applied by Koltun (25) to (p, 2p) reactions for a number of nuclei without corrections for distortions and recoil effects, with good agreement. In a more recent analysis of the $^{12}C(e, e'p)$ reaction (6) (including these corrections) a large disagreement was found: 4.0 MeV binding energy per proton from equation 4.8 compared to 6.9 MeV from mass tables. Similar deviations were found (7) for heavier nuclei.

Physically the most interesting explanation for the discrepancy would be the presence of true three-body forces in the nuclear Hamiltonian, in which case the relation 4.6 and thus the sum rule 4.7 no longer hold. Before such a conclusion can be drawn one must be certain that the right-hand side of equation 4.8 has been correctly extracted from experiment. We note, for example, that in the analysis mentioned above no corrections were made for multiple scattering. However, because such processes correspond to events in which protons have lost energy, these corrections would tend to increase the discrepancy.

Another possible explanation is that in the construction of the sum rule some of the single-particle strength has been missed, because it occurs either with a strength too weak and/or structureless to be separated from the background, or above the experimental cut-off. This is suggested particularly by the fact that, although one expects contributions due to short-range two-body correlations from states above the Fermi surface, so far only knock-out from normally occupied states has been identified for most nuclei.

A reliable theoretical estimate of the behavior of $P(\mathbf{k}, E)$ for large k and E, and therefore of the contribution of high-energy states to $\langle T \rangle$ and $\langle E \rangle$, requires a detailed calculation of the dynamics of the correlations (28). However, a crude estimate of the magnitude of the contribution of 2p2h components in the ground state wave function of the target to the right-hand side of equation 4.7 (but not to the distribution of the strength) can be obtained as follows. Taking the mre for $\alpha > F$ to be given by the unperturbed energy of the residual 1p2h state, using a free spectrum for the particle states, one finds that on the average $\langle \frac{1}{2}(k_\alpha^2/2m + \varepsilon_\alpha) \rangle_{\alpha > F}$ is equal to $\langle \varepsilon_\alpha \rangle_{\alpha < F}$, the mean (occupied) single-particle energy. One thus has

$$\Delta E_A/A = 1/2A \sum_{\alpha > F} \rho_\alpha(k_\alpha^2/2m + \varepsilon_\alpha) \approx (1 - \langle \rho_\alpha \rangle_{\alpha < F}) \langle \varepsilon_\alpha \rangle_{\alpha < F}. \qquad 4.10$$

Using $\langle \rho_\alpha \rangle_{\alpha < F} \sim 0.85$ (cf section 4.3) this results (including the reduction of the contribution from $\alpha < F$) in roughly 2–3 MeV extra binding per nucleon.

In any case some theoretical understanding of the large E region is desirable, for this is one place where one expects the reaction mechanism to break down. In particular, one expects events in which the detected nucleon was not the one struck by the electron (but rather a nucleon correlated with it). In this case the cross section is no longer proportional to the spectral function, and thus it is doubtful that the large E contribution to the sum rule 4.8 could ever be obtained directly from experiments. So far only rough estimates exist for these effects (15, 17).

Finally we note that because E_A and the kinetic energy T are finite (for all commonly used nucleon-nucleon potentials), it follows from equation 4.9 that the $\sum_\alpha \rho_\alpha \varepsilon_\alpha$ is bound from below; therefore all $\rho_\alpha \varepsilon_\alpha$ must be finite even for a system interacting through hard-core two-body forces. In contrast, the analogous mean addition energies, defined as $\varepsilon_\alpha^+ = (1 - \rho_\alpha)^{-1} \langle \psi_A | [a_\alpha, H] a_\alpha^+ | \psi_A \rangle$, which are, for example, probed in stripping reactions, would be infinite for singular hard-core interactions.

4.3   Mean Removal Energies and Self-Consistent Theories

Self-consistent calculations of nuclear ground state properties yield not only direct observables like the binding energy and charge distribution, but also, usually as a byproduct, single-particle energies and wave functions. The problem discussed in this section is how one can make a proper comparison between these theoretical single-particle energies and the mre defined above. This problem can be studied most conveniently by examining the linked-cluster perturbation expansion for the mre. This makes it possible, first to show how one should calculate the mre given an approximate theory for the target ground state, and second to make a comparison with the single-particle energies defined in self-consistent theories.

For the perturbation expansion one divides (28) the Hamiltonian H into an unperturbed one-body Hamiltonian H_0 and a perturbation H': $H = H_0 + H'$, with $H_0 = T + U$ and $H' = \sum_{i < j} V_{ij} - U$, where U is an auxillary one-body potential. The unperturbed shell model Hamiltonian H_0 has a set of single-particle eigenstates $|\alpha\rangle$ with energies W_α: $H_0 |\alpha\rangle = W_\alpha |\alpha\rangle$. The ground state energy may then be written as

$$E_A = E_0 + \Delta E_A, \qquad\qquad 4.11$$

where

$$E_0 = \sum_{\alpha < F} W_\alpha = \sum_{\alpha < F} (T_\alpha + U_\alpha). \qquad\qquad 4.12$$

The energy shift ΔE_A can be obtained diagrammatically from the Goldstone linked-cluster expansion (28).

Using equation 4.3 one sees that the mre can be expressed as

$$\varepsilon_\alpha = T_\alpha + U_\alpha + \Delta\varepsilon_\alpha/\rho_\alpha, \qquad\qquad 4.13$$

with

$$\Delta\varepsilon_\alpha = -\langle\psi_A|[a_\alpha^+, H']a_\alpha|\psi_A\rangle. \qquad\qquad 4.14$$

It has been shown by Koltun (24) and Schäfer (26) that in the linked-cluster perturbation expansion $\Delta\varepsilon_\alpha$ is given by the sum of all diagrams that contribute to ΔE_A but that have each line in the orbital α in turn held fixed, summing all other lines. The diagrams for the occupation probabilities ρ_α can also be obtained (24) from those of ΔE_A in the standard way. By combining the linked-cluster expansions of $\Delta\varepsilon_\alpha$ and ρ_α one thus obtains the expansion for the mre ε_α. Therefore, if a particular approximation to the ground state energy E_A is considered by specifying which diagrams are included in ΔE_A, ε_α can directly be calculated to the same approximation.[3]

Some of the lowest-order diagrams that contribute to the mre are shown in Figure 5. The point to note is that all diagrams that contribute to the mre also appear in the linked-cluster expansion for the target ground state energy shift ΔE_A (after closing the external lines α). Also, shown in Figure 6 is an example of a diagram that does not contribute to the mre, because it does not occur in the ΔE_A expansion.

This diagrammatic expansion makes possible a direct comparison between mre and the single-particle energies of self-consistent theories. For example, keeping only the lowest-order diagram (Figure 5a), in correspondence with the Brueckner-Hartree-Fock (BHF) theory, one finds $\varepsilon_\alpha = T_\alpha + \sum_{\beta<F}\langle\alpha\beta|G|\alpha\beta\rangle$ (where G represents the reaction matrix), and thus because of the BHF self-consistency condition, $U_\alpha = \sum_{\beta<F}\langle\alpha\beta|G|\alpha\beta\rangle$, the mre are the same as the single-particle energies $W_\alpha^{\mathrm{BHF}} = T_\alpha + U_\alpha$. Furthermore, to this order one has $\rho_\alpha = 1\,(0)$ for $\alpha < F\,(\alpha > F)$, and thus the BHF expression for the binding energy [neglecting center-of-mass, i.e. $O(1/A)$, corrections],

$$E_A^{\mathrm{BHF}} = E_A^0 - \tfrac{1}{2}\sum_{\alpha<F} U_\alpha = \tfrac{1}{2}\sum_{\alpha<F}(T_\alpha + W_\alpha^{\mathrm{BHF}}), \qquad\qquad 4.15$$

is consistent (to this order) with the exact sum rule 4.9.

It has been noted (3, 29) that when one inserts the experimental removal energies instead of W^{BHF} in equation 4.15, one finds a total binding energy E_A too low by several MeV per nucleon. This is similar to the present experimental situation of the sum rule 4.9; the strength that has been observed so far appears to be due only to occupied states and leads via the sum rule to an underbinding. This indicates that theories

---

[3] We note that the energy-weighted sum rule 4.9 that interrelates ρ_α, ε_α, and E_A can be considered as an application of the above.

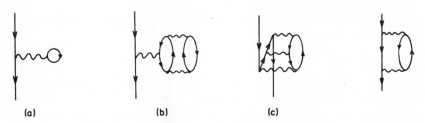

(a)   (b)   (c)

Figure 5 (a–c above)   Some diagrams that appear in the expansion for the mean removal energy.

Figure 6 (far right)   The lowest-order diagram which contributes to the separation energy but not the mean removal energy.

such as BHF, for which the relation 4.15 between the binding energy and the mre holds, are not satisfactory.

Essentially two methods to improve the situation have been applied: in the first, following Brandow (30, 31), some higher-order diagrams are explicitly included in the single-particle potential (32–35), and in the second, higher-order many-body effects are simulated by making the residual interaction H' density dependent (36, 37).

In Brandow's formulation (31) the development is in terms of self-energy insertions $M_\alpha(\omega)$ in particle or hole lines in diagrams (ω is an energy parameter). These insertions can be separated into two classes: the so-called on-shell, M_α^{on}, and the off-shell, M_α^{off}, insertions. The M_α^{on} insertions are on-the-energy-shell in the sense that the value of ω is determined only by the energies of the external lines α, and not by the rest of the larger diagram. They can thus be summed so as to give the effect of a generalized instantaneous single-particle potential (therefore the effect on M_α^{on} can be incorporated into an independent-particle model). The effect of M_α^{off} cannot be represented by such a potential, and therefore M_α^{off} insertions give rise to deviations from the IPSM. Diagrammatically, insertions on a single-hole line α in which the external lines overlap in time (e.g. those shown in Figure 5) are classified as M_α^{on}; if there exists a time at which neither external line is present the insertions are instead classified as M_α^{off}.

Following Brandow (31) the self-consistent single-particle energies and occupation probabilities (for $\alpha < F$) are defined by

$$W_\alpha^{Br} = T_\alpha + M_\alpha^{on}(W_\alpha^{Br}),\qquad\qquad 4.16$$

and

$$\rho_\alpha = \left[1 - \frac{\partial M_\alpha^{on}(\omega)}{\partial \omega}\right]_{\omega = W_\alpha^{Br}}^{-1}$$

It can easily be seen that M_α^{on} and $\Delta\varepsilon_\alpha$ have essentially the same structure. Because the external lines of the diagrams that contribute to M_α^{on} overlap in time, they can be closed; thus these diagrams also contribute to ΔE_A and therefore to $\Delta\varepsilon_\alpha$. A more careful analysis (24) shows that for the normally occupied states $M_\alpha^{on} - U_\alpha$ and $\Delta\varepsilon_\alpha/\rho_\alpha$ have the same perturbation expansion. The self-consistent single-particle energies W_α^{Br} are thus identical with the mre ε_α for the normally occupied states. For the normally empty states ($\alpha > F$) no such simple relation exists. Because it is necessary to have correlations in the ground state wave function in order to knock out nucleons from above the Fermi sea, the mre in this case correspond (assuming 2p2h correlations) to (minus) a 1p2h excitation energy, and therefore are very different from normal single-particle energies.

It should be noted that Brandow's theory is rather formal and has not been applied in its full complexity. A simplified version that has been used in practical calculations is the renormalized Brueckner-Hartree-Fock (RBHF) theory (32–35). The essential improvement of RBHF over the BHF theory is that the effect of the depletion of the occupied states ($\rho_\alpha < 1$) on the single-particle potential is taken into account explicitly. This corresponds to the inclusion of Diagram 5b (so-called third-order rearrangement) resulting in less negative single-particle energies for the occupied states: $W_\alpha = T_\alpha + \sum_{\beta < F} \rho_\beta \langle\alpha\beta|G|\alpha\beta\rangle$. Thus to this order the mre and single-particle energies are again the same. In this theory the binding energy is given by

$$E_A^{RBHF} = E_A^0 - \tfrac{1}{2}\sum_{\alpha < F} \rho_\alpha U_\alpha = \tfrac{1}{2}\sum_{\alpha < F} [\rho_\alpha T_\alpha + (2-\rho_\alpha)W_\alpha], \qquad 4.17$$

which (unlike in BHF theory) differs in form from the sum rule 4.9. This difference arises from the fact that in equation 4.17 the sum is restricted to states below the Fermi surface. Using the estimate 4.10 for the contribution from states above the Fermi surface, however, we see that the sum rule 4.9 agrees with equation 4.17. We note that although equation 4.17 is valid for the total binding energy, the apparent kinetic energy contribution $\sum_{\alpha < F} T_\alpha$ is smaller than the true kinetic energy $\langle T\rangle = \sum_{\alpha\beta} \rho_{\alpha\beta} T_{\alpha\beta}$.

Another generalization of BHF that seems to improve the relation of the removal energies to the total binding energy is the density-dependent version (DDHF). In this approach higher-order many-body effects are simulated by the inclusion of a density-dependent term in the two-body interaction. This approach leads to good agreement with experimental binding energies and charge distributions (36, 37), but is somewhat less fundamental because the interaction is usually adjusted to fit nuclear matter properties. The single-particle energies W_α^{DDHF} can again be interpreted as mean removal energies (36) and the total binding energy is now expressed as (36)

$$E_A^{DDHF} = \tfrac{1}{2} \sum_{\alpha < F} (T_\alpha + W_\alpha^{DDHF}) + E_R, \qquad\qquad 4.18$$

where the correction term E_R comes from the density dependence of the interaction.

Because of the large number of self-consistent calculations that have been published we do not attempt to give a comprehensive list of results, but rather mention some general trends. In the BHF approach for most two-body interactions not enough binding is obtained; the RBHF theory gives about the same binding but less negative values for the single-particle energies, and in the DDHF good agreement for binding energies is obtained; the single-particle energies of the RBHF and DDHF theories tend to be less negative than the experimental removal energies. In DDHF calculations the energy of the 0s proton state seems to saturate around 50 MeV (36). Recent (e, e′p) experiments on ^{40}Ca (7, 9) seem to favor a somewhat higher $|\varepsilon_{0s}|$ than the predicted 43 MeV (37) and 47 MeV (36). The occupation probabilities of the occupied states obtained in RBHF are on the order of 0.85 rather independent of the particular state (34, 35).

5   FRAGMENTATION OF SINGLE-PARTICLE STRENGTH

As has been pointed out, the knock-out process, ideally an instantaneous reaction, will not in general leave the residual nucleus in an eigenstate. Therefore it is necessary to go beyond the IPSM to interpret the observed spectral function. As was shown in the previous section, some of the effects of the residual interaction—basically the effects represented by M_α^{on} insertions that give rise to an instantaneous single-particle potential—could be incorporated into a generalized independent-particle model, characterized by the zeroth and first moments of the spectral function. The higher moments characterize effects such as fragmentation that cannot be incorporated in such a model.

5.1   Second Moment and Rearrangement Energy

The second moment of the spectral function is expressed as (24, 38)

$$\int dE \, E^2 P_\alpha(E) = \langle \psi_A | a_\alpha^+ (E_A - H)^2 a_\alpha | \psi_A \rangle$$
$$= \langle \psi_A | [a_\alpha^+, H][H, a_\alpha] | \psi_A \rangle, \qquad\qquad 5.1$$

in terms of which the dispersion, or mean square deviation, σ_a^2, of the single-particle strength for an orbital a is given by

$$\sigma_a^2 = 1/\rho_a \int dE\, E^2 P_a(E) - (\varepsilon_a)^2. \qquad 5.2$$

A related quantity that can also serve as a measure of deviations from the IPSM is the single-particle rearrangement energy. For the highest occupied shells the dominant part of the single-particle strength is in general located in one state (usually the lowest state in the $A-1$ excitation spectrum with appropriate spin and parity), which permits a meaningful definition of single-particle separation energies: $\varepsilon_a^{sep} = E_A - E_{A-1}^a$ (39). Then the single-particle rearrangement energy, ε_a^R, can be defined[4] naturally as the difference between the mre and the separation energy $\varepsilon_a^R = \varepsilon_a^{sep} - \varepsilon_a$ (27, 39). Clearly the rearrangement energy is positive, and in principle observable for the valence shells. For deeper bound shells the definition of ε_a^{sep} (and therefore ε_a^R) becomes increasingly arbitrary, and probably meaningless for the most deeply bound shells because of the short lifetimes of the nuclear states.

5.2   Green Function Theory

Theoretically the more detailed structure of the spectral function has often been discussed (40, 41) in the framework of Green function theory. The spectral function is closely related to the single-particle Green function $S_\alpha(\omega)$ (28) (for simplicity taken to be diagonal)

$$P_\alpha(E) = 1/\pi \lim_{\eta \downarrow 0} \text{Im}\, S_\alpha^{hole}(-E-i\eta), \qquad 5.3$$

where

$$S_\alpha^{hole}(\omega) = \langle \psi_A | a_\alpha^+ (\omega + H - E_A)^{-1} a_\alpha | \psi_A \rangle. \qquad 5.4$$

For $E > (E_{A-1}^0 - E_A)$, the physical region, the imaginary part of S_α^{hole} is equal (41) to the imaginary part of the full Green function S_α, which can be expressed as (28)

$$S_\alpha(\omega) = [\omega - W_\alpha - \Sigma_\alpha(\omega)]^{-1}, \qquad 5.5$$

where $\Sigma_\alpha(\omega)$ is the Dyson irreducible self-energy operator, which can be calculated by standard methods of many-body theory (28). In the following W_α will be taken to be the BHF single-particle energy.

The Σ_α insertions that appear in equation 5.5 form an irreducible sub-

---

[4] It should be noted that there exist several other definitions of the rearrangement energy. For example, the so-called third-order rearrangement diagram (Figure 5b) is sometimes also referred to as rearrangement energy (43, 45). The latter, however, should be considered as a purely theoretical quantity that cannot be compared to experiment.

set of the M_α insertions discussed in section 4.3, namely those insertions that cannot be cut into two separate parts by cutting one hole line (e.g. the diagrams of Figures 5 and 6). Consequently, if one makes a separation of Σ_α into Σ_α^{on} and Σ_α^{off} analogously to that for the M_α insertions, one finds that for M_α's consisting of more than one Σ_α, both Σ_α^{on} and Σ_α^{off} can, in general, occur. Thus Σ_α^{off} also contributes to M_α^{on} and therefore (unlike M_α^{off}) to E_A and the mre. However, taking the simplest approach by restricting oneself in equation 5.5 to insertions of one class, one again finds (24, 27, 40) that Σ_α^{on} produces shifts and changes in the single-particle strength whereas Σ_α^{off} contributes to the spreading of the single-particle strength. By selecting the appropriate set of Σ_α's the spectral function can in principle be calculated to any desired approximation.

The lowest-order Σ_α^{off} insertion (shown in Figure 6) is given by (28)

$$\Sigma_\alpha^{(2)}(\omega) = \tfrac{1}{2} \sum_{ph_1h_2} \langle \alpha p |G| h_1 h_2 \rangle^2 /(\omega + W_p - W_{h_1} - W_{h_2}),\qquad 5.6$$

where G represents the two-body reaction matrix and the sum runs over all possible intermediate 1p2h states. From equations 5.3–6 one finds that

$$\sigma_a^2 = \frac{1}{\pi} \lim_{\eta \downarrow 0} \int d\omega\, \mathrm{Im}\, \Sigma_\alpha^{(2)}(\omega - i\eta) = \tfrac{1}{2} \sum_{ph_1h_2} \langle \alpha p |G| h_1 h_2 \rangle^2,\qquad 5.7$$

and

$$\varepsilon_a^R = \Sigma_\alpha^{(2)}(W_\alpha) = \tfrac{1}{2} \sum_{ph_1h_2} \langle \alpha p |G| h_1 h_2 \rangle^2 /(W_\alpha + W_p - W_{h_1} - W_{h_2}).\qquad 5.8$$

In practical analysis of knock-out processes σ_a has not yet been used very often, because of the difficulty in identifying all pieces of single-particle strength in the tail of the spectral function. In particular, for deeply bound hole states a more common quantity [suggested by the experimental results on the 0s hole distribution (see Figure 2) which resembles the shape given by a quasi-particle pole] that has been used to characterize the fragmentation, is the spreading width (40, 43). Theoretically such a quasi-particle pole is obtained by averaging the Green function $S_\alpha(\omega)$ over a small energy interval to eliminate the fine structure of the states. The use of a Lorentzian averaging function with width I leads to a spreading width (40):

$$\Gamma_a^{\downarrow} = 2\,\mathrm{Im}\,\langle \Sigma_\alpha^{(2)}(\omega) \rangle_I = 2\pi/d \langle \alpha p |G| h_1 h_2 \rangle^2,\qquad 5.9$$

where $1/d$ is the density of the background states $(d \ll I \ll \Gamma)$. This quantity is somewhat less fundamental than σ_a because of the model-dependent assumptions that must be made; moreover the predicted symmetric shape of the distribution seems questionable.

Only a few calculations of σ_a, ε_a^R, or $\Gamma_a^{\downarrow}$ have been performed with realistic interactions. One of the problems with the calculation of $\Sigma_\alpha^{(2)}$ is the slow convergence of the sum over intermediate 1p2h states with increasing excitation energy (42). Also, because the convergence of the many-body perturbation approach has been questioned recently (51), it is not clear whether the lowest-order estimates can be trusted.

Typical values obtained for ε_a^R are in the range of 2–6 MeV [estimated from the lowest-order contribution for the least bound shells in light nuclei (35, 43)]. On the other hand, the large fragmentation implied by such values of ε_a^R for these shells has not been observed (3).

In the above calculations, the effect of the continuum that gives rise to an escape width $\Gamma^\uparrow$ for states above the threshold for particle emission and that thus contributes to the total width has been neglected. We note that in a recent (d, ^{3}He) reaction on the tin isotopes (44) a quasi-particle-type pole for the g9/2 proton orbital (one major shell below the Fermi surface) has been observed below the threshold. This strongly indicates that in some cases there is a considerable fragmentation in the absence of continuum effects.

5.3   Methods to Calculate the Spectral Function

It was shown in the previous sections that the gross properties of the spectral function (momentum dependence, mean removal energies) could be calculated in the framework of the self-consistent single-particle model, and that the second moment could also be obtained rather directly. Several attempts to calculate a more detailed energy dependence of the spectral function $P_\alpha(E)$ from first principles using microscopic nuclear models have been made by Lipperheide et al (5, 41, 47), Faessler et al (48), and Becker (39, 46).

In the simplest approach one takes a configuration space consisting of the shell model hole state α coupled through residual interactions to 1p2h states (usually restricted to two major shells). The particle can be in a continuum orbit, making possible particle decay of all eigenstates that are coupled to these particular basis states, resulting in a finite escape width $\Gamma^\uparrow$ of the peaks in the removal spectrum. For a more complete description one can take into account the mutual coupling between the 1p2h states through two-body interactions, and also the effect of ground state correlations (41).

Actual calculations have been carried out both on the basis of perturbation approach using the Green function techniques and by shell model diagonalization in a selected configuration space. The perturbation approach (5, 46) in general leads to removal energy spectra that exhibit many rather narrow peaks spread over a large energy region. The continuum model approach (47), in which the coupling to the continuum is treated more accurately, yields smoother energy dependence, although it still predicts more structure, for example, in the 0s distribution in ^{12}C, than is seen experimentally. The calculated positions of the peaks are also quite sensitive to the choice of the single-particle energies for the 1p2h states that are quasidegenerate with the 0s hole state.

6   CENTER-OF-MASS EFFECTS

So far we have considered only theories that, because they are based on the IPSM, violate translational invariance. The difficulties that arise when one tries to incorporate the latter in microscopic nuclear theories are well known (49). At the simplest stage of approximation, however, there is one model for which this center-of-mass problem can be treated exactly (50), namely the harmonic oscillator shell model (HOSM). The single-nucleon structure of the intrinsic wave function is very similar to that for the IPSM; however, because the motion of the particles is no longer independent, some intuitive concepts developed by working with shell model-based theories fail (for example, one can have intrinsic occupation probabilities larger than unity).

In passing from the shell model to the intrinsic wave function one finds (20) that the single-particle wave functions, and thus the momentum distributions, are modified by a typical $O(1/A)$ correction, represented by $\phi_\alpha(\mathbf{k}) \to \phi_\alpha\{\mathbf{k}[A/(A-1)]^{1/2}\}$. The effect on the occupation probabilities and the removal energies, however, can be more pronounced. For example, in ^{16}O the 0s hole state is 20% spurious (20), i.e. the 0s hole state wave function, representing a $1\,\hbar\omega$ excitation with respect to the ground state, has a component in which the center-of-mass wave function is in the 0p state:

$$\psi_{(0s)^{-1}}^{HOSM}\,\mathbf{r}_i = \alpha\phi_{(0s)^{-1}}^{int}\,\mathbf{r}_i'\,\chi_{0s}^{CM}(\mathbf{R}_{A-1}) + \beta\phi_{(0p)^{-1}}^{int}(\mathbf{r}_i')\chi_{0p}^{CM}(\mathbf{R}_{A-1}),$$ 6.1

where $\beta^2 = 1 - \alpha^2 = 0.2$. For the spectroscopic proton strength in the intrinsic frame, one thus has $N_{0s} = 2 \times 0.8 = 1.6$ and $N_{0p} = 6 + 2 \times 0.2 = 6.4$. In general, for 0p shell nuclei one finds $N_{0p} = Z - N_{0s} = [A/(A-1)](Z-2)$ rather than $N_{0p} = Z-2$ as in the shell model (20). It is interesting to note that the predicted corrected ratio of spectroscopic strength for ^{12}C, $N_{0s}/N_{0p} = \frac{3}{8}$, seems to be consistent with the results of recent (e, e'p) experiments by the Saclay group (6).

The spuriousness of the 0s state has also an important effect on the calculation of single-particle energies; namely if one takes the expectation value of the Hamiltonian in the shell model 0s hole state one is including a spurious contribution to the potential energy from the 0p hole state U_{0p} (because of the excitation of the center-of-mass motion in the spurious component, such an effect does not occur for the kinetic energy). Correcting for this contribution one finds for ^{16}O (20)

$$\varepsilon_{0s}^{int} = \varepsilon_{0s}^{SM} + \tfrac{1}{4}(V_{0s}^{SM} - V_{0p}^{SM}), \qquad\qquad 6.2$$

which gives roughly a 5 MeV increase in the binding of the 0s state. A similar correction for the spuriousness of the 0s hole state has recently been included by Becker et al (35) in an RBHF calculation for ^{16}O.

7   CONCLUSION

In this article the present status of knock-out experiments and the interpretation of their results in terms of nuclear models have been reviewed. Special attention has been given to the mean removal energies that could constitute an important test of various self-consistent nuclear structure theories like RBHF and DDHF.

The experimental situation has improved considerably over the last few years mainly as a result of the recent $(e, e'p)$ studies. Although the results of these experiments tend to be in reasonable agreement with theoretical predictions, further improvement is necessary for a good quantitative analysis.

As has been mentioned, the method used to extract the strengths for knock-out from the various orbitals from the momentum distributions involves some fundamental difficulties (e.g. expansion in an overcomplete set). Because the interpretation of the momentum distribution for knock-out from a particular orbital is much cleaner, it would thus be desirable to improve the resolution so as to separate the final states of the residual nucleus.

The exploration of the region of large E and k is of obvious interest. Though not yet observed, some strength is expected in this region from short-range two-body correlations and final state interactions. This is also suggested by the failure of the binding energy sum rule 4.8. As a large region of phase space is available, one expects this strength to be spread rather thinly, which could be the reason that it has not been seen. Obviously, theoretical calculations are needed to predict what kinematical conditions would be most favorable for the observation of these effects.

A point that definitely needs more study is that of obtaining a better understanding of the reaction mechanism itself. So far this problem has been treated in a semiphenomenological fashion, using, for example, experimentally determined optical potentials and simple models for multiple scattering to calculate the corrections to PWIA. In addition very little work has been done on corrections for processes in which the detected nucleon is not the one struck by the electron. Although these problems are mainly theoretical in nature they can also be investigated experimentally. This could, for example, be done by making a series of measurements under different kinematical conditions, but with E and k fixed. Because in PWIA the results are the same (aside from kinematical factors), the deviations would give a check on whether the corrections to PWIA have been calculated correctly.

ACKNOWLEDGMENTS

The authors would like to thank Dr. J. Mougey, Dr. I. Sick, and Dr. G. J. Wagner for stimulating discussions. Thanks are also due to Mrs. Cocky van Bueren-Kooi and Mrs. Marijke Oskam-Tamboezer for typing the manuscript.

This work is part of the research program of the Institute for Nuclear Physics Research (I.K.O.) made possible by financial support from the Foundation for Fundamental Research on Matter (Z.W.O.) and the Netherlands Organization for the Advancement of Pure Research (Z.W.O.).

Literature Cited

1. Jackson, D. F. 1971. *Advan. Nucl. Phys.* 4:1
2. Jacob, G., Maris, Th. A. J. 1966. *Rev. Mod. Phys.* 38:121; 1973. *Rev. Mod. Phys.* 45:6
3. Wagner, G. J. 1973. *Lecture Notes in Physics,* Vol. 23. Berlin: Springer-Verlag
4. Tyrén, H. et al 1966. *Nucl. Phys.* 79:321
5. Gross, D. H. E., Lipperheide, R. 1970. *Nucl. Phys. A* 150:449; Wille, U., Lipperheide, R. 1972. *Nucl. Phys. A* 189:113
6. Bernheim, M. et al 1974. *Phys. Rev. Lett.* 32:898
7. Bernheim, M. et al 1974. *Proc. Int. Conf. Nucl. Struct. Spectrosc.* II:412. Amsterdam: Scholar's Press
8. Hiramatsu, H. et al 1973. *Phys. Lett. B* 44:50
9. Nakamura, K. et al 1974. *Phys. Rev. Lett.* 33:853; Nakamura, K. 1974. *Rep. Univ. Tokyo* UPTN–36
10. Köbberling, M. et al 1974. *Nucl. Phys. A* 231:504
11. Antoufiev, Y. P. et al 1972. *Phys. Lett. B* 42:347
12. Hood, S. T. et al 1973. *Phys. Rev. A* 8:2494; Hood, S. T. et al 1974. *Phys. Rev. A* 9:260; McCarthy, I. E. 1973. *J. Phys. B* 6:2358
13. Epp, C. D., Griffy, T. A. 1970. *Phys. Rev. C* 1:1633
14. Devanathan, V. 1967. *Ann Phys.* 43:74
15. de Forest, T. Jr. 1967. *Ann. Phys.* 45:365
16. de Forest, T. Jr. 1971. *Nucl. Phys. A* 163:237
17. Pittel, S., Austern, N. 1974. *Nucl. Phys. A* 218:221
18. Viollier, R. D., Alder, K. 1971. *Helv. Phys. Acta* 44:77
19. Borie, E., Drechsel, D. 1971. *Nucl. Phys. A* 167:369
20. Dieperink, A. E. L., de Forest, T. Jr. 1974. *Phys. Rev. C* 10:543
21. Elton, L. R. B., Swift, A. 1967. *Nucl. Phys. A* 94:52

22. James, A. N. et al 1969. *Nucl. Phys. A* 133:89, 138:145
23. Kobe, D. H. 1969. *J. Chem. Phys.* 50: 5183
24. Koltun, D. 1974. *Phys. Rev. C* 9:484
25. Koltun, D. 1972. *Phys. Rev. Lett.* 28: 182
26. Schäfer, L. 1973. *Nucl. Phys. A* 217: 361
27. Dieperink, A. E. L., Brussaard, P. J., Cusson, R. Y. 1972. *Nucl. Phys. A* 180:110
28. Fetter, A. L., Walecka, J. D. 1971. *Quantum Theory of Many-Particle Systems.* New York: McGraw-Hill
29. Becker, R. L., Patterson, M. R. 1971. *Nucl. Phys. A* 178:88
30. Brandow, B. 1967. *Rev. Mod. Phys.* 39:771
31. Brandow, B. 1970. *Ann. Phys.* 57: 214
32. Becker, R. L. 1970. *Phys. Rev. Lett.* 24:400
33. McCarthy, R. J., Davies, K. T. R. 1970. *Phys. Rev. C* 1:1644
34. Davies, K. T. R., McCarthy, R. J., Sauer, P. U. 1972. *Phys. Rev. C* 6: 1461; Davies, K. T. R., McCarthy, R. J. 1971. *Phys. Rev. C* 4:81
35. Becker, R. L., Davies, K. T. R., Patterson, M. R. 1974. *Phys. Rev. C* 9:1221
36. Vautherin, D., Brink, D. M. 1972. *Phys. Rev. C* 5:626
37. Campi, X., Sprung, D. W. L. 1972. *Nucl. Phys. A* 194:401
38. Dieperink, A. E. L., Brussaard, P. J. 1973. *Z. Phys.* 261:117
39. Becker, R. L. 1972. *Proc. Int. Symp. Present Status Novel Develop. Nucl. Many-Body Probl.,* Rome 2:205
40. Engelbracht, C. A., Weidenmüller, H. A. 1972. *Nucl. Phys. A* 184:385
41. Fritsch, W., Lipperheide, R., Wille, U. 1972. *Nucl. Phys. A* 198:515
42. Dieperink, A. E. L. 1974. *Proc. Int. Conf. Nucl. Struct. Spectrosc.,* Amsterdam 1:33
43. Padjen, R. et al 1973. *Phys. Rev. C* 8:2024
44. van der Werf, S. Y. et al 1974. *Phys. Rev. Lett.* 33:712
45. Müther, H., Faessler, A., Goeke, K. 1973. *Nucl. Phys. A* 215:213
46. Becker, R. L. 1973. *Oak Ridge Nat. Lab. Phys. Div. Ann. Rep. 1972,* No. ORNL–4844, p. 10
47. Fritsch, W., Lipperheide, R., Wille, U. 1973. *Phys. Lett. B* 45:103
48. Faessler, A., Kusuno, S., Strobel, G. L. 1973. *Nucl. Phys. A* 203:513
49. Lipkin, H. J. 1958. *Phys. Rev.* 109: 2071, 110:1395
50. Elliott, J. P., Skyrme, T. H. R. 1955. *Proc. Roy. Soc. A* 232:561
51. Barrett, B. R., Kirson, M. W. 1973. *Advan. Nucl. Phys.* 6:219

NEUTRON STARS[1] ✳5558

Gordon Baym
Department of Physics, University of Illinois, Urbana, Illinois 61801

Christopher Pethick
Department of Physics, University of Illinois, Urbana, Illinois 61801,
and Nordita, Copenhagen

CONTENTS

1 INTRODUCTION

Léon Rosenfeld, at the 1973 Solvay meeting, recounted that on the
day in 1932 that word came to Copenhagen from Cambridge telling of
the identification of the neutron, he, Bohr, and Landau spent the evening

[1] This work has been supported in part by NSF Grants GP 40395 and
GP 37485X, and by an A. P. Sloan Foundation Research Fellowship to C.P.

discussing possible implications of this great discovery; it was then that Landau first suggested the possibility of cold dense stars composed primarily of neutrons, "unheimliche Sterne," as he described them (see 1). In 1934 Baade & Zwicky (2) independently proposed the idea of neutron stars, observing that they would be of very small radius and extremely high density, and could be much more highly bound than ordinary stars. They also advanced the suggestion that neutron stars would be produced in supernova explosions, the endpoint in the evolution of more massive stars.

The first calculation of models of neutron stars was carried out by Oppenheimer & Volkoff (3) in 1939, who assumed the matter to be composed of a dense free neutron gas. Further theoretical work proceeded slowly in the next two decades, largely because it was believed that thermal radiation from neutron stars would, as a consequence of their small radii and hence small surface areas, be too weak to be observed at astronomical distances. Then in 1967 pulsars were discovered (4), and Gold shortly thereafter (5) proposed the now generally accepted view that they are rotating neutron stars. The period since 1968 has been one of intense theoretical activity on the properties of neutron stars. Further stimulus to their study has come from the conclusion that certain of the recently observed compact X-ray sources in our galaxy are neutron stars in close binary orbits with more ordinary stars (6).

Neutron stars have a range of possible masses M with a lower limit of ~ 0.1 solar masses ($M_\odot$) and an upper limit somewhere between 1.5 and $\sim 2M_\odot$. The uncertainty in the upper mass limit reflects our lack of knowledge of the properties of matter at high densities. The typical radius of a neutron star is ~ 10 km, and thus the typical densities in the interior range up to and beyond the density ρ_0 of nuclear matter ($\sim 2.8 \times 10^{14}$ g cm^{-3} or 0.17 nucleons per fm^3). In a sense neutron stars are giant nuclei, with $A \sim 10^{57}$.

The gravitational acceleration $g = GM/R^2$ at the surface of a neutron star is about 10^{11} times as strong as that on earth; the gravitational binding energy GmM/R of a particle at the surface is about one tenth of its rest energy mc^2. The energy emitted in the compact X-ray sources comes directly from the release of gravitational energy of matter being accreted onto the neutron star from the companion star. The energy source of pulsars, however, is the kinetic energy of rotation E_{rot} of the neutron star. This energy also comes from the release of gravitational binding energy; in a supernova, as the core of the star collapses with conserved angular momentum L to form a neutron star, the gravitational potential energy is converted into kinetic energy of rotation $L^2/2I$, where I is the stellar moment of inertia.

The matter inside neutron stars is cold by microscopic standards; within a few days after their birth they cool by neutrino emission to interior temperatures of less than 1 MeV $\approx 10^{10}$°K, and throughout most of their early life T is ~ 10–100 keV (7, 8). Furthermore, nuclear equilibration times are sufficiently rapid compared with cooling times (9) that, to a good approximation, the matter inside can be considered to be in its absolute ground state, allowing for beta as well as strong and electromagnetic interactions. Figure 1 shows a cross section of a typical neutron star. The outer layers are a solid crust ~ 1 km thick, consisting, except in the outer few meters, of a lattice of bare nuclei immersed in a degenerate electron gas. As one goes deeper into the crust, the nuclear species become, because of the rising electron Fermi energy, progressively more neutron rich, beginning (ideally) as ^{56}Fe, through ^{118}Kr at mass density $\rho \approx 4.3 \times 10^{11}$ g cm^{-3}. At this density,

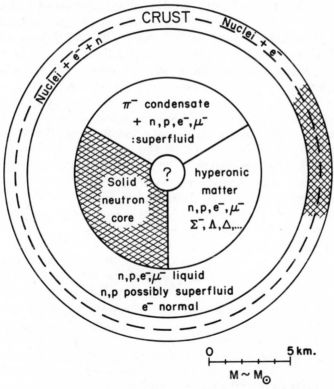

Figure 1 Schematic cross section of a neutron star, showing the outer crust, the neutron drip line (*dashed line*), the liquid interior, and the alternatives of a solid neutron core, a pion condensed core, or a hyperonic matter core.

the "neutron drip" point, the nuclei have become so neutron rich that with increasing density the continuum neutron states begin to be filled, and the lattice of neutron-rich nuclei becomes permeated by a sea of neutrons. By densities on the order of nuclear matter density the nuclei are practically touching, and beyond this they merge into a liquid composed primarily of neutrons, some protons, electrons, and a few muons. Most likely the neutrons are superfluid and the protons super-conducting; the electrons, however, are normal.

There are several possibilities proposed for the states of matter at the high pressures in the deep interior. One is that as the density increases, and the baryon and electron chemical potentials rise, the matter remains fluid and various hyperons are formed, which live stably in the star. A second is that the matter, primarily neutrons, again solidifies, now with a single baryon at each lattice site. A third is that the pion field develops a macroscopically occupied mode, or condensate. This latter "pion-condensed" state would be superconducting. Finally, the nature of matter at ultrahigh densities, when the baryon cores are strongly overlapping, as may occur in the centers of more massive neutron stars, remains a completely unsolved problem.

In this article we review recent developments in the understanding of the properties of condensed matter at high densities, and their implications for the structure of neutron stars, trying to emphasize the important points of physics; because space does not permit us to go into detail on all topics, we have had to be somewhat selective. Other reviews that the reader may find useful are as follows: on relativistic astrophysics, an overview including discussion of properties of matter at high densities, supernova and neutron stars (10); on pulsars (11); on earlier work on neutron stars (12); on matter at extreme densities and high temperatures (13); on dense matter, a collection of papers describing recent work (14); on the equation of state at high densities with detailed tabulations (15, 16).

2   NUCLEAR PHYSICS OF THE CRUST

In this section we first describe the structure of matter in the crust, assuming it to be in complete nuclear equilibrium at zero temperature in zero magnetic field; we consider separately effects due to strong magnetic fields, and possible departures from the ground state.

2.1   Nuclei Before Neutron Drip

2.1.1   EQUILIBRIUM NUCLIDES   Below mass densities $\rho \sim 4.3 \times 10^{11}$ g cm^{-3}, the matter at zero temperature consists of nuclei in a lattice, together, as in a normal metal, with a sea of electrons providing electrical neutrality. The most energetically favorable nucleus at low densities is ^{56}Fe, the endpoint of thermonuclear burning. At densities below $\sim 10^4$ g cm^{-3}, a fraction of the electrons are bound to the nuclei. The major difficulty in deriving the equation of state is the calculation of the electronic energy. The calculations of Feynman, Metropolis & Teller (17), who used the Thomas-Fermi method, provide a sufficiently accurate description of this regime for constructing zero temperature model stars. However, the very low density region has but small influence on the interior structural properties of the stars, whereas finite temperature and magnetic field effects modify considerably the properties of the matter in this region.

When the spacing between nuclei becomes (with increasing density) small compared with the Thomas-Fermi radius r_{TF} of an isolated neutral atom, the nuclei become effectively completely ionized. Because $r_{\text{TF}} \sim a_0 Z^{-1/3}$, where a_0 is the Bohr radius and Z the number of protons in the nucleus, one may assume complete ionization (18) for $\rho \gg 6.3 A Z$ g cm^{-3}, where A is the mass number of the nucleus; this density is about 10^4 g cm^{-3} for ^{56}Fe.

Above $\rho \sim 10^4$ g cm^{-3} the electrons are essentially free. The electron kinetic energy at the Fermi surface rises with ρ, and by $\rho \sim 10^5$ g cm^{-3} it is generally $\gg k_B T$; the electrons then form a degenerate plasma, which is fully relativistic above $\rho \sim 10^7$ g cm^{-3}. In the relativistic regime electron interaction energies are $\sim e^2/\hbar c$ times electron kinetic energies, and may be completely neglected. The electron plasma frequency $\omega_p = (4\pi n_e e^2 c^2/\mu_e)^{1/2}$ is $\sim 3.3 \times 10^6 (Z/A)^{1/3} \rho^{1/3}$ °K, where ρ is here the mass density in g cm^{-3}, n_e is the electron density, and μ_e is the electron Fermi energy including the electron rest mass. In the relativistic electron regime $\mu_e = (3\pi^2 n_e)^{1/3} \hbar c$, and $\hbar \omega_p = \mu_e/17.97$. Above $\rho \sim 6 \times 10^7 (T/10^9)^3$ g cm^{-3} the plasma frequency exceeds the ambient temperature T (measured here in °K) and consequently there is negligible thermal radiation present.

For $\mu_e \gtrsim 1$ MeV, or $\rho \gtrsim 8 \times 10^6$ g cm^{-3}, ^{56}Fe is no longer the most energetically favorable nucleus, since the energy of the matter can be lowered if ^{56}Fe nuclei capture energetic electrons from the top of the Fermi sea and rearrange themselves into ^{62}Ni nuclei. In the process of electron capture, protons (in nuclei) are converted into neutrons via the reaction

$$\text{e}^- + \text{p} \to n + \nu_e; \qquad\qquad 2.1$$

the neutrino escapes, lowering the energy of the system. The equilibrium nuclide present becomes increasingly neutron rich with increasing ρ;

the nuclei cannot decay via the inverse beta reaction

$$n \to p + e^- + \bar{\nu}_e, \qquad 2.2$$

because all final electron states allowed by energy conservation lie within the already filled Fermi sea. To find the ground state of matter below neutron drip one must, in computing the ground state energy at given ρ, also determine the nuclide present. This was originally done by Salpeter (19), and later by Baym, Pethick & Sutherland (20) who included also the effects of the lattice Coulomb energy on the equilibrium nuclide. The ground state energy per unit volume before neutron drip, assuming matter to be composed of nuclei of nucleon number A and charge Z, is

$$E = n_N[W_N(A, Z) + W_L] + E_e(n_e), \qquad 2.3$$

where $n_N[=n_e/Z]$ is the number of nuclei per unit volume, W_N is the total energy of an isolated nucleus, including rest masses of the nucleons and the internal nuclear Coulomb energy but not including any electron energy, and $E_e(n_e)$ is the total electron energy per unit volume. The term W_L is the lattice Coulomb energy per nucleus; for a given density of nuclei of charge Z the lattice energy is minimized for a body-centered cubic lattice, and is given by (21)

$$W_L = -1.819620 Z^2 e^2/a, \qquad 2.4$$

where the lattice constant $a = (2/n_N)^{1/3}$. For completely relativistic electrons, $E_e = \frac{3}{4} n_e \mu_e$. Given a knowledge of $W_N(A, Z)$ for various A and Z one simply searches for the A and Z that give the lowest E, for fixed mean baryon density $n_b = n_N A$, and electrical neutrality $n_e = n_b Z/A$. The main problem in the regime 10^9–4.3×10^{11} g cm^{-3}, the neutron drip point, is the lack of knowledge of nuclear masses far from the valley of beta stability. Considerable work has been done on this problem in connection with the search for superheavy elements in the laboratory (see 22) and also in connection with the origin of the elements (23). Useful extrapolations have been given by Myers & Swiatecki (24), Garvey et al (25), and Seeger (26).

Table 1 shows the equilibrium nuclei present below neutron drip as found by Baym, Pethick & Sutherland (20), using the mass tables of (24). In Table 1, ρ_{max} is the maximum density at which the nuclide is present, b is the binding energy per nucleon in the nucleus, defined by writing $W_N = m_n c^2(A - Z) + m_p c^2 Z - bA$. At the transition from one nuclide (Z, A) to the next (Z', A') at constant pressure (the physical condition in a star), the electron density remains roughly continuous, because the electrons supply essentially all the pressure in this regime; thus the

Table 1   Equilibrium nuclei below neutron drip[a]

Nucleus	b (MeV)	Z/A	ρ_{max} (g cm^{-3})	μ_e (MeV)	$\Delta\rho/\rho$ (%)
^{56}Fe	8.7905	0.4643	8.1×10^6	0.95	2.9
^{62}Ni	8.7947	0.4516	2.7×10^8	2.6	3.1
^{64}Ni	8.7777	0.4375	1.2×10^9	4.2	7.9
^{84}Se	8.6797	0.4048	8.2×10^9	7.7	3.5
^{82}Ge	8.5964	0.3902	2.2×10^{10}	10.6	3.8
^{80}Zn	8.4675	0.3750	4.8×10^{10}	13.6	4.1
^{78}Ni	8.2873	0.3590	1.6×10^{11}	20.0	4.6
^{76}Fe	7.9967	0.3421	1.8×10^{11}	20.2	2.2
^{124}Mo	7.8577	0.3387	1.9×10^{11}	20.5	3.1
^{122}Zr	7.6705	0.3279	2.7×10^{11}	22.9	3.3
^{120}Sr	7.4522	0.3166	3.7×10^{11}	25.2	3.5
^{118}Kr	7.2002	0.3051	(4.3×10^{11})	(26.2)	—

[a] b is the binding energy per nucleon, ρ_{max} is the maximum density at which the nuclide is present, μ_e is the electron chemical potential (including electron rest mass) at that density, and $\Delta\rho/\rho$ is the fractional increase in the mass density in the transition to the next nuclide. The value of $\rho_{max} = 4.3 \times 10^{11}$ g cm^{-3} is the density at which neutron drip begins.

baryon density and hence mass density must be discontinuous as Z/A decreases at each transition, with $\Delta\rho/\rho \approx (Z/A)/(Z'/A') - 1$. Values of $\Delta\rho/\rho$ are given in Table 1. In the sequence of nuclides in this table, only the first three, ^{56}Fe, ^{62}Ni, and ^{64}Ni, are stable in the laboratory; the fourth, ^{84}Se, has a laboratory half-life of 3.3 min. The masses of the remaining nuclei are determined by extrapolating from laboratory nuclei. The nuclei from ^{84}Se to ^{76}Fe have 50 neutrons, while beyond this the nuclei from ^{124}Mo to ^{118}Kr have 82 neutrons. Both neutron numbers are magic for laboratory nuclei; that they turn out to be favorable here reflects the shell structure assumed in the extrapolations of W_N for neutron-rich nuclei. Shell model calculations by Negele & Vautherin (27) (see section 2.2) indicate that 82 neutrons remains a magic number for nuclei as neutron rich as ^{118}Kr. It is worth remarking that the masses of predrip nuclei calculated by Negele and Vautherin are in good agreement with the extrapolations of (24).

While the lattice contribution to the energy density is small, $\sim Z^{2/3}(e^2/\hbar c)E_e$, it plays a significant role at higher densities in the determination of the equilibrium nucleus (20, 28). Omitting the lattice energy one finds (20) a sequence of nuclei ^{56}Fe(8.5×10^6), ^{62}Ni(1.3×10^8), ^{58}Fe(2.8×10^8), ^{64}Ni(1.1×10^9), ^{66}Ni(1.3×10^9), ^{68}Ni(1.9×10^9), ^{84}Se($6.2 \times$

10^9), ^{82}Ge(1.7×10^{10}), ^{80}Zn(3.8×10^{10}), ^{78}Ni(1.3×10^{11}), ^{76}Fe(2.2×10^{11}), ^{74}Cr(3.1×10^{11}), and until neutron drip at 3.7×10^{11} g cm^{-3}, ^{118}Kr; the numbers in parentheses are the maximum density in g cm^{-3} at which the nuclide is found. Comparing these nuclides with those in Table 1 one sees that at a given density, within a given neutron shell sequence, the nucleus calculated by including the lattice has a higher A than the nucleus found by omitting the lattice. The reason is that the equilibrium species of nucleus present is determined by a competition between the nuclear surface energy, favoring nuclei with large A, and the Coulomb energy, favoring nuclei with small A. At a density of 2×10^{11} g cm^{-3}, for example, the nuclei occupy $\sim 10^{-3}$ of space, and the effect of the lattice energy, inversely proportional to the distance between nuclei, is to reduce the net Coulomb energy by $\sim 15\%$.

2.1.2 CHEMICAL COMPOUNDS IN THE CRUST   Dyson (29) has suggested that at low densities compound structures in the crust, such as He and Fe nuclei in a NaCl structure, may be chemically stable, having lower energy than they would if separated into He and Fe crystals. Witten (30) has carried out detailed lattice energy calculations including zero-point energy and some electron screening effects, and finds the electronic plus Coulomb energies of such compounds to be favorable by a few tenths of a percent. However, these calculations omit the difference in nuclear binding energy per nucleon of He and Fe; the compounds do not have *total* energy density less than that of pure Fe, at the same mean baryon density. Such compound structure might be relevant if the outer portion of the crust does not come into overall nuclear equilibrium, either because of inadequate cooking at early times in the stellar history, or else through accretion or conceivable spallation processes at the stellar surface (see section 2.3).

2.1.3 MATTER IN STRONG MAGNETIC FIELDS   On the basis of current models of pulsars (11) and pulsating compact X-ray sources (6) one expects magnetic fields $B \sim 10^{12}$ gauss at the neutron star surfaces. Fields of such strength are also expected from scaling arguments, if the total magnetic flux threading a neutron star were the same as for an ordinary star or a magnetic white dwarf. Such high fields dramatically modify the properties of the matter in the lower-density regions of a neutron star. A clear account of these effects is given by Ruderman (31). We summarize here the important features.

   Classically an electron in a strong magnetic field B moves in a spiral about the field. Quantum mechanically an electron moves on a quantized Landau orbit, which corresponds to spiral motion around a cylinder of radius $r_n = (2n)^{1/2}\hat{\rho}$, where

$$\hat{\rho} = (\hbar/m_e\,\omega_c)^{1/2} = 2.6 \times 10^{-4}B^{-1/2}\ \text{cm} \qquad\qquad 2.5$$

($\omega_c = eB/m_e\,c$ is the cyclotron frequency and $n = 0, 1, 2, \ldots$); the width of the orbit is $\sim\hat{\rho}$. In the numerical expression in (2.5) B is in gauss. Consider then an atom in a strong B field. The basic point is that if $\hat{\rho}$ is small compared with the Bohr radius a_0/Z of the most tightly bound atomic electron (in zero field) then the magnetic field will dominate the electron motion perpendicular to the field, causing the atom to become a compact cylinder with axis along the field. Weaker fields can also have an appreciable effect on the atomic structure. In the absence of the Coulomb field the energies of the Landau orbits are given by (32)

$$E = (n+\tfrac{1}{2})\hbar\omega_c + (e/m_e\,c)\,\boldsymbol{\sigma}\cdot\mathbf{B} + (p_z^2/2m_e), \qquad\qquad 2.6$$

where p_z is the momentum along the field. [Generally we assume B to be in the z direction and take the nucleus to be at the origin.] The axes of the spirals, the guiding centers of the motion, are distributed uniformly in the x-y plane with a density $(2\pi\hat{\rho}^2)^{-1} = (e/hc)B$. [Note that hc/e is the flux quantum.] Now with a Coulomb potential in addition to the magnetic field, the guiding centers revolve in the x-y plane in circles about the z axis, but the density of guiding centers remains $(2\pi\hat{\rho}^2)^{-1}$.

To illustrate the effect of a strong magnetic field on an atom, let us first consider hydrogen. In the ground state, the electron is essentially in the $n = 0$ Landau orbital, centered on the proton, with its spin antiparallel to $\mathbf{B}$. However, its wavefunction for the z motion is not a plane wave but rather is localized by the Coulomb field of the proton. Roughly speaking the electron probability distribution is a uniform cylinder of radius $\hat{\rho}$, whose length we denote by l. The energy of the electron may be estimated as $E \sim \hbar^2/(2m_e\,l^2) - (e^2/l)\ln(l/\hat{\rho})$. The first term is the kinetic energy associated with the z motion; the second term is the electrostatic energy evaluated for $l \gg \hat{\rho}$. Minimizing with respect to l one finds $l \sim a_0/\ln(a_0/\hat{\rho}) \ll a_0$ and $E \sim -(\hbar^2/2ma_0^2)[\ln(a_0/\hat{\rho})]^2$, which is proportional to $\ln^2 B$; this will, for large fields, be much larger in magnitude than the binding energy of a hydrogen atom in zero field.

In a multi-electron atom in its ground state in a high magnetic field the electrons, Z in number, fill the Z orbitals with $n = 0$ and nodeless z-wavefunctions that lie closest to the nucleus. All electron spins are antiparallel to the field; to flip a spin in 10^{12} gauss costs ~ 10 keV. The last filled orbital is at a distance $\hat{\rho}_Z \sim (2Z)^{1/2}\hat{\rho}$ from the nucleus, and has an energy $E_Z \sim -Z\,e^2/(2Z)^{1/2}\hat{\rho}$. This way of filling single particle states will give the lowest energy atomic state so long as the single particle states with guiding centers close to the nucleus, but with (excited) z-wavefunctions having a node at $z = 0$, have energies $> E_Z$. Such states with nodes are weakly bound and have energies comparable to that of the lowest Bohr state in the absence of the field, and will therefore be filled only if $-Z^2 e^2/a_0 \lesssim E_Z$ or $\eta \equiv a_0/Z\hat{\rho}_z = (B/4.6 \times 10^9 Z^3)^{1/2} \lesssim 1$.

Analysis of the filling of electron levels as a function of η leads to the picture of atoms in strong fields shown in Figure 2, which illustrates schematically the cross-section of the electron distribution in the x-z plane. In Figure 2, a corresponds to the higher field regime, $\eta \gg 1$ discussed above (see also reference 33). In b, some of the electrons occupy states near the nucleus with nodes at $z = 0$. For $1 \gg \eta \gg Z^{-3/2}$ no electrons are in states with the quantum number $n > 1$, but the inner orbitals contain many electrons in states having nodes in the z part of the wavefunction (34, 35). In c, the field is so weak that it can be treated by perturbation theory; here states with many n values are occupied, and the atom is slightly compressed in the direction perpendicular to **B**.

The atoms in the surface of a neutron star with a strong magnetic field bind together to form a rather unusual solid. Since the atoms have large quadrupole moments, electrostatic effects alone can produce binding. Along the field, however, quantum mechanical covalent bonding of atoms is much more effective. Consider first a hydrogen molecule. In the absence of a magnetic field it is unfavorable for two hydrogen atoms with electrons of the same spin to bind, since if the Pauli principle is to be satisfied, the electrons can approach each other only if one of them is excited from the $1s$ state to a state in the next shell; this costs an energy comparable to the atomic binding energy. In a high magnetic field, the Pauli principle can be satisfied with little cost in

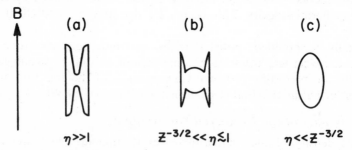

Figure 2   Cross sections of atoms in strong magnetic fields, showing the electronic distribution in the x-z plane, with the field in the z direction. (a) corresponds to a very strong field, (b) to intermediate, and (c) to a weak field. η is defined in the text.

energy by aligning the two protons along the z axis and putting one electron into the lowest orbital, whose axis passes through the positions of the protons, and a second electron in the next orbital, which passes within $\sim \hat{\rho}$ of the protons and therefore has an energy little different from the lowest orbital. The binding of such molecules is considerable (36). Similar arguments can be made for atoms with $Z > 1$.

By adding more atoms one can make covalently bonded chains, in which electrons can move easily along the length of the chain, but are confined in the perpendicular direction by the magnetic field. Ruderman (37) estimates that in the limit $\eta \gg 1$ the energy per atom of such a chain in its lowest energy state is $E_a \sim -0.5(Z^3\,e^2/a_0)\eta^{4/5}$, which greatly exceeds that for a single atom (38) $E_a^{(1)} \sim -(9/8)(Z^3\,e^2/a_0)\ln^2 \eta$; the spacing L between nuclei is $\sim 2.4a_0/(Z\eta^{4/5})$, and the diameter of the electron sheath is approximately the same. Neighboring chains are bound together by electrostatic attraction; in the state of lowest energy adjacent chains are displaced a distance $L/2$ along the z axis relative to each other. At zero pressure the mass density ρ of matter formed of these magnetically polymerized chains is much higher than that of conventional matter; for $\eta \gg 1$,

$$\mu \sim [(AZ^3 m_n)/(12a_0^3)]\eta^{12/5}$$
$$\sim 4 \times 10^3 (A/56)(26/Z)^{3/5}(B/10^{12})^{6/5} \text{ g cm}^{-3}. \qquad 2.7$$

More detailed calculations (39) show that at the surface of a star with $B = 10^{12}$ gauss the matter ends abruptly at a density 2.7×10^3 g cm^{-3} for ^{56}Fe, compared with the zero field density 7.86 g cm^{-3}. For a calculation including exchange effects, but assuming the nuclei to form a uniform positive background, see (40).

The structure of the matter is affected by the field up to densities such that the mean spacing between electrons is small compared with the Larmor radius v_f/ω_c of an electron at the Fermi surface; v_f is the electron Fermi velocity. This occurs for densities $\gg 2 \times 10^4 (B/10^{12})^{3/2}$ g cm^{-3}.

Finally, it should be noted that the electronic band structure, and hence the electrical conduction properties of matter in strong fields, remains an interesting problem. Consequences of the structure of the matter for pulsar emission mechanisms are discussed in (39).

2.2   Nuclei in the Free-Neutron Regime

As the density of the matter increases it becomes more and more neutron rich, as a consequence of the Pauli principle. Beyond a density $\rho_{\text{drip}} \approx 4.3 \times 10^{11}$ g cm$^{-3} = 2.6 \times 10^{-4}$ baryons per fm^3, the neutron drip point, the bound neutron states in nuclei become filled and it becomes energetically favorable to populate the neutron continuum states. Thus the nuclei become immersed in a neutron gas as well as an electron gas. The equilibrium nuclei in this regime were first calculated by Langer et al (41) and then by Bethe, Börner & Sato (42). In these calculations the energy of the nuclei was estimated from a standard semi-empirical mass formula for ordinary nuclei, and the energy of the neutron gas was taken from the calculations of Weiss & Cameron (43) (in 41) and Németh & Sprung (44) (in 42). On the basis of these calculations it was found that the nuclei dissolved at $\rho \sim 5 \times 10^{13}$ g cm^3, and at higher densities the favored state was a mixture of uniform neutron, proton, and electron gases.

Baym, Bethe & Pethick (BBP) (45) attempted to improve upon these calculations by employing better expressions for the neutron gas energy, but found it impossible to give a thermodynamically consistent description of the phase transition between the state with nuclei and the uniform liquid state. The difficulty was traced to a number of inadequacies in the basic model. First, the nuclei present in the free-neutron regime are very neutron rich, with $Z/A \lesssim 0.3$; the matter in them is little different from the free-neutron gas outside the nuclei. However, in the early calculations based on a semi-empirical mass formula, the energies of the matter inside nuclei and the free neutrons were obtained in quite different ways—the nuclear matter energy from a mass formula which is fitted to nuclei having $Z/A \approx 0.5$, and the neutron gas energy from a calculation starting from a neutron-neutron interaction—and it is unlikely that they will give similar results when applied to dilute solutions of protons in neutrons. Second, the nuclear surface energy was assumed to be that of a nucleus with no matter outside. However, the neutron gas outside reduces the nuclear surface energy appreciably, as is clear from the fact that when the matter inside the nucleus and that outside become identical with each other the surface energy must vanish. Third, the lattice Coulomb energy was neglected. BBP overcame the inconsistencies in earlier calculations by developing a "compressible liquid drop" model of the nuclei that used a single expression for the energy of nuclear matter, as a function of density and proton concentration, to calculate the bulk energies of nuclear matter and the neutron gas. This expression was consistent with empirical data for laboratory nuclei for $Z/A \approx 0.5$, and was made to agree for zero proton concentration with Siemens & Pandharipande's (46) nuclear matter theory calculations of the energy of a pure neutron gas (see section 3.1). The surface giving the energy per particle as a function of proton concentration and density is shown in Figure 3 for the parameters used by BBP. To take into account the reduction of the surface energy, BBP constructed an expression for the energy per unit area, which tended to zero when the matter on the two sides of the surface became identical, and which gave the empirical value for ordinary nuclei. As BBP were at pains to point out, the expression was only a first approximation and was the most uncertain feature of their calculation. The Coulomb lattice energy was taken into account using the Wigner-Seitz approximation, in which a unit cell of the crystal is replaced by a sphere of the same volume.

The equilibrium nucleus was found by minimizing the total energy density E for fixed nucleon number and zero charge. This leads to the beta equilibrium condition

$$\mu_n = \mu_p + \mu_e \qquad\qquad 2.8$$

relating the neutron chemical potential $\mu_n \equiv \partial E/\partial n_n$ (at fixed densities of nuclei n_n and protons n_p, where n_n is the neutron density), the proton chemical potential $\mu_p \equiv \partial E/\partial n_p$ (at fixed n_n and n_p), and the electron chemical potential. It also leads to the condition that the chemical potentials of neutrons inside and outside a nucleus must be identical. A further condition obtained from the minimization, on taking the compressibility of nuclear matter into account, is that the pressure

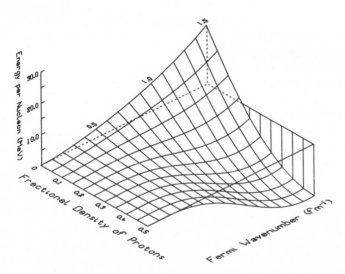

Figure 3 Energy per nucleon in bulk nucleon matter, plotted as a function of the fractional concentration of protons, x, and the mean Fermi wavenumber, k, related to the density of nucleons by $n_b = k^3/1.5\pi^2$. The minimum for $x = 0.5$ at $k \sim 1.4$ fm^{-1} corresponds to bound symmetric nuclear matter; pure neutron matter does not bind. From (45).

of the neutron gas outside a nucleus must balance the total pressure of a nucleus, given by the bulk pressure of the matter inside plus the surface tension pressure and pressure due to Coulomb repulsion. For a given concentration of protons inside the nucleus, the densities of neutrons inside and outside nuclei must be adjusted to achieve equality of chemical potentials and pressures.

Now let us describe some features of the results obtained by BBP. At neutron drip the pressure P is provided almost entirely by the electrons, but as the density increases the free neutrons become increasingly important and give 20% of the total pressure at $\rho = 1.5 \times 10^{12}$ g cm^{-3}, and over 80% for $\rho > 1.5 \times 10^{13}$ g cm^{-3}. At neutron drip the adiabatic index $\Gamma \equiv \partial \ln P / \partial \ln n_b$, where n_b is the mean baryon density, has a value very close to 4/3, the value for a relativistic free Fermi gas. Just above neutron drip Γ drops as $(\rho - \rho_{\mathrm{drip}})^{1/2}$, essentially because the low density neutron gas contributes appreciably to ρ but gives only a small contribution, $\sim (n_{\mathrm{free\ neutron}})^{5/3}$, to the pressure. This implies a small increase in P relative to ρ, and hence a small Γ. As a consequence of the fact that matter in the lower density part of the free neutron regime is very compressible, no stable stars can have central densities in this range (see section 5).

The nuclear A and Z are determined by a competition between the Coulomb and surface energies. One finds

$$Z = w_{\mathrm{surf}}/2w_c\, x, \qquad\qquad\qquad 2.9$$

where $w_{\mathrm{surf}} A^{2/3}$ is the surface energy of the nucleus (w_{surf} depends on the proton concentration $x \equiv Z/A$) and $w_c Z^2/A^{1/3}$ is the Coulomb energy, including the lattice contribution. A and Z, which are important for calculations of transport coefficients and elastic moduli of the crust, are therefore rather sensitive to the surface and Coulomb energies. The equation of state, on the other hand, is very insensitive to the surface and Coulomb energies, since they make up but a small part of the total energy and change little with density. BBP found Z values of over 100 at densities close to nuclear matter density, at about which point the transition to the uniform liquid state takes place; but these values were very uncertain due to the crude estimate they made of the surface energy.

Efforts to treat the nuclear surface better took two different directions. On the one hand Buchler & Barkat (47, 48), Barkat, Buchler & Ingber (49), and Arponen (50) used differential Thomas-Fermi theory to calculate the energy of complete unit cells of the crystal lattice. These methods have the advantage of being able to treat the rather thick nuclear surfaces one finds close to the transition to the uniform state. Numerically, however, they are rather complicated calculations, since no separation is made between bulk and surface energies. Thus, since one is particularly interested in Z, which depends sensitively on surface effects, one must calculate the total energy with high precision if the effects of the rather small surface energy contributions are to be accurately estimated. Buchler, Barkat and Ingber solved the differential equations of Thomas-Fermi theory and Arponen used a variational method, but both methods gave $Z \approx 35$ over most of the density range, falling at higher densities.

The second approach to the surface energy problem by Ravenhall, Bennett & Pethick (RBP) (51) was closer in spirit to BBP's work than were the Thomas-Fermi calculations described above. RBP calculated the energy per unit area of plane interfaces between nuclear matter and a neutron gas in the Hartree-Fock approximation, using the Skyrme interaction (52, 53), which is essentially a velocity-dependent contact interaction. They found that the surface energy dropped more rapidly than BBP's surface energy as Z/A decreased, and consequently the equilibrium Z's they found were smaller than BBP's, typically ≈ 37, and in good agreement at lower densities with the values obtained from the Thomas-Fermi calculations of Buchler & Barkat.

The most detailed investigation of nuclei in this regime has been carried out by Negele & Vautherin (NV) (27) who performed Hartree-

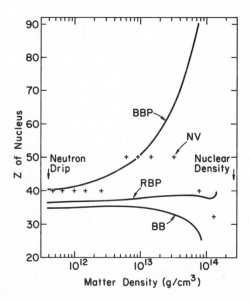

Figure 4   Proton number of nuclei in the crust beyond neutron drip, as a function of the average mass density of the matter. The notations BBP, RBP, BB, and NV refer to (45, 51, 48, 27) respectively.

Fock calculations of unit cells, using their density matrix expansion approach (54), which leads to an effective interaction of the Skyrme type, but with parameters directly related to nuclear matter theory calculations. Their results for the equation of state are in close agreement with BBP. The Z's they found are given by the crosses in Figure 4, where plots of Z's found by other workers are also indicated. As a consequence of shell effects, which are not taken into account in any of the earlier calculations, Z equals either 40 or 50, with the sole exception of the highest density point. NV's calculations shows that $Z = 40$ and 50 are magic numbers even for very neutron-rich nuclei immersed in a neutron gas. The average Z value, ~ 45, is somewhat higher than RBP's, ≈ 37. NV showed how this could be understood in terms of a surface curvature contribution to the energy of a nucleus ($\sim A^{1/3}$), which was not included in the compressible drop model calculations (45), but is implicitly taken into account in their work. In summary, there is now good agreement between the results of different methods for calculating Z; while the full quantum-mechanical calculations like NV's are necessary for investigating the fine details of the properties of the matter, both the compressible liquid drop model, with a suitably chosen surface energy, and the Thomas-Fermi method provide useful descriptions of the gross features.

In Figure 5 we show NV's calculation of the proton (lower curve) and neutron (upper curve) density distributions along the line joining the centers of adjacent cells. The very large A values here of the nuclei include the free neutrons outside the nucleus. (The A computed by BBP included only nucleons within the nucleus.) By ^{1100}Sn one can see the free neutron gas in the figure, while by ^{982}Ge at $\rho \sim 0.5\rho_0$, the highest density computed, one sees that the neutron density is almost uniformly spaced out with a small, nearly sinusoidal variation. The protons always remain clumped together.

At slightly higher densities, $\sim \rho_0$, the neutron and proton density variations smooth out with a first order transition (45), and the matter becomes a uniform liquid. Properties of the liquid regime are reviewed in section 3.1.

2.3   Departures from the Ground State

In the calculations described above, matter was always assumed to be in its absolute ground state. However, in a supernova, matter is heated to extremely high temperatures, and as a consequence the nuclei in a neutron star immediately after formation will not be those corresponding to the lowest energy state. At finite temperatures one expects fluctuations of the nuclei from the most favorable species. In general one expects that before neutron drip the nuclei will, through interactions with the

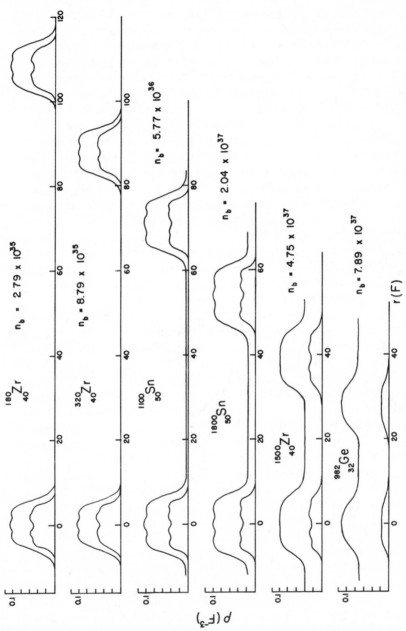

Figure 5 Density profiles of lattice unit cells in the crust for various average densities. The upper curve is the neutron number density; the lower is the proton density; n_b denotes the average nucleon density, measured in nucleons/cm^3. The horizontal axis is distance in fm. From (27). (The density of nuclear matter is 1.7×10^{39} nucleons/cm^3.)

electron sea, tend towards the most favorable Z value for their given A; odd-even effects may, however, inhibit the approach to the most favorable Z. Processes that change A are much more unlikely. Estimate of rates of nuclear transitions in the crust, and the approach to equilibrium, have been given by Tsuruta & Cameron (9). After neutron drip it becomes easy to change A via the free neutron sea. Still, processes that change the total number of nuclei occur very slowly, especially because of large Coulomb (lattice) energies (see also 55).

Bisnovatyi-Kogan & Chechetkin (56, 57) have suggested that the nuclear energy that is released when nuclei in the free neutron regime make transitions to lower energy states may be an important source of energy in neutron stars. Apparently, in these calculations the fact that nuclei can exist in equilibrium with a neutron gas having a positive chemical potential was not taken into account, and more detailed calculations are needed to decide just how important this energy source is.

Another nuclear energy source which could play a role in certain compact X-ray sources is nuclear fusion in the outer layer of a neutron star, through accretion of hydrogen-rich matter onto the surface. Rosenbluth et al (58) suggested that, even though for an accreting neutron star the gravitational energy available (~ 100 MeV/nucleon) greatly exceeds the nuclear energy (~ 7 MeV/nucleon), the nuclear energy could be the dominant source of radiation emitted in soft X rays by the pulsating X-ray sources. In the accreting neutron star model for these sources, the accreting matter is focused toward the magnetic poles by the star's magnetic field and most of the energy is radiated in keV or higher energy X rays from a relatively small portion of the stellar surface. Rosenbluth et al argued that nuclear burning would occur only when the density of accreting matter reached $\sim 10^7$ g cm^{-3}, when neutrons would be produced by electron capture by protons, followed by deuterium production via the reaction $p + n \rightarrow {}^2H$, and higher burning processes. Because the nuclear energy would be released in a part of the star where the thermal conductivity is very high, it would heat the whole interior of the star and not be lost directly through the envelope, which has a poor conductivity. Estimates of the surface temperature due to this process are $\sim 10^6$ °K (or 100 eV). Even though the total nuclear energy release is much less than the gravitational energy release, the nuclear energy, which is emitted from the whole surface of the star, could give more radiation at soft X-ray energies than the hot spots of small surface area produced by gravitational energy release at the magnetic poles.

Hansen & Van Horn (59) have recently pointed out that the analysis of (58) is incomplete. Because the accreting matter at the stellar surface is hot, thermonuclear reactions will take place at densities $\sim 10^3$–10^5 g

cm^{-3}, much less than that required to produce electron capture on protons. They find that the layers in which nuclear burning occurs are unstable, on time scales varying from milliseconds to months. Such instabilities may perhaps be related to the observed time variability of compact X-ray sources, but more detailed calculations are needed.

3 NEUTRON STAR INTERIORS

3.1 The Liquid Phase

Beyond the density $\sim \rho_0$ where the nuclei dissolve, the matter becomes a uniform liquid of neutrons, protons, electrons, and a small fraction of muons, the latter occurring because the electron Fermi energy begins to exceed the muon rest mass here. Beta equilibrium requires $\mu_n = \mu_p + \mu_e$ (cf section 2.8), where μ_p is the proton chemical potential, while charge neutrality requires that the number densities of the species obey $n_p = n_e + n_\mu$. For noninteracting particles $\mu_p \ll \mu_e$, and therefore the condition $\mu_n \simeq \mu_e$ implies that $n_p/n_n \simeq (\mu_e/2mc^2)^{3/2}$; thus at $\rho \sim \rho_0$ the matter is $\sim 99\%$ neutrons. Nuclear interactions lower μ_p much more than μ_n [by ~ 50 MeV at nuclear density (45, 54)], since the mean proton-neutron interaction is more attractive than the neutron-neutron interaction at the relative particle energies in the matter; this has the effect of increasing the proton concentration to ~ 3–4%. Again, the neutrons are stable because the beta decay $n \to p + e^- + \bar{\nu}_e$ is forbidden by the Pauli principle.

3.1.1 NUCLEAR MATTER THEORY   The matter up to densities $\lesssim 2\rho_0$ is well described by the techniques of the Brueckner-Bethe-Goldstone nuclear matter theory. This theory is reviewed in (60); references to early work on the pure neutron gas can be found there and in (15). Recent calculations of pure neutron matter using nuclear matter theory with two body clusters are given by Siemens & Pandharipande (46), based on the Reid potential, and Buchler & Ingber (61), based on Ingber's potential (62); see also (63–65). The energy curve in Figure 3 for pure neutron matter (zero proton fraction) shows essentially the results of the calculations of (46); note that pure neutron matter does not bind. Sjöberg (66) has done a nuclear matter calculation with the Reid potential of the energy and chemical potentials of nuclear matter for arbitrary proton concentration, also retaining only two-body clusters.

Three-body clusters are much less important in pure neutron matter than in symmetric nuclear matter since these arise primarily from tensor forces acting in $T = 0$ states, while pairs of neutrons are always in $T = 1$ states. However, three-body clusters play a role in determining the chemical potential of protons in neutron matter. To simulate effects of higher order clusters Sjöberg has also calculated asymmetric nuclear matter with the $T = 0$ interaction increased by 65%, an amount that enables him to reproduce the correct binding energy of symmetric nuclear matter at density ρ_0. This lowers μ_p in pure neutron matter by ~ 20 MeV for $\rho \gtrsim \rho_0$, and hence increases somewhat the relative proton fraction above the results of (45).

Calculations of Landau Fermi-liquid parameters, and hence compressibility, effective mass at the Fermi surface and low temperature specific heat, and spin susceptibility have been given by Bäckman et al (67) (see also 68). The possibility that neutron star matter becomes ferromagnetic at densities above ρ_0 was suggested in (69–71). More detailed calculations (64, 67, 72–78) rule out this possibility (see also 79).

3.1.2 VARIATIONAL CALCULATIONS As the density increases, many-body clusters become more and more important, and beyond $\rho \sim 2\rho_0$ nuclear matter theory can no longer be expected to give valid results. At higher densities variational calculations have proven useful. These have been based on a Jastrow trial wavefunction of initially unsymmetrized form

$$\psi(\mathbf{r}_1, \mathbf{r}_2 \ldots \mathbf{r}_N) = \prod_{i<j} f_{ij}(|\mathbf{r}_i - \mathbf{r}_j|) \prod_i \phi_i(\mathbf{r}_i), \qquad 3.1$$

where for a liquid the $\phi_i(\mathbf{r})$ are plane waves $\exp(i\mathbf{k}_i \cdot \mathbf{r})$. The functions f_{ij}, the Jastrow factors, take into account correlations between the particles, in particular preventing them from approaching too close to one another. The energy is minimized by varying the f_{ij}. The detailed form of the ground state energy in terms of the ϕ_i and f_{ij} involves rather complicated multiple integrals, and can be found in (80, 81).

Pandharipande (82) has given a very simple "lowest order constrained variation" technique for evaluating the f's in which he requires $f(r) = 1$ for $r > d$ and $\partial f/\partial r = 0$ at $r = d$, where the healing distance d is chosen so that $n_b \int_0^d d^3 r f^2(r) = 1$, where n_b is the particle density; this condition ensures that there is only one particle on the average within a distance d of any particle. Hence with this constraint on f, only two-body clusters become important, and for $r < d$, each f is determined by the Schrödinger-like equation:

$$[-(\hbar^2/m)\nabla^2 + v(r)]\psi_{ij}(r) = [\lambda_{ij} + (\hbar^2/m)\kappa_{ij}^2]\psi_{ij}(r), \qquad 3.2$$

where $\psi_{ij}(r) = \exp(i\boldsymbol{\kappa}_{ij} \cdot \mathbf{r}) f_{ij}(r)$, $\boldsymbol{\kappa}_{ij} = (\mathbf{k}_i - \mathbf{k}_j)/2$, $\mathbf{r} = \mathbf{r}_i - \mathbf{r}_j$, $v(r)$ is the two-body potential and λ_{ij}, the contribution of the two-body interaction to the average field, is adjusted so that f_{ij} satisfies the boundary condition at d. The expectation value of the energy is then

$$E = \sum_i k_i^2/2m + \tfrac{1}{2}\sum_{ij} \int d^3r \psi_{ij}^*[v - (\hbar^2/m)(\nabla^2 + \kappa_{ij}^2)]\psi. \qquad 3.3$$

For neutron matter in practice, $d = 1.2 \; r_0$, where $(4\pi/3)r_0^3 n_b = 1$. Pandharipande & Garde (83) computed in this way the equilibrium composition of neutron star matter, allowing for n, p, e$^-$, and μ^- (and later for hyperons; see section 3.1.4 below), and found that for $\rho \gtrsim 1$ fm^{-3} the matter becomes pure neutrons.

Pandharipande & Bethe (81) have gone on to give a cluster expansion for evaluating higher order corrections in the absence of the constraint on f. Contributions of exchange diagrams decrease rapidly as the number of particles exchanged increases. Results of this method for liquid ^{3}He and ^{4}He agree extremely well with experiment; for neutron matter the lowest order constrained results appear to be within ~ 5–10% of the more exact results, and in excellent agreement at $\rho \sim \rho_0$ with nuclear matter theory (46). Pandharipande & Bethe have calculated the equation of state for pure neutron matter up to $\rho \sim 5$ fm^{-3} using the Reid potential. Bethe & Johnson (84), using their new phenomenological potentials which are more consistent with meson exchanges than is Reid's, have applied the lowest order technique to compute the equation of state for pure neutron matter, as well as for matter with hyperons.

Other variational calculations based on equation 3.1, treated in the two-body approximation and also treated by the Wu-Feenberg (85) method which sums higher order clusters, have been given by Miller et al (86). These calculations agree well with each other, indicating that higher order clusters are not important in neutron matter; they also agree well with Pandharipande & Bethe's results below $\rho \sim 0.9$ fm^{-3}, but deviate significantly beyond this, probably due to an inadequate treatment of the Pauli principle in reference (86).

3.1.3 EFFECTS OF ISOBARS The intermediate range attraction in the nucleon-nucleon interaction is believed to arise primarily from two pion exchange processes, where the two nucleons scatter into intermediate states in which they can be NN, NΔ, or $\Delta\Delta$; here Δ (or isobar or N*) is the 1236 MeV $T = 3/2$, $J = 3/2$ π-N resonance. As Green & Haapakoski (87) pointed out, such attractive two pion exchange processes will tend to be suppressed in matter, due to modification of intermediate state energies, and to a lesser extent to the Pauli principle; thus the equation of state of matter should be harder than that calculated using a matter-independent description of the intermediate range attraction, as, for example, in the Reid potential. [In symmetric nuclear matter this effect has the good result of lowering the calculated saturation density (87), whereas in hypernuclei a similar effect in scattering of NΛ into intermediate $N\Sigma$ states decreases the Λ binding energy (88).] To illustrate the importance of this effect in neutron matter, Green & Haapakoski present a model NN, $N\Delta$ coupled channel calculation of the energy contribution from $^1S_0$ states, using Pandharipande's lowest order constrained variational method. Further work here is clearly needed. See also (89) and section 3.2.

3.1.4 HYPERONIC MATTER   As the density increases in the liquid state, the neutron and electron chemical potentials μ_n and μ_e increase until it becomes favorable for $e^- + n \to \Sigma^- + \nu_e$. In this way (12, 90, 91) one begins to build up a Fermi sea of degenerate Σ^- in "chemical equilibrium" with the neutrons and electrons. As long as the energy of a Σ^- is less than $\mu_n + \mu_e$, decay of the Σ^- is forbidden by the exclusion principle, and the Σ^- will live stably in the matter. Similarly, other baryons, such as Σ^0, Σ^+, Λ, and Δ (in four possible charge states) can be formed. One expects Σ^- to appear before Λ even though they are more massive, since $e^- + n \to \Sigma^- + \nu_e$ removes a high energy e^- as well as a neutron.

The problem in computing the properties of such "hyperonic matter" is that one knows very little about the forces between hyperons; much beyond $\rho \sim 5$–$10\ \rho_0$ one no longer expects the forces to be describable in terms of static two-body potentials. Various two-body potential models of interactions of hyperons and nucleons have been proposed. Langer & Rosen (92) and Libby & Thomas (93) assumed a universal force to act between all pairs of baryons. Pandharipande (82) examined a similar model with the (central) Reid $T = 1$ state interaction assumed between all baryons, plus the one-pion exchange potential between nucleon pairs. In general one expects that if the nucleon core repulsion is primarily due to (isoscalar) ω exchange, then the repulsive core should be the same between all baryons. Figure 6 shows the composition of hyperonic matter in this model; note that Δ's are the most prevalent species at high densities due to their higher spin multiplicity $(2j+1 = 4)$. Since this choice of interaction is too attractive for many pairs, including neutron-proton, he also examined a model with the intermediate range attraction arbitrarily reduced by 10% between all pairs except n-n and p-p. Somewhat more realistic is the model of Pandharipande & Garde (83) in which they include the full Reid interaction between pairs of nucleons, and two models for the n-hyperon interaction; in the first it is taken to be the n-p interaction, while in the second it is argued on the basis of SU(3) symmetry that the $\Sigma^- - n$ interaction in triplet states is better replaced by other components of the Reid potential that are less attractive than the Reid $T = 0$ triplet interaction. In the first case Σ^-, but no other hyperons, make a brief appearance for $\rho \sim 0.5\ \text{fm}^{-3}$, whereas in the second, because of the weakened hyperon-nucleon attraction, it is never energetically favorable for hyperons to appear. The possibility of Δ's was not allowed for.

Bethe & Johnson (84), in computing the equation of state and com-

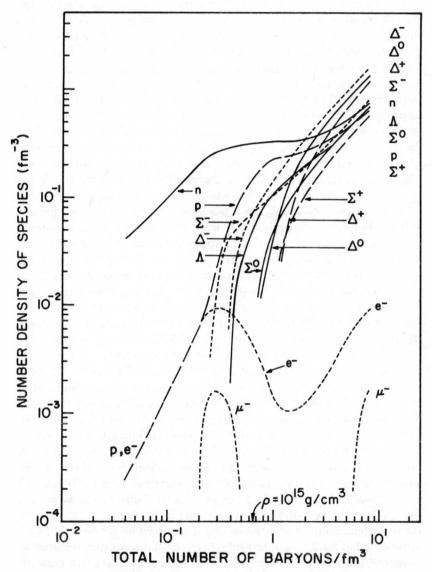

Figure 6 Number densities of the various components of hyperonic matter, as a function of total baryon density. After (82).

position of hyperonic matter, chose the hyperon-nucleon and hyperon-hyperon interaction to be the $^1D_2$ nucleon-nucleon potential in all even states, for all hyperons, while in all odd states the potential was chosen to be the spin-isospin average of the p-state nucleon-nucleon potential, minus the one pion exchange contribution. Such a choice is consistent with the result from hypernuclei that the hyperon-nucleon attraction is slightly less strong than that of two nucleons. Bethe & Johnson find the result that the inclusion of hyperons in their models produces only a slight softening of the equation of state.

A further application of SU(3) symmetry arguments to the determination of the effective hyperon potentials is due to Moszkowski (64), who assumes the same short range repulsion between all baryon pairs, but modifies the attraction as follows. A universal attraction is used for all like baryon pairs, and a second for all unlike pairs, except that the Λ-nucleon and Σ-nucleon couplings are reduced by a factor 2/3. Similarly the coupling between pairs of strange hyperons is reduced by 4/9. These factors are introduced on the basis of the SU(3) quark model under the assumption that the attraction arises from exchange of scalar mesons with the baryon-scalar coupling constant proportional to the number of nonstrange quarks contained in the baryon. Four model interactions are assumed, and equations of state are computed for both pure neutron and hadronic matter. Neutron star matter calculated with these interactions contains relatively few hyperons, a result consistent with that of (83).

Further discussions of the treatment of Δ's as elementary particles in dense matter are found in (94, 95).

3.1.5   SUPERFLUIDITY AND SUPERCONDUCTIVITY   The neutrons in the crust and interior of neutron stars, as well as the protons in the interior, are most likely superfluid (96, 97). It is very unlikely that the electrons will ever be superconducting (97). One can estimate, from the BCS weak coupling theory (98), the transition temperature $T_{c,e}$ for electron superconductivity to be $T_{c,e} \sim T_{f,e}\, e^{-1/N(0)V}$; $T_{f,e}$ is the electron Fermi temperature expressed in terms of the electron Fermi momentum p_e by $kT_{f,e} = cp_e$ for fully relativistic electrons. The density of states (for one spin population) at the electron Fermi surface is $N(0) = p_e^2/2\pi^2\hbar^3 c$, since the electron effective mass is p_e/c. The mean net attraction V between electrons is a quantity of order $e^2(\hbar/p_e)^2$, so that $N(0)V \sim e^2/\hbar c$. Thus $T_{c,e}$ is essentially zero; at stellar temperatures the electrons are normal.

Superfluidity of neutrons, as well as protons, arises from pairing interactions, as in nuclei and laboratory superconductors. Estimates of neutron and proton energy gaps for $^1S_0$ pairing have been given by many authors using a variety of nucleon-nucleon potentials. The most careful recent calculations are those for neutrons and protons by Chao, Clark & Yang (99, 100) who use the method of correlated basis functions (80) extended to include pairing correlations; and those for neutrons by Takatsuka (101), who employs soft-core effective potentials. These authors' results for the neutron $^1S_0$ gaps are in reasonable agreement with the early calculations of Hoffberg et al (102) using effective pairing interactions derived from phase shifts.

As Ruderman pointed out (103, 102), at the relative neutron-neutron scattering energies in neutron star matter above nuclear densities, the $^3P_2$ is the dominant attractive phase shift, as a consequence of the tensor force, and thus $^3P_2$ pairing should be favored at such densities. Detailed calculations of anisotropic $^3P_2$ pairing in neutron matter have been done by Takatsuka (101), Tamagaki (104), and Tamagaki & Takatsuka (105). It should be pointed out though that all these calculations, including those for $^1S_0$ pairing, should be regarded as first estimates; because the gaps depend exponentially on the strength of the interaction, the results depend sensitively on small effects such as modification of the proton-proton interaction due to polarization of the neutron liquid, and single particle self-energies and lifetimes (106).

Figure 7 shows the neutron superfluid transition temperatures computed by Hoffberg et al (102), together with the proton superfluid transition temperatures, from Chao et al (100). The superfluid transition temperatures for neutron and for proton pairing may be estimated from the BCS result $T_c \approx \Delta/1.76$ for $^1S_0$ pairing, where Δ is the corresponding $T = 0$ energy gap. For the $^3P_2$ gap computed in (102) $T_c \approx \Delta/2.4$, where Δ is the maximum gap (as a function of direction on the Fermi surface). The transition temperatures are generally well above the expected ambient temperatures in all but very young neutron stars. The picture of superfluidity in neutron stars that emerges is that the neutron fluid in the crust should be paired in $^1S_0$ states. The transition from $^1S_0$ to $^3P_2$ pairing occurs (coincidentally) at about the density where the nuclei dissolve, and thus in the liquid interior one expects $^3P_2$ neutron pairing and $^1S_0$ proton pairing.

One consequence of neutron superfluidity is that because neutron stars rotate, the neutrons should form an array of quantized vortices, analogous to the vortices in rotating superfluid helium II. The quantum of circulation is $h/2m$, where $2m$ is the mass of a pair of neutrons. For a rotation period of $\frac{1}{30}$ sec (that of the Crab pulsar) the vortex spacing is $\sim 10^{-3}$ cm, small compared with characteristic stellar dimensions. By contrast the vortices are widely spaced compared with the neutron coherence length $\xi_n = \hbar^2 k_n/\pi\Delta_n m \sim 10^{-12}$ cm, where Δ_n is the neutron energy gap, and k_n the neutron Fermi wavenumber. The number of vortices present is always large enough that the moment of inertia of the neutron superfluid, in the case of $^1S_0$ pairing, is not reduced from the classical value. In general a $^3P_2$ superfluid is anisotropic with complicated flow properties. Ginzburg-Landau equations for describing these properties have been derived by Richardson (107) and Fujita & Tsuneto (108).

Let us consider now the properties of proton superconductivity (109). The critical magnetic field H_c (or H_{c1} for type II superconductivity) is of order 10^{16} gauss (110). [Note that the scale of critical fields is $(hc/2e)/4\pi r_p^2$, where r_p is the mean proton separation; for $r_p \sim 10^{-13}$ cm, this is $\sim 10^{17}$ gauss.] Since $B \sim 10^{12}$ gauss in a neutron star one would normally expect a Meissner effect in which the magnetic flux is expelled from the regions of superconducting protons. Such an expulsion would have a serious effect on the electromagnetic properties of pulsars. However, because of the enormous electrical conductivity of the normal state (111) ($\sigma \sim 10^{29}$ sec^{-1}) the time (55) $\tau \approx (4\pi\sigma/c^2)R^2(B/2H_c)$ required to

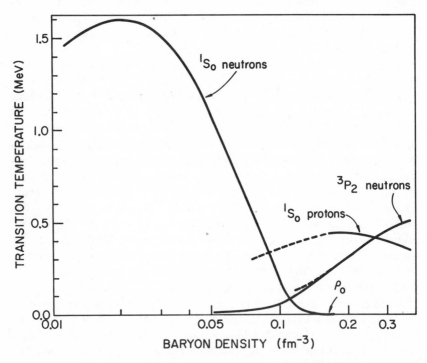

Figure 7 Superfluid transition temperatures as a function of total baryon density. The neutron transition temperatures are taken from (102), while the proton transition temperatures are from (100), using the BCS relation $k_B T_c = \Delta/1.76$.

expel flux from a region of size $R \sim 1$ km is $\gg 10^8$ years. The super-conductivity, rather than waiting so long, simply nucleates with the field present. Although this is technically a metastable situation, the actual lifetime of the state can be shown to be orders of magnitude longer than the time τ.

If the protons are a type I superconductor they will learn to live with the field by forming an intermediate state configuration, with alternate layers of superconducting material, which are field free, and normal material containing the field. On the other hand, for type II super-conductivity, which is more likely, the magnetic flux will be contained in an array of quantized vortices, each of a single flux quantum $\phi_0 = hc/2e$. This is the same state as occurs in laboratory type II superconductors for $H_{c1} < B < H_{c2}$, only now $B \ll H_{c1}$.

The criterion for type II superconductivity is that the penetration depth λ should be greater than ξ_p, the proton coherence length, divided by $(2)^{1/2}$. Estimating λ from the London result (112) $\lambda = (mc^2/4\pi n_p e^2)^{1/2}$, where n_p is the proton number density, and using $\xi_p = \hbar^2 k_p/\pi \Delta_p m$, where k_p is the proton Fermi wavenumber, we find

$$\xi_p/(2)^{1/2}\lambda = [(8/3\pi^2)(e^2/\hbar c)(\hbar k_p/mc)]^{1/2}\varepsilon_p/\Delta_p; \qquad 3.4$$

here $\varepsilon_p = \hbar^2 k_p^2/2m$. From the calculations of (100), ε_p/Δ_p is ~ 10 from the outer surface of the liquid interior to $\rho \sim 5 \times 10^{14}$ g cm^{-3}. In this range $\xi_p > (2)^{1/2}\lambda$, indicating that the protons should be a type II super-conductor.

At temperatures well below the transition temperature the reduction of the specific heat (113) and neutrino emission rates (114, 115) below their values in the normal state affects the cooling of neutron stars (8).

3.2 Solidification of Neutron Matter

The overall similarity—a short range repulsive core with an intermediate range attraction—of the interatomic potential between rare gas atoms and the neutron-neutron interaction, raises the possibility that, as in the rare gases, neutron matter under sufficient pressure may solidify with a single nucleon at each lattice site. If this occurs then neutron stars would have a solid inner core, in addition to the solid outer crust. Such a solid core could have several observational consequences. For example, a solid core in the Vela pulsar could contain sufficient elastic energy to be seismically active enough to explain, via the "starquake" model (116, 219), the observed speedups in the Vela pulsar's pulse repetition frequency and hence in its rotation rate. Dyson, at the Sixth Texas Symposium on Relativistic Astrophysics, pointed out that the (transverse) rumblings of a solid core would, because of the high density and sound velocity, be an efficient source of gravitational radiation. Furthermore, a neutron star with an oblate solid core could have a relatively rapid wobble. Wobble, due either to solid crust or core (117), could be the clock mechanism for the 35^d cycle observed in the accreting close binary X-ray source Hercules X-1, if, say, the accretion of matter onto the neutron star in the binary, the source of the X rays, depended on the orientations of the magnetic moment and rotation axis of the neutron star with respect to the plane of the binary orbit (118).

The possibility of solidification of nuclear matter was discussed relatively early by Cazzola et al (119), Coldwell & Wilets (120), and Coldwell (121). Recent interest in the problem was stimulated by the work of Banarjee et al (122), whose calculations suggested that in the absence of quantum zero point effects a neutron solid could be stable (though not necessarily of lower energy than a liquid at the same density), and by the calculations of Anderson & Palmer (123), based on de Boer's theory of corresponding states (124) which has been so successful for rare gas liquids and solids.

In this latter method the nucleon-nucleon interaction (suitably averaged over angular momentum) is replaced by a state-independent central potential of the Lennard-Jones form, but with an inwardly shifted core; by comparison with rare gas data, particularly that of helium, Palmer & Anderson (125) find that neutron matter should solidify at $\rho \sim \rho_0$. Clark & Chao (126) attempted to allow for density dependence in the choice of the effective Lennard-Jones potential, and found a solidification density $\rho_s \sim 2\rho_0$.

The "corresponding state" results are inconclusive, however, because the nucleon-nucleon interaction has strong state dependence, in addition to having a much softer core ($\sim r^{-1}$ at very short distances) than that $[\sim (r+a)^{-12}]$ of the (shifted) Lennard-Jones potential, where a is the core shift distance. Solidification of the rare gases at high densities is due primarily to localization of the atoms by the hard core (127); on the other hand, it is not evident that the (soft) Yukawa nucleon core is capable of overcoming the zero point motion sufficiently to localize the nucleons. Cochran & Chester (128) and Cochran (129) have done extensive Monte-Carlo calculations to study the validity of the corresponding state approach and conclude that the softness of the Yukawa nucleon core leads to quite different solidification behavior, compared with results of calculations in which one models the Yukawa interaction by a hard-sphere (130) or Lennard-Jones nucleon interaction. The uncertainties in the determinations of the strength of the nucleon core present a further complication in assessing the role of the core in producing solidification. The state dependence becomes significant because the relative angular momentum of two localized nucleons in a solid can be rather high, up to $l \sim 2\pi/\theta$, where θ is the relative angle of localization of the two nucleons. The s-wave component is generally very small and thus much of the s-wave attraction is lost by localization; the higher partial wave forces, particularly the d-state, are more important. In addition the tensor force can play a large role in producing spatial order (89). One must resort to detailed calculations to determine if and when solidification occurs.

Quantum calculations of the ground state energy of solid neutron matter have been performed by several groups starting from either a variational or t-matrix formalism. We describe first the variational calculations. These are based on the Jastrow trial wavefunction of the initially unsymmetrized form (3.1), where for a solid the single particle wavefunction $\phi_i(\mathbf{r}_i)$ is taken to be a Gaussian centered on lattice site $\mathbf{R}_i$, and whose width is regarded as a variational parameter.

Nosanow & Parish (131) evaluated both solid and liquid energies of a neutron system taking the potential to be the Reid $^1S_0$ for all even l states and the central part of the Reid $^3P_2$-$^3F_2$ for all odd l states. They used a Monte-Carlo technique to compute the energy, and included effects of exchange by means of the leading term in a cluster expansion (85, 132). The solidification density, where the solid ground state energy falls below the liquid energy, was found to occur at the rather low value $\rho \approx \rho_0$. As we have already discussed, such a choice of potential overestimates the amount of attraction in the solid, lowering the energy of the solid relative to the liquid, where s-wave interactions are more important. Since their solid has a large zero point motion, the lowest order treatment of exchange effects is open to question (133).

Pandharipande (133) chose as a potential instead the Reid central $^1D_2$ in all even l states, except $l = 0$, where he uses the Reid $^1S_0$; the odd state potential is again the Reid central $^3P_2$-$^3F_2$. The energy was evaluated in the solid by his lowest order constrained variational method (section 3.1.2), and compared with the liquid energies computed by Pandharipande & Bethe (81) in their "hypernetted-chain" approximation, using central Reid potentials. The result was that the liquid energies remained below the solid energies up to the highest density considered, $\rho = 3.6$ fm^{-3}, and thus there was no evidence for solidification using central potentials alone.

The t-matrix approach (134–136) is more analogous to the familiar Brueckner-Bethe-Goldstone nuclear matter theory (60), where one solves for the motion of a pair of particles in the presence of the background sea of the other particles. The self-consistent single particle potential is taken to be a harmonic crystal potential; in the absence of pair correlations the wavefunction is a product of Gaussians. Three-body clusters and further higher order correlations can be calculated, but in a solid are expected to be small because of the localization of particles near lattice sites. The Bethe-Goldstone equation for the correlated two particle wavefunction is solved by an angular momentum decomposition, which gives a set of coupled integral equations for the various partial wave components of the pair wavefunction. In the solid this approach has the advantage of being able to take into account much of the angular momentum dependence of the nucleon-nucleon interaction, and can be expected to be useful up to rather high densities (at which nuclear matter theory for the liquid is unreliable, due to the importance of higher-order clusters and exchange effects).

Canuto & Chitre (137, 138) applied this technique to the solid, using the Reid potential for $l \leq 2$, the Reid $^1D_2$ for even $l \geq 4$, and the central Reid $^3P_2$–$^3F_2$ for odd $l \geq 3$. Comparing their computed neutron solid energies with the liquid energies computed by Pandharipande (139), they conclude that solidification should occur at $\rho \simeq 1.0 \, \text{fm}^{-3}$. They also examine the *stability* of the solid (138), and find that below $\rho \sim 1$–2 fm^{-3}, the elastic constant C_{44} becomes negative, indicating that the solid is unstable against infinitesimal shears.

The calculations described above were based on a range of different methods and used a variety of potentials, and it was not clear whether differences between the results obtained were due to inadequacies of some of the methods, or differences between the potentials. At the Urbana Workshop on Dense Matter in April 1973 it was agreed that the various groups should all calculate a simple model problem, proposed by Bethe, to check the reliability of the calculational methods. Since the repulsive, short range part of the potential plays an important role in causing solidification, the interaction between neutrons in Bethe's "homework problem" was chosen to be the repulsive part of the Reid $^1S_0$ potential, $V_{HW}(r) = 6484.2 \, \text{e}^{-7x}/x$ (in MeV, where $x = m_\pi r$), which was assumed to act in all partial waves. To avoid complications with exchange, the "neutrons" were assumed to obey Boltzmann statistics.

There is now nearly a consensus that the homework neutrons do not solidify. Takemori & Guyer (140) have recently carried out t-matrix calculations which differ in detail from those of Canuto et al (137, 138, 141); they find that at the three densities examined ($\rho \leq 3 \, \text{fm}^{-3}$), there is no minimum in the solid energy as a function of the width of the single particle Gaussians; this indicates that there is no self-consistent solid wavefunction and hence no stable solid configuration. They also find the same result when calculating with the Reid $^1S_0$ potential, including attraction, and indicate that introduction of further components of the Reid potential and exchange corrections does not modify this result. Their calculations were done for a body-centered cubic (bcc) lattice (with an anti-ferromagnetic spin order in the more realistic calculations). Similar results were reported by Østgaard (142). These results are in distinction to those of Canuto, Lodenquai & Chitre (141) who find a stable homework solid configuration at all densities examined $(0.2 \leq \rho \leq 1 \ \text{fm}^{-3})$, with energies lying below the homework liquid energies computed by Pandharipande (143). The interested reader can find in Takemori & Guyer's paper (140) a critical comparison of their work with that of Canuto et al.

Chakravarty, Miller & Woo (144), in a Jastrow variational calculation, called the "quasi-crystal approximation" (145), found solid configurations and energies in close agreement with those of Canuto et al (141). However, their solids have a sufficiently large zero point motion to cast doubt on the validity of their method of calculating the energy. Their liquid energies, computed in a low order Jastrow procedure, lie above their solid energies (indicating in this order of approximation that homework neutrons solidify). Shen, Sim & Woo (146) also examined effects of three-particle correlations on the liquid energies and found only small corrections, while Shen & Woo (147) calculated liquid energies using the hypernetted chain and Percus-Yevick equations (80), and found results essentially equivalent to those of Chakravarty et al. Recently, however, Lowy & Woo (148) have reexamined the calculation of the solid phase from the point of view of the Kirkwood-Monroe theory (149), in which one looks for states havng a spatially periodic density distribution, and conclude that no such distributions are possible, and thus that the homework solid does not exist.

Further evidence against the solidification of homework neutrons was provided by the Monte-Carlo calculations of Cochran & Chester (128). For a repulsive potential $V(r) = 5725 \, e^{-4.0r}/r$ (in MeV, with r in fm) they find only weak evidence for a possible solid configuration, and then always at greater energy than the liquid. With an added attraction $-1974 \, e^{-2.9r}/r$, local minima associated with solidification disappear. The liquid calculated with repulsion alone agrees well with calculations of Shen & Woo (147) for the same potential.

Pandharipande (143) has applied the Pandharipande-Bethe method (81) to calculation of the homework liquid, and finds for the $5725 \, e^{-4.0r}/r$ potential the results are indistinguishable from the Cochran-Chester Monte-Carlo calculations. His lowest order constrained variational calculations lie ~ 10–15% above these more exact results. He does not find, however, any stable solid configurations.

The result of the homework problem has been to produce general agreement on the calculated energies of the liquid and on the inability of the soft repulsive core to produce solidification in the density range 1–10 ρ_0. Given the disparity of the work of (141) on the homework solid with that of other workers, the description of (137, 138) of the solidification of real neutron matter is open to question.

All the calculations described above have been done in the framework of nucleons interacting via two-body potentials. Pandharipande & Smith (89) have described a possible mechanism for solidification utilizing the fact that interacting nucleons spend time in intermediate states as Δ's or isobars (see section 3.1.3). They consider, as a model, a solid configuration constructed by stacking plane layers of neutrons; in any layer the spins are all parallel and aligned perpendicular to the layer, the spin direction reversing from one layer to the next. In this arrangement the tensor force is attractive. Such a configuration, with a nonvanishing divergence of the $(T = 1)$ spin density has a nonvanishing expectation value of the π° field, and thus has a "π° condensate" (see section 3.3.3). The π° field excites the nucleons to intermediate Δ states, and this leads to an enhancement of the effective tensor force in the matter sufficient to produce a solid of lower energy than the liquid for $\rho \gtrsim 2\rho_0$. (The normal Reid tensor force is too weak to lead to such solidification.) While Pandharipande & Smith have not shown that a solid configuration, as opposed to a layered "liquid crystal" arrangement, is more energetically favorable, their calculations do indicate how enhancement of the tensor force can lead to spatial ordering in dense matter; such models deserve further study.

3.3   Pion Condensation

The question of whether pions appear in dense neutron star matter has been of substantial theoretical interest, because of their possible effects on the neutrino cooling of neutron stars and the equation of state, and because of the relation of the question to the understanding of the properties of pions in laboratory matter. Neglecting for the moment the strong interactions of pions with the matter, one sees that π^- will form via $n \rightarrow p + \pi^-$ once the neutron-proton chemical potential difference $\mu_n - \mu_p = \mu_e$ exceeds the π^- rest mass $m_\pi = 139.6$ MeV. In neutron star matter in beta equilibrium, $\mu_n - \mu_p$ reaches ~ 100 MeV at nuclear matter density (45), and one might expect the appearance of π^- at slightly higher densities.

One cannot, of course, neglect the interaction of the pion with the background matter. As noted by Bahcall & Ruderman (see 92), the strong repulsive s-wave pion-nucleon interaction tends to increase the pion effective mass and therefore the pion threshold density, the first point at which pions can appear. At low π^- energies the increase $\Delta E_{\pi^-}^s$ in the energy of a single π^- due to s-wave interactions with the nucleons in the matter can be estimated in terms of measured pion-nucleon scattering lengths from

$$\Delta E_\pi^s = -(2\pi\hbar^2/m_\pi)[a_3 n_n + \tfrac{1}{3}(2a_1 + a_3)n_p] \approx 217(n_n - n_p)\ \text{MeV}, \qquad 3.5$$

where the densities n_n and n_p are measured in fm^{-3}, and a_3 and a_1 are the scattering lengths in isospin $3/2$ and $1/2$ channels; experimentally $(150, 151)$, the isospin-odd scattering length $a_1 - a_3 \approx 0.262 m_\pi^{-1} \approx 0.370\,\text{fm}$, while the isospin-even scattering length $2a_3 + a_1 \approx -0.01 m_\pi^{-1} \approx 0$. In symmetric nuclear matter ($n_n = n_p$), ΔE_π^s is close to zero, while in neutron star matter at nuclear density ΔE_π^s raises the pion energy by ~ 38 MeV. Calculations of neutron star matter indicate that $\mu_n - \mu_p$ is always less than $m_\pi + E_\pi^s$, and thus if the only pion-nucleon interaction were s-wave, pions would never appear in neutron star matter. However, Migdal $(152–154)$, and independently Sawyer and Scalapino $(155–157)$ pointed out that the attractive p-wave pion-nucleon interaction, which leads to the $T = 3/2$, $J = 3/2$ pion-nucleon scattering resonance Δ, would greatly reduce the energy of pions in the medium. Such p-wave pion self-energy processes are shown in Figure 8, where a corresponds to the processes $\pi^- + p \leftrightarrow n$ and $\pi^+ + n \leftrightarrow p$, and b corresponds to $\pi^- + n \leftrightarrow \Delta^-$ and $\pi^- + p \leftrightarrow \Delta^0$, as well as $\pi^+ + n \leftrightarrow \Delta^+$ and $\pi^+ + p \leftrightarrow \Delta^{++}$. The best recent calculations $(158, 159)$, which take into account effects due to nuclear forces, indicate that negative pions can indeed appear at densities not much beyond nuclear matter density.

Because pions are bosons, if they do appear in the ground state they will macroscopically occupy the lowest available mode, i.e. form a condensate as in ordinary Bose-Einstein condensation (see, e.g. 160). Such a state corresponds to a classical or coherent excitation of the pion field in the medium, or alternatively, to a nonvanishing ground

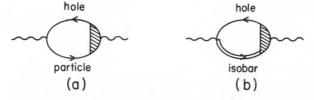

hole   hole

particle   isobar
(a)   (b)

Figure 8 Feynman diagrams for pion self-energy processes in the medium. The wavy lines are pions, the solid lines nucleons, and the hatched regions are the p-wave pion-nucleon vertices in the medium. (a) is the p-wave coupling of the pion to a nucleon particle-hole state, while (b) is the coupling of the pion to a nucleon hole-isobar (Δ) state; the isobar is represented by a double line.

state expectation value of the pion field; in the normal ground state, with no pions, this expectation value vanishes.

3.3.1 THE π^- CONDENSATION THRESHOLD

The appearance in the ground state of charged pions, or more precisely, pion-like modes of excitation, leading to a nonvanishing expectation value of the charged pion field, could actually take place through several different physical mechanisms. To see this, consider the excited states of the matter with the quantum numbers of the π^+ or π^-, relative to the ground state. These include, first, the π^+ and π^- states, which reduce to the free $\pi^\pm$ states in the absence of matter. The properties of these states are of course modified by the matter. Also there is a continuum of proton (quasi)-particle–neutron (quasi)hole states, with minimum energy $\mu_p - \mu_n < 0$, and the quantum numbers of the π^+; in matter with some protons present there is in addition a continuum of neutron particle–proton hole states with the quantum numbers of the π^-. Furthermore there are continua of nucleon hole–isobar (Δ) states.

Due to p-wave interactions which couple the "free" pion states to the particle–hole states, neutron star matter may develop, as Migdal (154) has shown, a neutron hole–proton particle 0^-, $T = 1$, collective mode, or spin-isospin sound, with the quantum numbers of the π^+. The range of wavevectors k for which this collective mode, which we denote by π_s^+, exists depends on the density. Where it exists its frequency lies *below* the continuum of proton particle–neutron hole states. This mode, a form of zero sound, analogous to the giant dipole mode, is an oscillation of the T_1 and T_2 components of the isospin density, together with a counteroscillation of the components of spin density parallel and anti-parallel to the wavevector.

In a pure neutron gas, condensation can take place as the density is raised, through two possible mechanisms. The first is when it becomes energetically possible for a neutron (quasi)particle to turn into a proton (quasi)particle plus a π^-; this transition would occur when the lowest energy π^- state in the system, of energy ω_-, has energy equal to $\mu_n - \mu_p$, the highest energy neutron particle–proton hole state. Before pions appear, ω_- is always $> \mu_n - \mu_p$. The second possibility is that the lowest energy π_s^+ state, of energy ω_s, becomes so low that $\omega_s + \omega_- = 0$. Then it becomes energetically favorable for the system spontaneously to create pairs of π^- and π_s^+. This latter mechanism appears the more likely in realistic calculations. [It is not necessary for the total momentum $\mathbf{q}$ of the pair π_s^+ and π^- (or in the first case that of the neutron particle–proton hole and π^-) to vanish for condensation to take place. Momentum conservation occurs by a transfer of the momentum difference, $\mathbf{q}$, to the system as a whole, with each particle recoiling by an infinitesimal amount $\mathbf{q}/N$, where N is the total number of particles; the energy of recoil $\sim q^2/2mN$ vanishes for large N.]

If the matter is initially in beta equilibrium, so that a small fraction of protons is initially present, then condensation can also occur through the mechanism $p \rightarrow n + \pi_s^+$, which is energetically possible when $\mu_p - \mu_n = \omega_s$. (When the π_s^+ spectrum begins to appear in the normal state it lies above $\mu_p - \mu_n$.) The existence of this additional mechanism for condensation means that matter in beta equilibrium can condense at a lower density than pure neutron matter. For example, in the simple model calculation (161, 162) in which one assumes a free nucleon gas with a nonrelativistic p-wave pion-nucleon interaction $H_p = -(f/m_\pi)N^\dagger\tau \cdot (\boldsymbol{\sigma} \cdot \mathbf{V})\Pi N$ [where $f \simeq (0.081 \times 4\pi)^{1/2}$ is the pion-nucleon p-wave coupling constant, N is here the nucleon annihilation operator, a spinor in isospin space, τ and $\boldsymbol{\sigma}$ are Pauli matrices for isospin and spin, and Π is the pion field operator; this interaction leads to a self-energy for the pion represented by the process in Figure 8a], together with the s-wave pion-nucleon interaction one finds that for pure neutrons the condensation threshold occurs at $\rho = 0.30$ fm^{-3}, while for neutrons, protons, and electrons in beta equilibrium, condensation occurs at $\rho = 0.08$ fm^{-3}. However, studies by Weise & Brown (163) indicate that with inclusion of short range correlations between nucleons, matter in beta equilibrium condenses at a density not very different from that for pure neutron matter.

The existence of the spin-sound mode can also affect the proton chemical potential in the normal phase of a pure neutron gas. For example, in a free neutron gas μ_p is, to a first approximation, zero, the energy of the lowest free proton state. However, beyond the density, call it ρ_E, where $\mu_p - \mu_n$ equals ω_s (and at which the system would condense were some protons initially present), it is more favorable when one adds a proton to the system to have it turn into $n + \pi_s^+$, even in the normal state. Thus beyond ρ_E, μ_p is lowered to $\mu_n + \omega_s < 0$. In general the proton chemical potential is continuous from the normal to the infinitesimally condensed phase.

The most realistic attempts to estimate the π^- condensation threshold in pure neutron matter have been made by Bäckman & Weise (158) and Bertsch & Johnson (159). These calculations include s- and p-wave pion-nucleon interactions, coupling to Δ states, and effects of nuclear forces. The latter modify the energies of the particle-hole states to which the pion field is coupled, as well as introducing vertex corrections in the p-wave pion-nucleon matrix element. At $\rho \sim 2\rho_0$, where condensation appears likely, the particle-hole energy shift is not as important as at nuclear densities, due to the velocity dependence of the nuclear forces. The vertex corrections in the pion channel are due to coupling of particle-hole states, as well as isobar–nucleon hole states, through the part of the effective nucleon-nucleon interaction of the form $g'(\sigma_1 \cdot \sigma_2) \times (\tau_1 \cdot \tau_2)$ [where σ_1 is the spin and τ_1 the isospin matrix element for the initial particle-hole state, and σ_2 and τ_2 the spin and isospin matrix elements for the final]; to a good approximation g' is a density- and momentum-dependent nuclear matter theory G-matrix element, written as a particle-particle interaction as $\langle \mathbf{p} - \mathbf{k}, \mathbf{p}' | G | \mathbf{p}, \mathbf{p}' - \mathbf{k} \rangle$ (with subtraction of the single pion exchange contribution in the direct channel). For $g' = \frac{1}{3}(f/m_\pi)^2$, the vertex corrections are equivalent to the Ericson-Ericson Lorentz-Lorenz effect (164), which is simply the removal, due to the short range particle correlations, of the effects of the attractive delta function in the one-pion exchange potential (165). For $g' > 0$ the vertex corrections make condensation more difficult. Migdal, in the framework of the Landau Fermi-liquid theory, estimates (166) that in symmetric nuclear matter $g' \approx 0.5(f/m_\pi)^2$. From nuclear matter theory calculations using the Reid potential, Bäckman & Weise (158) estimate that $g' \approx 0.43(f/m_\pi)^2$, while Bertsch and Johnson (cited in 167) find $g' \simeq 0.55 - 0.65(f/m_\pi)^2$, reasonably independent of density and momentum. The detailed value of g' depends critically on an accurate description of the short range correlations; the difference between these results reflects differences in the correlations assumed.

Bäckman & Weise find condensation at $\rho_c = 0.34$ fm^{-3}, with $\mu_\pi = 147$ MeV and wavevector $k = 2.05$ fm^{-1}, while Bertsch & Johnson find the somewhat lower values $\rho_c \approx 0.225$ fm^{-3}, $\mu_\pi = 119$ MeV, $k = 1.50$ fm^{-1}. In both calculations condensation occurs due to spontaneous creation of π^- with the collective mode π_s^+ [lying ~ 25 MeV below the particle-hole continuum (159)]. These results for ρ_c lie above Migdal's earlier estimate $\rho_c \lesssim 2\rho_0$ (154).

3.3.2 PROPERTIES OF THE CONDENSED STATE   The charged-pion condensed system is characterized by a nonvanishing expectation value $\langle \Pi(\mathbf{r}, t) \rangle$ of the charged pion field. In the normal state this expectation value vanishes from charge conservation, as well as parity conservation, since the pion field is pseudoscalar. The π^- condensed phase is a state of broken symmetry with a complex order parameter $\langle \Pi \rangle$, and it is thus superconducting. It is instructive to compare this state with a neutral Bose superfluid such as liquid ^{4}He and with a BCS superconductor. In ^{4}He the nonvanishing complex order parameter is $\langle \psi(\mathbf{r}, t) \rangle$, where ψ is the ^{4}He field operator. In a BCS superconductor one has pairing of particles of opposite spin and momenta, and $\langle\psi_\uparrow(\mathbf{r},t)\psi_\downarrow(\mathbf{r},t)\rangle$ is the order parameter. One can equivalently regard the π^- condensed phase as arising from a pairing of neutron particles with proton holes, and proton particles with neutron holes, with order parameter $\langle\psi_n^\dagger(\mathbf{r},t)\sigma\psi_p(\mathbf{r},t)\rangle$, the divergence of which is the nonrelativistic source of the pion field $\langle\Pi\rangle$.

From another point of view one can look at the π_s^+ mode in the normal state as the fluctuations in isospin space of the neutron isospin vectors. The neutrons normally point along the negative T_z axis; the mode π_s^+ is a small oscillation about zero of the angle α that the vector makes with respect to the negative T_z axis. At the point of condensation this mode becomes "soft" and the system beyond this prefers a non-zero value of α. The Hamiltonian is then diagonalized (156, 157, 161) by nucleon state vectors that are linear combinations of neutrons and protons.

The equilibrium conditions (168) of the π^- condensed state are first that $\langle\Pi(\mathbf{r},t)\rangle = [\exp(-i\mu_\pi t)][\langle\Pi(\mathbf{r})\rangle$, where $\mu_\pi = \mu_n - \mu_p$ is the π^- chemical potential. Even though the pions may condense into a mode $\exp(i\mathbf{k}\cdot\mathbf{r})$ with a non-zero wavevector $\mathbf{k}$, or perhaps a more complicated mode with spatially varying modulus (169), the spatial average of the electromagnetic current must vanish in the ground state. So also must the spatial averages of the baryon current, and the electron-like and muon-like lepton currents vanish. The net charge density associated with the condensed field (in the absence of interactions dependent on $\dot{\Pi}$) is $-2\mu_\pi|\langle\Pi\rangle|^2$ (see 168); since generally $\mu_\pi > 0$, this is a negative charge density. The overall system is electrically neutral.

Model calculations of the condensed state were first given by Sawyer and Scalapino (155–157) (see also 168, 170, 171). These were improved by Baym & Flowers (161) and Au & Baym (162), who included "minimal effects" of nuclear forces via the Lorentz-Lorenz effect, and pi-pi interactions via Weinberg's chirally symmetric phenomenological pion-nucleon Lagrangian (172). A formulation of pion condensation in terms of the σ model and the general relation of pion condensation to chiral symmetry, together with methods for inclusion of Δ's in the condensed phase, are discussed by Campbell, Dashen & Manassah (173, 174), who also consider excitations of the condensate, and the influence of pion condensation on neutrino emission from hot neutron stars (see also 175 and 176). A simple pedagogical model calculation of the condensed phase is given in (177). There have been to date no calculations of the condensed phase including full effects of nuclear forces, isobars, and pi-pi interactions; and so, a realistic assessment of the effect of condensation on the equation of state is not yet possible. A preliminary study of the effects of condensation on neutron star models has been carried out by Hartle et al (178).

3.3.3 π° CONDENSATION Migdal (154), as well as Sawyer & Yao (169) (see also 179), have considered possible π° condensation of neutron star matter. Such condensation is described by a nonvanishing real expectation value $\langle \Pi_0(\mathbf{r}) \rangle$ of the neutral pion field. Because the order parameter is real the state is therefore not superconducting. Rather, since the (nonrelativistic) source of the neutral pion field is the divergence of the $T = 1$, $T_z = 0$ nucleon spin polarization, the neutral condensed mode $\langle \Pi_0(\mathbf{r}) \rangle$ will have a finite wavevector, thus varying in magnitude in space and leading to a spatially nonuniform spin-ordered phase. One such possible state is the Pandharipande-Smith solid (89) with a non-zero $\langle \Pi_0 \rangle$ described in section 3.2. Migdal (166) (see also 180) has estimated onset of infinitesimal π° condensation in neutron matter at $\rho \sim 0.7\rho_0$, and wavevector $k \sim 3$ fm^{-1}; he and co-workers have also discussed possible π° condensation in symmetric nuclear matter (166), and in real nuclei (181, but see also 182).

3.4   Ultrahigh Densities

At ultrahigh densities, $\gg 1$ fm^{-3}, one knows very little about the properties of matter. The main problem is not that one has little idea how to calculate in this density regime, given the forces between particles; rather the question is whether matter here can be described in terms of phenomenological forces between distinct asymptotic particles. At these densities the meson clouds of the baryons are always strongly overlapping. The situation is somewhat analogous to that of one's being given the complete phenomenology of low energy hydrogen atom-hydrogen atom scattering and being asked to describe the properties of hydrogen gas compressed to a mean interparticle spacing of 10^{-2} Å.

The property of high density matter most important for the structure of neutron stars is the equation of state $P(\rho)$, relating the pressure to the mass density $\rho = E/c^2$, where E is the energy density including rest mass (see section 5). Several general statements can be made about the asymptotic behavior of $P(\rho)$ independent of detailed models. In a system of relativistic free particles, the energy per baryon E/n_b is proportional to $n_b^{1/3}$, where n_b is the baryon density; then $P = n_b^2 \, \partial(E/n_b)/\partial n_b$ equals $\rho c^2/3$. This result also holds in any theory of elementary particles where the strong interactions obey "scaling" (183), i.e. have no characteristic length scale (as for electromagnetic interactions), for then the energy per baryon is still proportional to $n_b^{1/3}$, the only characteristic wavenumber. More generally though, as Zel'dovich first pointed out (184), the pressure can exceed $\rho c^2/3$. If the energy per baryon becomes asymptotically proportional to the baryon density, as in the field theory models described below, then P approaches ρc^2.

On grounds of causality one may argue that asymptotically P cannot exceed ρc^2, for if it did the sound velocity of the matter, $s = (\partial P/\partial \rho)^{1/2}$, would at some density exceed the velocity of light. However, this argument is not rigorously true. As Bludman & Ruderman (185) have shown, $s > c$ does not imply a violation of causality in the case of an "active" or amplifying medium (as in a laser); it is always valid, however, for matter in its ground state. Bludman & Ruderman also suggest that baryon-antibaryon production processes will generally serve to keep $P \leqq \rho c^2$ (see also 186, 187).

Specific attempts to describe matter in this ultradense regime have been made from statistical models and also from the point of view of field theory. Frautschi et al (188–190), Lee, Leung & Wang (191), and Leung & Wang (192) have presented a description of dense matter as a system of noninteracting zero width baryon resonances, with baryon number unity, in statistical equilibrium. The spectrum of masses of the resonances is assumed to be given by $dn(m) \sim m^a \exp(m/m_0)\, dm$ where $dn(m)$ is the number of baryon resonances of mass between m and $m+dm$. Such a spectrum (190) is suggested by Hagedorn's statistical bootstrap model and the Veneziano model, with $m_0 \approx 160$ MeV and $-7/2 \lesssim a \lesssim -5/2$. Only resonances with $mc^2 < \mu_n$, the neutron chemical potential, are present in the matter. By a simple calculation one finds that in the asymptotically high density limit the baryon density n_b is $\sim \exp(\mu_n/m_0)$, the mass density ρ is $\sim n_b \mu_n$, while the pressure P is only $\sim n_b m_0 \sim \rho/\ln\rho$. This equation of state is much softer than the limiting form $P = \rho c^2$. The reason for the softness of the equation of state is that the only interactions included in the model are attractive baryon-meson interactions, and these only insofar as they lead to baryon resonances. No repulsive interactions, such as those leading to the hard core of the nucleon-neuclon interaction, are included. Neutron star models (193, 194) based on such an equation of state at high densities are very sensitive to the details of how this equation of state is joined on to that used at low densities ($\lesssim 5 \times 10^{14}$ g cm^{-3}). See also Section 5.

Recently Canuto & Lodenquai (195) have applied Landau's hydrodynamical model of elementary particle interactions (196) in an attempt to deduce the sound velocity of ultradense matter from high-energy p-p collisions.

Another approach to the properties of ultradense matter has been to describe it in terms of the elementary constituents of matter, "bare" baryons, or quarks, interacting through boson fields. One of the first such models was given by Zel'dovich (184), who considered baryons interacting through a classical vector field of mass m_v. In the limit of high densities in such a model, correlations between baryons become unimportant and the interaction energy density is given by the Hartree expression

$$E_{int} = \tfrac{1}{2}g_v^2 \int d^3r' \left[\exp\left(-m_v |\mathbf{r}-\mathbf{r}'|\right)/(4\pi |\mathbf{r}-\mathbf{r}'|)\right]n_b(r)n_b(r')$$
$$= \tfrac{1}{2}(g_v/m_v)^2 n_b^2 \qquad\qquad 3.6$$

where g_v is the vector coupling constant. As $n_b \to \infty$ the potential energy dominates the kinetic energy ($\sim n_b^{4/3}$ per unit volume), so $P \to \rho c^2$.

Recently, Walecka (197) has extended this model by including interactions of the baryons with a scalar field σ of mass m_s, as well as a vector field. The vector field produces a repulsion between the nucleons, while scalar exchange leads to attraction, so that for $g_v^2 > g_s^2$, $m_v > m_s$, this model provides a first description of the nucleon-nucleon force. Effects of pions are not included (but see 198). The coupling of the baryons to the scalar field is of the form $g_s \sigma \bar{\psi}\psi$. When field fluctuations are ignored, the baryon effective mass is $m^* = m - g_s\langle\sigma\rangle = m - g_s \bar{n}_b/m_s^2$, where $\bar{n}_b = \langle\bar{\psi}\psi\rangle = \sum_{p \leq p_f} m^*/\varepsilon_p$, $\varepsilon_p = (p^2 + m^{*2})^{1/2}$, and m is the bare baryon mass; the total energy density is

$$E = \tfrac{1}{2}(g_v^2/m_v^2)n_b^2 + \tfrac{1}{2}(g_s^2/m_s^2)\bar{n}_b^2 + \sum_{p \leq p_f} \varepsilon_p. \qquad\qquad 3.7$$

Also $n_b = \langle\psi^\dagger\psi\rangle = \sum_{p \leq p_f} 1$. Asymptotically $P \to \rho c^2$, while at lower densities, equation 3.7 applied to nuclear matter implies saturation. Chin & Walecka (199) fit the parameters g_v^2/m_v^2 and g_σ^2/m_σ^2 to reproduce the binding energy and density of nuclear matter, and so derive an equation of state for neutron star matter. While this is only a very rudimentary description of matter at nuclear densities, where correlations are important, it is a promising approach to the properties of ultradense matter. See also the scalar field model in (200).

A model similar to Walecka's containing in addition nonlinear interactions of the σ field has been explored by Källman (201). This approach is closely related to Lee & Wick's theory (202) of a possible abnormal state where $m^* \to 0$ in symmetric nuclear matter. It is not clear whether one should expect such abnormal states in neutron star matter. In this context, we mention the work of Bodmer (203) and Ne'eman (204) (see also 205, 84, and 193), who explored the possibility of a collapsed state of nuclear matter due to an attractive "heart" core in the nucleon-nucleon interaction, arising, for example, from exchange of massive spin-2 mesons.

Recent "bag" descriptions of the nucleon in terms of quarks interacting via vector gluons (206, 207) may also provide a useful basis for constructing models of ultradense matter. A quark-like model of dense matter with integrally charged quarks and massless vector gluons has been given by Iachello et al (208), who also consider the possibility that the "quarks" form a solid.

4 TRANSPORT PROPERTIES

The large magnetic fields in neutron stars, which are crucial to all pulsar emission mechanisms, are maintained by electrical currents flowing within the interiors of the stars. As a result of ohmic dissipation these currents decay on a time scale $\tau_D = (4\pi\sigma/c^2)(R/\pi)^2$ for the fundamental mode, where σ is the electrical conductivity (assumed to be homogeneous) and R is the stellar radius (209). Baym, Pethick & Pines (111) calculated the electrical conductivity of the neutron-proton-electron liquid in neutron stars and showed that it is so high that τ_D is typically greater than the age of the universe, and consequently decay of magnetic fields by ohmic dissipation in neutron star interiors is negligible. The high electrical conductivities ($\sim 10^{29} s^{-1}$ under typical neutron star conditions compared with $\sigma \sim 10^{17} s^{-1}$ for Cu at room temperature) are a consequence of the fact that both the electrons, which are the important current carriers, and the protons from which they scatter, are degenerate; this severely restricts the phase space into which an electron and proton can scatter. The electrical conductivity of matter, if it forms a neutron solid, is an open question, but it is likely to be high. One should further note that if high density matter is a superconductor, times for removing magnetic flux from the star will greatly exceed τ_D (see section 3.1.5). Extensive calculations of the electrical and thermal conductivities of matter at densities of the order of ρ_0 and below, which occur in the outer parts of neutron stars, have recently been carried out by Flowers & Itoh (210), who give an account of earlier work. In the crust of a neutron star a number of solid-state effects, such as scattering by dislocations, can be important. These have recently been discussed by Ewart, Guyer & Greenstein (211), who conclude that it is rather unlikely that they will affect decay times substantially. Values of τ_D for early models of light neutron stars have been determined by Chanmugan & Gabriel (212); these do not differ significantly from the earlier results (111).

Immediately after formation, the temperature of a neutron star is $\sim 10^{12} \, °K \sim 100$ MeV and neutrino emission is the chief source of energy loss from the neutron star (7). The neutrinos have mean free paths much greater than typical neutron star radii and freely escape from the star. At central temperatures $\lesssim 1$ MeV thermal conduction processes begin to give more heat loss than does neutrino emission. The interior of a neutron star has a very high thermal conductivity, due to the dense degenerate electron gas there, and most of the temperature drop between the center of the star and the surface occurs within a few meters of the surface, where the thermal conductivity is relatively low. If one neglects the effect of magnetic fields and possible superfluidity in the interior of the star, one finds surface temperatures (7) $\sim 10^6\,°K$ at an age of $\sim 10^6$ yr, a typical pulsar age. Interior temperatures are typically one or two orders of magnitude larger than the surface temperature. A strong magnetic field increases the thermal conductivity in the surface region, and superfluidity of the interior reduces the heat capacity. Both these effects reduce the surface temperature at a given age, and Tsuruta estimates surface temperatures in the range 10^3–$10^5\,°K$ at an age of 10^6 yr (8, 212, 214).

The viscosity of dense matter is important for determining the coupling between various portions of a neutron star. This also has been calculated by Flowers & Itoh (210), as well as in (213). The second viscosity, resulting from the finite time for the composition of hadronic matter to reach equilibrium after compression, can play an important role in damping vibrations of neutron stars (216–218).

5   NEUTRON STAR MODELS

The ranges of masses and moments of inertia of neutron stars are of direct astrophysical interest; in addition, more subtle details of their structure, such as the rigidity of the crust and a possible solid core, may have observational consequences (219). A knowledge of the equation of state at densities beyond 1 baryon/fm^3 is required to have a good estimate of the maximum mass of a neutron star. Recent attempts to discover black holes in our galaxy rely on one being able to distinguish a black hole from a neutron star by its mass. For example, observations (220) of the Cygnus X-1 compact X-ray source suggest a model of an accreting compact object of mass $\geq 6M_\odot$ orbiting a more normal star; the argument is that since it does not seem possible that a neutron star, even if rapidly rotating, could have such a large mass, the object in Cyg X-1 (given the correctness of the model) would have to be a black hole.

Certain bounds can be placed on the maximum mass, independent of detailed models of the equation of state. Sabbadini & Hartle (221) present a simple argument that, given the correctness of the equation of state (45) up to $\rho \sim 5 \times 10^{14}$ g cm^{-3}, the maximum mass of a neutron star cannot exceed $\sim 5M_\odot$, whatever the equation of state at higher densities. (See also 231.) If one imposes the condition of causality, $P < \rho c^2$ (see section 3.4), on the equation of state, then, as Rhoades & Ruffini show (222), the maximum mass is reduced to $3.2M_\odot$; similar bounds have been found by Nauenberg & Chapline (223).

Given an equation of state $P(\rho)$, one constructs models of neutron stars by integrating (numerically) the general-relativistic equation of hydrostatic balance, the Tolman-Oppenheimer-Volkoff equation (see e.g. 224)

$$\frac{\partial P(r)}{\partial r} = -G\frac{[\rho(r)+P(r)/c^2][m(r)+4\pi r^3 P(r)/c^2]}{r^2[1-2Gm(r)/rc^2]} \qquad 5.1$$

to determine the pressure $P(r)$ and mass density $\rho(r)$ at radius r in a spherical star. Here $G = 6.67 \times 10^{-8}$ erg cm g^{-2} is the gravitational constant and $m(r) = \int_0^r d^3 r' \rho(r')$ is the mass contained within radius r. If one lets $c^2 \to \infty$, equation 5.1 reduces to the familiar nonrelativistic equation of hydrostatic balance. The pressure P and density ρ are related by the equation of state. The central density $\rho_c = \rho(r = 0)$, the integration constant for equation 5.1, uniquely specifies the star, at zero temperature.

Figure 9 shows the masses of cold, equilibrium white dwarfs and neutron stars as functions of their central density, as computed in (20). Stars to the left of the maximum at $\rho_c = 1.4 \times 10^9$ g cm^{-3} are stable white dwarfs, while stars to the right of the minimum at $\rho_c = 1.55 \times 10^{14}$ g cm^{-3} are neutron stars. The dashed portions of the curves are calculated for two different equations of state, both due to Pandharipande (82); the lower curve assumed hyperons to be present, while the upper curve was calculated for pure neutrons. These two curves give one some idea of the range of uncertainty in knowledge of the structure of massive neutron stars. Neutron stars beyond the maximum are again unstable. Stars with $M > M_{max}$ haven't sufficient pressure to resist gravitational collapse. The maxima here are at $1.41M_\odot$ and $1.66M_\odot$. Generally, the stiffer the equation of state, the higher will be the mass of the most massive stable neutron star, the lower its central density, and the higher its radius and moment of inertia.

In order for a star to be stable against radial perturbations it is necessary, nonrelativistically, that the adiabatic index, $\Gamma = \partial \ln P/\partial \ln n_b$ averaged (weighted with respect to P) over the star, be greater than 4/3. For an ideal monatomic gas Γ equals 5/3, and for an ideal relativistic gas such as photons or relativistic electrons it equals 4/3. Thus, before neutron drip, Γ is greater than 4/3, but it is very close to 4/3 by the point of neutron drip. However, as remarked in section 2.2, just above neutron drip, Γ falls well below 4/3. As the density of free neutrons begins to grow, the matter becomes stiffer and Γ begins to rise, becoming greater than 4/3 again when the neutron Fermi pressure becomes on the order of the electron pressure; this occurs at about 7×10^{12} g cm^{-3}.

One can see now the reason why there are two classes of dense, cold stars—white dwarfs and neutron stars—and not a continuum in between. Basically, most of the matter in a star is at densities close to the central density, and thus when $\Gamma(\rho_c)$ is 4/3, the star will not be stable. There are stable stars with central densities in the region below neutron drip—these are white dwarfs—and stable neutron stars with high central densities, well above the point where Γ becomes greater than 4/3. The microscopic phenomenon of the nuclei becoming so neutron rich that they begin to drip neutrons leads to the lower limit on the branch of stable neutron stars.

The fact that the white dwarfs made of equilibrium zero temperature

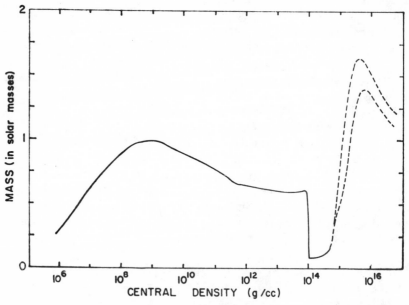

Figure 9   Mass versus central density for zero-temperature nonrotating stars in nuclear equilibrium. Stars to the left of the maximum at $\rho_c = 1.4 \times 10^9$ g cm^{-3} are stable white dwarfs, while stars to the right of the minimum at 1.55×10^{14} g cm^{-3} are neutron stars. The lower dashed extension of the curve was constructed from Pandharipande's hyperonic equation of state C (82), and the upper dashed extension from Pandharipande's equation of state for pure neutrons (82). Neutron stars beyond the maximum are unstable. From (20).

matter become unstable at the central density 1.4×10^9 g cm^{-3} is an interesting reflection of nuclear shell structure. Just below this central density the stars are made primarily of ^{56}Fe, ^{62}Ni, and ^{64}Ni, with a small core of ^{84}Se. As mentioned in section 2.1, as the nuclide present in the matter changes to one with a smaller Z/A, it must be accompanied by a discontinuous change in the density for the pressure to remain continuous. In the ^{64}Ni-^{84}Se transition there is a particularly large change, $\Delta\rho/\rho = 7.9\%$. That ^{84}Se is the next nucleus in the sequence is a consequence of the fact that it has a closed shell structure with 50 neutrons, which is energetically quite favorable. Now the presence in the matter of phase transitions with density discontinuities makes the matter effectively softer. To see this, imagine a closed vessel containing a liquid on the bottom in equilibrium with its vapor on top; even if the liquid and gas are each separately incompressible, the system in the vessel would be compressible, since as one reduces the volume of the vessel, gas is converted to denser liquid. By the point where ^{84}Se enters, $\Gamma - 4/3$ is ~ 0.01; once there is a large enough ^{64}Ni-^{84}Se interface present the matter becomes soft enough for the effective Γ to drop below 4/3, and the stars become unstable. The limiting cold white dwarf mass here is exactly $1.00 M_{\odot}$. Actual white dwarfs never reach complete nuclear equilibrium and hence have higher maximum possible masses.

The most important observational handle on the equation of state and hence interactions of matter at high densities is the possibility of determining the masses of neutron stars in close binaries. Recently, Middleditch et al (225) have determined the mass of the compact object (almost certainly a neutron star) in the compact X-ray source Her X-1 to be $1.33 \pm 0.2 M_{\odot}$. This appears to eliminate the models of neutron stars computed by Leung & Wang (193), using an equation of state constructed to agree with the soft noninteracting baryon resonance equation of state, which neglects repulsion, at ultrahigh ρ; they find a maximum neutron star mass between 0.27 and $0.50 M_{\odot}$ depending on the details of the interpolation. A second observational handle is based on a comparison of the luminosity of the Crab nebula with the rate of energy loss of the rotating neutron star in the center. The energy emitted by the nebula that must be steadily supplied by the pulsar is at the least 0.7×10^{38} erg sec^{-1} (11). On the other hand, the energy loss due to rotational slowdown of the neutron star is $I\Omega\dot{\Omega}$ where I is its moment of inertia, $\Omega = 189.65$ sec^{-1} is the angular frequency, and $\dot{\Omega} \approx -\Omega/2483$ yrs is its time derivative. This implies that I is $\gtrsim 1.5 \times 10^{44}$ g cm^2 (about one-fifth that of the earth). By contrast, the models of (193) all have moments of inertia less than 1×10^{44} g cm^2, which is yet another indication that the noninteracting baryon resonance equation of state is too soft. On the other hand, models generated from equations of state based on interactions including baryon repulsion (226) have masses and moments of inertia consistent with the observational bounds.

A number of properties of neutron star models are sensitive to the equation of state at nuclear densities and above. In particular, the inclusion of Δ's leads to a significantly harder equation of state, which, among other things, increases considerably the amount of crust material in the models (89). An extensive comparison of neutron star models based on recent equations of state is given by Arnett & Bowers (226) [see also Cohen & Börner (227)]. We mention too the recent models calculated by Moszkowski (228), and Malone et al (229); and also by Wagoner & Malone (230), who study models of neutron stars in the parametrized post-Newtonian approximation of metric theories of gravity.

CONCLUSION

The equation of state of dense matter is now reasonably well understood up to about nuclear-matter densities. At densities greater than that there are a number of uncertainties, and the effects of Δ's, possible other states such as pion condensation, solidification, abnormal states, and the properties of matter at ultrahigh densities remain interesting areas for future research. A number of properties of neutron star models are rather sensitive to the properties of matter at densities above that of nuclear matter, and therefore there exists the possibility of obtaining information about the properties of such matter from observations of pulsars and compact X-ray sources.

Literature Cited

1. Rosenfeld, L. 1974. In *Astrophysics and Gravitation, Proc. Solvay Conference on Physics, 16th.* Brussels: Editions de l'Université de Bruxelles, p. 174
2. Baade, W., Zwicky, F. 1934. *Proc. Nat. Acad. Sci. USA* 20:259
3. Oppenheimer, J. R., Volkoff, G. M. 1939. *Phys. Rev.* 55:374
4. Hewish, A., Bell, S. J., Pilkington, J. D. H., Scott, P. F., Collins, R. A. 1968. *Nature* 217:709
5. Gold, T. 1969. *Nature* 221:25
6. Blumenthal, G. R., Tucker, W. H. 1974. *Ann. Rev. Astron. Astrophys.* 12:23
7. Tsuruta, S., Cameron, A. G. W. 1966.

   Can. J. Phys. 44:1863
8. Tsuruta, S., Canuto, V., Lodenquai, J., Ruderman, M. 1972. *Ap. J.* 176:739
9. Tsuruta, S., Cameron, A. G. W. 1965. *Can. J. Phys.* 43:2056
10. Zel'dovich, Ya. B., Novikov, I. D. 1971. *Relativistic Astrophysics.* Chicago: Univ. Chicago
11. Ruderman, M. 1972. *Ann. Rev. Astron. Astrophys.* 10:427
12. Cameron, A. G. W. 1970. *Ann. Rev. Astron. Astrophys.* 8:179
13. Kirzhnits, D. A. 1971. *Usp. Fiz. Nauk* 104:489. Transl. 1972. *Sov. Phys. Usp.* 14:512
14. Hansen, C. J., ed. 1974. *Physics of*

Dense Matter, IAU Symp. No. 53.
Dordrecht: Reidel
15. Canuto, V. 1974. *Ann. Rev. Astron. Astrophys.* 12:167
16. Canuto, V. 1975. *Ann. Rev. Astron. Astrophys.* 13:335–80
17. Feynman, R. P., Metropolis, N., Teller, E. 1949. *Phys. Rev.* 75:1561
18. Landau, L. D., Lifshitz, E. M. 1969. *Statistical Physics,* Chap. XI. Reading, Mass: Addison-Wesley. 2nd ed.
19. Salpeter, E. E. 1961. *Ap. J.* 134:669
20. Baym, G., Pethick, C., Sutherland, P. 1971. *Ap. J.* 170:299
21. Coldwell-Horsfall, R. A., Maradudin, A. A. 1960. *J. Math. Phys.* 1:395
22. CERN. 1970. *Proc. Int. Conf. Properties Nucl. Far From Reg. Beta-Stabil., Leysin, Switzerland.* Geneva: CERN
23. Schramm, D. N., Arnett, W. D., eds. 1973. *Explosive Nucleosynthesis.* Austin: Univ. Texas Press
24. Myers, W. D., Swiatecki, W. J. 1966. *Nucl. Phys.* 81:1
25. Garvey, G. T., Gerace, W. J., Jaffe, R. L., Talmi, I., Kelson, I. 1969. *Rev. Mod. Phys.* 41:51
26. Seeger, P. A. 1970. See Ref. 22, p. 217
27. Negele, J. W., Vautherin, D. 1973. *Nucl. Phys. A* 207:298
28. Volodin, V. A., Kirzhnits, D. A. 1971. *Zh. Eksp. Teor. Fiz. Pisma Red.* 13:450; 1971. *JETP Lett.* 13:320
29. Dyson, F. J. 1971. *Ann. Phys. NY* 63:1
30. Witten, T. A. 1974. *Ap. J.* 188:615
31. Ruderman, M. A. 1974. See Ref. 14, p. 117
32. Landau, L. D., Lifshitz, E. M. 1965. *Quantum Mechanics.* Reading, Mass: Addison-Wesley
33. Kadomtsev, B., Kudryavtsev, V. 1971. *Zh. Eksp. Teor. Fiz. Pisma Red.* 13:15; 1971. *JETP Lett.* 13:9
34. Kadomtsev, B. 1970. *Zh. Eksp. Teor. Fiz.* 58:1765; 1970. *Sov. Phys. JETP* 31:945
35. Mueller, R., Rau, A. R., Spruch, L. 1971. *Phys. Rev. Lett.* 26:1136
36. Kadomtsev, B., Kudryavtsev, V.
1971. *Zh. Eksp. Teor. Fiz. Pisma Red.* 13:61; 1971. *JETP Lett.* 13:42
37. Ruderman, M. A. 1971. *Phys. Rev. Lett.* 27:1306
38. Cohen, R., Lodenquai, J., Ruderman, M. 1970. *Phys. Rev. Lett.* 25:467
39. Chen, H.-H., Ruderman, M. A., Sutherland, P. G. 1974. *Ap. J.* 191:473
40. Kaplan, J. I., Glasser, M. L. 1972. *Phys. Rev. Lett.* 28:1077
41. Langer, W. D., Rosen, L. C., Cohen, J. M., Cameron, A. G. W. 1969. *Ap. Space Sci.* 5:529
42. Bethe, H. A., Börner, G., Sato, K. 1970. *Astron. Astrophys.* 7:279
43. Weiss, R. A., Cameron, A. G. W. 1969. *Can. J. Phys.* 47:2171
44. Németh, J., Sprung, D. W. L. 1968. *Phys. Rev.* 176:1496
45. Baym, G., Bethe, H. A., Pethick, C. 1971. *Nucl. Phys. A* 175:225
46. Siemens, P. J., Pandharipande, V. R. 1971. *Nucl. Phys. A* 173:561
47. Buchler, J. R., Barkat, Z. 1971. *Astrophys. Lett.* 7:169
48. Buchler, J. R., Barkat, Z. 1971. *Phys. Rev. Lett.* 27:48
49. Barkat, Z., Buchler, J. R., Ingber, L. 1972. *Ap. J.* 176:723
50. Arponen, J. 1972. *Nucl. Phys. A* 191:257
51. Ravenhall, D. G., Bennett, C. D., Pethick, C. J. 1972. *Phys. Rev. Lett.* 28:978
52. Skyrme, T. H. R. 1959. *Nucl. Phys.* 9:615
53. Vautherin, D., Brink, D. M. 1970. *Phys. Lett. B* 32:149
54. Negele, J. W., Vautherin, D. 1972. *Phys. Rev. C* 5:1472
55. Baym, G. 1970. *Neutron Stars.* Copenhagen: Nordita
56. Bisnovatyi-Kogan, G. S., Chechetkin, V. M. 1973. *Zh. Eksp. Teor. Fiz. Pisma Red.* 17:1; 1973. *JETP Lett.* 17:437
57. Bisnovatyi-Kogan, G. S., Chechetkin, V. M. 1974. *Astrophys. Space Sci.* 26:25
58. Rosenbluth, M. N., Ruderman, M., Dyson, F., Bahcall, J. N., Shaham, J., Ostriker, J. 1973. *Ap. J.* 184:907
59. Hansen, C. J., Van Horn, H. M.

1975. *Ap. J.* 195:735
60. Bethe, H. A. 1971. *Ann. Rev. Nucl. Sci.* 21:93
61. Buchler, J. R., Ingber, L. 1971. *Nucl. Phys. A* 170:1
62. Ingber, L. 1968. *Phys. Rev.* 174:1250
63. Bäckman, S.-O., Sjöberg, O. 1973. *Acta Acad. Abo. Ser. B* 33:1
64. Moszkowski, S. A. 1974. *Phys. Rev. D* 9:1613
65. Haensel, P. 1975. *Nucl. Phys. A* 245:528
66. Sjöberg, O. 1973. *Nucl. Phys. A* 209:363
67. Bäckman, S.-O., Källman, C.-G., Sjöberg, O. 1973. *Phys. Lett. B* 43:263
68. Nitsch, J. 1972. *Z. Phys.* 251:141
69. Brownell, D. H., Callaway, J. 1969. *Nuovo Cimento B* 60:169
70. Silverstein, S. D. 1969. *Phys. Rev. Lett.* 23:139
71. Rice, M. J. 1969. *Phys. Lett. A* 29:637
72. Clark, J. W., Chao, N.-C. 1969. *Lett. Nuovo Cimento* 2:185
73. Clark, J. W. 1969. *Phys. Rev. Lett.* 23:1463
74. Pearson, J. M., Saunier, G. 1970. *Phys. Rev. Lett.* 24:325
75. Pandharipande, V. R., Garde, V. K., Srivastava, J. K. 1972. *Phys. Lett. B* 38:485
76. Pfarr, J. 1972. *Z. Phys.* 251:152
77. Dabrowski, J., Haensel, P. 1973. *Phys. Rev. C* 7:916
78. Haensel, P. 1975. *Phys. Rev. C.* 11:1822
79. Østgaard, E. 1970. *Nucl. Phys. A* 154:202
80. Feenberg, E. 1969. *Theory of Quantum Liquids.* New York: Academic
81. Pandharipande, V. R., Bethe, H. A. 1973. *Phys. Rev. C* 7:1312
82. Pandharipande, V. R. 1971. *Nucl. Phys. A* 178:123
83. Pandharipande, V. R., Garde, V. K. 1972. *Phys. Lett. B* 39:608
84. Bethe, H. A., Johnson, M. 1974. *Nucl. Phys. A* 230:1
85. Wu, F. Y., Feenberg, E. 1962. *Phys. Rev.* 128:943
86. Miller, M., Woo, C. W., Clark, J. W., Ter Louw, W. J. 1972. *Nucl. Phys.*

A 184:1
87. Green, A. M., Haapakoski, P. 1974. *Nucl. Phys. A.* 221:429; 1974. *Proc. Int. Conf. High Energy Phys. Nucl. Struct., 5th, Uppsala, 1973,* p. 44. Stockholm: Almqvist & Wiksell
88. Bodmer, A. R., Rote, D. M. 1971. *Nucl. Phys. A* 169:1
89. Pandharipande, V. R., Smith, R. A. 1975. *Nucl. Phys. A* 237:507
90. Ambartsumyan, V. A., Saakyan, G. S. 1960. *Astron. Zh.* 37:193; 1960. *Sov. Astron.* 4:187
91. Tsuruta, S., Cameron, A. G. W. 1966. *Can. J. Phys.* 44:1895
92. Langer, W. D., Rosen, L. C. 1970. *Astrophys. Space Sci.* 6:217
93. Libby, L. M., Thomas, F. J. 1969. *Phys. Lett. B* 30:88
94. Sawyer, R. F. 1972. *Ap. J.* 176:531. Erratum 1972. *Ap. J.* 178:279
95. Dashen, R. F., Rajaraman, R. 1974. *Phys. Rev. D* 10:708
96. Migdal, A. B. 1959. *Zh. Eksp. Teor. Fiz.* 37:249; 1960. *Sov. Phys. JETP* 10:176
97. Ginzburg, V. L., Kirzhnits, D. A. 1964. *Zh. Eksp. Teor. Fiz.* 47:2006; 1965. *Sov. Phys. JETP* 20:1346; Ginzburg, V. L. 1969. *J. Stat. Phys.* 1:3
98. Bardeen, J., Cooper, L. N., Schrieffer, J. R. 1957. *Phys. Rev.* 108:1175
99. Yang, C.-H., Clark, J. W. 1971. *Nucl. Phys. A* 174:49
100. Chao, N.-C., Clark, J. W., Yang, C.-H. 1972. *Nucl. Phys. A* 179:320
101. Takatsuka, T. 1972. *Prog. Theor. Phys.* 48:1517
102. Hoffberg, M., Glassgold, A. E., Richardson, R. W., Ruderman, M. 1970. *Phys. Rev. Lett.* 24:775
103. Ruderman, M. 1967. In *Proc. Ann. East. Theor. Phys. Conf., 5th, 1966,* ed. D. Feldman, p. 25. New York: Benjamin
104. Tamagaki, R. 1970. *Prog. Theor. Phys.* 44:905
105. Takatsuka, T., Tamagaki, R. 1971. *Prog. Theor. Phys.* 46:114
106. Itoh, N., Alpar, M. A. 1974. *J. Phys. A* 7:1970
107. Richardson, R. W. 1972. *Phys. Rev. D* 5:1883

108. Fujita, T., Tsuneto, T. 1972. *Prog. Theor. Phys.* 48:766. Erratum 1973. *Prog. Theor. Phys.* 49:371
109. Baym, G., Pethick, C., Pines, D. 1969. *Nature* 224:673
110. Yang, C.-H., Clark, J. W. 1972. *Lett. Nuovo Cimento* 4:969
111. Baym, G., Pethick, C., Pines, D. 1969. *Nature* 224:674
112. London, F. 1950. *Superfluids,* Vol. I. New York: Wiley
113. Itoh, N. 1969. *Prog. Theor. Phys.* 6:1478
114. Wolf, R. A. 1966. *Ap. J.* 145:834
115. Itoh, N., Tsuneto, T. 1972. *Prog. Theor. Phys.* 48:1849
116. Ruderman, M. A. 1969. *Nature* 223:597
117. Lamb, F. K., Pethick, C. J., Pines, D., Shaham, J. 1974. Unpublished manuscript
118. Pines, D., Pethick, C. J., Lamb, F. 1973. In *Proc. Tex. Symp. Relativistic Astrophys., 6th, Ann. NY Acad. Sci.* 224:237
119. Cazzola, P., Lucaroni, L., Scaringi, C. 1966. *Nuovo Cimento* 43:250
120. Coldwell, R. L., Wilets, L. 1967. *Bull. Am. Phys. Soc. II* 12:498
121. Coldwell, R. L. 1969. PhD thesis. Univ. Washington. Unpublished
122. Banarjee, B., Chitre, S. M., Garde, V. K. 1970. *Phys. Rev. Lett.* 25:1125
123. Anderson, P. W., Palmer, R. G. 1971. *Nature Phys. Sci.* 231:145
124. de Boer, J. 1957. *Progr. Low Temp. Phys.* 2:1. Amsterdam: North-Holland
125. Palmer, R. G., Anderson, P. W. 1974. *Phys. Rev. D* 9:3281
126. Clark, J. W., Chao, N.-C. 1972. *Nature Phys. Sci.* 236:37
127. Brandow, B. 1973. In *Proc. Texas Symp. Relativistic Astrophys., 6th, Ann. NY Acad. Sci.* 224:232
128. Cochran, S., Chester, G. V. 1975. *Phys. Rev.* In press
129. Cochran, S. G. 1974. PhD thesis. Cornell Univ. Unpublished
130. Schiff, D. 1973. *Nature Phys. Sci.* 243:130
131. Nosanow, L. H., Parish, L. J. 1973. In *Proc. Texas Symp. Relativistic Astrophys., 6th, Ann. NY Acad. Sci.* 224:226
132. Nosanow, L. H. 1966. *Phys. Rev.* 146:120
133. Pandharipande, V. R. 1973. *Nucl. Phys. A* 217:1
134. Iwamoto, F., Namaizawa, H. 1966. *Prog. Theor. Phys. Suppl.* 37:243
135. Guyer, R. A., Zane, L. I. 1969. *Phys. Rev.* 188:445
136. Guyer, R. A. 1969. *Solid State Physics,* ed. F. Seitz, D. Turnbull, H. Ehrenreich, 23:413. New York: Academic
137. Canuto, V., Chitre, S. M. 1973. *Phys. Rev. Lett.* 30:999; 1973. *Nature Phys. Sci.* 243:63
138. Canuto, V., Chitre, S. M. 1974. *Phys. Rev. D* 9:1587
139. Pandharipande, V. R. 1971. *Nucl. Phys. A* 174:641
140. Takemori, M. T., Guyer, R. A. 1975. *Phys. Rev.* In press
141. Canuto, V., Lodenquai, J., Chitre, S. M. 1974. *Nucl. Phys. A* 233:521
142. Østgaard, E. 1973. *Phys. Lett. B* 47:303
143. Pandharipande, V. R. 1975. *Nucl. Phys. A.* To be published
144. Chakravarty, S., Miller, M. D., Woo, C.-W. 1974. *Nucl. Phys. A* 220:223
145. Woo, C.-W., Massey, W. E. 1969. *Phys. Rev.* 177:272
146. Shen, L., Sim, H.-K., Woo, C.-W. 1974. *Phys. Rev. D* 10:3925
147. Shen, L., Woo, C.-W. 1974. *Phys. Rev. D* 10:371
148. Lowy, D. N., Woo, C.-W. 1975. To be published
149. Kirkwood, J. G., Monroe, E. 1941. *J. Chem. Phys.* 9:514
150. Carter, J. R., Bugg, D. V., Carter, A. A. 1973. *Nucl. Phys. B* 58:378
151. Pilkuhn, H. et al 1973. *Nucl. Phys. B* 65:460
152. Migdal, A. B. 1971. *Zh. Eksp. Teor. Fiz.* 61:2209; 1972. *Sov. Phys. JETP* 34:1184
153. Migdal, A. B. 1971. *Usp. Fiz. Nauk* 105:775; 1972. *Sov. Phys. Usp.* 14:813
154. Migdal, A. B. 1973. *Phys. Rev. Lett.* 31:247
155. Sawyer, R. F. 1972. *Phys. Rev. Lett.* 29:382

156. Scalapino, D. J. 1972. *Phys. Rev. Lett.* 29:386
157. Sawyer, R. F., Scalapino, D. J. 1973. *Phys. Rev. D* 7:953
158. Bäckman, S.-O., Weise, W. 1975. *Phys. Lett. B* 55:1
159. Bertsch, G. F., Johnson, M. B. 1975. *Phys. Rev.*
160. Baym, G. 1969. In *Mathematical Methods in Solid State and Superfluid Theory,* ed. R. C. Clark, G. H. Derrick, p. 121. Edinburgh: Oliver & Boyd
161. Baym, G., Flowers, E. 1974. *Nucl. Phys. A* 222:29
162. Au, C.-K., Baym, G. 1974. *Nucl. Phys. A* 236:500
163. Weise, W., Brown, G. E. 1974. *Phys. Lett. B* 48:397
164. Ericson, M., Ericson, T. E. O. 1966. *Ann. Phys. N Y* 36:323
165. Barshay, S., Brown, G. E. 1973. *Phys. Lett. B* 47:107
166. Migdal, A. B. 1973. *Nucl. Phys. A* 210:421
167. Bertsch, G. F., Johnson, M. Personal communication from G. F. Bertsch
168. Baym, G. 1973. *Phys. Rev. Lett.* 30:1340
169. Sawyer, R. F., Yao, A. C. 1973. *Phys. Rev. D* 7:953
170. Migdal, A. B. 1973. *Phys. Lett. B* 45:448, 47:96; 1973. *Zh. Eksp. Teor. Fiz. Pisma Red.* 18:443; 1973. *JETP Lett.* 18:260
171. Kogut, J., Manassah, J. T. 1972. *Phys. Lett. A* 41:129
172. Weinberg, S. 1968. *Phys. Rev.* 166:1568
173. Campbell, D. K., Dashen, R. F., Manassah, J. T. 1975. *Phys. Rev.* In press
174. Campbell, D. K., Dashen, R. F., Manassah, J. T. 1975. *Phys. Rev.* In press
175. Markin, O. A., Mishustin, I. H. 1974. *Zh. Eksp. Teor. Fiz. Pisma Red.* 20:497; *JETP Lett.* 20:226
176. Migdal, A. B., Markin, O., Mishustin, I. 1974. *Zh. Eksp. Teor. Fiz.* 66:443; *Sov. Phys. JETP* 39:312
177. Baym, G., Campbell, D. K., Dashen, R. F., Manassah, J. T. 1975. *Phys.*

    Lett. B. In press
178. Hartle, J. B., Sawyer, R. F., Scalapino, D. J. 1975. *Ap. J.* In press
179. Barshay, S., Vagradov, G., Brown, G. E. 1973. *Phys. Lett. B* 43:359
180. Palmer, R. G., Tosatti, E., Anderson, P. W. 1973. *Nature Phys. Sci.* 245:119
181. Migdal, A. B., Kirichenko, N. A., Sorokin, G. A. 1974. *Zh. Eksp. Teor. Fiz. Pisma Red.* 19:326; 1974. *JETP Lett.* 19:185; 1974. *Phys. Lett. B* 50:411; Migdal, A. B. To be published
182. Barshay, S., Brown, G. E. 1973. *Phys. Lett. B* 47:107
183. Bjorken, J. D. 1969. *Phys. Rev.* 179:1547
184. Zel'dovich, Ya. B. 1961. *Zh. Eksp. Teor. Fiz.* 41:1609; 1962. *Sov. Phys. JETP* 14:1143
185. Bludman, S. A., Ruderman, M. A. 1970. *Phys. Rev. D* 1:3243
186. Bludman, S. A., Ruderman, M. A. 1968. *Phys. Rev.* 170:1176
187. Ruderman, M. A. 1968. *Phys. Rev.* 172:1286
188. Frautschi, S., Bahcall, J. N., Steigman, G., Wheeler, J. C. 1971. *Comments Astrophys. Space Phys.* 3:121
189. Wheeler, J. C. 1974. See Ref. 14, p. 77
190. Frautschi, S. 1971. *Phys. Rev. D* 3:2821
191. Lee, H., Leung, Y. C., Wang, C. G. 1971. *Ap. J.* 166:387
192. Leung, Y. C., Wang, C. G. 1972. *Ap. J.* 181:895
193. Leung, Y. C., Wang, C. G. 1971. *Ap. J.* 170:499
194. Ruffini, R. 1973. In *Black Holes,* ed. C. DeWitt, B. S. DeWitt, 446–89. New York: Gordon & Breach
195. Canuto, V., Lodenquai, J. 1975. *Phys. Rev. D* 11:233
196. Landau, L. D. 1953. *Izv. Akad. Nauk SSSR Ser. Fiz.* 17:51. Transl. 1965. In *Collected Papers of L. D. Landau,* ed. D. ter Haar, p. 569. Oxford: Pergamon
197. Walecka, J. D. 1974. *Ann. Phys. N Y* 83:491
198. Bowers, R. L., Campbell, J. A.,

Zimmerman, R. L. 1973. *Phys. Rev. D* 7:2278, 2289
199. Chin, S. A., Walecka, J. D. 1974. *Phys. Lett. B* 52:24
200. Kalman, G. 1974. *Phys. Rev. D* 9: 1656
201. Källman, C.-G. 1975. *Phys. Lett. B* 55:178
202. Lee, T. D., Wick, G. C. 1974. *Phys. Rev. D* 9:2291
203. Bodmer, A. R. 1971. *Phys. Rev. D* 4:1601
204. Ne'eman, Y. 1974. See Ref. 14, p. 111
205. Feenberg, E., Primakoff, H. 1946. *Phys. Rev.* 70:980
206. Chodos, A., Jaffe, R. L., Johnson, K., Thorn, C. B., Weisskopf, V. F. 1974. *Phys. Rev. D* 9:3471
207. Bardeen, W. A., Chanowitz, M., Drell, S. D., Giles, R., Weinstein, M., Yau, T. M. *Phys. Rev. D* 11: 1094
208. Iachello, F., Langer, W. D., Lande, A. 1974. *Nucl. Phys. A* 219:612
209. Lamb, H. 1883. *Phil. Trans.* 174: 519
210. Flowers, E., Itoh, N. 1975. To be published
211. Ewart, G. M., Guyer, R. A., Greenstein, G. 1975. To be published
212. Chanmugan, G., Gabriel, M. 1971. *Astron. Astrophys.* 11:268
213. Tsuruta, S. See Ref. 14, p. 209
214. Tsuruta, S. 1975. In *Proc. Cardiff Symp. Solid State Astrophys.*, 1974. To be published
215. Heintzmann, H., Nitsch, J. 1972. *Astron. Astrophys.* 21:291
216. Jones, P. B. 1970. *Astrophys. Lett.* 5:33
217. Jones, P. B. 1971. *Proc. Roy. Soc. London A* 323:111
218. Langer, W. D., Cameron, A. G. W. 1969. *Astrophys. Space Sci.* 5:213
219. Pines, D., Shaham, J., Ruderman, M. A. 1974. See Ref. 14, p. 189. Also in *Proc. 6th Texas Symp. Relativistic Astrophys., Ann. NY Acad. Sci.* 224:190
220. Giacconi, R. 1974. In *Gravitational Radiation and Gravitational Collapse, Proc. IAU Symp. No. 64*, ed. C. DeWitt-Morette, p. 147. Dordrecht: Reidel
221. Sabbadini, A. G., Hartle, J. B. 1973. *Astrophys. Space Sci.* 25:117
222. Rhoades, C. E., Ruffini, R. 1974. *Phys. Rev. Lett.* 32:324
223. Nauenberg, M., Chapline, G. 1973. *Ap. J.* 179:277
224. Weinberg, S. 1972. *Gravitation and Cosmology.* New York: Wiley
225. Middleditch, J., Nelson, J., Mast, T. 1975. In *Proc. Texas Symp. Relativistic Astrophys., 7th, Ann. NY Acad. Sci.* In press
226. Arnett, W. D., Bowers, R. L. 1974. *Neutron Star Structure—A Survey.* Univ. Texas Publ. Astron. No. 9
227. Cohen, J. M., Börner, G. See Ref. 14, p. 237
228. Moszkowski, S. A. 1975. To be published
229. Malone, R. C., Johnson, M. B., Bethe, H. A. 1975. To be published
230. Wagoner, R. V., Malone, R. C. 1973. *Ap. J.* 189:L75
231. Zwicky, F. 1939. *Phys. Rev.* 55:726

STATUS AND FUTURE   ✶5559
DIRECTIONS OF THE WORLD
PROGRAM IN FUSION
RESEARCH AND DEVELOPMENT

Robert L. Hirsch
US Energy Research and Development Administration, Washington DC 20545

CONTENTS

1   INTRODUCTION

Fusion research has made major strides in recent years, and the outlook
for the ultimate success of developing practical fusion power is now considered to be excellent. Indeed, most fusion scientists no longer doubt
that fusion power production will be physically possible. As they go
forward to experimentally demonstrate this belief, they are simultaneously

beginning to delve into the problems of building practical, reliable, economical fusion systems. This is not to imply that all of the central physics problems associated with fusion have been solved, because they most assuredly have not. Rather, it is that physicists believe that present physical foundations, technologies, diagnostics, trained manpower, and program directions will lead to a number of feasible techniques to generate fusion power. In a sense, fusioneers feel that they have reached or passed the knee of the "S" curve of basic knowledge, and, as often happens in research, this feeling comes prior to actual physical demonstration of the phenomena under study.

As a result of this confidence, significant programs were initiated about five years ago to begin to examine the fusion reactor embodiments of the concepts being studied in the laboratories. These studies are now in their second or third phase of development, and they are beginning to provide useful guidance to physics research programs, to infant fusion reactor technology programs, and to program decision making. In the new field of fusion reactor technology, the situation is presently one of innovation, discovery, and change, in many ways reminiscent of the "golden" years of fission technology, when new reactor types were being conceived and studied for a variety of purposes.

In the following, the present state of fusion research and development is described for the reader who is presumed to be familiar with general nuclear technology but who has had relatively little exposure to the details of fusion. First, the basics of fusion are briefly reviewed. Next, the often interrelated approaches to fusion power are described, including descriptions of present status, near term directions, and expectations. The emphasis in each major national or multinational program is then briefly presented, and finally, highlights of fusion reactor engineering studies are described.

The emphasis herein is on general concepts and problems. Details of plasma physics and engineering can be obtained from any of a variety of excellent references, some of which are cited. One reference of particular note is the IAEA World Survey of Major Facilities in Controlled Fusion Research, 1973, Edition I (1), which describes the characteristics and objectives of many major fusion experiments as well as the general character of the various theory and technology programs.

2   FUSION BASICS

Fusion is the process that produces energy in the sun and the stars. It can proceed at useful rates only at extremely high temperatures—of the order of 5 keV (50,000,000°C) or more. At these temperatures fusion fuel atoms are stripped of their electrons and exist in the form of a gaseous mixture of ions and electrons called a plasma. Because energy is obviously required to heat a gas to such high temperatures, it is necessary to contain a suitably dense fusion plasma for a minimum length of time to allow as much energy to be produced from fusion as was consumed to do the initial heating. This "break-even" condition is usually expressed as a density-confinement time product $n\tau$.

The fuel cycle that will be used in first generation fusion reactors is based on the fusion of two isotopes of hydrogen: deuterium and tritium (DT). The plasma temperature required for net energy production using this reaction is the lowest and the energy gain per reaction is the highest. Some important features of this reaction are shown in Figure 1. Deuterium is obtained from ordinary water where it naturally exists in essentially infinite supply. Lithium is used to create tritium through the absorption of neutrons from the DT reaction. Lithium is now available from land reserves and can be extracted from the oceans, where it is available at low cost in quantities adequate for an estimated millions of years of fusion power. Because the basic fusion fuels are hydrogen isotopes, ordinary hydrogen is usually used in magnetic confinement fusion experiments to simulate the behavior of DT plasmas. Hydrogen does not fuse under the same conditions that DT does, and so its use simplifies many experimental studies.

Confinement of a hot fusion plasma is accomplished in the sun and stars by gravitational forces. On earth that option is not available and

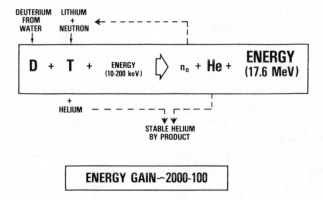

Figure 1 General characteristics of the deuterium-tritium fusion fuel cycle. In an operating reactor deuterium and tritium nuclei must be given energies of the order of 10 keV or more. The DT reaction produces a neutron, helium, and a large amount of energy. The neutron can be utilized to breed tritium from lithium. The energy gain from this reaction can be as high as about 2000.

of course solid materials are unable to contain plasmas at such extreme temperatures. The most attractive approaches to fusion power utilize magnetic fields or inertia to confine a fusion plasma. Magnetic confinement utilizes the ability of magnetic forces to hold individual charged particles in circular or helical orbits around magnetic field lines. Inertial confinement depends upon the fact that it takes a finite time for a rapidly heated, high pressure mass to disassemble because of its inertia.

Consider first the situation in magnetic confinement, which has been underway since 1952. In the 1950s, plasma production and heating techniques were developed, and reactor level plasma densities were produced quickly. In the early 1960s, thermonuclear temperatures were achieved for the first time in a magnetic confinement experiment. Since then, these temperatures have been reproduced in a variety of different experimental systems, whenever the appropriate financial investment was made.

The major problem in magnetic confinement has been plasma instabilities that cause premature and excessive escape of plasma. Plasma instabilities fall into two categories: magnetohydrodynamic instabilities, which involve gross motions of plasmas, and microinstabilities, which occur on the microscale and which involve relatively small plasma masses. These instabilities can appear and grow under a variety of conditions; they can saturate, causing small or large losses; or they can develop to the point that they destroy the plasma catastrophically. Whether these instabilities exist at all depends upon such factors as the magnetic field strength and/or the configuration, the ion and/or electron distribution function, the presence or absence of impurities, the distribution and strength of the plasma current (if one exists), the neutral gas density and distribution, etc. To model plasmas analytically a number of simplifying assumptions must always be made. In earlier years these assumptions often masked important instabilities, and so a great deal of experience needed to be developed in order to pose an analytic problem in plasma theory that had some semblance of reality. Experimentally, it was often found that a number of instabilities were present simultaneously, and it was and sometimes still is a laborious job to isolate and eliminate these complex phenomena.

These problems required a major theoretical and experimental effort to develop an appropriate understanding of plasma physics, which was a relatively undeveloped science prior to the advent of fusion research. As plasma instabilities were identified and understood, their deleterious effects were overcome and plasma confinement began to improve dramatically beginning about 1965. By 1971, a variety of confinement concepts had closely approached the maximum confinement time theoretically possible and, in effect, the confinement problem was broken.

Table 1   Fusion requirements and achievements to date for magnetic confinement

Quantity	Minimum Levels Required	Date First Achieved
Density	greater than 10^{13} cm^{-3}	1953
Temperature	greater than 5 keV	1962
Confinement time	greater than $\left(\dfrac{10^{13} \text{ cm}^{-3}}{\text{density}} \right)$ sec	1969 (for near-classical confinement)

Table 1 shows the minimum required levels of plasma density, temperature, and confinement time for DT fusion using magnetic confinement. The table also shows the dates when each of these minimum level conditions was achieved.

A word of explanation is necessary with respect to confinement time. Classical plasma confinement time is the theoretical upper limit on the length of time that a plasma can be confined for a given finite volume of magnetic field, strength of magnetic field, and plasma energy state. The classical confinement time in present-day, relatively small experiments is smaller than that required for a reactor. Larger experiments will be required to achieve reactor-level confinement times. The significance of achieving near-classical confinement time in smaller systems is that it showed that magnetic confinement can indeed be utilized to contain hot plasmas as had been predicted.

In magnetic confinement research, reactor-level temperatures and densities and near-classical confinement times have been achieved in combinations of at most two in a variety of different experiments. In order to have some perspective on these achievements, it is worthwhile to review some important facts that are central to fusion research. First, plasma pressure, which is proportional to the product of plasma density and plasma temperature, is high for fusion plasmas. In principle, magnetic, electromagnetic, electrostatic, and inertial forces are all capable of exerting a sufficiently large counter-pressure on plasmas to contain them. For magnetic confinement of fusion plasmas, relatively high magnetic field strengths are required. Two additional requirements necessary to reach fusion plasma conditions are large plasma volume to achieve long confinement times and long-time duration of the experiment. These conditions lead to the need for a large volume of long-lived and strong magnetic field, which is expensive. These facts dictated a logical approach to fusion by magnetic confinement. This approach was first to do as much as possible in small, low cost experiments, and then, when the results justify them, to build larger, more expensive experiments to further study important questions of physics and to demonstrate that plasma conditions scale with size, volume, and magnetic field, as theory predicts. When fusion experiments are large enough, all three fusion reactor conditions of temperature, density, and confinement time can then be demonstrated simultaneously.

In inertial confinement the basic idea is as follows: A very large amount of energy is delivered in a very short period of time to a spherical pellet containing deuterium and tritium fuel, usually conceived to be in a solid (frozen) form. The energy input could come from a very short pulsed laser, electron beam, or ion beam source. Upon interacting with the surface of the pellet, the input energy vaporizes and ionizes the surface material, causing a large part of it ($\gtrsim 70\%$) to rapidly expand outward. The resulting reaction force causes the remainder of the pellet to be imploded inward by an effect that has aptly been described as an "implosion rocket." Implosion can cause heating and also compression to extremely high densities. At the peak of the implosion the fuel can be so hot and so dense that in principle large amounts of fusion energy can be released before the compressed matter explodes. The "containment" is then very short in duration and is determined by the inertia of the pellet mass, which determines the explosive disassembly time.

The conditions necessary for practical fusion power by inertial principles are shown in Table 2. Also indicated are the conditions that are believed to have been achieved to date using laser implosion techniques. The conditions are further from the required levels than those in magnetic confinement largely because much more energetic lasers than presently exist will be necessary to approach practical inertial fusion conditions. In addition, the laser fusion program is relatively new and experimental conditions have not yet been optimized. The degree of certainty associated with the experimentally measured parameters is relatively modest due to the very severe diagnostic problems of making measurements of phenomena that occur in times measured in picoseconds (10^{-12} sec) to nanoseconds (10^{-9} sec) in pellets whose sizes are measured in tens to hundreds of microns.

Table 2 Fusion requirements and achievements to date for inertial confinement

Quantity	Minimum Levels Required	Levels Achieved to Date Using Laser Techniques
Density	10^{26} cm^{-3}	10^{23} cm^{-3}
Temperature	10 keV	1 keV
Confinement time	10^{-11} sec	10^{-11} sec

In recent years there have been dramatic developments related to the energy requirements for fusion by magnetic confinement. As mentioned earlier, an ion temperature[1] above 5 keV is necessary to provide a useful fusion rate. The use of the term *temperature* implies the existence of a Maxwell-Boltzmann equilibrium distribution function. This is a reasonable expectation in a plasma that is not having energy continuously added because the thermalization times due to interparticle coulomb collisions are much shorter than the fusion collision time. At equilibrium the $n\tau$ value required for break-even in deuterium-tritium is about 5×10^{13} cm^{-3} sec. This condition is often called the Lawson criterion.

In a Maxwell-Boltzmann plasma the majority of the fusions occur between the ions in the energetic tail and the lower energy ions that compose the bulk of the plasma. Since the early days of the program it was recognized that the fusion rate in a plasma could be dramatically increased by adding high energy ions to give the ion energy distribution a larger population at high energies than would exist for a Maxwell-Boltzmann plasma. This concept was revived in 1971 by Dawson, Furth & Tenney (2) and recently has been developed as the basis for some of the upcoming first generation DT burning experiments in magnetic confinement systems. Using this so-called two-component concept, break-even conditions can be achieved at $n\tau \approx 10^{13}$ cm^{-3} sec. This concept is discussed further below.

3   APPROACHES TO FUSION

3.1   General

The majority of the world fusion program has been and is presently concentrated on magnetic confinement concepts, but recent theoretical and experimental progress has led to the initiation of significant programs in inertial confinement. Magnetic confinement concepts are characterized as follows:

1. Closed or open, depending upon whether the configurations are generally toroidal or linear in shape;
2. Low-β or high-β [2] depending upon whether the plasma pressure is much less than or approximately equal to the magnetic field pressure;

---

[1] Under many conditions of practical interest, the ion and electron temperatures or energies in a plasma can be quite different. Since the ions fuse, their temperature is more important.

[2] Beta is defined as the ratio of the plasma pressure to the pressure exerted by the magnetic field.

3. Pulsed or steady-state, depending upon whether they are "batch-burn" or can be continuously refueled.

The majority of magnetic confinement research is concentrated on toroidal concepts, because these concepts have proven to be the best plasma containers and because it has proved difficult to inhibit the natural end losses in linear, open systems.

3.2   Low-β Toroidal Systems

Consider low-β closed concepts first. These are usually circular, although some can be race-track shaped. All low-β closed systems of potential reactor interest utilize a strong toroidal magnetic field, and all utilize a secondary poloidal magnetic field to provide a helical shape (rotational

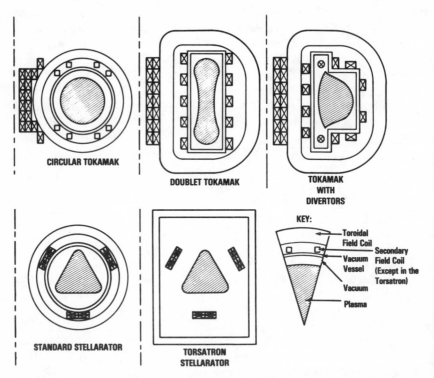

Figure 2   Cross sections of important low-β toroidal systems. Four of the five important low-β configurations utilize a toroidal magnetic field produced by toroidal field coils. Inside the coils are conductors to produce appropriate vertical shaping fields. Inside the conductors is the vacuum vessel containing the plasma. The Torsatron configuration utilizes three sets of coils individually wound to produce the combination of fields generated by separate windings in the other configuration.

transform) to the resultant total field. This helical field prohibits a natural tendency towards charge separation (and therefore instability) that occurs in simple toroidal fields. Beyond these simple thoughts, matters become complicated because some plasmas have circular cross sections, whereas others are triangular, kidney-shaped, elliptic, D-shaped, etc. The so-called stellarator gets its poloidal field from specially shaped helical coils located inside the toroidal field coils, whereas the tokamak uses a current through the plasma to produce its poloidal field. Because it is important to maintain a high plasma purity, some closed systems "divert" magnetic field lines that exist near the chamber wall into special auxiliary chambers outside the confinement region in order to decrease the flux of energetic plasma on the wall and to capture ionized impurities from the wall before they diffuse into the plasma. There thus exists a continuum of low-β closed configurations that are very much interrelated. Figure 2 shows the basic cross-sectional shapes and components of a few of the most important closed concepts under study today.

The β values of tokamaks and stellarators can vary from 0.1–20%, but β values above 1–5% are needed for reactors. All of the variations shown in Figure 2 have power reactor potential; all would operate at relatively low density $(5 \times 10^{13}–5 \times 10^{14} \ cm^{-3})$ as reactors; all would operate in either the long pulse (minutes-hours) or steady state mode; and so all would require refueling during operation. Tokamak plasmas have currents flowing in them, which, besides providing a necessary secondary magnetic field, also serve to preheat the plasma to the kilo-electron volt temperature range. Heating is needed in stellarators not only to create the initial plasma but to heat it to reactor levels.

To date, tokamaks have achieved plasma parameters closer to those required for a fusion reactor than have stellarators. Because of this success, the lower cost and simplicity of tokamaks, and their very attractive reactor potential, low-β toroidal research emphasizes the tokamak worldwide. Present-day tokamak experiments typically exhibit electron temperatures in 1–3 keV range, ion temperatures in the 0.2–1 keV range, densities in the $10^{13}–10^{14} \ cm^{-3}$ range, and confinement times between 5 and 20 msec. Some of the most important tokamak experiments worldwide are the T-4[3] (Figure 3) and T-6 in the Soviet Union, the TFR in France, the Cleo in the United Kingdom, the Pulsator in the Federal German Republic, the JFT-2 in Japan, and the ATC (Adiabatic Toroidal Compressor), ORMAK, Doublet II, and Alcator in the United States.

---

[3] Fusion experiments have names of varying derivations. For the characteristics, achievements, and objectives of each experiment see (1).

Figure 3 The T-3 tokamak in the I.V. Kurchatov Institute, Moscow, USSR. In the 1968–1969 period, this tokamak produced the data that convinced the world of the superior performance characteristics of tokamaks. T-3 was subsequently modified and is now called T-4.

Tokamak plasma confinement is theoretically and experimentally dominated by diffusive losses, which means that confinement time generally increases as the square of plasma size, assuming other parameters are held constant. Experimentally observed confinement times are about ten times lower than the very favorable classical law but are more than adequate to meet reactor requirements, if they continue to hold in larger systems. Critical tests of diffusion losses will take place in 1976 when two large new tokamaks come into operation. These are the Princeton Large Torus (PLT) at Princeton University and the T-10 at the I.V. Kurchatov Institute in Moscow. Both experiments are about 2–3 times larger in minor diameter than most previous tokamaks, and both are expected to yield confinement times of about 0.1 sec at densities near 10^{14} cm^{-3}. Improved confinement due to the increased size should also allow the achievement of higher temperatures and, therefore, the investigation of the diffusion behavior in a plasma with parameters much more closely approximating those of a fusion reactor.

The plasma physics problems of principal concern in large, high temperature tokamaks are associated with magnetohydrodynamic (MHD) equilibrium and so-called trapped particle microinstabilities. The MHD problem is associated with the need to keep the plasma current distribution within stable bounds. This requires careful attention to establishing the proper initial current distribution, complicated by the fact that the magnitude of the current density determines the magnitude of the local heating, which varies as a function of temperature, which in turn determines the plasma conductivity, which in turn determines the rate at which current can diffuse through the plasma. Experience in smaller tokamaks and computer modeling provides a good guide as to how to handle this problem. Experimental flexibility is provided by controlling the rise time of the plasma current and by neutral beams, which can be used for tailoring the temperature, and therefore the current profile, if necessary. One uncertainty is associated with the effects of impurities that can cool the outer regions of the plasma, causing an undesirable shrinkage of the current channel, but a number of techniques to inhibit this effect are already in hand, and other techniques are under investigation.

With respect to trapped particle instabilities, linear theory indicates the conditions under which they will be excited and nonlinear theory indicates roughly at what level they should saturate. It is recognized that trapped particle instabilities will degrade plasma confinement but the level of degradation is not predictable to better than about a factor of three. These instabilities are present to varying degrees in existing tokamaks and some research devices, but detailed study must await the new large, high temperature tokamaks.

The natural heating mechanism in a tokamak, namely the plasma current, is not sufficient to raise the plasma temperature to fusion reactor values. This is because the resistivity of a plasma decreases rapidly as temperature increases, so that ohmic heating becomes progressively less effective as fusion temperatures are approached. Thus, supplementary heating methods must be utilized, and in the last two years significant progress has been achieved in this area of research.

Supplementary heating methods generally involve introducing or creating some very energetic charged particles in the plasma. These gradually lose energy to the remainder of the plasma by simple collisions and sometimes by wave-particle interactions. A different method is adiabatic compression, which has been shown to work according to theory in the ATC experiment at Princeton (3).

Probably the most popular means for heating low-β closed systems is neutral beam heating. In this technique energetic charged particles are created outside the plasma in a plasma source from which they are extracted and accelerated electrostatically. The accelerated ions are then neutralized by charge exchange so they can cross the magnetic field that insulates the plasma from the outside world. Once inside the plasma the energetic neutrals become trapped because they are quickly ionized, adding both energy and new particles to the plasma. This technique has been successfully demonstrated in the ATC (4) and ORMAK (5) experiments in the United States.

Radio frequency (RF) fields can be utilized to resonantly or non-resonantly heat a plasma. Resonance can occur at the ion cyclotron frequency or at upper harmonics. This type of heating affects the ions directly and requires frequencies in the 50 MHz range. Another choice exists at the electron cyclotron frequency, of interest in cases where the confinement is sufficient to allow the electrons to equilibrate with the ions; frequencies here are in the 100 GHz range. Yet another choice is a hybrid resonance of the electrons and ions in the 1000 MHz range.

In the USA, the ATC and the ORMAK devices have each achieved 50% increases in ion temperature to ~ 500 eV with neutral beam power in the 200 kW range. Such power levels are comparable to the power dissipated in the plasma by the plasma current. At this writing, neutral injection at the 500 kW level is underway on both devices. Presently, the Soviet T-11 tokamak (similar to the ORMAK in size and magnetic field) is in the early stages of experimentation with 100 kW of neutral power. Similar experiments are planned for the TFR device in France. Theory predicts no instabilities at these levels of injection and none have been experimentally observed.

RF heating in tokamaks has been tested recently on the Princeton ST (ion cyclotron resonance) (6) and the TM-3 (electron cyclotron resonance) in the Soviet Union (7). The ST experiments showed effective coupling of energy to the plasma but difficulty in depositing a significant amount of energy due to an influx of gas from the wall caused by desorbed, partially stripped oxygen, which happened also to be resonant with the heating frequency.

RF heating experiments are planned for the TFR in France in the near future, and high power RF will be tested in PLT in two or three years. The Soviet program has significant RF heating efforts on TM-3 and R-03 and will use high power RF on T-10 in one to two years. Finally, the entire fusion laboratory at Grenoble, France, is dedicated to pursuit of RF heating on tokamaks and stellarators.

Fusion reactor plasmas must be very pure, because impurities cause enhanced radiative energy losses and replace reacting nuclei with non-fusible nuclei. Impurities are present in today's experiments; their source is the wall surrounding the plasma. The mechanisms for the liberation of impurities from the wall are principally 1. sputtering by fast neutrals and ions, and 2. incidental contact of energetic plasma with the wall. Impurity take-up by a tokamak plasma can occur during the initial formation of the discharge and also during steady operation. Until recently, all tokamaks showed both kinds of take-up, but now the ATC and the Alcator have means to minimize start-up–associated impurities. In ATC part of the vacuum chamber was covered with a titanium getter, which apparently collects impurities before they have an opportunity to diffuse into the plasma. The resultant Z-effective (effective atomic number) of the plasma is very close to the pure hydrogen value of unity, rather than the values of four or five observed previously (8). In Alcator the same effect seems to have been achieved by discharge cleaning the chamber walls with unusually high energy discharges. The Alcator plasma then seems to stay clean for tens of confinement times (9). Both experiments thus seem to have found techniques for producing clean initial plasmas. These techniques should be useful in other experiments and should provide plasmas better suited to the study of long-term impurity take-up.

Independent of these very promising recent results, the creation and migration of impurities is an important area of study. Several experiments are now under construction to study and reduce impurity problems. Included are the PDX (Poloidal Divertor Experiment) at Princeton, the ASDEX (Axi-Symmetric Divertor Experiment) at the Garching Laboratory in West Germany, and the T-9 and the TBO at the Kurchatov Institute in Moscow. These devices will operate in 1977–1979 and will use diverted magnetic field lines to move ionized impurities at the plasma periphery to remote regions where the impurities can be trapped or pumped from the system. In Japan, the DIVA tokamak is now operating with a divertor. At Princeton a divertor on the FM-1 spherator[4] has operated according to a simple theory and demonstrated an 80% efficiency for sweeping diffusing plasma away from the walls.

For any fusion system to be of practical value, it must contain a reasonably stable plasma. This means that the plasma must be stable against gross, fluid-like motions, called magnetohydrodynamic (MHD) instabilities, of which there are many. Once MHD stability has been assured, then a large class of fine-scale instabilities, called micro-instabilities, must be guarded against. The amount of tolerable instability is very much a function of the confinement configuration. For instance,

---

[4] A Spherator is a tokamak-like closed system that uses a solid conductor inside the plasma to conduct the toroidal current. Because the poloidal field is highest at the internal ring, plasma does not come in contact with it.

tokamaks have very long classical plasma confinement times at reactor scale, and so a significant amount of plasma instability would be permissible in a tokamak reactor. On the other hand, magnetic mirrors (see below) must be almost perfectly stabilized against all instabilities to be of reactor interest. Theta pinches (see below) are most sensitive to MHD problems and appear to be relatively, indeed anomalously, insensitive to microinstabilities. Finally, inertial systems are sensitive to a more limited class of both types of instabilities (also see below).

In low-β closed systems, the most worrisome instabilities have generally been theoretically and experimentally identified and regions of stable operation have been established, although further study is required to fully understand all aspects of both the stable and unstable regimes. These stable regimes for low-β closed systems with circular plasma cross sections are characterized by relatively small poloidal fields, i.e. relatively small plasma current and relatively small β. Reactors will be more economical the higher the β, because high-β corresponds to a more efficient use of the main magnetic field which, with auxiliaries (structure, power supplied, coolant, etc), represents one of the most expensive elements of a magnetic confinement reactor. Theory (10) and small-scale experiments (11) both indicate that much higher beta values are achievable when certain noncircular plasma cross sections are used. Configurations of particular interest are those with the Doublet (see Figure 2), "D," and elliptical shapes. The Doublet III experiment, scheduled to begin operation in 1978 at the General Atomic Company, is aimed at testing most of the important noncircular tokamak configurations at near reactor-level conditions. The very large Euratom Joint European Torus (JET) experiment will utilize a "D"-shaped cross section in its attempt to achieve reactor-level conditions when it begins operation in 1979–1980.

Although the various major world fusion programs have somewhat different philosophical approaches to fusion research and development, all agree that the next major step in fusion research must be the construction of very large new tokamak experiments. The Europeans under Euratom began their plans ahead of everyone else. Their motivations were apparently as follows: (a) in their judgment a sufficiently sound scientific basis existed to justify such a large step; and (b) new larger tokamak experiments are clearly needed to precisely establish how tokamak confinement scales with size at reactor scale.

The proposed Euratom JET experiment is well into preliminary design and is scheduled to be funded in 1975 and to operate in 1979–1980. Its site has not yet been chosen. Basically, JET will use conventional water-cooled copper coils to produce a 35 kG field. It will employ a "D"-shaped cross section, which has been theoretically shown to exhibit a higher β

than the more conventional circular type. If problems arise, a somewhat smaller circular plasma can be utilized with only minor adjustment. Plans are to use neutral beam heating, but RF could be added relatively easily, if it proves necessary or desirable. Primary emphasis will be on studying hydrogen plasmas, but after two or three years DT mixtures will be substituted to study burning plasmas. JET will not be built for extensive DT operation, so extensive maintenance and modification will not be possible after initial DT fueling.

The next, more advanced, very large tokamak will be the US Tokamak Fusion Test Reactor (TFTR) to be started in 1975 and completed in 1980–1981 at a total cost of $215 million. The purpose of the TFTR will be to study very large hydrogen tokamak plasmas and then to relatively quickly move to the study of DT plasmas. With DT, the goal will be to create power densities equal to those of a tokamak power reactor. TFTR will be built with extensive remote handling and maintenance equipment

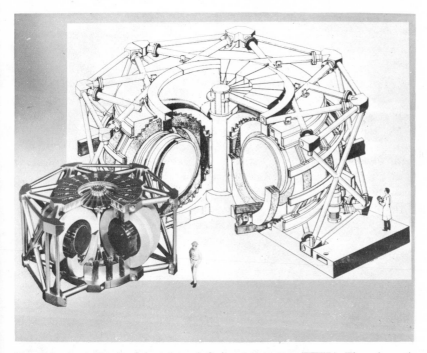

Figure 4 A schematic of the tokamak fusion test reactor (TFTR). The schematic shows the general configuration of the TFTR, a very large DT burning experimental system to be built at the Princeton Plasma Physics Laboratory for operation in 1980–1981. The insert shows a model of the Princeton Large Torus (PLT) presently nearing completion. In many respects, PLT is a prototype of the TFTR.

so that extensive DT operation will be possible and so that engineering experience can be gained with a radioactive fusion power system. A schematic of TFTR alongside a scaled picture of a model of PLT is shown in Figure 4.

Like JET, TFTR will use conventional water-cooled copper coils. Whereas JET is designed according to nominal scaling laws, TFTR is designed to meet its objectives based on the most conservative scaling assumptions. If confinement scaling turns out to be better, the facility will be capable of taking full advantage of the improved conditions. One means by which TFTR will attempt to meet its objectives will include the use of neutral beams to create a nonequilibrium plasma of the two-component type described in Section 2. In this way additional insurance is provided to insure the achievement of TFTR's DT burning objectives.

The Soviet Union is planning a new tokamak that is larger than either JET or TFTR. It is called T-20 and is scheduled to operate in 1982. If experiments in the next few years on T-10 and PLT indicate very favorable scaling, then T-20 will be configured as the core of an experimental electrical power reactor with a thick prototypical blanket. If scaling is less favorable, the blanket will be reduced or eliminated, and the additional space used for plasma. T-20 will be aimed at reaching Lawson conditions and will be built to operate repeatedly with DT, again, using water-cooled coils.

The Japanese are also designing a new, very large tokamak which they hope to operate in 1980. It will be similar in size to JET and TFTR, but will have a circular cross section and will not be designed for DT. Besides being a vehicle for hydrogen plasma physics studies, the JT-60 will serve as a near-term focus for a massive scale-up of the Japanese program, which has been relatively small in size.

In many ways stellarators are quite similar to tokamaks. Experimentally, confinement appears to be classical in some low temperature, low density devices, notably the Saturn in the Soviet Union and the Wendelstein in Germany. In larger devices operating at higher field and density (Wendelstein IIb and the Uragan I at the Kharkov Institute in the Soviet Union) confinement appears to be similar to that in tokamaks, but experimental uncertainties cloud this conclusion in part because the difference between classical and anomalous behavior is not very large in those stellarators. Heating with RF power near the ion cyclotron frequency is common and has been shown to be effective in stellarators. Neutral beam injection heating has not yet been tested on stellarators, but experience in tokamaks suggests that it should be very effective.

Present-day stellarators are limited fundamentally by small plasma size, which is a direct result of the engineering difficulties associated with the very complex forces produced by the helical windings. This smaller size coupled with weaker poloidal fields means that present stellarators have smaller confinement times than most present-day tokamaks. New, larger, higher-field stellarators are in various stages of construction or initial operation, and the future of the stellarator approach depends very much on the success of such new machines at the CLEO (United Kingdom), WEGA (France), Wendelstein VII (Federal German Republic), and L-2 (Soviet Union). If results are favorable, the stellarator may begin to compete with the tokamak, in spite of the well-recognized difficulty in handling the very large magnetic forces on the helical windings. The major impetus for stellarators is their inherent ability to operate in a steady state mode, which, if possible at all, will be difficult in tokamaks, which now utilize transformers to maintain the plasma current.

3.3   High-β Systems

High-β systems are of particular interest because they make better use of the applied magnetic field than do low-β systems. Because magnetic field coils and power supplies will represent a significant fraction of the cost of a magnetic confinement fusion reactor, a significant economic advantage might accrue, if indeed other aspects are comparable. Most high-β concepts are characterized by high plasma density. Because power output is a direct function of the density-confinement time product $n\tau$, high-density, high-β systems allow shorter confinement time, and indeed almost all such systems operate in a short pulse, batch-fuel burn mode that does not require refueling during the power pulse.[5] To some degree these advantages would be offset by the fact that rapid pulsed operation would require the efficient storage, handling, and circulation of very large quantities of electrical energy.

Pulsed high-β concepts have densities in the range 10^{15}–10^{19} cm^{-3}, well above the 5×10^{13}–5×10^{14} cm^{-3} characteristic of low-β systems. These systems utilize rapidly rising magnetic fields to shock heat and adiabatically compress plasma, with the ratio of shock heating to compressional heating varying according to the concept and, to a degree, to the state of development of pulsed power technology. Of the various pulsed high-β concepts, the most important are the theta pinch, the belt pinch, the Z pinch, and the imploding liner, the listing not being in order of world-wide emphasis. The theta and Z pinches and the imploding liner concept can and have been built in both the open and closed geometry, but present efforts are concentrated on the closed or toroidal embodi-

---

[5] The one exception is the EBT (Elmo Bumpy Torus), which is a steady-state, high-β exploratory concept in the United States (1).

ments. The belt pinch is always toroidal. The theta and belt pinch, and most imploding liner work, make use of strong pulsed toroidal fields with secondary poloidal fields for MHD stability. The main fields in the Z pinch come from the strong current that flows around the torus in the plasma and an applied toroidal field. In a number of respects the Z pinch is thus similar to the tokamak. Theta, Z, and belt pinches all typically operate with magnetic fields in the 30–100 kG range, which can be reasonably obtained using what is today conventional coil and capacitor storage technology.

On the other hand, the imploding liner concept aims at using megagauss magnetic fields, which cannot be created using conventional coils because of strength of materials limitations. To achieve megagauss fields a relatively cool plasma in a relatively weak magnetic field is compressed by a metal liner that is driven inward by an external, rapidly rising field. Because the compression of the liner is much faster than the skin time

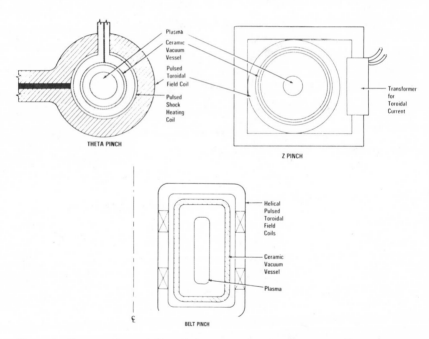

Figure 5 Cross sections of three high-β concepts. The primary approach in high-β systems is the theta pinch, here shown in the configuration necessary for separated shock heating and compression. Both the Z pinch and the belt pinch utilize relatively strong currents flowing in the plasma, which make them somewhat similar to the tokamak, with the important exception being that they operate at much higher β.

of the inner magnetic field in the liner, both the plasma and trapped field can in principle be compressed to fusion temperatures at very high densities and megagauss magnetic field levels for short but, in principle, adequate times. Figure 5 shows the general configurations of the three major high-density, high-β fusion concepts.

Early theta-pinch research was focused on linear systems and demonstrated the following: (a) ion heating to thermonuclear temperatures (3–5 keV) could be routinely achieved in the theta-pinch geometry, and (b) loss of plasma from the ends of the theta pinch was the limiting factor on confinement. These results were first obtained in the United States (12). Concurrent experiments in Germany (13) and England (14) confirmed the US observations, and experiments on an 8-m linear pinch in England established that cross-field diffusion in a linear theta pinch is essentially classical (15).

The decay time for particle losses through the ends of 1–10-m linear theta-pinch experiments is only 2–20 μsec, so the extrapolated length of a break-even linear theta-pinch experiment is about 1 km, and the length of a reactor is about 10 km, a length generally regarded as impractically long. Therefore, in the mid-1960s physicists began to consider toroidal theta pinches in which the "ends" and hence the "end losses" would be eliminated.

It was recognized from the beginning of toroidal theta-pinch research that the curvature of the field lines would require supplementary magnetic fields to provide toroidal equilibrium. As in low-β stellarators, helical magnetic fields were to be superimposed on the basic toroidal magnetic field to produce the required field patterns. It was recognized that once the plasma is brought to its equilibrium position, some additional forces might be necessary to stabilize the plasma motion and maintain the plasma in toroidal equilibrium. It was thought that these "shimming" forces could be supplied by either a feedback or dynamic stabilization system. The world's largest toroidal theta pinch—the Scyllac at Los Alamos—has been testing these concepts since it first came into operation in a toroidal configuration in April 1974. The results from Scyllac will have a major effect on the future of theta-pinch research. If long-term confinement is possible as predicted, then a new, larger, reactor-grade theta-pinch experiment may be warranted.

Present high voltage theta pinches, including Scyllac, are devices in which the energy for ion heating by both fast implosion and adiabatic compression is derived from a single large capacitor bank. But high voltage capacitor banks of sizes larger than that of Scyllac would be too large and complex to be practical, and so future theta pinches must utilize other technologies that dictate a separation of these two processes. This concept is called staging. It involves the use of a high quality, fast rising (microseconds), shock-heating field to preheat the plasma, followed by a slower (milliseconds) adiabatic compression to bring the plasma to reactor temperatures and densities (16). Although the shock heating apparatus is expensive in cost per unit of energy, it need supply only a few percent of the energy needed to heat the plasma, so its total cost can be comparatively low. The slower adiabatic compression system can be of much lower unit cost so that the overall cost can in principle be within practical bounds.

Another advantage of the staging approach is that the compression ratio in a staged system would not need to be as extreme as in present systems, in which the plasma is compressed to a radius of 0.1–0.2 of the discharge tube radius. Compressions to 0.3–0.6 radius ratio would mean that the plasma would be in better magnetic "communication" with the conducting wall, which could then help to stabilize a number of MHD instabilities, eliminating the need for fast response, expensive feedback stabilization apparatus. The basic concepts of staging and wall stabilization of "fat" theta-pinch plasmas are now being tested in the Staged Theta Pinch experiment, which began operation in early 1975 at Los Alamos.

Z pinches were among the first concepts pursued in fusion research in the early 1950s. They were abandoned in the United States because of MHD instability problems but were pursued in England principally in their toroidal configuration. Recently, a new approach to the MHD instability problem was conceived and this is now being tested on the Culham, U.K., HBTX (High Beta Toroidal Experiment), and the Los Alamos ZT-1 (17). The new concept involves providing a field reversal at the edge of a rapidly formed plasma. Initial results from both devices indicate that theory and experiment seem to agree (18). The problem is that the electron temperature in both experiments is low (10–20 eV), so that the plasma is very resistive and the reversed field is lost in times of the order of 10 μsec, thereby destroying the equilibrium. Future directions are toward higher electron temperature and therefore more highly conductive plasmas capable of holding the reversed field for a much longer period. The attractiveness of the Z pinch has always been that the main confining field comes from the plasma current rather than from an external magnet, which has mechanical and electrical limitations that are more restrictive than those of a high current in a plasma.

A belt pinch is rather like a combination of a theta and Z pinch but with a plasma whose cross section is elongated in a height-to-width ratio of 5–10. In principle, betas approaching unity are possible. To date β's of 30% have been achieved in the BP II at the Garching, FGR, laboratory at temperatures of about 40 eV (19). Because significant plasma currents flow in belt pinches, they are in some ways high-β versions of very noncircular tokamaks.

The principal program in imploding liner research is underway at the Kurchatov Institute in Moscow; a much smaller effort exists at the US Naval Research Laboratory. At present the Soviet program is aimed at developing the technology required to achieve the necessary liner compressions. Emphasis is on inductive energy storage development and studies of the fluid stability of imploding liners of different materials and configurations (spinning or not). Preliminary plasma experiments are aimed at producing suitably preheated, stable plasmas for subsequent liner compression. Significant plasma experiments with liners are due in 2–3 years and should provide a good insight into the practicality of the concept. Proponents of imploding liner reactors believe that their concept will not suffer from the impurity problems that could be important in longer pulsed and steady state systems, and they believe that liner replacements can be accomplished with relative ease using ideas that are now in the conceptual stage.

3.4   Open Systems

Open systems, or magnetic mirrors as they are often called, are generally linear in shape. To inhibit the natural tendency of confined plasma to "slip" along the field lines out the ends, stronger fields are usually added at the ends to reflect or "mirror" plasma back into the system. The original mirror concept employed two simple ring magnets, but it was found to be MHD unstable. By adding additional current-carrying bars parallel to, and symmetrically arranged about, the axis of the system, an MHD-stable geometry was obtained. This concept was experimentally pioneered in the USSR in the early 1960s (20). The resulting magnet configuration is still being utilized in the USSR. However, analysis shows that a more efficient way of obtaining the same configuration is through the use of a coil shaped like the seam of a baseball, or through the use of two curved, flattened, interlocking coils called a "yin-yang" set.

The principal world program in mirror research is in the United States, and a more basically oriented effort exists in the USSR. Significant mirror programs were underway in the Euratom countries for many years, but they were terminated in the early 1970s due to a desire to shift to tokamak research, a general lack of progress, and the theoretically small energy gain predicted for basic mirror reactor concepts.

The above remarks apply to the "classical" approach to mirror research. Two more unconventional approaches merit note also. One is the linear theta pinch described in the previous section. Because its β is near unity, its plasma excludes most of the applied magnetic field, and it naturally generates its own mirrors at the ends, where the plasma density (and beta) drop to very small values. The other notable variant is the multiple mirror configuration being pursued in Novosibirsk, USSR (21). Here the magnetic field is used primarily to insulate the plasma from the wall, i.e. to reduce the perpendicular thermal conductivity, while the multiple mirrors are aimed at limiting axial plasma flow. This multiple mirror concept eventually is aimed at a β of about 100, i.e. it would be a wall-confined, magnetically insulated system rather than a usual magnetic confinement system. An essential feature of the concept is the use of an intense, relativistic electron beam to rapidly heat the plasma to fusion temperatures. Because the concept is in an early phase of research, it is not described further. All three concepts are shown in Figure 6.

The principal experiment in mirror research has been and is still the 2X-II at the Lawrence Livermore Laboratory. This device has regularly produced plasmas with ion temperatures of 1–10 keV. Plasma densities are in the 10^{13}–10^{14} cm^{-3} range; the initial plasma β is approximately 0.5; and confinement has been shown to be classical (22). The achievement of classical confinement (the theoretically predicted maximum) in 2X-II was a major achievement and essential to the continuation of this line of

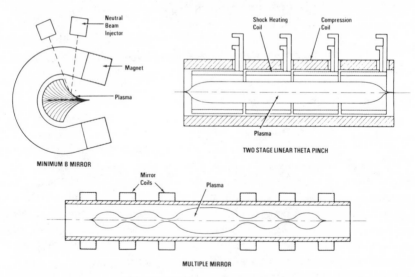

Figure 6 Schematic of the various magnetic mirror concepts. Main-line mirror research is concentrated on the minimum-B magnetic mirror concept, here shown with a magnetic coil shaped like the seam of a baseball and fed by neutral beams. Also shown is the two-stage theta pinch and the very high pressure multiple mirror concept. Both of these variants are pulsed systems, while the minimum-B mirror is steady state.

research. This is because the basic mirror concept offers very little margin below classical confinement, if it is to be of practical interest.

2X-II was built as a pulsed machine[6] to minimize initial cost and because the first key questions of mirror confinement physics could be addressed easily in a pulsed system. A practical mirror reactor must be a steady state system, however. Thus, the next questions to be addressed in 2X-II are associated with maintaining the plasma density with neutral beams and also verifying the theoretical prediction (23) that particle confinement increases as $E^{3/2}$ ($E =$ the mean ion energy). Because the 2X-II plasma provides a well-behaved, well-understood starting point, the experiment was extensively modified in 1974 to enable it to test heating and refueling by neutral beams and to test ion energy scaling. Specifically, an unprecedented 12 MW of pulsed 20 keV neutral beams (12 beam modules) were added for this next phase of the research.[7] At this writing the formerly achieved good confinement of its basic target plasma has been reestablished and experiments with the beams have begun. Thus far, a 25% decrease in plasma decay has been achieved with about 1 MW of beam power in accordance with expectations, and no deleterious instabilities have been observed (24). The data are preliminary, however, and the full 12 MW of power will be required for a definitive test.

The method of plasma formation in 2X-II would not be easily adaptable to a reactor, and 2X-II uses pulsed copper coils, which also would not be useful in a reactor. These technologically related problems are being addressed in a second experiment, the Baseball II (BB II), which utilizes a steady state superconducting magnet. BB-II is now testing creation of an initial target plasma through the use of a short pulsed, energetic CO_2 laser to vaporize, ionize, and heat material introduced into the vacuum magnetic field in the form of a small pellet of NH_2. Because the nitrogen would quickly leave the plasma, a pure, energetic hydrogen plasma should result. This laser-initiated target plasma concept was conceived and studied at the United Aircraft Research Laboratories, where it is now being further developed in a new, smaller (than BB-II) experiment called LITE (Laser-Initiated Target Experiment) (25).

In summary, 2X-II is aimed at demonstrating further heating and refueling by neutral beams, and it will test the theoretically predicted

[6] An energetic, dense plasma is created in a very short time, and this plasma then decays in density, because it has no feed to compensate for the natural plasma loss out the ends.

[7] It is worthy of note that the operational experience gained on 2X-II with twelve neutral beams has direct applicability to future tokamaks that will also utilize multiple neutral beams.

$E^{3/2}$ confinement scaling. BB-II and LITE are aimed at developing the technologically important laser-target concept for eventual use in a superconducting mirror system, similar in technology to BB-II.

The US mirror program benefits greatly from the physics understanding that has been and is being developed in the less applications-oriented mirror program at the Kurchatov Institute in the USSR. Specifically, recent experiments on the Soviet PR-7 have exhibited the theoretically predicted drift loss-cone, flute instability late in time in a pulsed mirror plasma (26). Theory and experiment were correlated, giving greater confidence in general mirror theory. Because this particular instability occurs outside the mirror reactor regime, its significance is more associated with basic understanding.

If mirror experiments in the next few years are successful, then a number of future options exist for this concept:

1. If confinement is exactly classical or if either serendipity or an invention make it better, the basic mirror concept could become the basis for a fusion power reactor.
2. Because the inherent difficulty with the mirror is its end losses, a toroidal linkage could conceivably be developed to connect the ends, thereby decreasing the end-loss energy drain and increasing the system energy gain. If fusion power were produced only in the mirror section, the toroidal link could be relatively inexpensive, and the resultant system would take advantage of the inherent ability of the mirror to contain energetic ions, which, of course, are what make fusion. If the toroidal link carried a plasma current, present-day tokamak research might be directly applicable.
3. In the future, the fusion program will require a large materials and component testing facility with easy access. A Fusion Engineering Research Facility (FERF) based upon a mirror might be an attractive option. Such a system might draw possibly 100 MW(e) to produce a steady state, fusion plasma environment for a variety of engineering studies. The advantage of using a mirror for such a FERF would be its inherent small size and easy radial access (27).

3.5   Inertial Confinement Concepts

Inertial confinement is based upon the fact that it takes a finite time for a mass of hot plasma to disassemble due to its internal pressure. Thus, if a suitable mass of DT is heated to fusion temperatures very quickly, it is theoretically possible to produce net energy in the short time that it takes for the resultant plasma to blow apart. When energy is rapidly deposited on the surface of a spherical pellet, a radially converging compression wave is created. This wave not only heats the solid, but also compresses it to high densities; spherical convergence magnifies these effects near the pellet center. Because the fusion reaction rate is proportional to the square of the density and because the disassembly time is roughly inversely proportional to the sound speed, which is insensitive to density, high compression is the key to efficient inertial confinement fusion concepts.

In the early 1960s the invention of short pulsed lasers led to the idea of using lasers to ignite DT pellets to make fusion. Calculations indicated that relatively high laser pulse energies would be required, so the idea of laser-pellet fusion lay essentially dormant while scientists developed higher powered, short pulse lasers. In the later 1960s lasers utilizing neodymium-doped glass were developed to the point of being able to deliver joules of energy in times of the order of nanoseconds (10^{-9} sec) or less. In 1968 Academician Basov from the Lebedev Institute in Moscow reported (28) observing neutron emission from a DD-loaded, lithium pellet irradiated with a short pulsed laser. This observation, coupled with then classified calculations showing that break-even (fusion yield equal to the laser-light-energy input) might occur at energies as low as a few hundred joules, prompted US researchers to mount a major program in laser fusion. The reason that part of the work was then classified was that the concept and techniques for developing high compression of fusion fuels underlie the hydrogen bomb. Subsequent world-wide interest in laser fusion and a reassessment of classification policy in the United States has led to the declassification of many of the important concepts of laser fusion so that today well over 90% of the work is unclassified.

In the late 1960s the laser-fusion problem appeared relatively straight-forward—a judgment that often prevails with a new idea about which little is known. (Note that this same kind of optimism prevailed in the early days of magnetic confinement research, when the complexities of plasma physics were unappreciated, largely because the field was essentially undeveloped.) Today, laser-fusion research has developed to the point of being a mature discipline wherein many of the difficult problems are being studied in a very sophisticated manner.

The problems of laser fusion can be classified as those related to the laser and those associated with the pellet. Consider first the laser. In the mid- to late 1960s, a variety of lasing media were identified and tested. The medium that showed the greatest promise for providing very short pulses in the picosecond (10^{-12} sec) to nanosecond (10^{-9} sec) range was glass doped with small quantities of neodymium and pumped with xenon-filled flash lamps. Great strides were made in developing Nd:glass lasers into the tens of joules output range, and it appeared relatively straightforward to develop these lasers into the kilojoule and tens of kilojoule range.

In the 1970–1971 period, researchers were predicting the near-term availability of hundreds to a thousand joules in the nanosecond and subnanosecond pulse range. This was to be accomplished by first selecting a single millijoule pulse from an oscillator and then amplifying it through progressively larger amplifiers. The early stage amplifiers were to be solid rods pumped by flash lamps around their periphery. Beyond a certain diameter (about 10 cm), rods cannot be uniformly pumped, so side-pumped glass disks were to be used to go to the larger diameters necessary to provide yet higher powers.[8] To go beyond the limits imposed by the fabrication of large disks, multiple-path lasers were conceived. The arrangement of an oscillator, amplifiers, and parallel chains is shown in Figure 7, along with schematics of a rod and a disk amplifier.

The development of higher laser energies turned out to be more difficult than anticipated. Some of the more significant problems that were encountered were as follows:

1. Production of large rods and disks of optical quality, uniformly doped, impurity-free glass represented a difficult problem. It took a number of years to develop appropriate fabrication techniques.

2. The disks in disk amplifiers have a high gain parallel to the face of the disk so that lasing can occur internally unless the disk edges are coated with a material with a very low reflection coefficient. It took $2\frac{1}{2}$ years to develop appropriate coatings to eliminate this problem.

3. Isolators had to be developed to protect the amplifiers. Laser amplifier chains have an extremely high optical gain—of the order of 10^6 and higher. This gain in the forward direction is basic to the achievement of a high energy output. A problem arises when the target reflects a small amount of the input light back into the high gain, not totally depleted amplifier chain, which then can amplify this small signal in the reverse direction back into the smaller amplifiers, which then can be damaged by the higher intensities thus generated. The required isolators had to have a low loss in the forward direction and a very high loss in the reverse direction. The biggest problem was to develop optical quality, radially uniform, large aperture isolators. Figure 7 shows how isolators are placed throughout a Nd:glass laser system to adequately protect it.

4. At high power, lasing media display a nonlinear index of refraction

---

[8] Laser glass damages above 5–10 J/cm^2 so large diameters are necessary in late and final stage amplifiers to achieve high energy output while maintaining good beam quality.

that can virtually destroy the coherence of the beam. This confused the interpretation of early pellet irradiations where there was insufficient knowledge of true beam spatial and temporal behavior and it was thought that much higher energies were provided to the target than was actually the case. Recent work at Livermore has helped to quantify this nonlinear effect (29). Although lower power densities in the laser glass are required to avoid this problem, it will apparently no longer be a source of confusion.

5. Laser alignment and diagnostics to verify beam coherence, energy, and position were difficult problems that required innovation and development.

In the future the optimum use of laser energy on a target will require the development of laser pulse shapers that can deliver a light pulse of specified duration and time history. The first of these has already been experimentally demonstrated by KMS Fusion Inc. in the United States

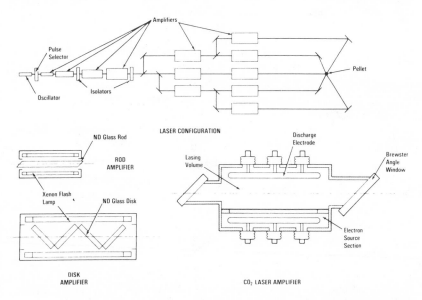

Figure 7 Schematics of a high power, short pulsed laser and some amplifier modules. High power lasers select a single pulse from an oscillator and increase its energy by passing it through a series of progressively larger amplifiers. The entire chain is protected by a number of isolators. Disk and rod amplifiers utilize the neodinium-doped glass in either the rod or disk form, pumped by xenon flash lamps. A CO_2 laser amplifier utilizes a pulsed relativistic electron beam to produce initial ionization in the gas. A secondary discharge then excites the molecules to the excited states necessary for laser amplification.

(30). These pulse shapers can deliver an almost arbitrarily determined pulse to the system for subsequent amplification. Since the gain of the amplifiers is nonlinear with power, careful computer calculations are necessary to optimize the total system so that the desired pulse appears at the output of the final amplifier.

The target pellet contains the DT fuel that is to be heated and compressed to fusion conditions. A variety of target configurations are possible. Most simply, the DT could be used in the form of a frozen solid sphere. Hollow spheres are desirable because they have a lower resistance to inward flow during the early phases of compression and because they prolong the duration of the implosion, allowing more laser energy to be coupled into the imploding pellet at a given peak laser power. Layers of various materials around the outside of the DT may be useful to improve laser energy absorption, to "tamp" the pellet to improve its compression characteristics, and to help to inertially confine the burning fuel at the culmination of the implosion.

Besides the optimization of the pellet configuration, a number of basic physics questions arise:

1. How efficient is the energy absorption process? If it is inefficient, larger laser energies would be required to achieve a given compression, and the overall energetics of the process would be less attractive. Involved in the absorption process is classical "inverse bremsstrahlung" plus a host of laser light-driven instabilities (31) that not only alter the overall absorption efficiency but also undesirably precondition the pellet (see below). Recent experimental data has been well correlated with theory, and it appears that laser light absorption can be 80–90% efficient up to an instability threshold that is wavelength dependent. At 10.6 μm this threshold is about 10^{12} W/cm^2 of laser energy on the pellet, while at 1.06 μm it is 10^{14}–10^{15} W/cm^2.

2. Will instabilities in the absorption process precondition the pellet in an adverse manner? In particular, one class of instabilities gives rise to a significant number of very energetic (100 keV) electrons. These electrons have a range of the order of the size of the pellet so that they can uniformly preheat the pellet, creating a back pressure that can subsequently inhibit high compression. X rays from these energetic electrons are observed in experiments (32), and so this problem must be avoided by appropriate pellet design or the use of shorter wavelength lasers.

3. Can Rayleigh-Taylor instabilities inhibit high compression? Initially the answer to this question appeared to be in the negative, but recent studies indicate it to be a problem (33). As a result, thicker layers of

tamper or fuel are required, and these require higher power inputs from the laser to achieve the required maximum compression. In addition, a high degree of pellet sphericity and dimensional uniformity is required to minimize this problem.

Presently, Nd : glass lasers are delivering hundreds of joules in nanosecond and subnanosecond pulses to target pellets of 50–100 μm diameter. Compression factors in the range of 10–100 have been achieved yielding neutron outputs of 10^6–10^7 neutrons (34). Diagnostics to resolve pellet behavior on picosecond time scales and tens of micron sizes are being developed. Lasers producing energies in the kilojoule range in subnanosecond pulses are under construction at the University of Rochester, the Lawrence Livermore Laboratory, and the Lebedev Institute. These lasers will produce higher compressions and begin to experimentally establish energy scaling laws.

Nd : glass lasers of 10 kJ pulse energy are now under construction at Livermore and the Lebedev for operation in 1977. These lasers will further extend the scaling laws and should produce a significant thermonuclear burn. Presently, break-even is calculated to require laser energies of 10–30 kJ (35) in a 0.1 nsec pulse, so it is not clear whether another generation of laser will be required to achieve this milestone. Energy gains of the order of 100 (fusion yield one hundred times the laser light energy input) will require laser energies roughly 100 times larger than those required for break-even, i.e. about 1 MJ.

Nd : glass lasers are about 0.1–0.2% efficient. Because the maximum energy gain from a laser pellet reaction is about 100, more efficient lasers will be required for future experiments and reactors. One possible candidate medium for this application is CO_2 gas. CO_2 lasers for laser fusion applications are under development at Los Alamos, where about 200 J have been generated in a nanosecond pulse at an efficiency about 1% (36). A CO_2 amplifier can be excited by an electrical discharge initiated and sustained by a relativistic electron beam. Figure 7 shows a schematic of such an amplifier, which could be arranged in chains similar to those used in Nd : glass systems. One problem with CO_2 is that it lases at a wavelength of 10.6 μm compared to 1.06 μm for Nd : glass. Recent experiments indicate that long wavelengths are less than optimum for laser fusion (32, 37). Therefore, frequency shifting of CO_2 laser light would be necessary or another gaseous medium might be required. For this reason, increased emphasis is being placed on the search for efficient short wavelength lasers.

Laser-fusion research has become a sophisticated discipline and considerable progress has been made in the last year in particular. Higher energy lasers will clearly pace future progress. The next really definitive experiments at significant burn or break-even should come late this decade and should provide a basis for evaluating this very interesting concept.

An alternate initiator for inertial confinement systems might be relativistic electron beam generators, which were developed originally by the military for weapons simulation studies. They produce intense bursts of electrons in the 0.1–10.0 MeV range in pulses of tens of nanoseconds duration and total pulse energies of 10–1000 kJ (38). Furthermore, these generators are very efficient—of the order of 30–50% of the stored energy appears in the beam. Relativistic electron generators were not of use for inertial systems initially, because the beams could not be focused onto the small pellets of interest and their pulse lengths were longer than optimum. In the early 1970s, methods of focusing short pulsed, relativistic electron beams were developed (39), and pulsed power technology progressed to the point where nanosecond pulses appeared feasible.

Programs on electron beam fusion are now being pursued at both Sandia Laboratories (40) and at the Kurchatov Institute. In 1974, experiments showed that relativistic electrons behave predictably when they impinge on high density solids (41) so that they might be usable to initiate implosions using tamped pellets. To date, compressions of hemispherical gold shells have been observed (42) and irradiation of spheres has been accomplished with one as well as two beams, showing good symmetry (43). The problem of efficiently transmitting and focusing a relativistic electron beam from the generator to the target is more difficult than the analogous problem in laser fusion. Near-term experiments will help to clarify these problems and should shed light on the feasibility of the pulsed electron beam approach.

4   THE MAJOR WORLD PROGRAMS

4.1   General

Major fusion research and development programs are underway in the Soviet Union, the United States, Europe, and Japan. The majority of the work is centered on magnetic confinement concepts with tokamaks receiving the greatest emphasis. The United States and the Soviet Union have the largest inertial confinement programs, with much smaller ones underway in Europe and Japan. The best measure of the size of fusion programs for purposes of comparison is believed to be based upon total manpower. From the World Survey (1), the distribution of effort in magnetic confinement research and development in 1973 was as follows:

Country	Percentage of World Effort
USSR	37.1
USA	15.8
Germany	13.0
Japan	9.3
France	7.2
United Kingdom	4.5
Other	13.1

Recently, rapid expansions have occurred in the United States and Japan and also in the Soviet Union to a somewhat smaller degree. The resultant redistribution might differ by only a few percentage points from that of 1973, with the US gaining in relative size and Europe diminishing somewhat.

Magnetic confinement research has been a notable example of effective international cooperation since the late 1950s. Numerous personnel exchanges have taken place, and excellent communication and open discussions have been the rule, particularly at triannual IAEA Conferences on Plasma Physics and Controlled Nuclear Fusion. A bilateral agreement between the US and the USSR in 1973 has led to a particularly close and effective cooperation between these two countries. The emphasis in each program is described in the following.

4.2 The Soviet Union Program

The Soviet program is the world's broadest in terms of magnetic confinement physics research. The Soviets conceived and developed the tokamak concept, which is now under study worldwide. In addition, they maintain a major program in the generically related stellarator concept. Their program in magnetic mirrors is much smaller, utilizing modest experiments for more basic studies. In the area of high-β research, the Soviets have a major program in imploding linear systems. This was initiated about three years ago and today is the sole activity of a specially created new branch of the Kurchatov Institute at Krasnaya Pachra—a fusion laboratory that is today as large as the largest US fusion laboratory at Princeton.

The relative distribution of effort in the Soviet magnetic confinement program now and that projected for the next five years is as follows:

Approach	Now (%)	Next Five Years (%)
Tokamaks	43	56
Stellarators	29	22
Open Systems (Mirrors)	14	11
Pulsed High-β Systems	14	11
	100	100

In Soviet tokamak research the T-4 is the largest operating experiment; it is presently being used to test ion cyclotron heating. Currently in fabrication is the world's first superconducting tokamak, the T-7, which is about the size of T-4 and which is scheduled to operate in 1976. Also in fabrication is the T-10, a larger experiment similar in size to the US PLT. In terms of very large tokamaks, the Soviets are designing T-20, which is larger than any of the other very large tokamaks (TFTR in the US, JET in Europe, and JT-60 in Japan). It will be designed to burn DT on a regular basis and could serve as the core of an experimental power reactor, if tokamak scaling laws turn out on the favorable side of present predictions. T-20 is to operate in 1982.

The Soviet stellarator program is the world's largest. A new, midsized stellarator called L-2 is beginning to operate at the Lebedev Institute. At Kharkov the Uragan has been upgraded in magnetic field to 25 kG, and an even larger experiment has been proposed. The observation of classical confinement and favorable scaling in the Uragan stellarator is of particular note.

In inertial confinement the Soviets were the first to announce the observation of neutron emission from a laser pellet experiment in 1968 (28). They are in the process of building a new, nanosecond pulse, 10 kJ laser, which is scheduled to operate in 1977. The exact size of the Soviet laser fusion program is not known, but it is estimated to be about the size of its US counterpart. Similarly, the size of the Soviet electron beam fusion program is believed comparable to that in the US.

4.3   The United States Program

In magnetic confinement the emphasis is about 60% on tokamaks and 20% apiece on mirrors and theta pinches. The US program more than doubled in budget from 1973 to 1975 and has begun serious programs in fusion reactor engineering, somewhat ahead of the rest of the world. A number of large new tokamaks are in fabrication, notably the circular cross-section PLT at Princeton, scheduled to operate in late 1975; the noncircular cross-section Doublet III at General Atomic, scheduled to operate in 1978; and the divertor tokamak PDX, scheduled to operate in 1977. The very large, DT burning TFTR will begin operation sometime between mid-1980 and mid-1981, depending largely on funding.

In theta pinches the Scyllac and Staged Theta Pinch are the critical experiments, with results projected for the next two years being of major importance to the future of this line of research. The future of the mirror program is heavily dependent upon results from the 2X-II, which should also yield critical data in the near term.

In technology, the US leads the world in neutral beam development and recently began major programs in superconducting magnet development for large tokamaks and mirrors. A significant program in fusion materials studies began in 1973 and now includes plans for the world's first high flux 14 MeV neutron generators.

The largest US fusion programs are at the Princeton Plasma Physics Laboratory, the Holifield National Laboratory, the Lawrence Livermore Laboratory, the Los Alamos Scientific Laboratory, and the General Atomic Company. About 70% of the program is carried out at national laboratories with 15% in universities and 15% in industry.

The government-sponsored inertial confinement program in the US is conducted within the Energy Research and Development Administration's military program, because there is close association of the technology with nuclear explosive technology and because applications in both military and civilian sectors are conceivable. Within this national program complementary efforts are conducted at Livermore and Los Alamos, with glass laser development being emphasized at Livermore and CO_2 laser development at Los Alamos. Electron-beam fusion is emphasized at the Sandia Corporation. In addition to limited Government sponsorship at other facilities, excellent privately sponsored programs are also underway at KMS Fusion, Inc., the University of Rochester, and Battelle Memorial Institute.

4.4   The Euratom Program

The largest European programs are in Germany, France, and England. Recently, a major effort was undertaken to unify under Euratom what were previously separate, independent programs, and today about 20–30% of European fusion is supported by Euratom funds. In general, the emphasis in Europe is largely on magnetic confinement plasma physics, but a significant interest in fusion reactor engineering is developing in Germany and England. The primary emphasis is on tokamaks with significant programs in stellarators (Germany and England) and high-β systems (Germany and England). Mirror programs in both England and France were terminated about two years ago, and a small Euratom program in laser fusion is being developed, largely to keep abreast of the programs in the US and the USSR.

The most significant undertaking under Euratom sponsorship is the very large JET tokamak. An excellent multinational design team was assembled in late 1973 in Culham led by the Frenchman Paul Rebut. In 1975, it is expected that a site for JET will be selected and major expenditures approved. If all goes well, JET could begin operation in late 1979 or early 1980.

In France the TFR tokamak at Fontenay-aux-Roses is presently the world's largest, measured in terms of plasma current. It is now the focus of the entire fusion program at the laboratory, which formerly was devoted to a broad range of research topics.

At the Garching laboratory near Munich, a small tokamak called Pulsator is in operation. Garching has a major program in high-β research involving a belt pinch and a toroidal theta pinch, which is smaller than the US Scyllac but which has provided information useful to the basic operation of these systems.

At the Culham Laboratory outside London, a number of small tokamaks, stellarators, and high-β experiments are under study. A decision in the late 1960s to reduce the size of the English fusion program to half its former size over a five-year period greatly inhibited progress. Recently, a decision to reverse this trend has led to a mild expansion at what has been one of the world's leading fusion centers.

4.5   The Japanese Program

The Japanese Program has been relatively small but recently has been the subject of major national attention. In 1975 the program is to be increased by about a factor of three as part of a scale-up centered around a very large new tokamak, JT-60, which is scheduled to operate in 1980. Of particular note in the past have been very competent programs in high-β and tokamak research. A small program in laser fusion is also receiving additional attention.

5   FUSION REACTOR ENGINEERING

5.1   Reactor Conceptual Design Studies

First generation fusion power plants will utilize the DT reaction, because it has the highest cross section at the lowest temperature and has the highest energy gain per reaction. Because 80% of the energy from DT fusion appears in the form of neutron kinetic energy and because neutron energy can most readily be converted by thermal processes, first generation fusion reactors will utilize thermal conversion cycles for electrical power generation.

The first conceptual design of a fusion reactor was performed for the stellarator concept in the mid-1950s (44). Three significant problems were identified in that study:

1. Plasma-wall interactions can lead to the liberation of significant amounts of impurities that can contaminate the plasma. (This revelation led to the invention of the magnetic divertor concept described previously.)
2. The fueling of a steady state or long pulse reactor will require special

attention because the plasma tends to shield itself from incoming fuel gas, beams, or pellets.

3. Plasma fusion power densities would have to be very high to provide a net power output in a system that utilizes large copper toroidal field coils. (This problem was essentially eliminated by the development of superconducting magnets.)

Plasma confinement progress in the later 1960s and early 1970s stimulated the development of conceptual designs for fusion reactors based on each of the four major approaches to fusion power (tokamak, theta-pinch, mirror, and laser-pellet systems). An excellent review of these reactor designs can be found in the proceedings of a 1974 IAEA workshop on fusion reactor engineering (45). Important characteristics and problems associated with each reactor concept are as follows:

1. For a tokamak reactor, (a) power generation of 2000–5000 MW(t) and 800–2500 MW(e); (b) a toroidal plasma of about 10-m major radius and about 3-m minor radius supporting a 10–15 MA discharge current in a 50–60 kG toroidal magnetic field. The plasma might be bounded by a "magnetic limiter" and a divertor through which leakage plasma might be exhausted. Except for the relatively narrow layers of diverted plasma, the discharge is surrounded by a reasonable vacuum.

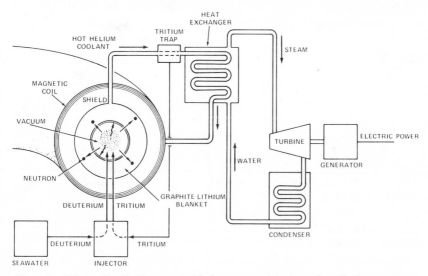

Figure 8   Schematic of helium-cooled fusion reactor. Neutrons from fusion reactions in the plasma are moderated in a graphite-lithium blanket. The blanket is cooled by helium, which produces steam in a standard heat exchanger. Tritium is bred in the blanket and removed from the helium coolant for use as a fuel.

The discharge current might be maintained for about 100 min followed by a 5-min recharge between pulses. Large superconducting magnets and high power neutral beams are among the largest development problems for tokamak reactors.

Figure 8 is a schematic of a toroidal DT fusion reactor that utilizes helium as a coolant, graphite as a moderator, and lithium as the fertile material for tritium production. Figure 9 is a drawing of a more detailed Princeton Plasma Physics Laboratory conceptual design (46). All tokamak power-reactor conceptual designs utilize superconducting magnets to produce the toroidal magnetic field. The Princeton reactor design proposes to fuel the reactor by the high energy injection of pellets of liquid DT. About 5% of the fuel is burned per pass through the reactor, so provision is made to separate the unburned fuel from the helium fusion product and subsequently to recycle it. Most impurities from wall bombardment are kept from entering the plasma by means of a divertor.

2. For a theta-pinch power reactor, (a) an output of 3000–4000 MW(t) and 1200–1800 MW(e), operating at one pulse every 10 sec; (b) a toroidal chamber major radius of 56 m; (c) a 0.5-m minor radius inside of which a theta-pinch plasma would be formed by shock heating and subsequent adiabatic compression to ignition. After a burn time of approximately 0.1 sec, a neutral gas blanket would be introduced around the plasma to cool it to avoid damage to the first wall. The reliable, rapid, and efficient transfer of large amounts of energy from an energy store to the compression coil is a major requirement for this concept and represents a sizable engineering challenge. Inherent to the theta pinch and all pulsed magnetic confinement reactor concepts is the need to cover the interior of the first wall and elements in the blanket with insulators capable of withstanding high voltages in the presence of 14 MeV neutrons and significant thermal cycling.

3. For a mirror fusion reactor, (a) an output of 350 MW(e), (b) a plasma approximately spherical in shape with a 130-m^3 volume; (c) a vacuum magnetic field of 50 kG with a 3:1 mirror ratio produced by a yin-yang pair of coils on a 7-m radius sphere; (d) 490 MW of neutral beams at 550 keV; (e) 590 MW of fusion power generated by the plasma and a fusion neutron power multiplication of two via nuclear reactions in the blanket. The large circulation of power inherent to a mirror reactor is due to the high rate of plasma leakage from ends of the mirror. To circulate this power efficiently converters will be needed to capture the energy of leakage plasma and transform it directly into electrical energy with efficiencies of the order of 70–85%.

4. Conceptual designs for a laser-fusion reactor have not been as

thoroughly developed as those for the magnetic confinement concepts, due to the relative newness of the laser-fusion program and large uncertainties in laser and pellet performance. Current projections are for pellet yields of 10–100 MJ, and cavities to contain such explosions appear feasible. The estimated time to restore necessary preexplosion vacuum conditions are somewhat uncertain, but multiple parallel cavities serviced by a single laser should provide needed flexibility. Laser systems capable of delivering megajoules in about a nanosecond with high repetition rates will be required and represent a large technological challenge. The fabrication of possibly complex pellets with stringent dimensional tolerances at low cost represents another problem. Finally, pellets must be repeatedly delivered to the focal spot of the laser to an accuracy of possibly tens of microns in the presence of vibrations from explosions in other chambers and the repeatedly pulsed laser.

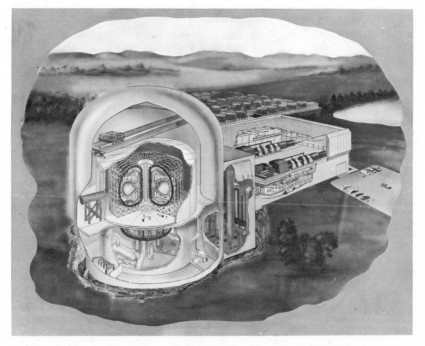

Figure 9 Conceptual design of a tokamak fusion reactor. Developed at the Princeton Plasma Physics Laboratory, this 2000 MW(e) reactor will operate in a pulsed mode with a power pulse of 97 min and a recharge time of 3 min. The reactor structure material is Nimonic PE-16, and the primary coolant is helium. The wall loading is 1.6 MW of DT neutrons per meter square.

5.2   Reactor Engineering Development and Testing Program

World fusion research to date has been justifiably centered primarily on solving plasma confinement problems. Reactor engineering problems until recently have been the subject of analytical study only. These studies have gone far toward defining the most important engineering questions, due in largu part to the extensive data and experience available from the fission reactor programs.

All fusion reactor designs have several features in common. A mixture of deuterium and tritium is burned. Tritium breeding is required and is achieved via neutron absorption in lithium placed in a blanket (approximately 1-m thick) surrounding the reacting plasma (see Figure 8). While some reactor designs use lithium or a lithium molten salt as a coolant and breeding material, favored designs use a solid compound of lithium (e.g. lithium aluminate or a lithium-aluminum alloy) in a blanket cooled by high pressure helium. Liquid lithium designs have more favorable heat transfer and tritium breeding characteristics, but they result in larger tritium inventories, and they require extensive technological development. All designs show satisfactory breeding ratios, although some utilize beryllium for neutron multiplication. Tritium doubling times can be as low as one month, which would allow an almost arbitrarily fast development of a fusion power economy from the standpoint of tritium supply.

Blanket/shield design and testing programs will be necessary in coming years to optimize the alternative blanket/shield/primary coolant configurations. These programs are being planned, primarily in the United States, but they have not been initiated. They must include consideration of such topics as blanket and shield neutronics, thermal hydraulics, coolant processing, materials compatibility, blanket and coolant corrosion and impurity control, fabrication and assembly techniques, remote maintenance, instrumentation and control, component development, and design optimization studies.

Another major task involves the development of large superconducting magnets. Existing niobium-titanium technology appears adequate for tokamaks, and niobium-tin technology appears to be developing rapidly enough to be practical for mirrors. Already large niobium-titanium superconducting magnet coils have been fabricated for bubble chamber applications, and these have operated reliably near the sizes envisioned for experimental tokamak power reactors. Development programs must provide magnets in large sizes with the following characteristics: (a) minimum refrigeration loads; (b) adequate stabilization against quenching; (c) ability to withstand large mechanical stresses; (d) ability to withstand nominal levels of radiation damage to conductor and insulator material without failure or excessive resistivity increases; and (*e*) adequate quality assurance and safety features to provide high plant safety and protection.

As discussed in Section 3, neutral beam injectors and other plasma heating devices are necessary to provide reactor-level plasma temperatures. Neutral beam technology programs in the US in particular have been very successful in developing neutral beam modules for successively larger plasma experiments. Future directions for neutral beam development are towards continuous operation and higher beam energies (200–600 kV). No "breakthroughs" will be required to achieve these goals but a substantial development program will be required.

Tokamak or mirror power reactors will operate in either a long-pulse or continuous mode so that continuous fueling will be necessary. Options include a fuel gas blanket or fuel beam or pellet injection. Investigations of these concepts are underway but a great deal remains to be learned, particularly about pellet fueling.

Long burn times can result in impurity buildup in the plasma from plasma-wall interactions. Excessive accumulations in the plasma can result in large radiation losses and resultant plasma cooling. As mentioned earlier, divertors have been proposed to prevent impurities from reaching the outer regions of hot, low-β toroidal plasmas. The degree to which divertors or other impurity prevention techniques can limit this inward diffusion of impurities will affect the burn time with shorter pulses resulting from poorer control. Short pulsed tokamak reactors could still be attractive, but the more rapid pulsing would require greater attention to the efficient switching of large amounts of energy.

One potentially attractive alternative to the divertor involves the idea of maintaining the tokamak discharge inside a neutral gas blanket that would serve the dual function of fueling the plasma while isolating it from the wall. Another concept would involve the use of mild plasma turbulence to inhibit impurity uptake by the plasma.

Tritium processing and control is a major engineering concern in all fusion power reactor designs. The problem may be conceptually divided as follows:

1. The tritium feed and exhaust system (exhaust vacuum pumping, impurity removal, isotopic enrichment, fuel storage, and reactor fueling).
2. The blanket tritium extraction system (extraction from the primary and secondary coolant and lithium replacement).
3. Tritium barriers to inhibit tritium migration through coolant piping, heat exchangers, etc.
4. Primary and secondary tritium containment and monitoring systems for both routine operation and fault conditions.

Conceptual designs to date have exhibited credible methods for attaining very low release rates for tritium. In the United States this was facilitated by the extensive body of data and experience developed in the nuclear weapons program.

The neutron activation products associated with 14-MeV neutron reactions with candidate reactor structural materials are nonvolatile in contrast to many important fission products. These activation products will be naturally and effectively contained in the structural materials so that routine releases of activation products to the environment are expected to be very small.

5.3   Materials Problems

All fusion reactor concepts will require structural materials that can withstand significant neutron damage. Fusion neutron fluxes on the first wall will be about a factor of ten lower than those in a fast breeder reactor. This will yield displacements per atom that are accordingly about a factor of ten lower. But the fusion neutron energy spectrum will run up to 14 MeV, resulting in hydrogen and helium production rates 10–100 times higher than those in a fast reactor. This will result in a different mix of important phenomena. This new mix will undoubtedly present new problems and some new opportunities. A significant amount of swelling will be allowable in coarse-tolerance fusion reactor structures, but excessive loss of ductility will not be tolerable. The outcome of this research will determine acceptable fluxes on the first wall, which will have a major impact on fusion power economics. Current reactor conceptual designs assume fluxes of 1 MW/m^2 or less, and this translates to projected capital costs of $400–$1000/kW electrical. These figures cannot be taken literally at this early phase of fusion reactor engineering; rather they serve to indicate that fusion reactor capital costs may be competitive with other energy sources. Note that fusion's fuel costs will be negligible— of the order of one one-hundredth of a mill per kilowatt hour.

In addition to the neutron damage problem, fusion reactor first walls must not yield a large amount of impurities. Because the first wall will be subject to X-ray, neutron, and charged particle bombardment, its surface characteristics must be optimized also. This second problem is not of concern in fission reactors, of course.

A major program in fusion materials research and development is underway in the United States, and the Soviets and the Germans are beginning to seriously address these problems also. In the United States part of the materials program is aimed at developing intense 14-MeV neutron sources both for surface and bulk irradiations. At present fission reactor radiations and ion bombardment simulations are being used to begin to study the fusion bulk damage problem. A 14-MeV neutron source at Livermore is now providing fluxes of 10^{12} neutrons cm^{-2} sec^{-1} for surface studies. A new rotating-solid-target, beam system is expected to yield fluxes of 10^{13} cm^{-2} sec^{-1} in 1977–1978, and a gas jet-beam system is proposed to yield reactor flux levels of 10^{14} cm^{-2} sec^{-1} in 1978–1979. Both of these sources are limited to small sample irradiations so that a large-volume, high flux facility for large sample irradiations and for component testing will be a clear need for the 1980s.

6   SUMMARY AND CONCLUSIONS

Experimental results from magnetic confinement experiments over the past five years have provided a basis for expecting that practical fusion power will be scientifically achievable. While there are a variety of attractive magnetic and inertial confinement concepts under study, the tokamak concept has emerged as the most promising. A number of new large tokamaks will come into operation in the next few years. These should achieve near reactor-level plasma conditions in hydrogen plasmas and should help to resolve a number of outstanding physics questions. Beyond these, the Europeans, the Americans, and the Soviets are all planning very large DT burning tokamaks to operate in the early 1980s to address many of the remaining physics problems and to produce reactor-level energy densities on a regular basis. After an engineering phase, all major world fusion programs are projecting commercial fusion power about the turn of the century.

Fusion reactor engineering is beginning in earnest throughout the world, with the United States the present leader. Although a variety of difficult engineering problems must be faced, the problem of developing suitable materials looms as the most significant, with the outcome probably having a first order effect on fusion power costs.

The outlook for fusion power is excellent, but a great many problems remain to be solved. The challenges are significant, but if success can be achieved as envisioned today, mankind will have a safe, environmentally attractive, reasonably priced energy source for essentially all time to come.

ACKNOWLEDGMENTS

The author is deeply indebted to members of the staff of the Division of Controlled Thermonuclear Research of the Energy Research and Development Administration for their assistance in collecting background material for this article. Particular thanks to Drs. Maurice Katz, David

Ignat, Kenneth Moses, F. Robert Scott, Bennett Miller, Robert Bingham, and Frank Coffman. Thanks are also due to Dr. Fred Tenney of the Princeton Plasma Physics Laboratory, who helped in the early formulation of the reactor engineering chapter.

Literature Cited

1. World Survey of Major Facilities in Controlled Fusion Research, International Atomic Energy Agency, Vienna, 1973
2. Dawson, J. M., Furth, H. P., Tenney, F. H. 1971. *Phys. Rev. Lett.* 26:1156
3. Bol, K. et al 1972. *Phys. Rev. Lett.* 29:1495
4. Bol, K. et al 1973. *Phys. Rev. Lett.* 32:661
5. Barnett, C. F. et al 1974. *Sixth European Conference on Controlled Fusion and Plasma Physics, Moscow, 1973*, II:330
6. Adam, J. et al. *Fifth Conference on Plasma Physics and Controlled Thermonuclear Research, Tokyo, 1974,* Paper A3-2. Vienna: Int. At. Energy Agency. In press; Hosea, J. C., Hooke, W. M. 1973. *Phys. Rev. Lett.* 31:150
7. Alikaev, V. V., Bobrovskij, G. A., Pozdnyak, V. I., Razumova, J. A., Sokolov, Yu. A. See Ref. 6, Paper A9-4
8. Stott, P. E., Daughney, C. C., Ellis, R. A. Jr. *Nuclear Fusion.* In press
9. Parker, R. R. et al 1975. Submitted for publication
10. Ohkawa, T. 1968. *Kakuyugo Kenkyu* 20:577
11. Jensen, R. H. et al 1975. *Phys. Rev. Lett.* 34:257
12. Jahoda, F. C., Little, E. M., Quinn, W. E., Ribe, F. L., Sawyer, G. A. 1964. *J. Appl. Phys.* 35:2351; Little, E. M., Quinn, W. E., Sawyer, G. A. 1965. *Phys. Fluids* 8:1168; Goldman, L. M., Kilb, R. W., Pollock, H. C. 1964. *Phys. Fluids* 7:1005
13. Andelfinger, C. et al 1966. Proceedings of the Second Conference on Plasma Physics and Controlled Nuclear Fusion Research, Culham, England I:256. Vienna: IAEA
14. Bodin, H. A. B. et al. See Ref. 13, p. 210
15. Bodin, H. A. B., McCartan, J., Newton, A. A., Wolf, G. H. 1969. *Proceedings of the Third International Conference on Plasma Physics and Controlled Nuclear Fusion Research, Novosibirsk, USSR* II:215. Vienna: IAEA
16. Freidberg, J. P., Morse, R. L., Ribe, F. L. 1974. AEC Symp. Ser. 31, 812–30. Washington DC: Technical Information Center, Office of Information Services, United States Atomic Energy Commission
17. Robinson, D. C. 1971. *Plasma Phys.* 13:439; Baker, D. A. et al 1971. *Fourth Conference on Plasma Physics and Controlled Nuclear Fusion Research, Madison, Wis.* I:203. Vienna: IAEA
18. Gowers, C. W. et al 1974. *Sixth European Conference on Controlled Fusion and Plasma Physics, Moscow, 1973* I:265; Baker, D. A. et al. *Fifth Conference on Plasma Physics and Controlled Nuclear Fusion Research, Tokyo.* Vienna: IAEA. To be published
19. Becker, G. et al. See Ref. 18, Fifth Conference
20. Baiborodov, Yu. T., Ioffe, M. S., Petrov, V. M., Sobolev, R. I. 1963. *Sov. At. Phys.* 14:459
21. Budker, G. I., Mirnov, V. V., Ryutov, D. D. 1971. *JETP Lett.* 14:320
22. Coensgen, F. H. et al. See Ref. 18, Fifth Conference
23. See for example, Rose, D. J., Clarke, M. Jr. 1965. *Plasmas and Controlled Fusion,* 380. Cambridge, Mass.: MIT Press
24. Fowler, T. K. Personal communication
25. Haught, A. F., Polk, D. H., Fader,

W. J. 1970. *Phys. Fluids* 13:2842
26. Gott, Yu. V. et al. See Ref. 18, Fifth Conference
27. Batzer, T. H. et al 1974. *Lawrence Livermore Lab. Rep.* UCRL 51617
28. Basov, N. G., Krynkov, P. G., Zakharov, S. D., Senatskii, Yu. V., Chekalin, S. V. 1968. *IEEE Quantum Electron.* 4:864
29. Boling, N., Glass, A., Owyoung, A. Lawrence Livermore Lab., Rep. UCRL 75628; Bettis, J., Glass, A., Guenther, A. 1974. *IEEE Region 6 Conference Record,* 43–55
30. Thomas, C. E., Siebert, L. D. 1975. *A Pulse Shape Generator for Laser Fusion,* Rep. KMSF-U-280. Ann Arbor, Mich.: KMS Fusion, Inc. To be published in *Applied Optics*
31. Brueckner, K., Jorna, S. 1974. *Rev. Mod. Phys.* 46:325
32. Kephart, J. F., Godwin, R. P., McCall, G. H. 1974. *Appl. Phys. Lett.* 25:108
33. Lindl, J. D., Meade, W. C. 1974. *2-D Simulation of Fluid Instability in Laser Fusion Pellets, Rep.* UCRL-76216. Livermore, Calif.: Lawrence Livermore Lab.
34. Charatis, G. et al. See Ref. 6, CN/33/F1
35. Nuckolls, J., Emmett, J., Wood, L. August 1973. *Physics Today,* 46
36. Boyer, K., Fenstermacher, C. A., Stratton, T. F. 1974. CO_2 *Oscillator Amplifier for Laser Fusion Research.* See Ref. 6
37. Lee, P., Giovanielli, D., Godwin, R. P., McCall, G. H. 1974. *Appl.*

Phys. Lett. 24:406; Godwin, R. P. 1974. *Laser Interaction and Related Phenomena,* Vol. 3. New York: Plenum
38. Bernstein, B., Smith, I. 1973. *IEEE Trans. Nucl. Sci.* NS-20:294
39. Morrow, P. L., Phillips, L. D., Stringfield, R. M. Jr., Deggett, W. V., Bennett, W. A. 1971. *Appl. Phys. Lett.* 19:444
40. Yonas, G., Poukey, J. W., Prestwich, K. R., Freeman, J. R., Toepfer, A. J., Clauser, M. J. 1974. *Nucl. Fusion* 14:731
41. Perry, F. C., Widner, M. M. Submitted to *J. Appl. Phys.*; Widner, M. M., Thompson, S. L. 1974. *Sandia Labs. Rep. SAND-74-351.* Albuquerque, New Mexico: Sandia Labs.
42. Toepfer, A. J. 1974. *Bull. Am. Phys. Soc.* 19:856
43. Chang, J. et al 1974. See Ref. 6
44. Spitzer, L. Jr., Grove, D., Johnson, W., Tonks, L., Westendorp, W. 1954. *New York Operations Office of the Atomic Energy Commission Rep.* No. NYO-6047
45. International Atomic Energy Agency. 1974. *Fusion Reactor Design Problems, Proceedings of the International Atomic Energy Agency Workshop in Culham, U.K., 29 January–15 February 1974.* Vienna: IAEA
46. Price, W. G. Jr. 1974. *The Princeton Reference Design Fusion Power Plant,* issued as Princeton Plasma Physics Lab. Rep. No. MATT-1082

PREEQUILIBRIUM DECAY ✻5560

Marshall Blann[1]

Department of Chemistry and Nuclear Structure Research Laboratory,[2]
University of Rochester, Rochester, New York 14627

CONTENTS

1   INTRODUCTION

This article explores a relatively new area of nuclear reaction models, an area dealing with the time-dependent evolution of reactions of nuclei excited in the 20–200 MeV range (1–8). In nuclear physics, as in many areas of physics and chemistry, it has been convenient to treat questions of reaction mechanisms by models that are at diametrically opposed

[1] Supported in part by the United States Energy Research and Development Administration.

[2] Supported by the National Science Foundation.

123

extremes in order to gain tractable results. On the one hand, models exist for "direct reactions" in which a single interaction between a projectile and one or several nucleons of the target nucleus is treated. At the other extreme, it is assumed that the projectile is captured by the target nucleus, and that the resulting "compound nucleus" attains statistical equilibrium without prior particle emission. The decay of the long-lived compound nucleus may then be treated by equilibrium statistical mechanics as was formulated for the case of nuclear reactions by Weisskopf (9) nearly 40 years ago.

Many review articles may be found on both direct reaction models (10–13) and on the compound nucleus model (14–16), and these articles illustrate that the dichotomy has been highly successful in many cases. However, in many spectra, continuous high energy components were observed that were consistent neither with predictions of the compound nucleus model nor with existing direct reaction models (17–19). As isochronous cyclotrons came into wide usage in the 1960s and higher projectile energies became common for nuclear reaction studies, these inexplicable spectral components became ever more obvious in a broad range of experimental results (20). In recent years these phenomena have been treated by classical models that formulate the decay into the continuum of a system with an initial partition of projectile energy between relatively few (intrinsic) degrees of freedom, progressing through more complicated configurations until an equilibrium distribution of energy is attained. A discussion of these models, the assumptions implicit and explicit to them, and their degree of success in reproducing experimental results is presented herein.

The first model capable of reproducing the shapes of the continuum spectra under discussion resulted from the pioneering works of Griffin in 1966 (1–4). Another important and powerful model result was due to Harp, Miller & Berne (5) in 1968, although not applied to the pre-equilibrium problem until 1971 (6). The latter model predicts both shapes and absolute cross sections of spectra. The hybrid model (7, 8) which combined aspects of both the Harp-Miller-Berne and Griffin models was suggested in 1971 and has been the most widely used of the formulations. The intranuclear cascade model (21, 22) was applied to this area in 1971 (6) and in later works (20, 23). This article, however, does not follow the historical sequence. Rather, the discussion of models follows a sequence based on the increasing numbers of assumptions and ease of calculation involved, in the expectation that in this way differences in the various formulations will be more clearly described.

These models have been called both *precompound* and *preequilibrium*. The latter term is used in this work (hereafter abbreviated PE) as the term *compound nucleus* is not unanimously used to refer to a system at equilibrium.

Several reviews have been written in recent years on the subject of PE decay (24–27); they contain greater detail on many aspects of the subject than is to be found in this review. No discussion of doorway state models is included in this review (28, 29). We briefly summarize areas of application of concepts of PE decay not reviewed in detail, and speculate as to probable directions for the future.

2 EXPERIMENTAL CHARACTERISTICS OF PREEQUILIBRIUM SPECTRA

Several examples may serve to characterize the properties of particle spectra that have been attributed to PE decay; these properties are summarized in this section. Much of the remainder of the article is devoted to a discussion of models that have been applied in support of the interpretation of these characteristics as due to decay of states of varying degrees of complexity occurring during the equilibration process.

Figure 1 shows ^{54}Fe(p, p') spectra at 35° at incident proton energies of 29, 39, and 62 MeV (20). In each case an evaporation-like component may be seen with a broad peak centered at about 5 MeV proton energy. This result is consistent with decay from an equilibrated compound nucleus. At fixed residual excitations of 0–10 MeV, sharp peaks are observed in the spectra corresponding to population of low-lying states in the residual nucleus, states that are most likely populated by a direct reaction mechanism. Between the direct and compound nucleus contributions to the spectra, there are continuous contributions that are explained neither by compound nucleus theory nor by existing direct reaction models.

Figure 2 shows angle integrated (p, p') spectra on several targets between ^{27}Al and ^{209}Bi for 62 MeV incident protons (20). The spectral contributions beyond the evaporation region may be seen to be rather independent of target mass with respect to shape, with a slight dependence ($\sim A^{1/3}$) as to cross section. A similar comparison is shown for a survey of (α, p) reactions at an incident ^{4}He energy of 55 MeV in Figure 3 (30). While the spectra beyond the evaporation region are similar in shape and cross section for the α-induced reactions, for all targets, they differ in shape from the nucleon-induced reactions at corresponding excitations. Those phenomena that have been attributed to PE decay are dependent on projectile mass and energy, but seem to be independent of target nucleus as to spectral shape, and independent of, or only slightly dependent on, characteristics of the target nucleus with respect to cross section.

Angular distributions for the ^{181}Ta(α, p) reaction are shown in Figure 4 (30). Characteristic of PE spectra, the results show the strongest forward peaking for the highest energy particles, with a fairly continuous decrease in degree of forward peaking with decreasing energy. Note that even low-energy particles show some forward peaking. While these features will be seen to be qualitatively consistent with the physical assumptions of PE models, no general formulation exists that will properly reproduce the experimental angular distributions. The intranuclear cascade approach, to be mentioned in Section 3, predicts angular distributions for nucleon-induced reactions only.

With these few examples of the experimental characteristics one wishes

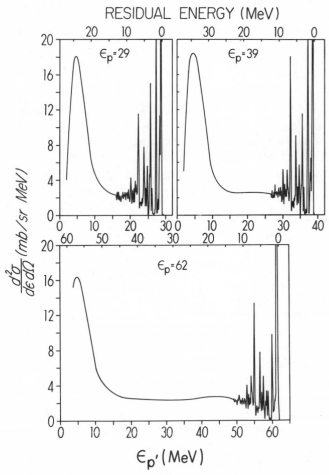

Figure 1   Inelastic ^{54}Fe(p, p′) spectra at 35° for three incident proton energies (20).

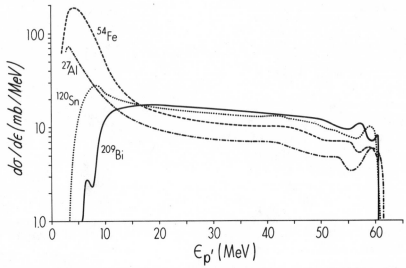

Figure 2 Angle-integrated (p, p') spectra on several targets for 62-MeV incident proton energy (20).

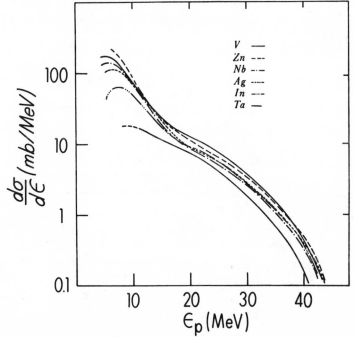

Figure 3 Angle-integrated (α, p) spectra from six targets for 54.8-MeV α particles (30).

to understand, we next consider models giving credence to the interpretation of these data as resulting from PE decay. A discussion of these models is followed by comparisons with experimental data.

3   CASCADE MODEL

Serber (21) pointed out that when incident nucleon wavelengths were short relative to internucleon distances within the nucleus (e.g. for incident nucleon energies in excess of 100 MeV), the interaction could be treated as a quasi-free scattering process. In this model, mean free paths and energy transfers in the assumed two-body scattering processes are based on experimental nucleon-nucleon (N-N) scattering cross sections and angular distributions, with collisions in which either partner has a final energy less than the Fermi energy being forbidden by the Pauli principle. For each scattering event, the position of collision within the nucleus and the energies and directions of each particle are followed explicitly in three-dimensional geometry. In formulations such as the code of Bertini (31) at Oak Ridge National Laboratory (hereafter referred to as ORNL) those particles reaching the nuclear surface with sufficient energy for emission are assumed to be emitted; in codes such as the Brookhaven Laboratory-Columbia University version (hereafter referred to as BNL),

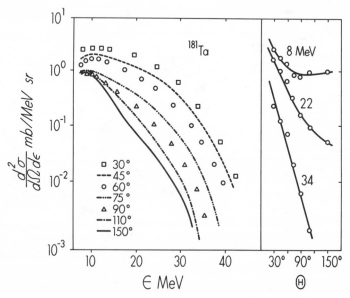

Figure 4   Spectra and angular distributions for the ^{181}Ta(α, p) reaction at 54.8-MeV incident α energy (30).

a multistep nuclear potential is used, with reflection and refraction effects at the boundaries (32–33a,b). Collisions are followed until all particle energies are below some arbitrary minimum. Greater detail on these models may be found in published articles (22, 31–34).

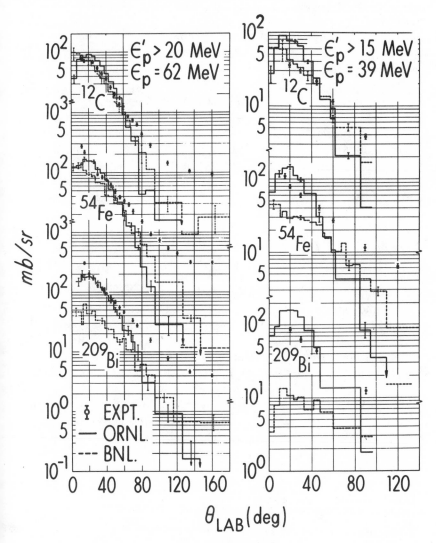

Figure 5 Experimental and calculated angular distributions for (p, p′) reactions on several targets at incident proton energies of 62 and 39 MeV (23). Cross sections are integrated beyond 20 and 15 MeV as indicated. Calculated results are due to cascade codes from Oak Ridge and Brookhaven-Columbia.

In principle, the cascade model was the first that might have been applied to PE phenomena, but it was not. This was in large part because of the belief that the assumption of successive two-body interactions becomes invalid at energies much below 100 MeV/nucleon. In any case, application of the cascade model to PE decay at energies below 100 MeV has for the most part come only in recent years (6, 20, 23, 35a,b) and has been limited to nucleon-induced reactions. Results of spectra calculated with the cascade model are presented in Section 8. Several calculated integral angular distributions are compared with experimental results in Figure 5. Results shown are for the cross sections for the (p, p′) reactions integrated for all proton energies beyond 15 MeV ($\varepsilon_p = 39$ MeV) or 20 MeV ($\varepsilon_p = 62$ MeV) (23). Detailed comparisons of differential yields may be found in (23). Figure 5 indicates that the major fraction of the PE reaction cross section has an angular distribution quite consistent with predictions of the cascade model. Detailed shortcomings can apparently be understood in terms of the approximate manner in which surface reflection and refraction were (BNL code) or were not (ORNL code) treated.

4   PHASE SPACE APPROACHES
TO PREEQUILIBRIUM DECAY

A great simplification results when the laboratory coordinate system of the cascade model is replaced by phase space (1–4). In this case it is only necessary to follow the partition of energy between particles and holes in time, and not the directions and position of nucleons in the nucleus. There is a great gain in ease and speed of calculation; the price paid is a loss of ability to calculate angular distributions in as explicit a manner as permitted by the cascade model.

All phase space models described herein make the assumption that energy transfer processes occur by binary interactions at all energies, as is done in the cascade model for high-energy interactions. All models assume incoherent processes, i.e. an implicit random phase approximation is made. Those models that can predict absolute cross sections on an a priori basis use free N-N scattering cross sections corrected for the Pauli exclusion principle or the imaginary part of the optical potential to evaluate the rate of intranuclear transitions. Use of the imaginary optical potential partially eases the assumption of binary energy transfer processes.

The model that makes fewest assumptions in its execution is a Boltzmann master equation approach that was applied to nuclear Fermi gas systems by Harp, Miller & Berne (5, 6). It is referred to as the HMB

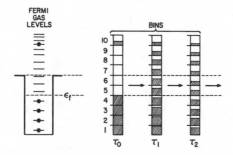

Figure 6   Representation of the equilibration process as formulated by the master equation of Harp, Miller & Berne (25). The shaded areas represent the occupied fraction of each bin, with occupations changing after each time interval.

model and is discussed in the following section. Next, the exciton model due to Griffin (1–4), which is physically most transparent and the first to be proposed, is described; this is followed by a discussion of models related to the exciton model but permitting a priori calculation of absolute spectral yields that is not possible in the exciton model (7, 8). Differences of formulations in use are presented and discussed; comparisons of several of the models with experimental results follow.

5   HARP-MILLER-BERNE MODEL

The physical description of the Harp-Miller-Berne (HMB) model consists of a set of coupled differential equations for computer solution, as illustrated in Figure 6. Consider the initiation of the reaction at an initial time τ_0, as shown on the left of the figure. Energy bins of some width, e.g. 1 MeV, are defined, and the number of available single-particle levels in each bin is computed and stored. The fractional occupation of each bin is followed in the calculation as a function of time. For the incident nucleon, the rate of allowed transitions with all nucleons in the nucleus is computed, as is the rate of reemission of the particle into the continuum. The particle flux is divided in proportion to the rate at which the particle may be emitted vs the rate at which it may make an internal transition. After computing the relative probabilities of scattering into and out of each bin and of emission from bins above the particle binding energies, populations of all bins are changed accordingly, as shown in the center of Figure 6. The calculation is repeated in units of time that are short with respect to the N-N collision time.

The master equations that describe the relaxation process in the one fermion model with a 1-MeV bin width may be represented (27) by

$$d(n_i g_i)/dt = \Sigma_{jkl} \omega_{kl,ij} g_k n_k g_l n_l (1 - n_i)(1 - n_j) g_i g_j$$

$$- \Sigma_{jkl} \omega_{ij,kl} g_i n_i g_j n_j (1 - n_k)(1 - n_l) g_k g_l - n_i g_i \lambda_c(i') \qquad 1.$$

where symbols are defined in Table 1 and i' is the energy outside the nucleus corresponding to the ith bin within the nucleus. The first term of equation 1 may be seen to represent the increase of particles in bin i from intranuclear N-N scattering processes, the second term the loss from the same process, and the third term the loss of particles from i due to emission into the continuum. A two-fermion (neutron and proton) gas formulation has been given by Harp & Miller (6) and is the code used in results to be presented in Section 8.

Most work published with this model has used single-particle levels computed with a Fermi gas spacing. An advantage of the model is that a realistic set of single-particle levels can be used with equal ease to reproduce effects due to nuclear structure; an example of this approach may be found in (27). The calculation can be performed for a nucleus initially in the ground state, as illustrated in Figure 6, or for any arbitrary initial configuration. The latter situation, applied to investigate the relaxation of a nucleus at very high excitation (~ 1 GeV) was the original motivation for this model and its initial application. The HMB model to date has been applied only to nucleon-induced reactions at moderate excitations and to high-excitation/relaxation problems such as the case just mentioned.

6   THE EXCITON MODEL

The similarities of spectral components beyond the evaporation maxima from different targets were observed long ago (17), as were the high temperatures of these components. Griffin was intrigued by the characteristics of the high-energy components of a large body of precise (p, n) spectra measured by the Wisconsin group (18, 19). To interpret these components he formulated a model in which it was assumed that equilibration between target and projectile was achieved by a succession of two-body interactions as represented pictorially in Figure 7. Each state is characterized by the number of excited particles (p) plus holes (h) (or excitons, $n = p + h$) defined with respect to the Fermi energy of the target. For each exciton number, some fraction of states have particles that are unbound, as shown in the lower part of Figure 7. Griffin made the ad hoc statistical assumption that every partition of energy for a given exciton number occurred with equal a priori probability during the equilibration process. He further assumed a predominance of transitions to more complicated states relative to those to simpler states, as the

Table 1   Definitions of symbols used in this review

Symbol	Definition
$n_i g_i$	Number of nucleons in the ith energy interval with g_i single particle states/MeV; intervals are 1 MeV wide, indexed from the bottom of the potential well
g_x; $g_x(\varepsilon)$	Single-particle state density for neutrons or protons in the equidistant model, or with an energy dependence $[g_x(\varepsilon)]$
$\omega_{kl,ij}$	Transition rate for a nucleon in one of the states i to collide with one in j such that the two nucleons go to the energy conserving states k and l; free N-N scattering cross sections for 90° collisions are used
m, s, σ	Nucleon mass and spin, and inverse cross section for channel energy ε
$\rho_c(\varepsilon)$	Density of states of a particle of energy ε in a volume Ω
$\varepsilon_f, \varepsilon$	Fermi energy, channel energy
$\lambda_c(\varepsilon)$	Emission rate into the continuum of a particle of channel energy ε; evaluated as $\sigma(2\varepsilon/m)^{1/2}\rho_c(\varepsilon)(2s+1)/g_x\Omega$
V, W	Real and imaginary potential depths; radius dependent in some uses
n	Exciton number, equal to number of particles (p) above the Fermi energy plus holes (h) below ε_f; subscript 0 represents initial value of n
$_n p_x$	Number of n excitons that are nucleon type x
$\langle \lambda_+^n(E) \rangle$	Average intranuclear transition rate of an n exciton state going to an $n+2$ exciton state; if subscript is minus, rate is to an $n-2$ exciton state
$\lambda_+(\varepsilon), \lambda_+^{NN}(\varepsilon), \lambda_+^W(\varepsilon)$	Intranuclear transition rate of a particle at energy $\varepsilon+V$; superscripts indicate whether evaluated from NN cross sections or the imaginary optical potential
$P(n, t)$	Fraction of nuclei in the original ensemble which are in an n exciton state at time t
τ_n	$1/\langle \lambda_+^n(E) \rangle$
$P_x(\varepsilon)$	Probability of emitting nucleon of type x with channel energy ε
$_n P_x(\varepsilon)$	Probability of emitting nucleon of type x with channel energy ε from an n exciton state
D_n	Fraction of reaction cross section surviving decay prior to reaching the n exciton configuration
$\rho_{p,h}(E) = \rho_n(E)$	Density of n exciton states
$\rho_{p,h}(U, \varepsilon)$	Density of n exciton states such that one, if emitted, would have channel energy ε; differs from $\rho_{n-1}(U)$ by factor $1/p$

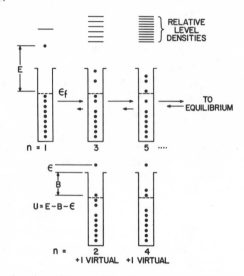

Figure 7   Representation of the equilibration process as formulated in the exciton model. Successive two-body interactions are indicated with some fraction of each exciton hierarchy having unbound particles (24).

exciton state densities are rapidly increasing functions of the exciton number. It is these assumptions that permit a simple closed-form expression for PE spectra in the exciton model, eliminating the need for the use of coupled differential equations as in the later HMB model. The decay probability is computed in the exciton model as a sum over contributions from states from some initial exciton number n_0 to the equilibrium value $\bar{n}$. Since all transitions are assumed to proceed by binary processes, each state has one particle and one hole more than the preceding state in the equilibration sequence (see Figure 7), and the sum over states is taken in units of $\Delta n = +2$. Although the "never come back" assumption implicit in the summation of terms becomes invalid as n approaches $\bar{n}$, the rate of particle emission drastically decreases before this point is reached, so the failure of the approximation in general causes no problem (e.g. see Figure 8).

The density of states of a system of p excited particles and h holes with constant single-particle level density g at excitation E was given by Ericson (36) as[3]

$$\rho_{p,h}(E) = \rho_n(E) = g(gE)^{n-1}/p!h!(n-1)!$$ 2.

The number of excitons that could be emitted at channel energy ε to

---

[3] Some of the shortcomings of this simple formula are discussed in Section 6.2.

$\varepsilon + d\varepsilon$, leaving the residual nucleus at excitation U with $p-1$ excited particles and h holes, is given by the ratio $\rho_{n-1}(U)\,d\varepsilon/\rho_n(E)$. Griffin formulated the model (1) for the probability of neutron emission in the range ε to $\varepsilon + d\varepsilon$ as

$$P(\varepsilon)\,d\varepsilon \propto \sum_{n=n_0}^{n=\bar{n}} \varepsilon^{1/2}\frac{\rho_{n-1}(U)}{\rho_n(E)}\,d\varepsilon \qquad\qquad 3.$$

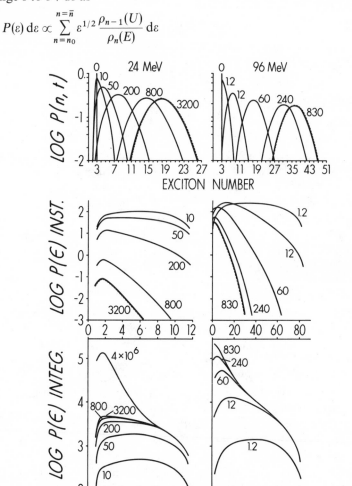

Figure 8 Equilibration process at two excitations in the $A = 100$ region in the exciton master equation approach. Top portion shows the exciton population as a function of time, initially a three-exciton state at $t = 0$. Central curves show instantaneous spectral emission rates at different times, and the bottom curves show the spectra integrated from $t = 0$ to t. Relative time scales at the two energies are not equal (43).

where the sum is taken for $\Delta n = +2$. These expressions were later extended to include internal reflection and barrier penetrability for charged particles as well as finite state lifetimes by Blann (37) and Blann & Lanzafame (38). Their result was

$$P(\varepsilon)\,d\varepsilon = \frac{(2s+1)}{\pi^2 \hbar^3}m\varepsilon\sigma \sum_{n=n_0}^{\bar{n}} \frac{\rho_{n-1}(U)}{\rho_n(E)}\tau_n\,d\varepsilon$$

$$= \frac{(2s+1)m\varepsilon\sigma}{\pi^2 \hbar^3 gE}\sum_{n=n_0}^{\bar{n}}(U/E)^{n-2}p(n-1)\tau_n\,d\varepsilon = \sum_{n=n_0}^{\bar{n}}P_n(\varepsilon)\,d\varepsilon. \qquad 4.$$

Most applications in the literature have basically used this form, with the lifetime $\tau_n = $ constant, or following work of Williams (39) with $\tau_n \propto n+1$. The latter result is from a consideration of the average number of states accessible for states of $\Delta n = \pm 2$, based on state densities as given by an equidistant spacing model. These results were derived by Williams in the hope that the transition rate of an n exciton state would be proportional to the average number of accessible final states ρ_n (where only a fraction of all final states are accessible to a single initial state) according to the Golden Rule, $\lambda_{n,n'} = 2\pi/\hbar|M|^2\rho_{n'}$ and in the hope that $|M|^2$ (the average squared two-body transition matrix) might be effectively constant over a broad range of excitations. Unfortunately, calculations based on N-N scattering in a nucleus do not support such an assumption at energies beyond ~ 25 MeV (40, 41). Rather, the constant lifetime assumption (37, 38) suggested earlier is more consistent with results based on N-N scattering. It was also demonstrated that a constant value of $|M|^2$ over a broad energy range leads to unrealistic results for PE decay cross sections (42). Ignorance as to means of evaluating $|M|^2$ on an a priori basis restricted the exciton model to calculation of relative spectral shapes rather than absolute spectral yields, with the assumption that PE particle emission was a small fraction of the compound nucleus cross section, or alternatively stated that the intranuclear transition rate was much greater than the rate of emission into the continuum at each stage of the equilibration process.

The main prediction of the exciton model that could be tested was that all targets bombarded by nucleons should give the same simple initial configuration characterized by the parameter n_0, within a one-unit uncertainty for pairing effects. If so, a graph of $\ln[d\sigma/d\varepsilon)/\varepsilon\sigma]$ versus $\ln U$ (hereafter referred to as the slope analysis method) should give a line of slope $n_0 - 2$ at low residual excitations U, where the leading term of equation 3 or 4 dominates. Griffin demonstrated that this was the case for (p, n) reactions (finding $n_0 = 2$-3) (1, 2) and that similar analyses of α-induced reactions gave $n_0 = 4$-5 (3, 4). These results were verified by analyses of many n-, p-, and α-induced reactions from a literature survey (43), and by analyses of many new experimental results at bombarding energies between 14 and 62 MeV (30, 42, 44–57) both by slope analysis and by the criterion of the n_0 value required to reproduce spectra or excitation functions. Analyses of this type have also been performed for d- and ^{3}He-induced reactions (58–63). The results at present comprise a convincingly large, consistent set of data to support the hypothesis of simple initial states decaying prior to the attainment of equilibrium with n_0 a function of projectile type. It has been shown that decay of states of greater complexity than the first is necessary to reproduce the complete experimental particle spectrum—i.e. a sum over terms as in equations 3 or 4 is required to reproduce experimental results; the first or "direct reaction" term alone is not adequate (24, 40, 45).

The result of a low value of n_0 dependent on projectile but independent of target is consistent with the observations of Section 2 of high-energy spectra having generally the same shape for a given projectile and energy, independent of target. The general change in shape of preequilibrium spectra with changing projectile energy is also reproduced well by equation 4, as has been illustrated (7, 24, 43).

6.1 Master Equation Exciton Model

The simple summation of terms of equation 4 is the most convenient formulation of the exciton model. Nonetheless, a pedagogically interesting result arises by writing a master equation (43) in which the population of any exciton number state may either increase or decrease with time. With symbols defined in Table 1, the set of equations may be represented by

$$dP(n,t)/dt = P(n-2,t)\langle \lambda_+^{n-2}(E)\rangle + P(n+2,t)\langle \lambda_-^{n+2}(E)\rangle -$$
$$P(n,t)[\langle \lambda_+^n(E)\rangle + \langle \lambda_-^n(E)\rangle] \qquad 5.$$

where the transition rates are evaluated on a relative basis as proportional to the average number of accessible final states (39).

Some illustrative results of the master equation approach are shown in Figure 8. The upper part of the figure shows the spreading in time of the distribution of excitation between states characterized by different exciton numbers (assuming $n_0 = 3$ at $t = 0$ for a nucleon-induced reaction) until a statistical distribution is approached and $P(n,t)$ becomes proportional to $\rho_n(E)$. The central portion of the figure shows the instantaneous emission spectra that would be observed at the indicated times from the corresponding exciton population distribution. It illustrates that as the energy becomes partitioned over an increasing number of degrees of freedom, the rate of high-energy particle emission decreases many orders of magnitude from the maximum value. This result is intuitively obvious in Figure 7. The bottom section of Figure 8 shows the integrated particle emission spectrum from $t = 0$ to time "t." At lower excitation energies (e.g. the result at 24 MeV) the spectrum could be expressed as a prompt component plus an equilibrium component because of the great difference in time scales for developing the "hard" and "soft" spectral components. But the prompt component in this case extends beyond the definition of a direct interaction, as states following interactions more complicated than a single two-body transition contribute. At higher excitations the distinction between prompt and equilibrium components is not as clear, and there is an implication that nuclei at higher excitations may well decay prior to the attainment of statistical equilibrium. This point is discussed in greater detail in the following sections.

It has been shown (6, 43) that in the limit of $t \to \infty$, the PE formulations of equations 4 and 5 reduce exactly to the Weisskopf equilibrium evaporation equation. Performing the calculation with equation 5 on a time-dependent basis, therefore, provides a single formulation that includes decay of the simple state following a single projectile-target interaction, generally called a direct reaction, decay of more complicated intermediate states, and decay of the long-lived equilibrium compound nucleus. For this reason, it is of pedagogical interest. It has been demonstrated that the simpler closed-form equation 4 gives essentially identical results, making it the practical choice for the calculation of spectra (43, 64).

6.2   Partial State Densities and Transition Rates

Much effort related to understanding the predictions of the exciton model has gone into investigating the partial state densities. It was first pointed out by Böhning that the enumeration of equation 2 allows configurations that violate the Pauli exclusion principle. Böhning presented elegant methods by which exact partial state densities that did not violate the Pauli principle could be computed for the case of an equidistant spacing model (65). Williams later published a simple closed-form result in which the excitation energy of equation 2 was reduced by an amount dependent on the number of excited particles and holes (66). Although these results were not exact as were those of Böhning, they have found wider usage because of their relative simplicity. It has meanwhile been shown that equation 2 is a quite satisfactory result to use in the range of excitation and exciton number where PE particle emission is likely to be important, and that the equidistant spacing model is a reasonable approximation of a Fermi gas model with respect to calculation of the partial state density (24, 26).

Computer codes have been written (and are available) permitting calculation of partial state densities using any arbitrary set of single particle levels. These results give an exact enumeration for a potential with finite well depth, with the Pauli exclusion principle considered. Results have been published using single particle levels due both to Nilsson (67, 68) and to Seegar (69). The results predict peaks in the spectra corresponding to quite high residual excitations (8–9 MeV) for nuclei near closed shells and large energy shifts resulting from shell structure at low excitations. These results indicate that equation 2 may be inadequate at low residual excitations for closed-shell and near-closed-shell nuclei. These effects, in fact, are probably a far greater limitation on the use of equation 2 than are neglect of the Pauli principle or the equidistant spacing approximation. It remains an open question as to whether or not these nuclear structure related aspects of PE particle spectra will be observed and identified experimentally. If so, the investigation of PE decay may yield information relevant to nuclear structure.

Much effort has been directed toward calculation of either relative or absolute partial state lifetimes for use in equations 4 or 5 or similar equations. Effects on rates due to limitation by the Pauli principle of densities of final states (70), or due to use of Fermi gas or harmonic oscillator potentials, have been studied (41). At excitations where PE emission is likely to occur, the latter considerations seem to be minor (41, 70).

As shown in Section 7, the use of partial state densities with a state lifetime, for example as in equation 4, may be erroneous for absolute calculations. If so, questions dealing with average partial state lifetimes are moot. Attempts to do absolute calculations within the framework of the exciton model formulation are described following the discussion of the hybrid model. This is done not so much for historical accuracy as to focus attention more clearly on the main point of difference between the two approaches in use.

7   HYBRID AND GEOMETRY-DEPENDENT HYBRID MODEL

The exciton model has an appealing simplicity and physical transparency in formulation. As originally formulated its shortcoming, relative to the cascade and HMB models, was that only relative spectral shapes, rather than absolute differential cross sections, could be calculated. This restriction resulted from ignorance of the partial state lifetimes of equation 4. Harp & Miller used the exciton model with partial state lifetimes evaluated from their master equation to compute an absolute spectral yield for the ^{181}Ta(p, n) reaction (6). Tying the lifetimes (or intranuclear transition rates) to a master equation calculation still did not provide a convenient model for use, nor did it provide a result in agreement with the experimental spectrum.

A formulation was suggested in which excited particle populations during equilibration were calculated by use of partial state densities as in the exciton model, and intranuclear transition rates of the excited particles were determined from calculations of the mean free paths of nucleons in nuclear matter (7). First formulations based mean free paths on free N-N scattering cross sections corrected for the Pauli principle (7); later formulations also used values based on the imaginary optical potential (40, 71). As this approach sought to combine the simplicity of formulation of the exciton model with the ability to predict absolute cross sections of the HMB model, it was called the "hybrid model." A later formulation was made in which effects of interactions in the diffuse nuclear surface could be included in the model, and was called the "geometry-dependent hybrid model" (GDH model) (8).

The PE decay probability in the hybrid model is given by

$$
\begin{aligned}
P_x(\varepsilon)\, d\varepsilon &= \sum_{n=n_0}^{\bar{n}} \left[{}_n P_x \frac{\rho_{p,h}(U, \varepsilon) g\, d\varepsilon}{\rho_{p,h}(E)} \right] \left[\frac{\lambda_c(\varepsilon)}{\lambda_c(\varepsilon) + \lambda_+(\varepsilon)} \right] D_n \\
&= \sum_{n=n_0}^{\bar{n}} {}_n P_x(\varepsilon)\, d\varepsilon.
\end{aligned}
\tag{6.}
$$

The expression in the first set of brackets uses partial state densities to give the number of particles of type x (neutrons or protons) in an n exciton state, that are in an unbound level with energy between ε and $\varepsilon + d\varepsilon$ in the continuum. The second set of brackets gives the fraction of those particles at energy ε that are emitted into the continuum rather than undergoing a transition to an $n+2$ exciton state. The D_n represents the population surviving particle emission from simpler states. Symbols are defined in Table 1.

7.1   Parameter Evaluation

Partial state densities and their shortcomings were discussed in the preceding section. The neutron or proton single-particle level densities have been defined by Fermi gas values, $g_n = (A-Z)/14$, $g_p = Z/14$, or by $g_n = g_p = A/28$, where A and Z are the mass and atomic numbers of the composite nucleus. However, as equation 6 has been applied at energies approaching 200 MeV, a more appropriate energy dependence is given (26) by $g_x(\varepsilon) = g_x [(\varepsilon + V)/\varepsilon_f]^{1/2}$.

The crucial quantity to evaluate in order to calculate absolute decay probabilities and cross sections is $\lambda_+(\varepsilon)$. Goldberger (72) and later Kikuchi & Kawai (73) calculated the average effective cross section $\langle\sigma\rangle$ for an excited nucleon to interact with nucleons having a Fermi gas momentum distribution. Free N-N scattering cross sections were used for this purpose, with all collision angles considered; collisions that left either particle with an energy below the Fermi energy were excluded by the Pauli principle. The mean free path (MFP) of the particle at ε is then given by MFP $(\varepsilon) = 1/\rho\langle\sigma\rangle$, where ρ is the density of nuclear matter. The intranuclear transition rate is given by $\lambda_+^{NN}(\varepsilon) = [2(\varepsilon+V)/m]^{1/2}/\text{MFP}(\varepsilon) = \rho\langle\sigma\rangle[2(\varepsilon+V)/m]^{1/2}$. Gadioli et al (41) have shown that transition rates based on N-N scattering with a harmonic oscillator potential give similar results to the square well used by Kikuchi & Kawai.

The mean free path may also be related to the optical potential (73): $\text{MFP}(\varepsilon) = (\hbar/2W)\{[\varepsilon+V+((\varepsilon+V)^2+W^2)^{1/2}]/m\}^{1/2}$ giving $(40)\lambda_+^W(\varepsilon) = [2(\varepsilon+V)/m]^{1/2}/\text{MFP}(\varepsilon) = 2W/\hbar$, if $\varepsilon+V \gg W$.

The intranuclear transition rates $\lambda_+^{NN}(\varepsilon)$ were parameterized by a binomial function for nucleon energies up to ~ 100 MeV for $N = Z$ nuclei. With this simplification the hybrid model may be written (7) as

$$\frac{d\sigma}{d\varepsilon} = \sigma_R/E \sum_{n=n_0}^{\bar{n}} \left[{}_n p_x (U/E)^{n-2}(n-1)\right] \left\{ D_n \sigma\varepsilon/g_x/[\sigma\varepsilon/g_x + 1890 \times \right.$$

$$\left. (\varepsilon+BE_x) - 8(\varepsilon+BE_x)^2] \right\}$$

6a.

where energies are in MeV and cross sections are in mb.

Present hybrid model codes calculate $\langle\sigma\rangle$ for the N and Z of the target nucleus under consideration, but many results presented in the literature and in this article have used the simpler result of equation 6a. When the imaginary optical potential is used to evaluate $\lambda_+^W(\varepsilon)$, the "best set" of parameters and form factors (surface and volume) of Becchetti-Greenlees are used (74). Further details of these calculations are to be found in (40).

A few comments on the accuracy to be expected from PE calculations would be appropriate. Unlike the equilibrium statistical model, where intranuclear transition rates do not enter the formulation, absolute cross sections calculated with PE models will have an error that to first order is linear with errors in $\lambda_+(\varepsilon)$. Energy-dependent errors in inverse reaction cross sections and in partial state densities are also likely to be of greater influence than in the equilibrium formulation. In consideration of these sources of error, it was suggested (40) that PE calculations performed without parameter adjustment should not be expected to reproduce experimental cross sections to better than the factor of two range (the equilibrium statistical model does no better for applications involving a comparable lack of parameter adjustment). Considering that PE spectra span several orders of magnitude in cross section, this still permits a quite stringent test of model predictions.

7.2 Alternate Formulations

The use of the transition rate of a particle at energy ε, $\lambda_+(\varepsilon)$, in equation 6 rather than the average transition rate of the n exciton state at excitation E, $\langle\lambda_+^n(E)\rangle$, as indicated in equation 4 is a significant deviation from the bookkeeping of the exciton model. This difference is of no consequence when only relative spectral shapes are calculated, but is very important when absolute spectral yields are computed. A formula for precompound decay using $\langle\lambda_+^n(E)\rangle$ based on N-N scattering results was presented by Blann & Mignerey (42) but was applied only to a few experimental cases for illustrative purposes (25). More recently the Milano group, in a series of publications that have made many contributions to a general understanding of PE decay, have greatly developed and refined this approach (41, 49–52, 75–77). In this formulation the calculated values of nucleon mean free paths are multiplied by an empirically determined constant, such that the calculated and experimental cross sections agree at one excitation energy. In the calculation of excitation functions, the multiplier parameter was determined for each target, and was found consistently to be within a range of 4 ± 1. The experimental systems that have been analyzed by this and other approaches described thus far, and the main basis of the calculations, are summarized in Table 2.

Proponents of the hybrid model lifetime dependence claim that use of average state lifetimes gives an incorrect influence of spectator particles and holes on λ_+, whereas followers of the exciton model dependence claim that multiple counting is involved if the hybrid model approach is taken. There is controversy on this question, and it must be considered open. It is an important question to answer, for much of the present effort in PE decay models has been directed toward exploring details of the partial state transition rates that have little significance if the bookkeeping implied by equation 4 or 5 with $\langle\lambda_+^n(E)\rangle$ is incorrect. Although a final answer to this question is not yet clear, several observations may be worthwhile.

Equation 6 has been used to analyze many n, p, d, ^{3}He, and α-induced reactions, in which particle spectra and/or excitation functions have been measured at excitations from 20 MeV to nearly 200 MeV (see Table 2). In all cases the experimental yields are quite satisfactorily reproduced when $\lambda_+^{NN}(\varepsilon)$ or $\lambda_+^W(\varepsilon)$ are used with no parameter adjustment.

If instead of $\lambda_+(\varepsilon)$ one uses a formulation with the average state transition rate $\langle \lambda_+^n(E) \rangle$, the mean free paths based on N-N scattering cross sections corrected for the Pauli principle no longer reproduce experimental yields; for nucleon-induced reactions, mean free paths must be multiplied by empirically determined factors of 4 ± 1, giving values of 15 to 30 fm (41). No comparisons of particle spectra from a single target at different bombarding energies have been published for this approach, so the question of the energy dependence of the multiplier is open.

It was shown in Figure 5 that the cascade model, using realistic nuclear density distributions, reproduces the main features of the angular distributions of PE spectra. The particle mean free paths of the cascade model are essentially the same values used in the hybrid (and GDH) model. It would seem to follow that if those mean free paths were increased by factors of three to five, as required when using $\langle \lambda_+^n(E) \rangle$, the forward peaking of the angular distributions could no longer be reproduced.

Analyses of experimental PE spectra by Cline (80) indicate that at a fixed excitation the nucleon mean free path multiplier for nucleon and α-induced reactions varies greatly, although recent results of Gadioli & Gadioli-Erba (77) are in contradiction to the results of Cline. When $\lambda_+(\varepsilon)$ is used, the same multiplier (unity) is used for all projectiles.

These observations are of a circumstantial nature and by no means answer the question as to which formulation is more appropriate. The important factors in PE decay are the number of particles to be found in each excitation energy range, and the length of time they are to be found there during the equilibration period. More precisely stated, it is the product of the number of particles at ε and their time of occupation that determine the PE spectrum. It is worthwhile to compare the two approaches under discussion using a method in which the mean free path does not enter as an additional free parameter; by this method it may be seen whether either approach gives a reasonable distribution of particle occupation probabilities during the equilibration sequence.

An attempt at such a comparison has been made by a computer experiment in which the numbers of particles versus excitation and the time during equilibration were calculated with the HMB master-equation model. This approach does not use partial state densities, and therefore makes no assumptions about their validity or bookkeeping. Rather the nucleon populations are directly followed on a time-dependent basis founded on N-N scattering kinematics. For the reaction $^{209}\text{Bi}(p, p')$ at an incident proton energy of 39 MeV, the product of particles and occupation time in each excitation interval during equilibration was

Table 2 Summary of basis and range of successful application of PE decay models for nucleon emission

Basis of Calculation	Angular Distribution	Surface Reactions	Free Parameters	Reactions	Projectile Energy Range (MeV)	Target Mass Range	References
Cascade Follow trajectory of each nucleon in 3 dimensions following binary collisions	Yes	Yes	None	Spectra (p,p')	39–62	12–209	23
HMB Phase space, solve master equation for equilibration by binary collisions	No	No	None	Spectra (p,p')	39–62	54–209	25, 27

				Reaction			
Hybrid/GDH	No	Yes (GDH)	None	Spectra			
Phase space, partial state densities with statistical assumption to follow exciton populations. Closed form expressions; uses excited particle lifetime $\lambda_+(\varepsilon)$, either N-N scattering or imaginary optical potential. All other models use only N-N or parameterize λ_+				(n, n')	14	28–93	78
				(p, p')	29–62	54–209	8, 25, 40
				(p, n)	18–45	40–208[a]	69
				(d, p)	24–45	57–197	62, 63
				(^{3}He, p)	24–26	57–117	61–63
				(^{3}He, n)	26	57–63	61
				(α, p)	30–59	51–197	7, 30, 42, 63
				(α, n)	18–20	56–124	47, 53
				Excitation Functions			
				(d, xnyp)	25–85	181–197	60, 63
				(^{3}He, xn)	20–80	181–197	63
				(α, xnyp)	20–160	51–197	25, 63, 79
Exciton	No	No	$\langle \lambda_+^n(E) \rangle$	Spectra			
Same basis as hybrid but using average state lifetimes $\langle \lambda_+^n(E) \rangle$				(n, p)	14	100–209	54, 75
				(n, n')	14	24–197	55, 56, 78
				(α, p)	55–59	51–66	57, 77
				Excitation Functions			
				(p, xn)	10–50	51–170	41, 49–51
				(α, n)	10–40	40–165	77

[a] A. Galonsky et al, private communication.

computed and is shown as a function of channel energy in Figure 9. Next a comparison was made to see if partial state densities with proper lifetimes would reproduce the distribution of the HMB model. The Fermi energy and nuclear densities were adjusted to the same values as were used in the HMB model, and the particle-sec distributions were computed (neglecting the small effect on lifetimes due to particle emission) as

$$\text{particle-secs } (\varepsilon) = \sum_{n=n_0}^{\bar{n}} \left[{}_nP_p \frac{\rho_n(U,\varepsilon)}{\rho_n(E)} \right] \left[\frac{1}{\lambda_+(\varepsilon)} \quad \text{or} \quad \frac{1}{\langle \lambda_+^n(E) \rangle} \right] g d\varepsilon.$$

These results are also shown in Figure 9, and, for this example, support the validity of using $\lambda_+(\varepsilon)$ in equation 6. This conclusion is independent of the nucleon mean free paths; if the same comparison were repeated with all mean free paths multiplied by, say, a factor of three, the same relative comparisons would result.

Comparisons among the HMB model, a similar model in which internal collisions are averaged over all angles (26) (A. Mignerey, personal com-

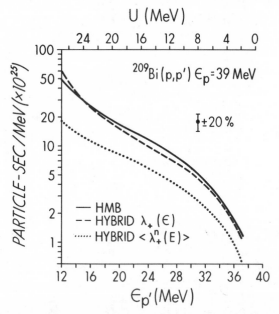

Figure 9 Proton bin populations during equilibration for the ^{209}Bi(p, p') reaction at 39-MeV incident proton energy. The ordinate gives the length of time one proton spends in each bin during the equilibration period, the abscissa the proton energy in the continuum. Results are calculated with the HMB model, and with partial state densities with $\lambda_+(\varepsilon)$ (dashed line) and with $\langle \lambda_+^n(E) \rangle$ (dotted line).

munication), and the two treatments under discussion have been made for
^{54}Fe(p, p') and ^{209}Bi(p, p') reactions at 39 and 62 MeV incident energy.
In each case the use of $\langle \lambda_+^n(E) \rangle$ gives spectral yields low with respect
to the other three calculations, consistent with the analysis shown in
Figure 9.

7.3   Results of Hybrid Model

Comparisons of experimental (α, p) spectra with results calculated with
equation 6 are shown in Figures 10 and 11. In Figure 10, differential
cross sections on ten targets for 55 MeV incident α-particles are compared with calculated results (30); the only parameter selected was the
initial excited particle number. Two values were used in the calculations:
$n_0 = 4$ (2 protons and 2 neutrons) and $n_0 = 5$ (3 protons and 2 neutrons).
The results show some evidence of an odd-even effect; odd proton targets
have (α, p) spectra that are better reproduced by the 5-exciton initial
state, even targets by $n_0 = 4$. However, a similar result can be achieved

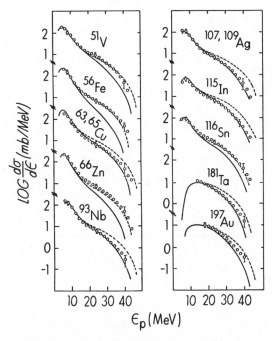

Figure 10   Experimental and calculated angle-integrated (α, p) spectra on ten targets
(30). Open circles represent experimental results. Dashed lines are hybrid model
results with $n_0 = 4$; solid lines with $n_0 = 5$. Equilibrium components are included in
the calculated spectra.

with $n_0 = 4$ for all spectra with an appropriate pairing correction for partial state densities (30). Earlier works sought to find an odd-even effect in nucleon-induced reactions at much lower bombarding energies (81, 82). These results had ambiguities as the targets were at or near shell closures where the latter effects were also likely to be present, as was emphasized by Grimes et al (82).

Agreement similar to that of Figure 10 between calculated and experimental (α, p) spectra on nine targets for 42-MeV incident particles has been reported (42). Figure 11 shows $^{93}Nb(\alpha, p)$ spectra at three bombarding energies between 30 and 55 MeV. Experimental results are from (30, 83, 84); calculations are unpublished results of Mignerey. The hybrid model with $n_0 = 5$ reproduces the experimental PE spectra over four orders of magnitude to within experimental uncertainties with no parameter variation. Results of tests at higher excitation energies may be seen in the $^{93}Nb(\alpha, xnyp)$ excitation functions of Bisplinghoff et al (79) in Figure 12; the calculated results were obtained with a code permitting multiple PE particle emission; the cross section not decaying prior to

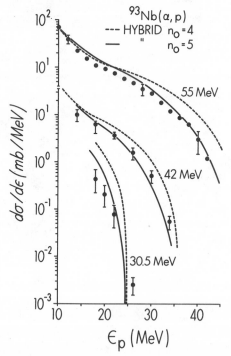

Figure 11   Experimental and calculated $^{93}Nb(\alpha, p)$ spectra at three incident α energies (A. Mignerey and M. Blann, unpublished).

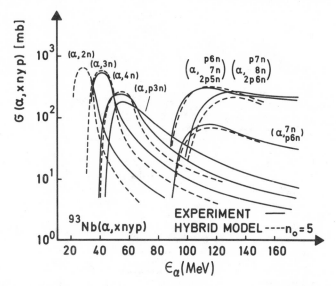

Figure 12 Experimental and calculated excitation functions for the reactions ^{93}Nb$(\alpha, xn yp)$ (79). For $x = 2, 3, 4$, the calculated results are multiplied by 0.6. For $x \geq 7$, the calculated results are multiplied by 1.5.

equilibration was treated by the Weisskopf-Ewing evaporation formula (85). Quite reasonable agreement is achieved at energies up to 160 MeV, as in previous cases with no parameter adjustment.

Results of equation 6 in reproducing proton and neutron spectra following ^{3}He bombardment (61) are shown in Figure 13. Chevarier et al have found $n_0 = 4$ required to reproduce the shape of spectra from ^{3}He-induced reactions on a large number of targets by applying the slope analysis method (Section 6). Having determined n_0 in this fashion, the spectra of Figure 13 were computed with no parameter adjustment. Results of application to deuteron-induced reactions (60) are shown in Figure 14. There is evidence in deuteron-induced reactions of an additional projectile breakup component that is not reproduced by PE decay models (62). Calculated results of Figure 14 were normalized as indicated.

Excitation functions for (p, n) reactions computed using the formulation of a Milano group (41) are shown in Figure 15. In computing these excitation functions, the mean free paths of nucleons in nuclear matter were multiplied by a constant as previously noted. The Milano group has gotten equally good fits to (p, n) and (p, xn) excitation functions for targets over a mass range of 50 to 210, at excitations up to 50 MeV. Recent calculations of (α, n) excitation functions in the same energy range have also appeared (77).

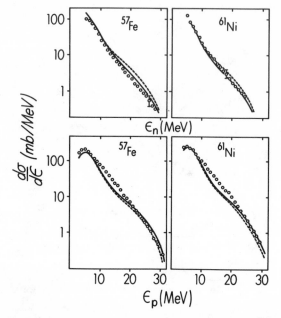

Figure 13   Angle-integrated neutron and proton spectra from ^{3}He bombardment of ^{57}Fe and ^{61}Ni targets. Open points are experimental results. The solid curve is the equilibrium plus hybrid model result for $n_0 = 2$ neutrons, 2 protons; the dashed curve for $n_0 = 1.5$ neutrons and 2.5 protons (61).

7.4   Geometry-Dependent Hybrid Model

Analyses of nucleon-induced reactions at incident energies of 18 (6) and 62 MeV (42) gave evidence of major spectral contributions from the nuclear surface. To investigate the importance of the nuclear density distribution on PE decay, the geometry-dependent hybrid (GDH) model was formulated (8)

$$\left(\frac{d\sigma}{d\varepsilon}\right) = \pi \lambda^2 \sum_{l=0}^{\infty} (2l+1) T_l \sum_{n=n_0}^{\bar{n}} {}_x P_n(\varepsilon) \qquad 7.$$

where density-dependent parameters of ${}_x P_n(\varepsilon)$ were averaged along the projectile paths corresponding to each partial wave. A Fermi density distribution function was used for this purpose. When $\lambda_+^W(\varepsilon)$ values were used (40) the volume and surface form factors and parameters of Becchetti & Greenlees (74) were applied.

In addition to changes in $\lambda_+^{NN}(\varepsilon)$ due to the density dependence of $\langle \sigma \rangle$, and change of velocity with position in the potential, there is a second effect of the dependence of the Fermi energy in a local density approximation. This is the most important effect, specifically that the hole energies cannot exceed the Fermi energy; partial state densities reflecting this limit will modify the exciton populations of equation 6. The first or direct interaction term of the equilibration sequence is the one for which the hole depth limitation is most important; partial state density expressions for this term were published (8), and more recently generalized expressions for all terms have been suggested (86).

Calculations with equations 6 and 7 indicate typical particle mean free paths of 3–6 fm (40). Such mean free paths would suggest that only the direct reaction term of equation 7 should be computed as a function of impact parameter, as nucleons following the first interaction should sample a large part of the nuclear volume. Consequently, calculations have also been performed in which the direct term ($n = 3$) is calculated with equation 7, and higher order terms ($n \geq 5$) are calculated

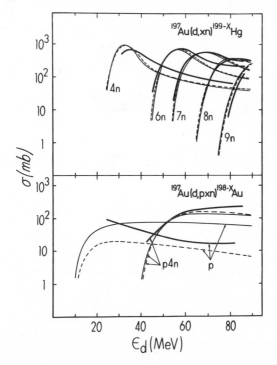

Figure 14 Excitation functions for deuteron-induced reactions of ^{197}Au. Heavy solid curves represent experimental results; thin solid curves represent hybrid model results with $n_0 = 2$; dashed curves represent hybrid model with $n_0 = 3$. The calculated xn results have been reduced by a factor 0.56; pxn calculated results have been increased by a factor 2.1 (60).

with the hybrid model (equation 6). Both approaches give similar results; the "direct plus hybrid" result provides the simpler calculation (40).

Results calculated with the GDH model are compared with ^{51}V(p, n) spectra at incident proton energies of 18 to 26 MeV in Figure 16 (69). The PE components calculated with both $\lambda_+^{NN}(\varepsilon)$ and $\lambda_+^W(\varepsilon)$ are in excellent agreement with experimental spectra, with the obvious exception of the peak corresponding to the isobaric analogue state. Results of spectra calculated by the "direct plus hybrid" method described above are displayed in Figure 17 for the reaction ^{209}Bi(p, p′); the direct reaction term (2p1h) gives the largest contribution to the PE spectrum but the contribution of higher order terms is also necessary to reproduce the entire spectrum. As in the previous figure, results using $\lambda_+^{NN}(\varepsilon)$ and $\lambda_+^W(\varepsilon)$ agree to well within the range in which the absolute accuracy of these models is meaningful. Since the use of the optical potential for evaluating $\lambda_+(\varepsilon)$ does not involve a binary collision assumption, the agreement of cross sections with results using $\lambda_+^{NN}(\varepsilon)$ tends to support the assumption.

Several calculations have been made using the GDH model for α-induced reactions. In separate results, Chevarier et al (30) and Bisplinghoff (63) both found some improvement in agreement with experimental results in using the GDH formulation rather than the simpler hybrid model result.

A calculation of the total fraction PE emission as a function of excitation

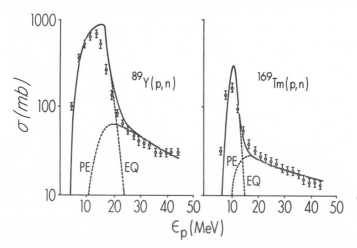

Figure 15 Experimental and calculated excitation functions for the reactions ^{89}Y(p, n) and ^{169}Tm(p, n). Points represent experimental yields. Calculated results for the equilibrium contribution (EQ) and PE component calculated with the formulation of (41) are indicated. The latter results used nucleon mean free paths multiplied by factors of 3 and 5 for ^{89}Y and ^{169}Tm, respectively.

energy for nucleon-induced reactions on a mass 60 nucleus is shown in Figure 18 (42). The points are experimental estimates based on measured results of the (p, n) (69) and (p, p') spectra (20), with estimated contributions for the nucleon channel not measured. Similar estimates for other initial exciton numbers have been published (42). In all cases, it is found that PE emission rises rapidly in importance beyond some minimum excitation (approximately the average particle binding energy per initial exciton), eventually accounting for the major fraction of the reaction cross section. Many, or most, nuclei decay prior to the attainment of equilibrium at excitations of several tens of MeV, according to these models (for light projectile-induced reactions).

Comment is appropriate on an apparent contradiction between the models under discussion and experimental evidence. At all but the lowest energies for which PE decay is observed, angular distributions show a

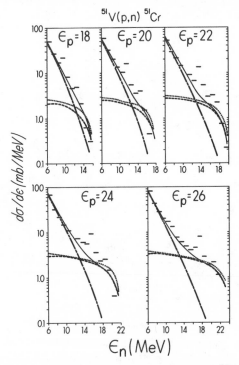

Figure 16 Calculated and experimental angle-integrated ^{51}V(p, n) spectra (69). Experimental results are given by horizontal bars. Calculated equilibrium components are given by long- and short-dash curves. PE results using $\lambda_+^{NN}(\varepsilon)$ with the GDH model are given by dashed curves; with $\lambda_+^W(E)$ by dotted curves. The thin solid curves are sums of equilibrium plus PE (N-N scattering) components.

forward peaking, often quite strongly so (e.g. Figures 4 and 5). Yet the models of Sections 5–7 rely on phase space arguments; those of Sections 6 and 7 refer to "intermediate states" in their formulation. This implies relatively long lifetimes for the intermediate states and angular distributions that are symmetric about 90° (87). Yet the evidence for the success of the hybrid/GDH model in reproducing angle-integrated experimental spectra, which include major contributions from "direct" or forward-peaked components, is quite compelling and suggests that there must be good reason for this.

It was shown in Figure 5 and in (23) that the intranuclear cascade model (abbreviated INCM), which does not employ phase space arguments (Section 3), reproduces the angular distributions and spectra of PE reactions quite satisfactorily (see also Figures 19 and 20 and the discussion in Section 8). It is clear from the discussion of the HMB model (Section 5) that the internal energy distributions following each nucleon-nucleon interaction in that model are identical to those that would be found in the INCM; the input kinematics of N-N scattering cross sections, energy distributions, and ground state nucleon momentum distributions are the same. It may be seen in Figure 9 that partial

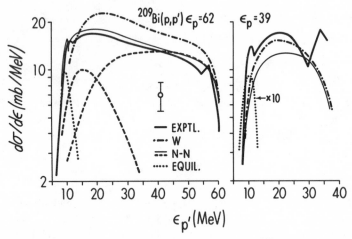

Figure 17 Experimental and calculated angle-integrated ^{209}Bi(p, p') spectra at 62- and 39-MeV incident proton energy. Experimental results (20) are indicated by the heavy solid curve. Dotted curves represent pure equilibrium components. Calculated results consist of a direct reaction ($n_0 = 3$) contribution computed with the GDH model, plus higher order terms ($n \geq 5$) computed with the hybrid model. The thin solid curves give results using $\lambda_+^{NN}(\varepsilon)$, the dot-dash curves represent results using $\lambda_+^{W}(\varepsilon)$. At 62 MeV the contributions of $n_0 = 3$ and $n \geq 5$ terms are shown by dashed curves.

state densities as used in the hybrid model provide a means of calculating the time-weighted spectrum of particles excited during the equilibration process. This result is mostly due to the near isotropicity of final energies following N-N scattering in the nuclear environment, and does not require an assumption of equilibrium; rather it is the result of the kinematics involved in the intranuclear cascade giving an energy distribution consistent with the statistical result of equation 2.

Phase space arguments are applied in estimating particle emission rates and branching ratios in the HMB, hybrid, and exciton formulations. This may well be an adequate estimate in the absence of true equilibrium. A large factor in $\lambda_c(\varepsilon)$ is the penetrability as given by the inverse reaction cross section, which carries with it a part of the dynamic and geometric information of the reaction.

It may be seen that the forward peaking of the angular distributions is consistent with particle emission following not just one, but several collisions. This follows from the mean free paths of 3–6 fm that are

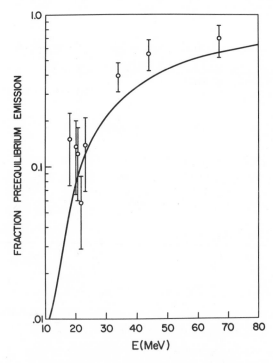

Figure 18   Fraction PE emission versus excitation for nucleon-induced reactions. The solid curve gives predictions of the GDH model for nuclei around $A = 60$; points are based on experimental spectra.

found in the INCM, HMB, and hybrid models, values that are short compared with the nuclear circumference.

From the observations of the preceding paragraphs, it seems that the hybrid model can be considered to be a simple closed form version of the INCM, and that it is not so strongly linked to phase space assumptions as it would initially appear to be. From the success of the hybrid, HMB, and INCM models in reproducing experimental results, it would appear that Serber's assumption was overly conservative. The N-N scattering approach would appear to be valid where the nucleon wavelength is short compared to its mean free path, rather than to the internucleon spacing. The Pauli principle extends this range over the energies of interest in PE decay.

8   COMPARISONS OF RESULTS
OF SEVERAL MODELS

Figures 19 and 20 present results of calculations of (p, p') reactions with the cascade, HMB, and GDH models at 39 and 62 MeV incident proton

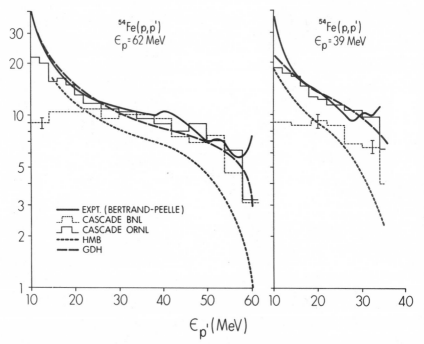

Figure 19   Angle-integrated ^{54}Fe(p, p') spectra compared with results calculated with several models. The curves are identified in the figures (25).

energy on ^{54}Fe and ^{209}Bi targets. The two cascade programs described in Section 3 were used.

In the case of ^{54}Fe(p, p') all model results are in satisfactory agreement with experimental spectra, with the HMB model and BNL cascade results being slightly low at 39-MeV incident proton energy. The results on ^{209}Bi show satisfactory agreement between experimental yields, and GDH and ORNL cascade calculations. The BNL cascade and HMB models give low results for both sets of spectra on ^{209}Bi. Shortcomings of the BNL cascade calculation at low energies (as in the ^{54}Fe case) and at high Z are apparently due to an overestimation of internal reflection and refraction at the multistep surface potential. The ORNL code has no provision for reflection or refraction, and this is not realistic. Reconciling these problems represents an open area for the intranuclear cascade calculation. The HMB model gives low results because of the failure to consider contributions from surface reactions.

Of the models discussed, only the hybrid/GDH model is capable of predicting on an a priori basis spectra and excitation functions due to PE decay for both nucleon- and complex particle-induced reactions. All comparisons to date indicate that it does so surprisingly well over a broad dynamic range in energy and cross sections. The cascade model can do this for nucleon-induced reactions, although very few cases have been explored

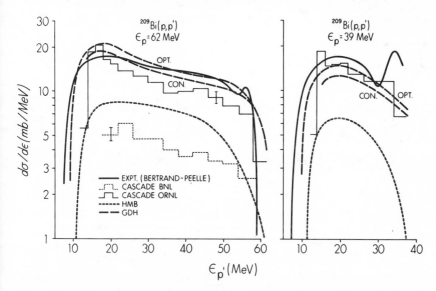

Figure 20 Spectra of ^{209}Bi(p, p') reaction compared with predictions of several models. The GDH model results were computed using both optical model and continuum model inverse cross sections (25).

to date in the lower energy regions of interest in this review. The exciton model as developed by groups in Milano has reproduced spectra and excitation functions of many nucleon- and some α-induced reactions, with an adjustment of the nucleon mean free path. This formulation has yet to be compared with data over as broad a dynamic range as has the hybrid model. Areas of application of these models have been summarized in Table 2.

9   OTHER AREAS OF INTEREST IN PREEQUILIBRIUM DECAY

9.1   Emission of Clusters

Considerable effort has been directed toward the interpretation of the emission of particles other than nucleons as due to PE decay. For all the assumptions made in formulating the models for nucleon emission, at least an order of magnitude is added to the uncertainties where clusters (e.g. α, d, t, ^{3}He) are concerned. This involves difficulties in calculating formation probabilities and numbers of levels per MeV for clusters, transition probabilities, and the basic question of whether the assumed mechanism is valid. What will be learned in this area remains to be seen. Some critical types of experiments, involving different particles in entrance and exit channels leading to the same final states, have only recently been completed (63).

The approaches taken may be grouped into three categories:

1.   The cluster is assumed to be preformed in the nucleus and is further assumed to scatter with the same kinematics as a nucleon; it is therefore treated as an exciton in an exciton model formulation. Such an approach has been taken to interpret (n, α) and (p, α) spectra, with reasonable success in reproducing the shape of the high-energy spectral component (88–91).

2.   It is assumed that of all the energy partitions allowed to excited particles and holes, some fraction will have just the right group of nucleons with correlated momenta to undergo cluster emission (92–95).

3.   A quasi-free scattering approach is taken for complex particles in either the entrance or exit channels; in this approach free nucleon/cluster elastic scattering angular distributions are used in a quasi-elastic approximation in the nuclear environment with application of Pauli exclusion principle to both scattered particles. This approach is still under development and few published results are available at the time of this review (26, 96). Attention in this subsection is therefore centered on categories 1 and 2 above.

Milazzo-Colli et al have used the form of equation 4 due to Williams (39) to calculate (p, α) and (n, α) spectra under the assumption that the incoming nucleon excites a preformed α particle (88–91). From analysis of experimental (n, p) and (n, n') spectra they evaluated lifetimes for excited states. A parameter was determined in this analysis to adjust the absolute (N, α) cross section of the calculated PE spectrum to the experimentally observed values. This parameter was interpreted as the probability that the incident nucleon interacts with a preformed α particle in the target nucleus.

In a related work, α decay half-lives of even-even nuclei were analyzed to extract an α preformation factor (97). The results of both sets of analyses are shown in Figure 21. There seems to be a general consistency between the parameter ψ (defined differently in the two cases) deduced from the two different types of experiments, which helps make the pre-formed α interpretation for (p, α) and (n, α) reactions seem reasonable.

The second approach to PE decay of clusters involves the assumption that partial state densities can be used to generate probabilities of the existence of clusters in the nucleus. Such an approach was first crudely formulated in (38) and was first applied to the case of α emission in (88). It was later formulated in greater detail in (92) and reformulated by the Bratislava group in the form that has had greatest success (93–95). Their expression for the decay rate for a cluster of β nucleons is mainly given by

$$W_\beta(\varepsilon)\,d\varepsilon = \gamma_\beta \, \frac{2S_\beta+1}{\pi^2\hbar^3} \, \mu_\beta \varepsilon_\beta \, \sigma_\beta(\varepsilon) \times \left[\frac{\rho(\beta,0,E-U)}{g} \frac{\rho(p-\beta,h,U)}{\rho(p,h,E)}\right]$$

$$\times \frac{\beta!(p-\beta)!}{p!}\right]\,d\varepsilon \qquad\qquad 8.$$

The quantity in the square brackets gives the number of p exciton

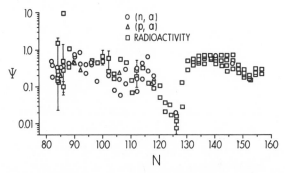

Figure 21   Preformation factor or probability of exciting a preformed α cluster versus target neutron number from radioactivity and induced reactions (88–91, 97).

configurations having β particles at excitation $E - U$ such that they could be emitted as a cluster of β nucleons at energy ε; the remaining $p - \beta$ particles and h holes share the residual energy U. It is assumed that some constant fraction of these combinations γ_β will actually have proper momenta to be a cluster; the value of γ_β is determined empirically for each cluster and for each target nucleus. The assumption of $\gamma_\beta = $ constant may be restated as follows: the greater the energy that β nucleons have to share between them, the larger the number of combinations of energy partition that will have an overlap characteristic of the cluster β. This assumption is open to question.

Results of equation 8 are shown in Figure 22, along with values of γ_β for reactions of [197]Au with 62-MeV protons (93). Results of an earlier calculation of Cline are also shown [approximately equation 8 without the factor $\gamma_\beta \rho(\beta, 0, E - U)/g$, and with an additional arbitrary multipli-

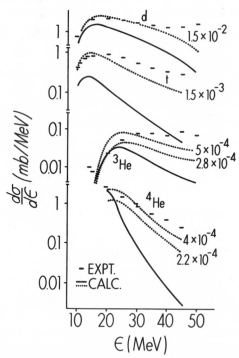

Figure 22   Calculated and experimental angle-integrated spectra of clusters emitted following 62-MeV proton bombardment of [197]Au. Experimental results are given by horizontal bars (20). Calculations with equation 8 are given by the dotted curves; preformation probabilities γ are indicated in the figure. The heavy solid curves are results of the formulation in (92).

cative factor of $\beta!$] (92). The earlier formulation reproduces neither the shapes nor the cross sections of experimental results in a satisfactory fashion.

It would seem that equation 8, or the earlier formulations, could be easily checked for internal consistency. Consider the reactions of Figure 22, which have also been measured at 29-MeV incident proton energy. If one takes the ratio of cross sections to be expected at excitation energies $E1$ and $E2$ such that the energy available to the clusters is equal ($\varepsilon1 = \varepsilon2$) then using equation 8 (and making the crude approximation that cross sections will be proportional to the emission rates independent of excitation energy), the expected yield ratio to first order is $P1/P2(\varepsilon) = (U1/U2)^{p-\beta-1} \cdot (E2/E1)^{n-1}$, all other terms canceling, including the assumed preformation fraction γ_β. If one considers the case of ^{197}Au(p, α) at 29- and 62-MeV incident proton energies, and considers the predicted cross-section ratios, e.g. for $\varepsilon_\alpha = 26$ and 29 MeV, respectively, one finds that in these two examples the calculated and experimental ratios disagree by factors of 400:1 and 500:1. These results do not support the mechanism assumed by equation 8 or earlier versions of this approach. Additional calculations of this type would be worthwhile either to show an internal consistency of the approach, or to eliminate it from further consideration.

The work that has been done with respect to complex particle emission must be regarded as bold explorations in an extremely difficult and evasive area. The reviewer is of the opinion that this area is still very much "open" and a fertile area for further work.

9.2   Other Areas in Preequilibrium Decay: Past, Present, and Future

Some application of PE decay models has been made in the interpretation of fission data. Since fission probably does not compete with the faster PE process, allowances must be made in interpretation of data. This was done by Bass et al (98) in the extraction of Γ_n/Γ_f from results of the proton-induced fission of uranium isotopes.

Calculations involving angular momentum-dependent partial state densities have been performed by the Bratislava group in an attempt to explain isomer yield ratios beyond excitation function maxima (48) and to investigate changes in emission rates due to angular momentum restrictions (99). Attempts to consider isospin effects in PE decay have been made (100). Concepts of PE decay have been applied to photonuclear reactions, where it was found that coherent excitation of the giant dipole resonance plays an important role (101). Kozlowski & Zglinski have reproduced the experimental n and p spectra following muon capture by application of the hybrid model with excellent results (personal communication, 1972).

PE concepts have been used to explain the isotopic yields of multi-nucleon transfer products in heavy ion-induced reactions (102). It may be shown that in high-energy, heavy-ion reactions, the theoretical equilibrium lifetimes of compound nuclei are less than relaxation periods and less than the fusion period. If so, PE decay must surely be considered in future analyses of heavy ion reaction products. Preliminary considerations have been given to the problem (103).

Structure has been observed at high residual excitations in particle spectra from reactions induced by several projectiles (104). It remains to be seen if these phenomena can be interpreted in a consistent and broadly successful fashion in terms of PE models using realistic partial state densities to replace the structureless results of the equidistant spacing model. If so, PE decay may become a useful tool in statistical nuclear structure studies of levels at high excitations.

Finally, an important area for the future is the question of reproducing angular distributions. One attempt in terms of a rotator with hybrid model decay periods has been made by Chevarier et al (105), with quite good results. However, no general formulation has yet appeared, and this must be considered an open and important area for further work.

10 CONCLUSIONS

Hopefully it is apparent that PE decay is an area of wide and increasing interest, having applications to the understanding of many aspects of nuclear reactions. The presence in many continuum spectra of a statistical component of a nonequilibrium nature seems clear. PE decay models formulated in terms of equilibration, predominantly by successive binary transitions, provide a consistently good description of the experimental results over a wide range of incident energies and projectile types. These models may be shown to provide a unified time-dependent description of nuclear reactions that includes direct, intermediate, and equilibrium components. Agreement between a priori calculations and experimental yields is generally much better than the factor of two that one might expect as a reasonable limit, and extends over a dynamic range of as many as four orders of magnitude. The broad success of these models is compelling a posteriori evidence that the basic physical assumptions are valid.

Yet it is true that the models described are simple classical models, derived without consideration of quantum mechanical results. In this regard they are analogous to the Weisskopf compound nucleus model in early years (which has proven to be extremely useful in nuclear physics). The past philosophy of model development in this area can be summarized by the quotation "Damn the torpedoes, full speed ahead." The time has come, however, for the vigor of the past to look to rigor in the future. In this regard Grimes et al (46) have shown the relationship between the exciton model and the model of doorway states (28, 29). Feshbach, Kerman & Koonin have indicated a relationship between long-lived intermediate states of the exciton model and a series of "chained" doorway states (87). Unfortunately the question of the forward peaked components that comprise the bulk of the spectra have yet to be treated in as rigorous a fashion. Other aspects of the theory are being investigated in the USSR (106, 107) and elsewhere.

The future will doubtless see many new applications of PE concepts in interpretation of a broad range of nuclear reaction phenomena, as well as a firmer theoretical understanding as to why these models work, and as to what circumstances may render them invalid. These investigations and results, like the models themselves, are time-dependent and deferred to the future.

ACKNOWLEDGMENTS

The author acknowledges with great pleasure many helpful discussions and communications with Professors A. K. Kerman, J. M. Miller, J. J. Griffin, E. Gadioli, L. Milazzo-Colli, and P. Oblozinsky, and with Drs. F. E. Bertrand and M. T. Magda during the course of this work. The author wishes to express his appreciation to Dr. J. R. Grover for helpful suggestions on this manuscript. Material included in this review was limited to documents in the literature or in the author's possession on January 15, 1975. Space limitations precluded thorough review of all works, and the author apologizes to those who may feel the emphasis of this work to be in improper balance.

Literature Cited

1. Griffin, J. J. 1966. *Phys. Rev. Lett.* 17:478
2. Griffin, J. J. 1968. *Intermediate Structure in Nuclear Reactions,* ed. H. P. Kennedy, R. Schrils. Lexington: Univ. Kentucky Press. 219 pp.
3. Griffin, J. J. 1967. *Int. Nucl. Phys. Conf., 1966, Gatlinburg, Tenn.,* ed. R. L. Becker, p. 778. New York: Academic
4. Griffin, J. J. 1967. *Phys. Lett. B* 24:5
5. Harp, G. D., Miller, J. M., Berne, B. J. 1968. *Phys. Rev.* 165:1166
6. Harp, G. D., Miller, J. M. 1971. *Phys. Rev. C* 3:1847
7. Blann, M. 1971. *Phys. Rev. Lett.* 27:337; 700E; 1550E
8. Blann, M. 1972. *Phys. Rev. Lett.* 28:757
9. Weisskopf, V. F. 1937. *Phys. Rev.* 53:295

10. Glendenning, N. K. 1963. *Ann. Rev. Nucl. Sci.* 13:191
11. Greider, K. R. 1965. *Ann. Rev. Nucl. Sci.* 15:291
12. Bethge, K. 1970. *Ann. Rev. Nucl. Sci.* 20:255
13. Tamura, T. 1969. *Ann. Rev. Nucl. Sci.* 19:99
14. Thomas, T. D. 1968. *Ann. Rev. Nucl. Sci.* 18:343
15. Fleury, A., Alexander, J. M. 1974. *Ann. Rev. Nucl. Sci.* 24:279
16. Bodansky, D. 1962. *Ann. Rev. Nucl. Sci.* 12:79
17. Sidorov, V. A. 1962. *Nucl. Phys.* 35:253
18. Holbrow, C., Barschall, H. 1963. *Nucl. Phys.* 42:264
19. Wood, R., Borchers, R., Barschall, H., 1965. *Nucl. Phys.* 71:529
20. Bertrand, F. E., Peelle, R. W. 1973. *Phys. Rev. C* 8:1045
21. Serber, R. 1947. *Phys. Rev.* 72:1114
22. Metropolis, N. et al 1958. *Phys. Rev.* 110:185, 204
23. Bertini, H. W., Harp, G. D., Bertrand, F. E. 1974. *Phys. Rev. C* 10:2472
24. Blann, M. 1972. *Proc. Summer School Nucl. Phys. 4th, 1971*, ed. W. Zych, 1–14.09. Warsaw, Poland: Warsaw Univ. Press
25. Blann, M. 1973. *Conf. Intermed. Processes Nucl. Reactions, Yugoslavia.* Heidelberg: Springer Verlag. 329 pp.
26. Blann, M. 1975. *Proc. Summer School on Nuclear Physics, 1974, Predeal, Romania.* Rep. COO-3494-13
27. Miller, J. M. 1973. *Proc. Int. Conf. Nucl. Phys., Munich,* ed. J. J. DeBoer, H. J. Mang, 2:597. Amsterdam: North-Holland
28. Feshbach, H., Kerman, A. K., Lemmer, R. H. 1967. *Ann. Phys. NY* 41:230
29. Mahaux, C. 1973. *Ann. Rev. Nucl. Sci.* 23:193
30. Chevarier, A. et al 1973. *Phys. Rev. C* 8:2155
31. Bertini, H. W. 1963. *Phys. Rev.* 131:1801
32. Chen, K. et al 1968. *Phys. Rev.* 166:949
33a. Chen, K., Friedlander, G., Miller, J. M. 1968. *Phys. Rev.* 176:1208
33b. Chen, K., Friedlander, G., Harp, G. D., Miller, J. M. 1971. *Phys. Rev. C* 4:2234
34. Hudis, J., Miller, J. M. 1959. *Ann. Rev. Nucl. Sci.* 9:159
35a. Porile, N. T. et al 1963. *Nucl. Phys.* 43:500
35b. Jackson, J. D. 1956. *Can. J. Phys.* 34:767
36. Ericson, T. 1960. *Adv. Phys.* 9:423
37. Blann, M. 1968. *Phys. Rev. Lett.* 21:1357
38. Blann, M., Lanzafame, F. M. 1970. *Nucl. Phys. A* 142:559
39. Williams, F. C. Jr. 1970. *Phys. Lett. B* 31:184
40. Blann, M. 1973. *Nucl. Phys. A* 213:570
41. Gadioli, E., Gadioli-Erba, E., Sona, P. G. 1973. *Nucl. Phys. A* 217:589
42. Blann, M., Mignerey, A. 1972. *Nucl. Phys. A* 186:245
43. Cline, C. K., Blann, M. 1971. *Nucl. Phys. A* 172:225
44. Demeyer, A. et al 1970. *J. Phys.* 31:847
45. Bimbot, R., LeBeyec, Y. 1971. *J. Phys.* 32:243
46. Grimes, S. M., Anderson, J. D., Pohl, B. A., McClure, J. W., Wong, C. 1971. *Phys. Rev. C* 4:607
47. Alevra, A. et al 1973. *Nucl. Phys. A* 209:557
48. Oblozinsky, P., Ribansky, I. 1972. *Nucl. Phys. A* 195:269
49. Birattari, C. et al 1971. *Nucl. Phys. A* 166:605
50. Birattari, C. et al 1973. *Nucl. Phys. A* 201:579
51. Birattari, C., Gadioli, E., Grassi Strini, A. M., Strini, G., Tagliaferri, G. 1973. *Nuovo Cimento Lett.* 7:101
52. Gadioli, E., Grassi Strini, A. M., LoBianco, G., Strini, G., Tagliaferri, G. 1974. *Il Nuovo Cimento* 22:547
53. Alevra, A., Duma, M., Lukas, I. R., Magda, M. T., Nistor, M. E. 1974. *Rev. Roum. Phys.* 19:55
54. Decowski, P., Grochulski, W., Marcinowski, A. 1973. See Ref. 27, 1:521
55. Hermsdorf, D., Sassonoff, S., Seeliger, D., Seidel, K. 1973. See Ref. 25, 1:521

56. Milazzo-Colli, L., Braga-Marcazzan, G. M. 1973. *Riv. Nuovo Cimento* 3:535
57. Bertrand, F. E., Peelle, R. W., Kalbach-Cline, C. 1974. *Phys. Rev. C* 10:1028
58. Chevarier, N., Chevarier, A., Demeyer, A., Duc, T. M. 1971. *J. Phys.* 32:483
59. Grimes, S. M., Anderson, J. D., McClure, J. W., Pohl, B. A., Wong, C. 1972. *Phys. Rev. C* 5:830
60. Jahn, P., Probst, H. J., Djaloeis, A., Davidson, W. F. Mayer-Boricke, C. 1973. *Nucl. Phys. A* 209:333
61. Chevarier, A. et al 1974. *Nucl. Phys. A* 231:64
62. Chevarier, A. et al 1975. *Nucl. Phys. A* 237:354
63. Bisplinghoff, J. 1974. *Zur Praecompoundemission Leichter Teilchen.* PhD thesis. Bonn Univ., Bonn, Germany. 74 pp.
64. Ribansky, I., Oblozinsky, P., Betak, E. 1973. *Nucl. Phys. A* 205:545
65. Böhning, M. 1970. *Nucl. Phys. A* 152:529
66. Williams, F. C. Jr. 1971. *Nucl. Phys. A* 166:231
67. Williams, F. C. Jr., Mignerey, A., Blann, M. 1973. *Nucl. Phys. A* 207:619
68. Albrecht, K., Blann, M. 1973. *Phys. Rev. C* 8:1481
69. Grimes, S. M., Anderson, J. D., Davis, J. C., Wong, C. 1973. *Phys. Rev. C* 8:1770
70. Oblozinsky, P., Ribansky, I., Betak, E. 1974. *Nucl. Phys. A* 226:347
71. Blann, M. *Winter Meet. Nucl. Phys., 9th, Villars, 1973.* ed. I. Iori. Milan: Univ. Milan. Reproduced copy
72. Goldberger, M. L. 1948. *Phys. Rev.* 74:1268
73. Kikuchi, K., Kawai, M. 1968. *Nuclear Matter and Nuclear Interactions.* Amsterdam: North-Holland
74. Becchetti, F. D., Greenlees, G. W. 1969. *Phys. Rev.* 182:1190
75. Braga-Marcazzan, G. M., Gadioli-Erba, E., Milazzo-Colli, L., Sona, P. G. 1972. *Phys. Rev. C* 6:1398
76. Gadioli, E., Milazzo-Colli, L. 1973. See Ref. 25

77. Gadioli, E., Gadioli-Erba, E. 1975. *Acta Phys. Slov.*
78. Kiessig, G., Reiff, R., eds. 1974. *Conf. Interact. Fast Neutrons with Nuclei. Tech. Univ. Dresden Rep. ZFK271.* 205 pp.
79. Bisplinghoff, J., Ernst, J., Machner, H., Mayer-Kuckuk, T., Schanz, R. *Nucl. Phys.* In press
80. Cline, C. K. 1973. *Nucl. Phys. A* 210:590
81. Lee, E. V., Griffin, J. J. 1972. *Phys. Rev. C* 5:1713
82. Grimes, S. M., Anderson, J. D., McClure, J. W., Pohl, B. A., Wong, C. 1973. *Phys. Rev. C* 7:343
83. Swenson, L. W., Gruhn, C. R. 1966. *Phys. Rev.* 146:886
84. West, R. W. 1966. *Phys. Rev.* 141:1033
85. Weisskopf, V., Ewing, D. H. 1940. *Phys. Rev.* 57:472
86. Betak, E. 1975. Personal communication
87. Feshbach, H. 1974. *Rev. Mod. Phys.* 46:1
88. Milazzo-Colli, L., Marcazzan-Braga, G. M. 1972. *Phys. Lett. B* 38:155
89. Braga-Marcazzan, G. M., Milazzo-Colli, L., Signorini, C. 1973. *Nuovo Cimento Lett.* 6:357
90. Milazzo-Colli, L., Braga-Marcazzan, G. M. 1973. *Nucl. Phys. A* 210:297
91. Milazzo-Colli, L., Braga-Marcazzan, G. M., Milazzo, M., Signorini, C. 1974. *Nucl. Phys. A* 218:274
92. Cline, C. K. 1972. *Nucl. Phys. A* 193:417
93. Ribansky, I., Oblozinsky, P. 1973. *Phys. Lett. B* 45:318
94. Betak, E. 1975. *Int. Symp. Neutron Induced Reactions, 1975, Czechoslovakia. Acta Phys. Slov.* In press
95. Oblozinsky, P. 1975. See Ref. 94
96. Mignerey, A., Scobel, W., Blann, M. 1974. *Bull. Am. Phys. Soc.* 19:1013
97. Milazzo-Colli, L., Bonetti, R. 1974. *Phys. Lett. B* 49:17
98. Bass, R. et al 1973. See Ref. 71
99. Oblozinsky, P., Ribansky, I. 1974. *Acta Phys. Slov.* 2:103
100. Kalbach-Cline, C., Huizenga, J. R., Vonach, H. K. 1974. *Nucl. Phys. A*

222:405
101. Lukyanov, V. K., Seliverstov, V. A.,
     Toneev, V. D. 1974. *JINR* P4-8001.
     Preprint
102. Gudima, K. K., Iljinov, A. S.,
     Toneev, V. D. 1974. *JINR* P7-7915.
     Preprint
103. Blann, M. 1974. *Nucl. Phys. A*
     235:211

104. Magda, M. T. et al 1973. *Nuovo
     Cimento Lett.* 8:396
105. Chevarier, N. et al 1973. See Ref. 71
106. Gudima, K. K., Ososkov, G. A.,
     Toneev, V. D. 1974. *JINR* P4-7821.
     Preprint
107. Rumyantzev, B. A., Heifetz, S. A.
     1974. Inst. Nucl. Phys. Joan, USSR,
     No. 74–50. Preprint

IDENTIFICATION OF NUCLEAR PARTICLES[1]

�×5561

Fred S. Goulding and Bernard G. Harvey

Lawrence Berkeley Laboratory, University of California,
Berkeley, California 94720

CONTENTS

1 INTRODUCTION

The past few years have seen a rapid development of techniques by which the identity of species produced in nuclear reactions may be established. Writing a nuclear reaction in the usual way—$X(a,b)Y$—we review methods for the identification of b by determining its atomic number Z

[1] This work was performed under the auspices of the United States Energy Research and Development Administration.

167

and mass number A. We consider only on-line techniques, ignoring methods based on radiochemistry, photographic plates, or track detectors. These latter have been reviewed by Price & Fleischer (1).

Generally the energy E of the particle must also be measured, and with the best possible resolution. It is sufficient, in many experiments, to measure the kinetic energy *differences* between more or less sharp lines in the energy spectrum of particle b, corresponding to the formation of discrete energy states of the residual nucleus Y. If the energies of the states of Y are well known from previous work, the energy differences need only be measured with sufficient precision to be sure that lines in the b spectrum are correctly associated with states of Y. Nuclear energy levels, however, are often closely spaced so only a small spread in the E measurement can be tolerated even though an absolute determination of E may not be important.

A complete determination of Z, A, and E requires the measurement of three quantities that are independent functions of Z, A, and E. No measurable quantities depend directly upon A, but for the non- or only slightly relativistic particles typically encountered in nuclear physics experiments, A is very nearly equal to the mass M of the particle (in atomic mass units), which is measurable. For nonrelativistic particles, M has only near-integral values. M and Z therefore need be determined only with enough accuracy to separate them from adjacent integral values. This is very easy for light particles (e.g. H and He isotopes) because the fractional differences between adjacent small integers are large. For heavy particles, the necessary resolution becomes difficult or even impossible to achieve.

Several types of measurements depend on independent functional combinations of M, Z, and E, but no one measurement uniquely determines these parameters. The value of E is nearly always required in an experiment but, fortunately, *individual* values of M and Z are often not needed. For example, the quantity MZ^2 can be obtained from a detector telescope that measures the energy loss of a particle passing through a thin detector into a second detector where its residual energy is deposited and measured. MZ^2 assumes unique values of 1, 2, 3, 12, and 16 for protons, deuterons, tritons, ^{3}He, and ^{4}He, respectively, so its value characterizes each of these isotopes unambiguously. For heavier ions more elaborate identification measurements are needed to discriminate between different ions. The following is a brief summary of the various methods and the information they yield:

(*a*) Total absorption in a detector (or detector telescope): Measurement of the total ionization produced in the detector(s) provides a linear measure of the particle energy E.

(*b*) Energy absorption in a thin detector: A thin transmission detector, included in a detector telescope, provides a direct measure of dE/dx for a particular segment of a particle's track. The rate of energy loss is approximately given by the simplified Bethe-Bloch (2) equation

$$- dE/dx = (aZ^2c^2/v^2)\ln\left[bv^2/(c^2-v^2)\right],\qquad\qquad 1.$$

where v is the particle velocity, c is the velocity of light and a and b are constants dependent only on the detector material. Although this equation is traditionally written in terms of the atomic number Z, the rate of energy loss actually depends on the rms charge state q_{eff} of the moving ion, which may not be fully stripped of atomic electrons (i.e. $q_{eff} \leqq Z$).

Since the logarithmic term varies only slowly with energy (or velocity) its effect will be neglected in this brief discussion. Also, for non-relativistic particles, $v^2 = 2E/M$. Therefore equation 1 can be simplified to

$$dE/dx \propto MZ^2/E.\qquad\qquad 2.$$

E can be computed by summing the detector telescope signals, so the measurement of dE/dx provides a measure of MZ^2.

(c) Time-of-flight measurement: Measurement of the time of flight (TOF) of a particle through a known flight path in vacuum determines the particle velocity v. We have

$$v^2 = 2E/M.\qquad\qquad 3.$$

If the value of E is known, then the TOF determines M. If this measurement of M is combined with a dE/dx determination, Z (or more accurately q_{eff}) can be determined.

In principle, therefore, a detector telescope measuring TOF, E, and dE/dx uniquely identifies a reaction product. However, limitations in the accuracy of determination of E, dE/dx, and v blur the results and prevent unique identification for the heavier isotopic species. Furthermore, an additional problem develops with heavy ions since q_{eff} is much less than Z for these ions.

(d) Bending in a magnetic field: Magnetic spectrometers provide yet another determination of a combination of the particle parameters M, Z, and E. In a fixed magnetic field B, the radius of curvature ρ of a particle is given by

$$B\rho \propto Mv/q,\qquad\qquad 4.$$

where q is the average charge of the ion. Since light particles, or heavy ions at high energies, emerge from a target fully stripped of electrons, their average charge q is equal to Z, and measurement of $B\rho$, TOF, and dE/dx is equivalent to a complete identification and energy measurement. For heavy ions at lower energies, $q \leqq Z$ and there may be ambiguities in the identification: these are discussed in Section 4.

Other physical effects that depend on M, Z, and E can also, in principle, be used for particle identification. The deflection of a particle in an electric field is one example, but it is so small for high-velocity particles that its use is not very practical.

2 IDENTIFICATION BY ABSORPTION IN A DETECTOR TELESCOPE

Basic ΔE,E Systems

As indicated earlier, a telescope of detectors can be used, sometimes alone or together with another measurement (e.g. time of flight), to

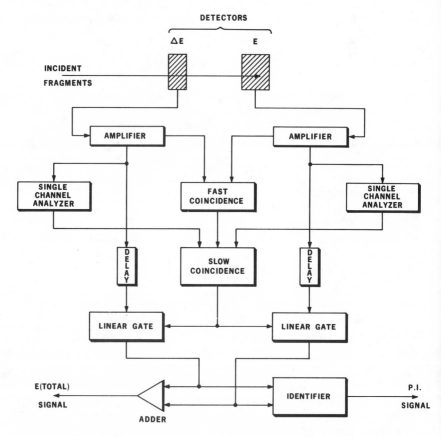

Figure 1   Block diagram of $\Delta E, E$ identifier system.

identify the fragments produced in a nuclear reaction. The simplest type of telescope and the associated electronics are shown in Figure 1. Detector thicknesses are so chosen that the particles of interest pass through the front ΔE detector, providing a ΔE signal, then stop in the rear E detector. The total energy signal is derived by adding the two signals, and particle identification is obtained from the sizes of the individual signals.

Referring to Figure 1, signals from the detectors are amplified to a convenient level and, providing that they satisfy certain energy (i.e. amplitude) and timing requirements, they are allowed to pass into the signal adder to generate the total energy signal. They also feed an identifier that develops an output signal whose size is ideally determined only by the type of particle—independent of its energy. The identifier function may be performed by a special-purpose analog computer unit, or by a digital computer, or sometimes by both. In general, analog calculations can be performed in a very short time, while digital operations are slower but offer more flexibility and accuracy. As shown in Figure 1, signals from the two detectors must be coincident in time and fall into energy windows defined by the single-channel analyzers. These two requirements reduce the chance that the summation of signals generated by two fragments will produce a particle identifier output that accidentally corresponds to a third type of ion. Narrowing the energy acceptance windows to the minimum values consistent with detecting the rare events, and reducing the resolving time of the fast coincidence circuit to its minimum tolerable value, reduces the chance of false identification.

Identification Algorithms

As shown in the Introduction, knowledge of dE/dx and E is adequate to establish the quantity MZ^2 for a fully stripped ion (i.e. $q_{eff} = Z$). In principle, therefore, a table look-up procedure can be used to identify the particle—or at least to identify its MZ^2 value. When a computer is available, this technique is sometimes used (3, 4), but it is often simpler, using the ΔE and E signals, to generate a function whose value is characteristic of a particular type of ion and independent of its energy. The purpose of the algorithms discussed here is to provide suitable analytical functions that come close to achieving this ideal.

The complete Bethe-Bloch equation that describes a charged particle's rate of energy loss by ionization is

$$- dE/dx = 4\pi n (e^4/m)(q_{eff}^2/v^2)\{\ln [2mv^2/I(1-\beta^2)] - \beta^2 - S - D\}, \qquad 5.$$

where n is the number of electrons/cm^3 of the absorber, e and m are the charge and mass of the electron, q_{eff} is the rms charge (in electronic units) on the ion, v is its velocity, $\beta = v/c$ where c is the velocity of light,

I is the mean ionization potential for the absorber, S is a correction for the fact that the electrons from different shells do not all equally participate in the ionization process, and D is a density correction.

The small corrections S and D will be neglected here. Furthermore, in most nuclear reaction experiments, the ions are nonrelativistic (i.e. $\beta^2 \rightarrow 0$) so equation 5 can be reduced to a simpler form.

The curves given in Figure 2, adapted from Northcliffe (5, 6), which show the behavior of various ions in an aluminum absorber, give a better perspective on the behavior of stopping power with energy and type of particle. The stopping power (see equation 5) is basically dependent on velocity (i.e. E/M) so it is convenient to plot the curves in terms of E/M. The stopping power is expressed in terms of $(1/Z^2)\,dE/dx$, where Z is the atomic number of the species. Equation 5 shows that the stopping power depends on q_{eff}, the rms charge state of the ion. For a proton, $q_{eff} = Z$ over most of the energy range shown in Figure 2, but for very heavy ions, the atom is fully stripped only at the very highest energies. Therefore, the curves of stopping power/Z^2 for heavy ions depart from those for protons except at the highest energies. At very low velocities, ions capture sufficient electrons to become neutral much of the time. In this region, collisions

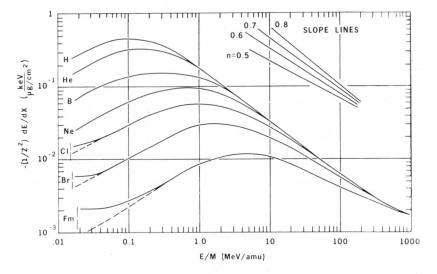

Figure 2 Stopping power (in aluminum) for various ions. (Adapted from Reference 6.) The stopping power axis is normalized in terms of $1/Z^2$ where Z is the nuclear charge of the particular ion. The energy scale is normalized in terms of the energy/nucleon of the ion. The slope lines at the top of the figure show the slopes corresponding to various power laws.

of the neutral atom with electrons and nuclei become the dominant energy loss mechanism.

For all ions of $Z < 10$ the curves of Figure 2 approximate a power law over a useful range of energies. Thus, we have

$$\text{stopping power}/Z^2 \propto (E/M)^n, \qquad\qquad 6.$$

where the value of n varies from about -0.75 for a proton to -0.5 for a heavy ion in the energy ranges commonly encountered in nuclear reactions. The power-law approximation for protons stems from the $1/E$ behavior of the main term in the right-hand side of equation 5, combined with the fact that the slow variation of the logarithmic term can almost be represented by a $E^{0.3}$ law. For heavy ions, charge exchange processes modify the energy dependence at low energies, resulting in a change in the exponent n.

Particle identifiers generally use one of two basic algorithms. The earliest identifiers (7–12) were based on simplified versions of equation 5. For nonrelativistic particles we may write

$$-dE/dx = a(q_{\text{eff}}^2 M/E)\ln(bE/M), \qquad\qquad 7.$$

where a and b are constants that depend on the absorber material. If a detector of thickness Δx absorbs a very small fraction of the incident energy of a particle to produce a signal proportional to the loss ΔE, and a second detector then absorbs the remainder of the energy E to produce a signal proportional to E, the signals are related as follows:

$$E\Delta E \propto \Delta x(M q_{\text{eff}}^2)\ln(bE/M). \qquad\qquad 8.$$

Neglecting the slow energy variation of the logarithmic term, this product is a measure of $M q_{\text{eff}}^2$. This can be computed to yield an output that depends, to first order, only on the value of $M q_{\text{eff}}^2$ for the detected particle.

Two modifications of this basic method are necessary where a broad range of energies and types of particle is to be analyzed. First, the assumption of constancy of the logarithmic term with energy is invalid, so the identifier output signal varies with energy, as well as with particle type, as shown schematically in curve a of Figure 3. It is common to add a term proportional to ΔE to the $E\Delta E$ product to partially correct for the fall-off at low energies. Because the ΔE signal rises at low energies, choice of an appropriate multiplying factor E_0 causes the resulting identifier output $(E + E_0)\Delta E$ to remain relatively constant with energy. This also provides some compensation for the change in q_{eff} at low ion velocities. It is also necessary to compensate for the fact that the energy loss in the ΔE detector is not infinitesimally small and that it

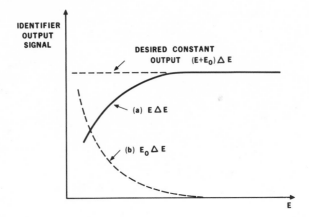

Figure 3   Behavior of the components of the output in a $E, \Delta E$ multiplier type of identifier.

may become quite a large part of the total energy E for some of the particles detected by the $E, \Delta E$ detector telescope. To allow for this in the identifier algorithm, the term $E \Delta E$ is modified to $(E + k \Delta E) \Delta E$ where the constant k is adjusted for best identification. If $\Delta E \ll E$, $k = 0.5$ is the correct value. The final form of the function used in *multiplier* types of identifiers is therefore

$$(E + E_0 + k\Delta E)\Delta E \propto M q_{\text{eff}}^2 \Delta x, \qquad 9.$$

where E_0 and k are parameters adjusted experimentally for optimum constancy of the identifier output as a function of energy.

The second basic type of identifier (13–18) largely avoids the need for experimental adjustment of parameters. Because a major difficulty with the multiplier type of identifier results from its dependence on measurement of an incremental energy loss, it appears that a more suitable method might be based on a range function. For a given ion, and over a limited range of energies, the curves of Figure 2 can be represented by the relationship

$$-\mathrm{d}E/\mathrm{d}x = E^n/a, \qquad 10.$$

where $n \approx -0.7$ for protons, becoming smaller for low-velocity ions. The value of a will be approximately proportional to $1/q_{\text{eff}}^2 M^{-n}$, or more roughly, $1/M q_{\text{eff}}^2$. The range R of an ion entering an absorber with energy E can be calculated by integrating the incremental elements of the path corresponding to incremental energy losses. Thus

$$R = \int_{E_1}^{E} (\mathrm{d}x/\mathrm{d}E)\,\mathrm{d}E + R_1, \qquad 11.$$

where E_1 represents the energy at which the simple relationship of equation 10 breaks down and R_1 represents the remaining range at energy E_1. For the high-energy particles commonly studied in nuclear reactions, $R_1 \ll R$ and $E_1 \ll E$. Neglecting these terms we therefore have

$$R \propto aE^{1-n}. \qquad\qquad 12.$$

For a proton, $n \approx -0.7$ so the range R_p is given by

$$R_p \propto aE^{1.7}. \qquad\qquad 13.$$

More generally, we may use an index $b = 1 - n$ that depends on the type of ion. Therefore

$$R \propto aE^b, \qquad\qquad 14.$$

where $a \approx 1/Mq_{\text{eff}}^2$.

If an ion deposits energy ΔE in the first detector of thickness Δx of a telescope, then stops in the second detector, depositing energy E, it is obvious that the range of the particle with energy $E + \Delta E$ is Δx longer than the range of the same particle with energy E. Therefore, from equation 14,

$$\Delta x/a \propto [(E + \Delta E)^b - E^b]. \qquad\qquad 15.$$

As a is approximately proportional to $1/Mq_{\text{eff}}^2$, the left side is roughly proportional to Mq_{eff}^2. The thickness Δx is a known quantity, whereas E and ΔE are determined by measuring the amplitude of the detector signals. Therefore, equation 15 provides a direct determination of Mq_{eff}^2 that involves no arbitrary constants and no limitation on the fraction of the incident energy deposited in the ΔE detector. These factors make identification based on equation 15 more generally useful than the method based on equation 9.

It is surprising how accurately the power-law relationship predicts the range of ions of widely differing types and energies. Skyrme (19) has shown that the errors amount to no more than a few percent over the energy range of 5–50 MeV for particles ranging from protons to alphas; as seen by inspection of Figure 2, the errors should not increase significantly up to much higher energies. Figure 2 indicates that the situation is not so favorable for heavy ions, and energy-dependent corrections (e.g. making b a function of ΔE) are often made. Examples of such corrections have appeared in the literature (20).

Chaminade et al (21) have used an approximation to the relationship of equation 15 that is more tolerant of imperfections in identifier circuits when ΔE is very small compared with E. Other useful approximations have been developed (22). Bird & Ollerhead (23) have extended the use of the range algorithm to low energies where the power-law approximation to the range-energy relationship is no longer valid. To achieve this they generalize the range-energy relationship to

$$R = aF(E).$$ 16.

Consequently equation 15 is replaced by

$$\Delta x/a = [F(E+\Delta E)-F(E)].$$ 17.

By storing range-energy tables [i.e. R v E] in a computer, and by using a table look-up method, they identify any particle registering in the detector telescope. Other computer methods (24) have appeared in the literature.

Identifier Circuits

The availability of medium-size on-line computers at accelerator laboratories has resulted in increasing use of digital calculations (by table look-up, or algorithms) for identifying particles. Depending on the speed of the computer, the processing of each event may require a few tens to a few hundreds of microseconds. Since the optimum shaping time in the signal paths prior to digitizing is only a few microseconds or less, computer processing of each event seriously reduces the data acquisition rate. If the ions of interest are rare ones in a large flux of less interesting particles, this rate limitation may be unacceptable. It is then convenient to use an analog identifier, which identifies particles in a few microseconds, to select only the interesting events for processing by the computer. The analog identifier may also be used alone without the help of a computer, whose main virtue is its ability to subject an event to more critical evaluation before deciding upon its type. For example, the computer might employ a modified form of the simple power-law algorithm to compensate for the change in b (in equation 14) that occurs for low-energy heavy ions.

Analog identifiers utilize circuit techniques to achieve the multiplication operation required by the function $(E+E_0+k\Delta E)\Delta E$ of equation 9, or the exponentiation required in the range algorithm represented by equation 15. The following basic methods have been employed:

1. Circuit elements whose output is proportional to the square of their inputs can be used to process two signals A and B to generate the functions $(A+B)^2$ and $(A-B)^2$. By taking the difference between these outputs, a final result proportional to AB is produced. This method, which used a special square-law tube, has now been superseded by some of the following techniques. Another element that exhibits almost a square-law characteristic is a field-effect transistor (FET). This has also been used (17, 25) as a function generator in identifiers.

2. By converting the amplitude of one of the signals to a time propor-

tional to amplitude, and then integrating the other signal for this time, an output proportional to the product AB is produced. Although this method has been employed in identifiers (8), its application is limited to the multiplier algorithm technique. Also the rate limitation caused by using time as an intermediate parameter in the calculation is not desirable.

3. A piecewise linear approximation using diode switches has also been used to approximate the various algorithms (26).

4. The most versatile method of performing the analog calculation required in an identifier uses elements exhibiting a logarithmic or exponential relationship between input and output. A semiconductor junction performs both these functions.

Figure $4A$ shows one implementation of a logarithmic function generator. The base-to-emitter junction of transistor Q forms the logarithmic element. The high gain operational amplifier forces the emitter voltage of Q to a value at which its collector current equals the input pulse current plus a very small standing current i_1. The standing current must be much smaller than any signal current of interest if the output voltage pulse is to be a reasonable approximation to the logarithm of the input current. Figure $4B$ shows an exponential function generator.

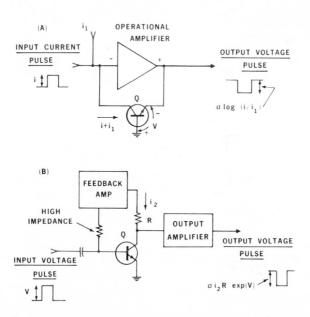

Figure 4   *A:* Basic circuit of a logarithmic function generator. *B:* Basic circuit of an exponential function generator.

Transistor Q is the exponential element whose standing current i_2 is stabilized by feedback to the base of Q. A voltage pulse V on the base of Q produces a change in the collector current proportional to $\exp(V)$ which, in turn, produces a proportional voltage across R. The high input impedance output amplifier allows the output to be scaled by an appropriate factor.

Logarithmic and exponential elements can be used to perform multiplication or exponentiation. Integrated-circuit four-quadrant multipliers, based on the logarithmic characteristics of semiconductor junctions, are now coming into use for identifiers (27). Figure 5A shows schematically the use of the logarithmic and exponential generators to perform the exponentiation operation. The variable gain amplifier permits adjustment of the power b in the range equation. When both a logarithmic element and an exponential element are used in sequence, as shown in Figure 5A, the temperature dependence of the output may be shown (13) to cancel out, resulting in a relatively drift-free function generator. The experimentally measured performance of a function generator based on these principles is shown in Figure 6.

Figure 5B shows one arrangement that can be used in identifiers. It has the virtue of requiring only a single function generator, thereby avoiding

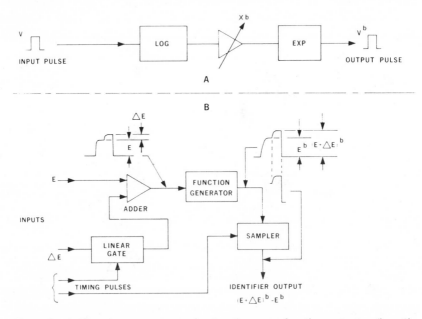

Figure 5 A: Function generator to develop the power function output $\propto$ (input)^b. B: Schematic of the system used in an identifier based on the range algorithm.

relative gain and zero drifts that would be present if two function generators were used. In this arrangement, the ΔE signal is mixed with the E signal after a short delay time. By measuring the height of the output step that occurs when the ΔE signal is allowed to enter the system, the required particle identifier output is produced.

These are a few examples of the circuits used in particle identifiers. This brief description omits many circuit details and does an injustice to the ingenuity employed in identifier circuits. Counting-rate performance, speed, linearity, stability, and accuracy are all important in these designs.

Detectors

Requirements on detectors and signal processing electronics for use in particle identification systems are generally similar to those imposed on detectors for nuclear spectroscopy. In the case of ΔE detectors, however, in addition to requiring good energy resolution, it is necessary to have near-zero dead layers at both the particle entrance and exit windows. Furthermore, the required ΔE detector thicknesses range from about 1 g/cm^2 to 0.1 mg/cm^2 depending on the types of ions being measured. The very thin detectors required for low-energy and/or heavy-ion measurements present serious problems in construction and handling.

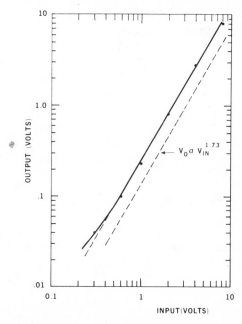

Figure 6   Measured response of a function generator of the type shown in Figure 5A.

Many early identifier experiments employed plastic or inorganic scintillation detectors. Because the requirement for thin windows in ΔE detectors eliminated the use of materials such as NaI that require protection from the atmosphere, CsI was used in some experiments. However, the rather poor energy-resolution capabilities of all scintillation detectors, and the nonlinearity of output signal as a function of the absorbed energy (particularly for the heavier ions), seriously limited the identification capabilities of scintillation detector telescopes. Furthermore, efficient light collection from thin ΔE scintillator foils presented serious difficulties. Despite these drawbacks, scintillation detectors were used in most of the early particle identification experiments and have been used in fission experiments quite recently (28).

The excellent energy resolution, linearity, and relative ease of fabrication of silicon detectors in the interesting range of thicknesses has improved the capability of particle identifiers to resolve adjacent isotopes up to the $Z = 10$ range. However, for low-energy heavy ions, even the thinnest available silicon ΔE detectors (5 μm) absorb too much energy; in these cases, gas ionization ΔE detectors have been used together with a silicon E detector (29, 30). At the present time, silicon is the dominant detector material for identifiers, but the increasing emphasis on heavy-ion physics is causing a revival of interest in gaseous detectors.

The interested reader is referred to one of the many treatises on semiconductor detectors (31–39). Here we emphasize those parts of the topic related to particle identification. Figures 7 and 8, which show the range of various ions in silicon, illustrate the wide variety of detector thicknesses

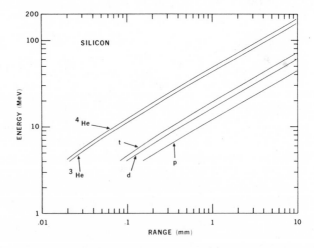

Figure 7 The range of hydrogen and helium ions in silicon.

required for particle identification in typical nuclear reaction experiments. E detectors for use with light ions having energies of up to 50 MeV/amu must have thicknesses as large as 1 cm, while those ΔE detectors intended to absorb only a fraction of the energy of heavy ions of 1 MeV/amu energy must have thicknesses well below 10 μm.

Thick detectors require very pure silicon, or the compensation of impurities by lithium-drifting. Therefore, either surface barriers on very high resistivity silicon or lithium-drifted silicon detectors may be used. In either case, 5 mm represents a practical limit to the thickness of silicon detectors. If thicker detectors are required, germanium (either high-purity or lithium-drifted) must be employed. These detectors require cooling near to liquid nitrogen temperature with the attendant problem of providing thin entry windows to the cryostat. Also, as cooled detectors require a very clean vacuum, it is usually necessary to provide a barrier between the scattering chamber and the detector. Thick silicon detectors may also require cooling to improve hole and electron mobilities and thereby to reduce the collection time and the signal rise-time.

At the other end of the range of detector thicknesses, very thin detectors are difficult to fabricate and to handle. Minor thickness variations across the detector area cause position-dependent variations in the

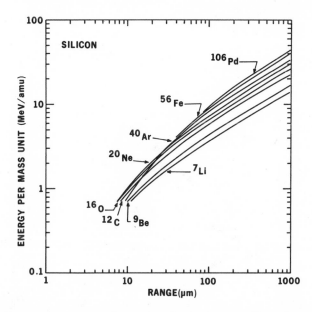

Figure 8   Range-energy curves for various heavy ions in silicon (note that the energy scale is in MeV amu^{-1}.

ΔE signals, resulting in poor particle identification. The most common fabrication techniques, which require etching of silicon wafers from both sides, make it difficult to provide detectors whose thickness is uniform to better than 0.5 μm over an area of 0.25 cm^2. In the future, epitaxial-growth and preferential etching (40, 41) may provide better detector thickness control.

Thin detectors exhibit large electrical capacities that limit the electronic resolution of the detector and its associated preamplifier. This fact, together with the problems of handling large-area, thin silicon wafers, causes a severe constraint on the area (and therefore efficiency) of a detector telescope. Consequently, it is common to use two or more independent telescopes to improve overall efficiency.

Another important aspect of detector performance, particularly for heavy-ion measurements, is the thickness of dead layers on either side of transmission detectors, or on the particle-entry side of E detectors.

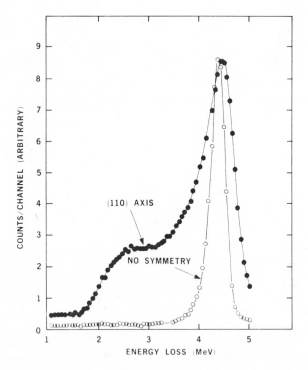

Figure 9   Energy lost by monoenergetic 40-MeV α particles passing through a 100-μm silicon crystal in two directions. (See Reference 46.)

Low-resistance evaporated-metal layer contacts (typically > 200 Å of Au) are essential where fast timing is required, and these constitute part of the dead layer. The total detector dead layer also includes a thin surface layer of silicon in which incomplete charge collection occurs. In diffused detectors the heavily-doped diffused layer partially acts as such a dead layer—for example, a 950°C, 30 min phosphorus diffusion in silicon produces an effective dead layer of about 0.3 μm of silicon. In surface barrier detectors the silicon dead layer is controlled by poorly understood parameters, such as recombination effects at the silicon-metal interface. A recent paper (42) details results on several types of surface-barrier detectors.

The whole question of dead layers is confused by the fact that results of measurement of dead-layer thickness depend upon the type of ion measured and on the applied detector voltage (43–45). Increasing the bias causes the plasma column of holes and electrons to erode more quickly, thereby reducing the probability of recombination in the dense plasma. Because the plasma density increases as dE/dx increases (i.e. for low-energy heavy ions), a detector capable of withstanding a bias voltage much greater than that required to accomplish full depletion should be used for heavy-ion identification. Also, in many surface-barrier detectors, a field inversion occurs just beneath the surface barrier; if sufficient bias can be applied, the resulting dead layer vanishes. Fields of about 50,000 V/cm just beneath the surface appear to be required to achieve this result.

Ion-channeling (46–48) causes anomalously low energy losses for those ions which become "channeled." The probability of ions following preferred channels increases for heavy slow-moving ions, and as the detector thickness is decreased, so this phenomenon can be important in heavy-ion experiments. Figure 9 shows the distribution of energy losses for 40 MeV α particles passing through a 100 μm silicon detector in two directions. The normal distribution of signals is produced when the beam is oriented essentially at random with respect to major axes of crystal symmetry, but when the beam is parallel to the $\langle 110 \rangle$ axis, a skewed distribution with many small signals is observed. The effect of channeling is illustrated by the identifier output spectrum seen in Figure 10 where the valley between the ^{3}He and ^{4}He peaks is filled in when the detector is cut normal to the $\langle 111 \rangle$ axis. It is therefore prudent to employ ΔE detectors made from silicon slices cut off-axis and to test for the effects of channeling.

Since the literature contains many discussions of signal processing electronics (34, 49–51) we do not dwell on this topic. Fast timing aspects of the electronics are discussed in Section 3.

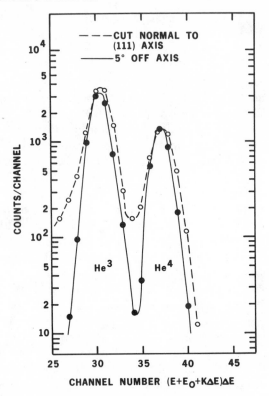

Figure 10   ³He and ⁴He identifier output distributions for detectors cut normal to the ⟨111⟩ axis and at 5° with respect to the previous cut. (See Reference 46.)

Resolving Power of ΔE,E Identifiers

While the median value of an identifier output should be constant and independent of energy for a given type of ion, we must also be concerned with fluctuations and errors in the measurement of ΔE and E and their effect on identification. These errors limit the ability of the system to resolve one type of ion from others having nearly the same value of MZ^2.

Figure 11 illustrates the accuracy required in the MZ^2 determination to unambiguously identify the known stable isotopes with $Z < 10$. In this figure, where a certain fractional error in determining MZ^2 represents a fixed vertical error, we note the existence of overlapping isotope series—for example, the MZ^2 range for known B isotopes (shown by dots) overlaps the range of Be and C isotopes. Also, note that the

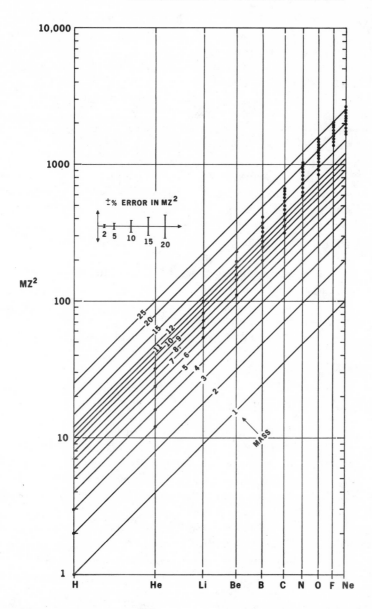

Figure 11 Plot of MZ^2 for ions of $Z \leq 10$ known to be stable against particle break-up. A fixed percentage error in determining MZ^2 constitutes a fixed vertical error in this plot.

percentage separation of the MZ^2 values of adjacent isotopes decreases for high Z so that, for the F and Ne isotopes (i.e. $MZ^2 \approx 1700$), the separation between adjacent stable isotopes is only 1% in MZ^2. This compares with 25% separation for the stable He isotopes. Therefore identification of heavy isotopes by measurement of MZ^2 alone demands very small fluctuations (or errors) in determination of ΔE and E.

The wide range of types and energy of particles makes a general statement of the measurement uncertainties impossible. The various factors responsible for errors or fluctuations in the ΔE and E signals include the following:

(a) Electronic noise causes a fluctuating signal whose effect on energy resolution depends on the pulse-shaping times that are employed. Speeding up the signal-processing system degrades the resolution almost in inverse proportion to the measurement time. The high capacitance of the ΔE detector generally makes the resolution of the ΔE system dominant over that of the E system. Fortunately, electronic noise is a major limitation in particle identifier experiments only in those cases where very little energy is deposited in the ΔE detector.

(b) Detector charge production statistics (52–54) result in a spread in the signals even when particles deposit a fixed amount of energy. The FWHM spread introduced by this effect is given by

$$\langle E \rangle = 2.35(FE\varepsilon)^{1/2}, \qquad\qquad 18.$$

where E is the energy deposited, ε is the average energy required to produce a hole-electron pair in the detector material ($\varepsilon \approx 3.7$ eV for silicon), and F is the Fano factor ($F \approx 0.12$ for silicon). Although the absolute value of the spread increases with absorbed energy, the fractional spread decreases as the energy increases. For this reason, this factor is rarely a serious limitation in identifier experiments where substantial amounts of energy are deposited in detectors. For example, the spread in a 20 MeV signal is only about 6 keV (FWHM) or 0.03%.

(c) Channeling effects in the ΔE detector may cause a significant fraction of the incident ions to deposit less than the normal amount of energy. This phenomenon must always be considered a potential source of fluctuations in ΔE signals, but can be largely avoided, as discussed in the previous section.

(d) Fluctuations in the charge state of heavy ions passing through a ΔE detector constitute a source of fluctuation in the energy loss in this detector. These fluctuations are zero for very high-velocity ions, which are fully stripped, but probably become important at energies where ions are only partly stripped, a typical situation in heavy-ion experiments. Although the average charge state of heavy ions has been studied (55,

56), very little work on the effect of fluctuations in charge state on ionization in detectors is available.

(e) Energy exchanges between the ion and electrons in the detector material occur as discrete events, and statistical fluctuations in the number of these events and in the nature of the energy-exchange process cause a spread in absorption between ions passing through a ΔE detector. If the number of exchanges along a track is large, details of the individual events are insignificant and an average value can be assumed for the interactions. In this regime, a Gaussian distribution of the energy losses results. However, when a detector is very thin and the incident ions are lightly ionizing, very few interactions take place and fluctuations in the individual interactions become dominant. Rare high-energy exchanges cause a skewed distribution of losses with a high-energy tail that is identified with the Landau (57) collision regime. Between these two extremes, a complex mixture of the two statistical processes occurs that can be analyzed using the approximate method of Symon (58) or the more exact approach of Vavilov (59). Experimental results (60, 61) agree closely with predictions of the Vavilov theory.

The spread in energy absorption can be calculated using tables (4, 19, 62) based on the Vavilov theory. When the number of interactions in the ΔE detector is large and a Gaussian distribution occurs (a typical case in identifiers), the width of the distribution can be calculated using Bohr's theory (63). According to this theory, a very thin absorber of thickness dx will introduce a spread given by

$$d(\sigma^2) = 4\pi e^4 q_{\text{eff}}^2 n \, dx \qquad\qquad 19.$$

(using the same nomenclature as in equation 5). Combining this with equation 5, and assuming $\beta = 0$, it can be expressed in the form

$$d(\sigma^2) \approx 2(m/ML)E \, dE, \qquad\qquad 20.$$

where $L = \ln(4mE/IM)$, which is a slowly varying function of E, dE is the energy loss in element dx, and m is the mass of the electron. Assuming that the total spread in energy absorption in a ΔE detector is much smaller than the average energy loss ΔE, equation 20 can be approximately integrated to yield a value for the spread

$$\sigma^2 = (m/ML)\{(E_0 + \Delta E)^2 - E_0^2\}, \qquad\qquad 21.$$

where E_0 is the energy of the ion on exit from the ΔE detector. If $\Delta E \ll E$, so that $E_0 \approx E$, this equation simplifies to

$$\sigma^2 = 2(m/ML)\Delta E E. \qquad\qquad 22.$$

For example, consider a 30 MeV α-particle beam losing an average of

3 MeV in a silicon ΔE detector. In this case, $L \approx 5$ and $m/M \approx 8000$; thus $\sigma \approx 70$ keV, and the FWHM spread will be almost 160 keV. This is a 5% spread in the ΔE measurement, clearly very significant. Fortunately, the fractional error decreases as M increases and as the energy loss ΔE increases. These fluctuations therefore become a serious problem only when thin ΔE detectors are used to identify lightly ionizing particles (4, 19).

(f) Thickness variations in the ΔE detector cause fluctuations in the ΔE and E signals. Fabrication procedures for ΔE detectors tend to produce a fixed range of thickness variations (~ 0.5 μm), so the resulting fractional ΔE spread decreases as the ΔE detector thickness increases. This problem becomes very important in experiments where heavy ions are identified with very thin ΔE detectors.

(g) The errors introduced by the effects discussed so far are primarily in ΔE signals. However, nuclear collisions occur especially near the end of a particle's track where the ion becomes neutral. Fluctuations in these collisions therefore mainly cause a spread in E signals. The FWHM spread due to this effect (64) is approximately given by

$$\langle E \rangle = 0.7 Z^{1/2} A^{4/3} \text{ keV.} \qquad 23.$$

This spread amounts to only 0.7 keV FWHM for protons, but it becomes quite large for heavy ions. When $Z \approx 10$, the contribution is well over 100 keV—by no means insignificant for those ions that just penetrate the ΔE detector and deposit very little energy in the E detector. This is common in heavy-ion experiments, since it is difficult to fabricate the adequately thin ΔE detectors.

(h) We have so far discussed fluctuations in signals from the ΔE and E detectors. Identification can also be influenced by nonlinearities in the relationship between the energy absorbed and the signal produced by a detector. The last two effects discussed here are of this nature. The first such effect is usually termed the *pulse-height defect* (45, 65–68), which is variously attributed to nuclear collisions near the end of the track of a heavy ion or to recombination losses in the dense plasma column existing near the end of the track. Its magnitude is difficult to assess since the range of ions encompasses cases in which the defect is negligible to cases in which it is very significant. Since the loss is associated with the end of a particle's track it is clear that the E signal becomes smaller than its anticipated value. Since this would cause the identifier to calculate a low value of MZ^2, a lower bound should be set on the E signals processed by the identifier.

(i) A further loss in the E signal results from dead layers on the back of the ΔE and on the front of the E detector. Since this energy loss is large for ions that barely penetrate through the ΔE detector, but is negligible for the longer-range particles, an energy dependence is caused in the identifier output. This is a serious effect for the heavier ions, and makes it essential to orient the two detectors to produce the thinnest possible total dead layer between them. The total can thereby generally be made less than 0.5 μm (silicon equivalent).

Multiple Detector Systems

In studies of relatively rare isotopic species accompanied by large numbers of more common particles, these identification errors become intolerable. Fortunately, the absolute cross section for a reaction is usually not important; rather, identification is required to select a meaningful fraction of the rare particles so that their energy distribution can be determined. A good example is the detection and, sometimes, mass determination of particle-stable neutron-rich isotopes near the boundary of stability. These experiments led to the first multidetector identifier system (14, 69). By allowing the ions to pass through a series of detectors, several simultaneous calculations can be made of the particle's identity (i.e. Mq_{eff}^2). Each individual calculation is subject to the errors discussed in the

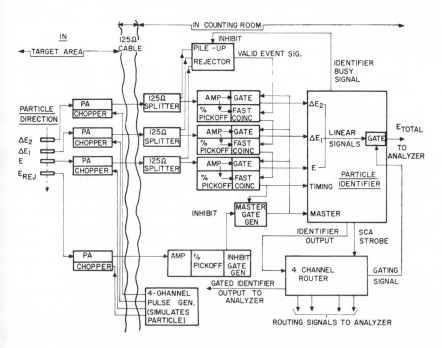

Figure 12   Block diagram of the triple-detector identifier system.

previous section, but a criterion is also placed on the agreement between the various answers before a particle is accepted for energy analysis. The rejection of all particles that deviate significantly from agreement between identifications reduces essentially to zero the chance of a neighboring isotope being falsely identified as the rare product.

A block diagram of one system commonly employed is shown in Figure 12. Here the detector thicknesses are chosen so that the particles of interest pass through two ΔE detectors, ΔE_2 and ΔE_1, then into the E detector. An E_{rej} detector behind the E detector permits rejection of all particles that pass through the E detector. For this to be efficient the dead layer on the rear of the E detector should be very small. As in the simpler identifier (Figures 1 and 5B), fast coincidence requirements and energy windows are set on all signals to reduce background to a minimum.

The logarithmic function generator of Figure 5 is convenient to use in a multidetector system, as the time-share principle can easily be extended to sequencing more than one ΔE signal. For the triple-detector identifier, the three-step waveform shown in Figure 13 is produced; measuring the height of the two steps on the top of the waveform developed by the function generator then yields two separate identifications:

$$(E+\Delta E_1)^b - E^b = T_1/a \qquad\qquad 24.$$

and

$$(E+\Delta E_2+\Delta E_1)^b - (E+\Delta E_1)^b = T_2/a, \qquad\qquad 25.$$

where a and b are the constants in the range-energy relationship (equation 14) and T_1 and T_2 are the thicknesses of the two ΔE detectors. (Note: T_1 is the thickness of the second detector in the telescope.) Using a simple logarithmic element, the ratio of the two results can be determined. It is a simple matter to accept only those events for which this ratio is almost equal to the ratio (T_1/T_2) of the two ΔE detector thicknesses.

As stated earlier, computer processing can be performed as an alternative to analog identification. It is often convenient to use the analog method to select interesting events for presentation to a computer, then to have the computer make a more rigorous investigation of the various signals to further check the particle's identity. In experiments designed to discover new isotopes, precautions must be taken to eliminate chance pile-up of common types of particle from being identified as one of the rare events. The computer, presented with signal amplitude- and time-information, is invaluable in this connection.

A logical extension is to use a multidetector telescope that permits

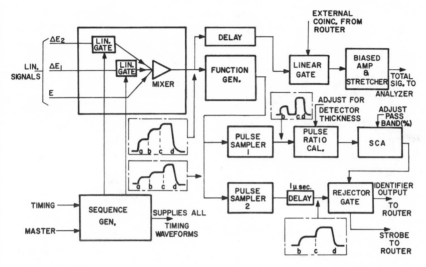

Figure 13   Diagram of the triple-detector identifier circuit showing waveforms.

recording the pattern of ionization along the track of a particle. This technique is limited to those higher-energy particles that penetrate most of the depth of a stack of detectors. Such telescopes are now being used (70) to identify high-velocity heavy ions. Because a detailed profile of ionization along a track is obtained, particles undergoing nuclear reactions in the detectors can be observed and rejected. This permits thick detector telescopes (e.g. high-purity germanium) to be used for high-energy light-ion experiments.

Experimental Results

Simple $\Delta E, E$ identifier telescopes have been used extensively since 1960, particularly in nuclear reaction experiments involving light ions. The first experiments using scintillation detectors and the multiplier algorithm achieved adequate identification of protons, deuterons, and tritons, but separation of ^{3}He from ^{4}He was only marginally possible and then only if the yield of ^{3}He was comparable to that of ^{4}He.

The improvement in identification resulting from the use of semiconductor detectors was quickly realized and exploited. The multiplier algorithm was used exclusively in these early experiments. Unfortunately, fabrication of very thin ΔE detectors was not then practical, so work on heavy ions required the use of other types of ΔE detectors. The first heavy-ion identification studies were carried out using a gas ionization detector for the ΔE detector with a silicon E detector. One result from these experiments (9) is shown in Figure 14.

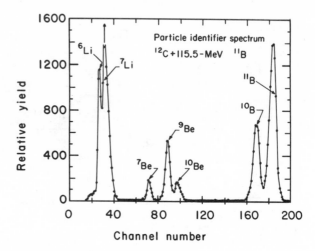

Figure 14   Heavy-ion particle identifier spectrum using $\Delta E, E$ multiplier algorithm. This experiment used a gas proportional ΔE detector (Reference 9).

The availability of thin silicon ΔE detectors quickly led to their use in identifier experiments. The first use of the range (power law) algorithm that imposed less restriction on the thickness of the ΔE detector resulted in a significant improvement in identifier performance. An early result achieved with this system is shown in Figure 15. As this system was applied to the search for rare neutron-rich isotopes, the need for better identifier resolution and lower backgrounds became evident; this led to the use of the triple-detector telescope. Results achieved in a specific reaction using this telescope as compared with those from a simple $\Delta E, E$ telescope are shown (14) in Figure 16.

This type of identifier has been used extensively for studies of neutron-rich isotopes near the boundary of stability. One example is the measurement of the mass of ^{8}He. Figure 17 shows the identifier output in an experiment in which 80 MeV α particles bombarded a ^{26}Mg target. This figure illustrates the large range of yields for the various isotopes; for example, a single ^{8}He particle was accompanied by 10^8–10^9 other particles. The small ^{8}He peak also includes some α-d coincident events in the telescope that were largely eliminated by detailed examination of the signals associated with each event. The final ^{8}He energy spectrum shown in Figure 18 contains ~ 25 events accumulated in several days of cyclotron operation.

Although experiments continue to use detector telescopes in this manner, the addition of time-of-flight measurement discussed in the following section has made studies possible on even rarer isotopes.

3   TIME-OF-FLIGHT IDENTIFIERS

General Considerations

A time-of-flight (TOF) measurement determines the velocity of a particle and hence the ratio E/M. If a separate E measurement is made, the mass of a particle can be determined. Sometimes this is adequate identification. On the other hand, when combined with a $\Delta E,E$ identifier, which determines both E and Mq^2_{eff}/E, TOF provides a complete determination of M, E, and q^2_{eff}.

Unfortunately, the basic simplicity of TOF methods is not matched by the hardware required to achieve the required timing performance. As long flight paths necessarily involve serious efficiency problems because

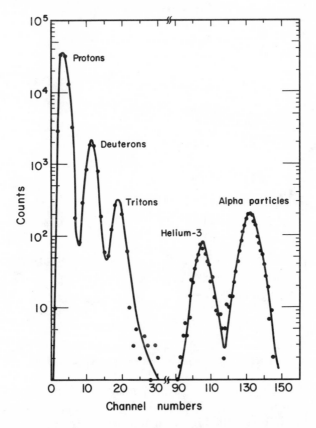

Figure 15   Early identifier spectrum using a silicon $\Delta E, E$ detector telescope and the range (power law) algorithm.

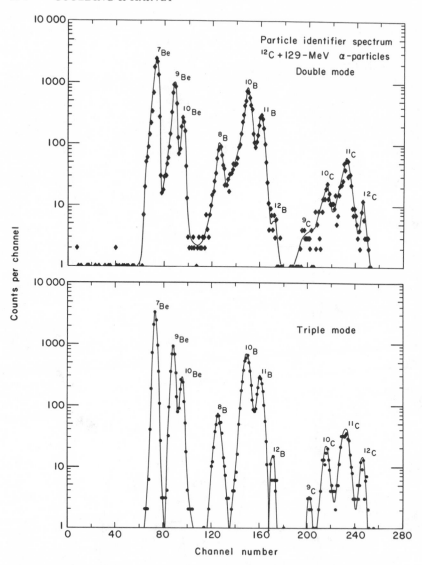

Figure 16 Comparison between the particle identifier spectrum achieved using the simple $\Delta E, E$ telescope and a triple-detector telescope and identifier.

of the poor collection geometry, short paths are desirable, and very fast timing circuits must be used. The velocity of an ion as a function of E/M is shown in Figure 19. It is given approximately by

$$v = 1.4(E/M)^{1/2} \text{ cm ns}^{-1}, \qquad\qquad 26.$$

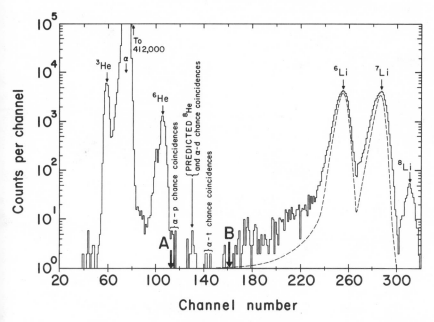

Figure 17 Identifier spectrum obtained in an experiment to measure the mass of ^{8}He. The reaction studied was ^{26}Mg(α, ^{8}He)^{22}Mg at an α-particle energy of 80 MeV.

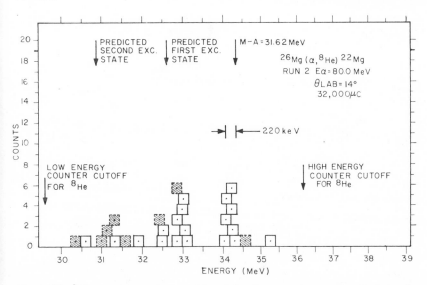

Figure 18 ^{8}He energy spectrum achieved in the experiment described in the caption to Figure 17.

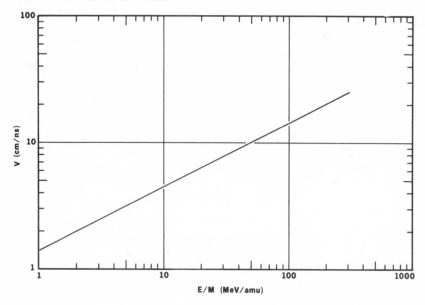

Figure 19   Plot of the velocity of ions as a function of E/M.

where 10 MeV/amu corresponds to about 4.5 cm ns^{-1}, and therefore a 2-ns flight time over a 10-cm flight path. If the accuracy of the time measurement is 100 ps—close to the best result yet achieved—the velocity measurement is accurate only to 5%, and the error in E/M is 10%. Achieving the required timing accuracy has been the major problem in applying TOF methods to identification of high-velocity ions.

It is convenient to rearrange equation 26 in terms of the mass:

$$M = 2Et^2/d^2,$$ 27.

where t is the flight time in nanoseconds and d is the distance in centimeters. If small statistical fluctuations δE, δt, and δd occur in measuring E, t, and d, the resulting fluctuation (δM) in mass determination is given by

$$(\delta M/M)^2 = (\delta E/E)+(2\delta t/t)^2+(2\delta d/d)^2,$$ 27.

where $\delta E/E$ is much less than 1% in most experiments, and $\delta d/d$ is usually very small. Therefore $\delta t/t$ is commonly the most important measurement error. In this case, we have

$$\delta M/M = 2.8(E/M)^{1/2}\,\delta t/d.$$ 29.

This is shown in Figure 20. We see that ^{16}O ions with an E/M of

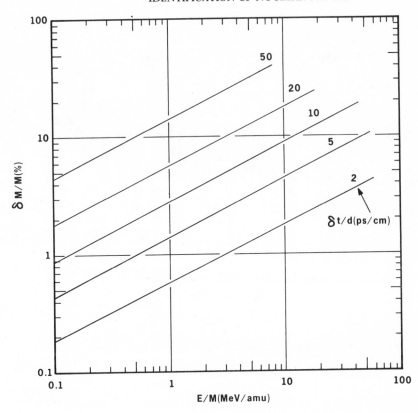

Figure 20   Plot of the mass resolution as a function of E/M for various resolutions (Δt) in the TOF measurement; d is the length of the flight path.

6 MeV/amu require a timing resolution of 9 ps cm^{-1} of flight path if they are to be resolved from other isotopes of mass 17 (i.e. $\delta M/M = 6\%$ or $\delta M \approx 1$). The flight paths used in actual experiments range from a few centimeters to a meter or more depending on the timing accuracy of the detectors used and on the accuracy required in determination of M.

Combining a TOF measurement with a $\Delta E,E$ particle identification produces a very useful two-dimensional result that is more tolerant of fluctuations in both the mass and MZ^2 determinations than is a single-parameter experiment. One representation of the two-dimensional data provided by a TOF and $\Delta E,E$ particle identifier is given in Figure 21. Allowing reasonable spreads both in the particle identifier output and in the mass determination due to timing errors, and assuming that all

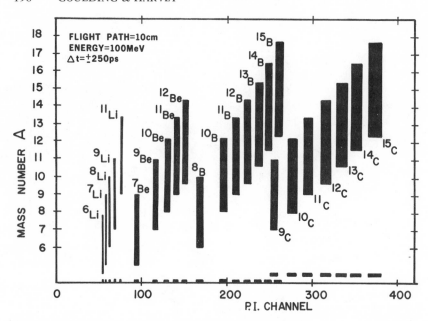

Figure 21   Two dimensional picture of MZ^2 particle identifier output combined with mass determination by measurement of time of flight. A 10-cm flight path, 100-MeV ions, and 250-ps timing accuracy are assumed.

particles have 100 MeV energy, the shaded regions, representing the errors in each determination, are well separated. Note that the series of carbon and boron isotopes overlap in the particle identifier dimension, so that the resolution of ^{9}C from ^{14}B and ^{15}B depends on the mass-identifying capability of the TOF measurement. Here we see very clearly the power of the combined system, for those events poorly resolved by one system are well resolved by the other.

Achieving the necessary time resolution for these experiments has taxed the limitations of both detectors and electronic circuits. A TOF measurement can be particularly useful for identifying low-velocity heavy ions for which very thin timing (ΔE) detectors must be used. They are of three basic types:

(*a*) Thin silicon detectors, which also provide a reasonably good ΔE signal but are relatively thick (> 1 mg cm^{-2}).

(*b*) Thin scintillator foils and photomultipliers that provide only a very crude ΔE signal but can be reasonably thin (~ 50–500 μg cm^{-2}).

(*c*) Thin secondary electron foils that emit electrons detected by either electron multipliers or scintillator-photomultiplier combinations. These foils provide almost no ΔE information but can be extremely thin (10–50 μg cm^{-2}). The thinner foils are naturally rather fragile.

In the first of these cases, the small detector signals must be amplified, and timing limitations are caused by the charge collection time in the detector, by the rise-time of the amplifier pulse, and by noise in the amplifier. In the latter two cases, the multiplier structures provide "noiseless" amplification. The dominant sources of time jitter are therefore statistics of emission from the foil and from the front end of the multiplier, and also the spread in transit time of electrons through the multiplier structure. The factors that limit time resolution are therefore quite different in the last two cases from those in the first. Once signals of suitable size are realized, all systems can use the same timing channels. In the case of semiconductor detectors, where energy-loss information is also derived, a slow signal-processing channel, designed to optimize pulse-amplitude measurements, parallels the fast timing channel.

We now briefly discuss the basic features of timing circuits before dealing separately with the various types of detector and front-end electronics. A fast timing channel includes the following items:

(a) A fast amplifier that exhibits the minimum possible rise-time and is capable of developing an output pulse in the 1 V range. Using the best high-frequency transistors, rise-times (10–90%) of about 1 ns are achievable in the amplifier. The shape of the rise of the output pulse is the result of multiple integrators in the amplifier causing a Gaussian— rather than exponentially—shaped rise. The design of fast amplifiers is discussed by Jackson (71).

(b) Pulse-shaping circuits used to shape signal pulses for optimum processing. The shaping includes differentiation to reduce pile-up probabilities; special operations, such as the bipolar pulse-shaping required by constant-fraction discriminators, may also be performed. Where charge collection in the detector dominates the overall signal rise-time, integration in the pulse shaper reduces the effect of noise produced in the input stages of the amplifier. This is usually not the case in the fast timing systems described in this paper.

(c) A fast discriminator picks off the best possible timing information from the signal pulse at the output of the amplifier. Fast discriminators may be designed to trigger at a fixed amplitude on the leading edge of signals (so-called leading-edge discriminators), or at a time related to a point on the leading edge where the signal reaches a fixed fraction of its final amplitude (constant-fraction discriminators or CFD). The latter type of discriminator, in principle, develops a signal whose timing is independent of the amplitude of the signal. Since particle identification experiments always involve widely varying signal amplitudes, CFDs are used.

Their design is discussed in many papers (72–76). Generally speaking, they use an input pulse shaped by double-differentiation to produce a zero-crossing point. The discriminator then triggers on the front edge and, when it retriggers on the zero-crossing point of the input signal, a timing signal is produced whose time is independent of signal amplitude. Delay-line differentiation is used to produce an asymmetrical bipolar signal with the negative undershoot constituting a preselected fraction of the initial positive portion of the waveform. Altering the amount of under-shoot changes the effective fraction of the signal rise at which the timing signal is developed. This must not be set at too low a value since triggering will then occur on the rounded portion of a signal where noise causes a large spread in the triggering point.

A useful feature of a constant-fraction discriminator is that the output pulse width (i.e. leading edge triggering to cross-over) is changed by pile-up pulses occurring within the width. Thus, pulse-width discrimination can be used to detect pile-up on a short time scale. This is an important supplement to slower conventional pile-up rejectors. It can be particularly important in those accelerator experiments where the beam is concentrated in intense short beam pulses that are separated by long intervals.

(d) A time-to-amplitude converter (TAC) is used to convert the time interval between timing pulses in two timing channels into a pulse amplitude linearly related to the time interval.

Measuring the velocity of a particle requires determination of its time of passage through two points. Sometimes the incident beam on a target is pulsed so that the time when secondary particles can leave the target is known. Detection of an ion a short distance away from the target is then adequate to determine its velocity. In some experiments, other radiation emitted from the target at the same time as the particle of interest can be detected and used to provide the fiducial signal. More generally it is necessary to interpose two or more detectors in the path of the particle, with at least the first detector thin enough for the particle to pass through it. Therefore, a wide variety of combinations of detectors is encountered in TOF identifiers. Some of these combinations are discussed in the following section.

Thin Scintillators

A number of plastic scintillators with decay times in the 1 ns time region are commercially available in the form of thin foils down to about 5 μm thickness (i.e. 500 μg cm^{-2}), and can be produced in the laboratory (77) down to about 50 μg cm^{-2}. These foils produce sufficient light output when densely ionizing particles pass through them to provide timing signals from photomultipliers observing the light. Two methods of coupling to the photomultiplier have been used. In the first (78), the foil is clamped between the two halves of a split lucite light pipe that couples to one or two phototubes. The second basic arrangement (79, 80) uses a hemispherical mirror to direct the light from a thin scintillator foil onto a phototube.

The time resolution of such detectors is limited partly by the electron emission statistics from the photocathode (which are largely due to the poor light emission and collection), and by the electron transit time spread through the multiplier structure. The time resolution that can be achieved is in the range of about 0.4–1 ns, becoming worse as the foil is made thinner.

Electron-Emitting Thin Foils

The emission of secondary electrons (δ rays) from solid surfaces when charged particles enter them is a well known phenomenon. Consequently electron-emitting thin foils interposed in the path of nuclear particles have long been considered as potentially fast detectors. Detection of the secondary electrons with the required speed is a nontrivial problem, but this method is now realizing its promise.

The secondary electron yield from a foil depends on the type of foil and the work function at its surface, and on the type, velocity and angle of incidence of the incident ion. Although low work-function surface coatings increase the electron yield, difficulties in preserving these surfaces have led to the use of uncoated plastic or carbon foils as thin as about 10 μg cm^{-2}. Electron yields from these foils range from approximately 10 for natural α particles (81) to about 100 for fission fragments. Much of the early work using these foils was in the field of fission studies (82–84), but it has recently been extended to nuclear reaction product analysis (81, 85–87).

Several types of electron detectors have been used to detect the secondary electrons. Early systems employed scintillation or semi-conductor detectors. These are insensitive to electrons having energies below 5–10 keV, so secondary electrons from the foil must be accelerated to this potential. However, positive ions in the detector region are also accelerated, striking surfaces in the vicinity and releasing electrons that are then attracted to the detector to produce spurious signals. Very careful design is therefore required to avoid very high background counting rates with this type of detector.

A better situation prevails when the secondary electrons from the foil are accelerated directly onto an electron multiplier structure. Only a low accelerating voltage (<1 kV) is required to achieve the full secondary emission ratio from the initial multiplying stage, and background is a much less serious problem. The first work using open electron-multiplier structures (77) employed a multiplier with Cu-Be-(BeO) dynodes (e.g. 56P17-2) that exhibit a time spread near to 1 ns. Besides this limitation, open-ended electron multipliers are sensitive to contamination that degrades their gain, and are affected by magnetic fields. More recently channel electron multipliers have been employed (86). Time resolutions in the 400–700 ps range were obtained using these devices, which are also much less sensitive to contamination than the conventional multiplier surfaces. Finally, microchannel plates have now been used as the electron detector in particle-timing experiments (87). These plates, about 1 mm thick and up to 3 inches in diameter, contain closely spaced micro-channels only 50 μm in diameter in which electron multiplication occurs. Because of the short distance traveled by the electron cloud advancing down a channel, only a very small time spread (< 100 ps) is introduced by the electron-multiplying process. Furthermore, the plates are rugged and rather insensitive to contamination problems. The electron gain of a single channel plate is limited by ion feedback effects to about 10^4, but higher gains are realized by using two plates in series, one with holes biased at a small angle. This *chevron* plate provides an electron gain of about 10^7 with a time spread below 100 ps. Channel plates actually image the impact point of electrons on their front face; if parallel field geometry is retained in the acceleration structure from the foil to the multiplier, the position of bursts of electrons emitted from the multiplier output side directly reflects the point of passage of the detected particle through the foil.

Figure 22A shows the arrangement employed in a foil-channel plate fast-timing detector. The use of plane-parallel field geometry reduces the electron transit time-spread and preserves the imaging capability of the detector. Position information can be obtained by splitting the anode to give several separate signal outputs or by replacing the final anode (collector) by a position-sensitive semiconductor detector as shown in Figure 22B. In this version it is necessary to accelerate electrons from the channel-plate onto the detector; also the inherent electron gain in the silicon detector permits the use of a single plate instead of the chevron plate, as in Figure 22A. In both arrangements, transit time-spreads in the electron paths between the foil and the front surface of the channel plate are reduced by using an open grid about 2 mm from the foil to accelerate the electrons very rapidly after their emission. A detector of the type shown in Figure 22A has exhibited a total timing spread (FWHM) below 150 ps for natural α particles passing through the foil. The combination of excellent timing and the small energy loss incurred by particles passing through 10 μg cm^{-2} carbon foils makes this type of detector a very

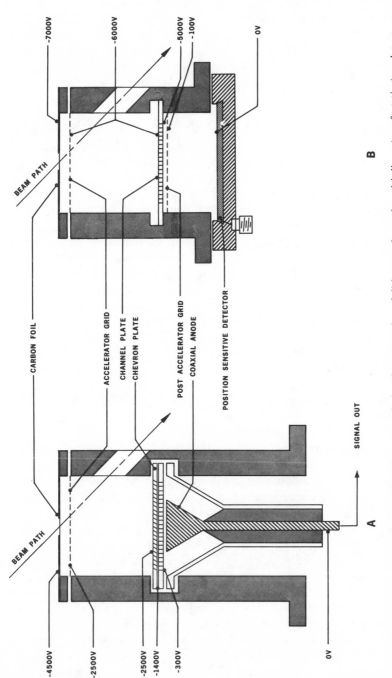

Figure 22   Diagram of a detector using electron emission from a carbon foil into a channel multiplier. *A*: a fast timing detector using a chevron plate. *B*: a system using a position-sensitive detector to provide an image of the emission pattern from the foil.

important new tool in mass identifiers. The imaging capability also offers the new potential of plotting the trajectory of particles entering a spectrometer magnet—in itself suggesting a simplification in the design of such spectrometers, as discussed toward the end of this paper.

Thin Silicon Detectors

Thin silicon detectors are commonly used to provide ΔE signals for use in $\Delta E,E$ identifier systems that determine Mq_{eff}^2. A natural extension of the technique is to separate the ΔE and E detectors by a suitable distance and to use the detector signals for timing purposes as well as energy measurements. A simpler version, in which the time was measured between a cyclotron beam pulse and the signal from the ΔE detector of a detector telescope, was used by Parkinson & Bodansky (88). More recently the full technique has been used by a number of authors (89–92). Since thick detectors exhibit intolerably long collection times, the technique is restricted in practice to applications where the two timing detectors (these could be the ΔE_2 and ΔE_1 detectors of a triple-detector identifier) have thicknesses below about 1000 μm. Practical limitations (see Section 2) restrict the minimum detector thickness to about 5 μm, although some experiments have been performed (89) with detectors as thin as 1.7 μm (i.e. about 400 μg cm^{-2}).

We will assume here that the electric field through the thickness of a detector is greater than that required to achieve the saturation velocity (10^7 cm sec^{-1}) of holes and electrons. At room temperature, this demands an overvoltage of approximately 2–3 V μm^{-1} of detector thickness; the required overvoltage decreases at lower temperature, becoming 1–2 V μm^{-1} at $-40°$C. The collection time is about 10 ps μm^{-1} of detector thickness. Therefore thin silicon detectors are potentially capable of extremely fast timing. Unfortunately, this simple picture is clouded by other factors. Slow erosion of the dense plasma produced in heavy-ion tracks significantly retards the signal produced by such particles. Furthermore, thin detectors necessarily exhibit large electrical capacitances, that degrade the signal/noise ratio of the detector amplifier system, and also make the effect of any series resistance or inductance in the input circuit very serious. Despite these effects, the timing capabilities of thin silicon detectors are essentially equal to those of electron-emitting foil detectors, while they also provide accurate ΔE energy signals. Their main drawbacks are their limited minimum thickness and area (< 0.5 cm^2).

Neglecting plasma erosion times and series resistance in the detector and input circuit, the time resolution of a thin silicon detector system is limited by the effect of electronic noise in the first stage of the amplifier that follows the detector. Noise fluctuations modulate the point on the signal rise where the timing discriminator fires, causing a jitter in the triggering time. As discussed earlier, the signal rise-time is determined mainly by the fastest achievable rise-time in the pulse amplifier. Commonly this is 1–2 ns. At high frequencies, noise is largely due to the random nature of electron flow through the input amplifier element (often called shot-noise).

The equivalent input noise voltage V_n at the input element is given approximately by the equation

$$\langle V_n^2 \rangle = 4kT\Delta f/g, \qquad\qquad 30.$$

where g is the transconductance of the device and Δf is the bandwidth of the system. The signal V_s developed at the input of the stage is given by

$$V_s = Q/(C_D + C_I), \qquad\qquad 31.$$

where Q is the charge released by an event in the detector, C_D is the detector capacitance, and C_I is the input capacity of the input amplifying stage. A rise-time τ (10–90%) is associated with a bandwidth $1/\pi\tau$; with the assumption, therefore, that the system is operated at room temperature, equation 30 can be simplified to

$$V_{n(\text{rms})} = 0.7 \times 10^{-4}/(\tau g)^{1/2}\ \text{V}, \qquad\qquad 32.$$

where τ is measured in ns and g in mA V^{-1}. Equation 31 can also be expressed in more practical terms:

$$V_s = 0.044E/(C_D + C_I)\ \text{V}, \qquad\qquad 33.$$

where E is the energy deposited in the detector in MeV, and C_D and C_I are expressed in pF. Assuming a linear signal rise, the fluctuation in timing caused by noise is therefore given by

$$\Delta t_{(\text{FWHM})} = 3.7(\tau/g)^{1/2}(C_D + C_I)/E\ \text{ps}. \qquad\qquad 34.$$

If $\tau = 2$ ns, $g = 50$ mA V^{-1}, $C_D + C_I = 500$ pF and $E = 10$ MeV— typical values in the ΔE detector channel—then $\Delta t_{(\text{FWHM})} = 37$ ps. Measured timing accuracies approach this value, but detector charge collection time and series resistance effects in the detector-input circuit can easily raise the effective time resolution of a pair of detectors, in the timing of actual particles, into the range near to 100 ps or more.

Equation 34 indicates that the time resolution is determined by the ratio $(C_D + C_I)/(g)^{1/2}$—the other factors in the equation do not depend on the input-amplifying element. Clearly the best timing would be achieved if an amplifying element with a very large ratio of g/C_I were used. Since this ratio is approximately 400 mA/V/3 pF for a bipolar transistor having a cut-off frequency f_T of 2 GHz, whereas the best field-effect transistors (FET) exhibit equivalent ratios of 50 mA/V/50 pF, it appears, at first sight, that a bipolar transistor should be used as the input amplifying element. However, the bipolar transistor, operating at a current of 10 mA to achieve the forementioned g/C_I ratio, exhibits an input impedance in the range of 30 ohms, so that the voltage pulse at the input decays quickly with a time constant $(C_D+C_I)30$—i.e. 15 ns, if $C_D+C_I = 500$ pF. Such a rapid decay makes difficult the slow signal processing required for energy measurement. Therefore, a field-effect transistor (FET) is commonly employed as the input amplifying element. The ratio g/C_I for an FET is determined by the ability of manufacturers to produce very small device structures; therefore although different values of g and C_I are available, their ratio tends to be constant for state-of-the-art devices. C_D is large for thin-silicon detectors, so its value nearly always exceeds C_I in these applications. Therefore an FET having the highest practical value of g (and therefore C_I) will give the best timing performance. At the present time 50 mA V^{-1} and 50 pF are representative of suitable devices. Equation 34 predicts that the best timing would result from using an FET exhibiting $C_I = C_D$, but this condition can rarely be satisfied in practice.

Figure 23 shows the input circuit used with a fast timing silicon detector. This diagram emphasizes the requirement for very low-inductance connections around the entire detector loop. A wire 1 cm long and 0.05 cm in diameter exhibits an inductance of 0.01 μH, which

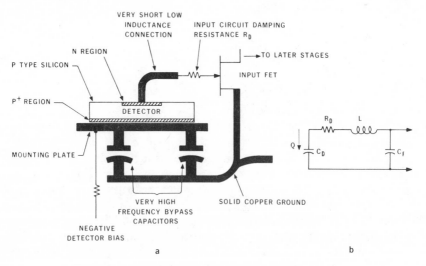

Figure 23   Input circuit of a silicon fast-timing detector.

will resonate at 250 MHz with a 300 pF FET input capacity. Clearly such a situation is intolerable if timing in the 100 ps range is to be achieved. Consequently, very short low-inductance connections should ideally be used in the detector circuit. To avoid ringing in the input circuit, a damping resistor ($\sim 10 \ \Omega$) must also be included in the circuit. A further timing limitation is imposed by the spreading resistance of the evaporated metal contact layers on the detector itself. If this resistance is not kept to a very low value (a few ohms), charge produced at different points across the detector area produces output signals that have been integrated to different degrees by the spreading resistance and detector capacitance combination. Since thick metal films constitute intolerably large dead layers on detectors used for heavy-ion measurements, a compromise must be made between their resistance and the dead layers they represent. Well-prepared gold films about 200 Å thick represent about the best compromise.

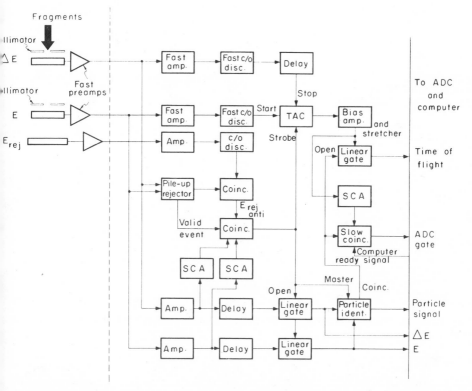

Figure 24   Block diagram of an identifier system using both MZ^2 and TOF measurements.

Typical Systems and Results

Simple TOF mass measurements have been made for many years in a broad range of experiments. These include some beam energy measurements (93), fission fragment mass distribution experiments (78), and some reaction experiments where the Z of the detected products was known or could be inferred. The more sophisticated systems, which depend on Mq_{eff}^2 particle identification using detector telescopes and on TOF measurement of mass, have been used in studies of heavy-ion transfer reactions (94, 95) and of the fragmentation products of high-energy bombardment of target materials (96–103). These products provide the

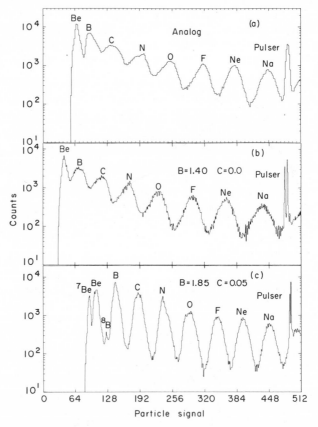

Figure 25   Identifier output spectrum for heavy ions covering a very broad energy range: (*a*) analogue calculated from $R = aE^b$ but with the value of $b \approx 1.4$. (*b*) computer processed (*b* also adjusted to 1.4). (*c*) computer processed with an energy-dependent exponent. (See equations 35, 36.)

opportunity for studies of nuclei far from the normal region of stability and permit tests of the various theoretical mass-stability relationships.

To illustrate particle identification using both the Mq_{eff}^2 and TOF methods we briefly review the results of Butler et al (20) and Bowman et al (103), who used essentially the same system but with improved data processing methods. Figure 24 shows a block diagram of the electronic system used by these authors. The particle identifier signal fed to the computer is calculated by an analog identifier performing the power-law calculation (i.e. equation 15). Events processed by the computer are selected on the basis of this signal, but a modified form of this algorithm is employed in the computer processing to make the identification less energy-dependent. The final identifier output P is given by

$$P = [\{(E+\Delta E)/k\}^n - (E/k)^n]^{1/2}, \qquad\qquad 35.$$

where k is a constant and

$$n = b - c\,\Delta E/T. \qquad\qquad 36.$$

The constants b and c are chosen empirically to realize the smallest energy dependence, and the square root of the whole expression is used to make the final output approximately proportional to Z rather than to Z^2, thereby producing a more convenient scale for output display.

The improved particle identifier behavior realized by this modification to the basic range algorithm is illustrated by the three cases shown in Figure 25. Using this method, elemental identification up to Argon

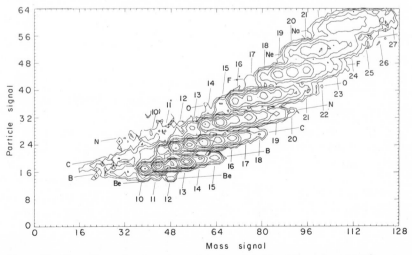

Figure 26   Two-dimensional contour plot of particle identifier output (MZ^2) versus mass (derived from TOF measurement).

($Z = 18$) has been achieved. In the experiments described by Butler et al, this was combined with a TOF measurement between a 22-μm ΔE and a 112-μm E detector. The time resolution achieved in these experiments was approximately 250 ps (FWHM). A two-dimensional contour plot of the resulting data is shown in Figure 26. In the more recent experiments by Bowman et al, amplitude-dependent time-walk effects were reduced by computer processing. By these methods the existence of several new particle-stable, neutron-rich isotopes of elements with $Z < 9$ has been demonstrated, and several other isotopes, whose stability was in question, have been shown to be unstable against particle breakup. These methods clearly provide a powerful new technique for exploring the boundary of the neutron-rich side of nuclear stability.

An approximate estimate of the isotopes that can be identified in simple $\Delta E, E$ systems can be made with the aid of Figure 27, taken from Butler et al. Depending on the type of ion, different values are assigned to the minimum energies considered essential in the E detector—these values are indicated on the horizontal axis. The curves then show the energy of those particles that pass through various ΔE detector thicknesses with enough energy remaining to satisfy this condition. The dotted line shows the boundary of the range of isotopes practically resolvable at the present time by $\Delta E, E$ detector identifiers without the help of TOF measurement. As explained earlier the two-dimensional picture achieved by adding TOF information extends the range of identifiable isotopes by separating the particle identifier spectra of adjacent elemental isotope series that overlap.

4   SYSTEMS INVOLVING MAGNETIC RIGIDITY

Fundamentals

The magnetic rigidity $B\rho$ is related by equation 4 to the mass, velocity, and charge state q of the particle. If the residual gas pressure in a spectrometer is sufficiently low ($\lesssim 10^{-5}$ Torr) the probability that a particle will undergo a charge-changing collision is small. The value of q as well as the nuclear charge Z and mass M (for nonrelativistic particles) is then restricted to integral values so that some of the measurements required for identification need only be made with sufficient precision to resolve adjacent integers. A precise measurement of $B\rho$ then yields the kinetic energy E with equal precision by setting M and q equal to their exact values and using

$$E = (B\rho)^2 q^2 / (2M).$$ 37.

This equation is, of course, nonrelativistic.

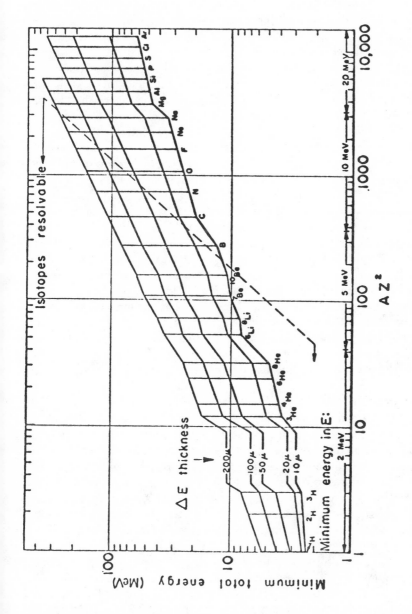

Figure 27   This figure indicates practical limits of MZ^2 identification using silicon detector telescopes.

If the spectrometer (deliberately or by accident) contains enough gas for a large number of charge-changing collisions to occur, $B\rho$ will be determined by the *average* charge state $\bar{q}$ of the ion, and of course $\bar{q}$ need not be integral. At intermediate pressure ($\sim 10^{-3}$–10^{-4} Torr), a few ions suffer a single charge-changing collision. The detection system then identifies most ions as having discrete integral charges, but there is a continuous background between one charge state and the next, as the position of a particle in the focal plane depends on where the charge-changing collision occurred in the spectrometer field.

No measurable quantity gives a direct determination of Z, but the energy lost in a suitable focal plane detector measures a quantity q_{eff}^2 which is a function of Z and of the velocity v. For light ions and high velocities, $q_{eff} = Z$, but for heavy ions of a few MeV per nucleon, $q_{eff} < Z$. The response of an energy-loss detector to ions of a given Z, therefore, depends for two reasons upon their position at the focal surface, first because v is higher at the high-energy end of the surface and second because there may be a significant change in q_{eff}. For example, for ^{16}O of ions of 50 MeV, $q_{eff}^2 \approx 55.2$, whereas for 35 MeV it is only 50.7, a drop of 8%. The energy loss signal for a given particle type therefore varies in a complicated manner as a function of focal plane position; fortunately, however, that position is known so that an empirical correction can be made. At the same time, the correction can include variations in the intrinsic response of the detector such as might be caused by variations in wire diameter (for a proportional counter) or by end effects upon the electric field. Because charge state equilibrium is reached in passage through just a few μg cm^{-2} of stopping material, the energy loss of a particle is virtually independent of the charge state with which it enters the detector.

In heavy-ion experiments, particles leave the target in a variety of charge states that are separated in the spectrometer magnet. The fraction in a given state depends upon Z, v, and the target element: this subject has been reviewed by Betz (56). To measure a reaction cross section, the fraction in the charge state detected must be known. Alternatively, all charge states can be converted into the average state $\bar{q}$ in a gas-filled spectrometer, but then the multiple scattering and energy loss straggling will damage the resolution of the $B\rho$ measurement.

Complete identification and energy measurement in a spectrometer thus requires that the five quantities, E, Z, A, q, and q_{eff} be determined; but, as mentioned above, q_{eff} is a function of E, A, and Z. To determine the remaining four variables, the following five quantities are in principle measurable: $B\rho$, v, dE/dx, E, and ρ_E (the radius of the orbit in an electric field). In fact, ρ_E has not so far been used because extremely strong fields are required to produce a useful deflection. Particle energy can be measured in silicon or germanium counters with excellent resolution, at least for short-range particles; but these detectors are available only with rather small areas.

Large-area plastic scintillators can be used to measure total particle energies with a resolution of a few percent. Because of the high degree of quenching, the light output for ions < 8 MeV per nucleon is approximately equal to a constant K times the range of the particle. The output is therefore approximately proportional to $E^{1.7}$. Bechetti (104) measured the signal from a 0.25″ thick disc of NE102 coupled through a short light pipe to a 6342A photomultiplier for several particles from protons to ^{12}C ions. He found that K varied with the atomic number of the particle as $K = 1.49\,Z + 0.18\,Z^2$ when the photomultiplier gain was adjusted so that a signal of 35 arbitrary units was obtained for 8.78 MeV ^4He ions and range was expressed in mg cm^{-2}. It is not known whether this equation properly describes the behavior of NE102 for ions with $Z > 6$.

The best available device for total energy measurement appears to be an ion chamber with sufficient stopping power to bring the particles of interest to rest within the chamber volume. DeVries et al (105) obtained a resolution of 1.2% with 38 MeV ^{16}O ions in a chamber that simultaneously measures particle position twice and energy loss. The system is described below in more detail.

The remaining quantities, $B\rho$, v, and dE/dx, can be measured with sufficient resolution (especially $B\rho$ and v), but four unknowns cannot be determined unambiguously by three measurements. Fortunately A, Z, and q have integral values so that the ambiguity in their determination is usually not troublesome; it is discussed later.

Magnetic spectrometers and focal plane detector systems have been reviewed by Hendrie (106).

Measurement of $B\rho$

The magnetic rigidity $B\rho$ is nearly always obtained by detecting the position at which the particle crosses the focal surface of a magnetic spectrometer. For adequate *identification* of the particle, the detection system does not need very good position resolution, but the requirement of good *energy* resolution for the particle spectra usually mandates that the position resolution shall be about 0.5–1 mm (FWHM).

Focal plane areas are typically 1–5 cm in the vertical dimension and up to 100 cm in the horizontal direction. For full exploitation of the momentum range of the focal surface, therefore, detectors of rather large area are required. Usually, only the position of a particle in the horizontal (radial) direction is measured, but several techniques permit a simultaneous measurement in the vertical direction that can be used to correct for a curved or inclined lineshape (107). Many spectrometers were designed for particle detection with photographic plates, for which it is an advantage to make the particle trajectories strongly inclined (typically at 45°) to the focal surface (106). Unfortunately, the non-normal entry often has a deleterious effect upon the position resolution obtained with nearly all types of detectors, especially in the measurement of particles with low specific ionization (see below). The same problem has been built into even quite recent spectrometer designs (108). When proper consideration (106) is given to the spectrometer system as a whole, it is evident that not only should the focal plane be normal to the trajectories, but it should also be flat, as most detectors cannot be constructed to follow a curved surface.

Virtually every type of detector used in nuclear and high-energy physics can form the basis of a position-measuring system provided that the units are cheap enough to be built into an array of large enough area. Matsuda et al (109) built arrays of 250 Si detectors and 200 individual proportional counters. The reliability of such a large array must be questioned. Cohen & Rubin (110) used an array of plastic scintillators, whereas Kobayashi & Takayanagi (111) put 1000 detectors on to 25 separate silicon wafers.

The multiwire proportional counter (MWPC) (112, 113) and spark chamber (112, 114) with sonic readout (115–117) have been used: they too are essentially arrays of detectors enclosed in a single large housing. They have the advantage over arrays of separated detectors in that there are no gaps along the sensitive length and in that they can in principle be built along a curve adapted to a focal surface that is not flat. The low counting rate (~ 100 events $\sec^{-1}$) of the spark chamber makes it unsuitable for use in nuclear physics spectrometers. Furthermore, high counting rates at discrete spots are known to cause insensitivity in these regions. The MWPC, however, offers many advantages, and it is somewhat surprising that its use is not more common now that readout systems are available that do not require an amplifier for each wire (118–120). But it has not yet been demonstrated that the MWPC is capable of high-resolution energy-loss measurement.

In a second class of counters, a single long detector, rather than an array, is used, and inspection of pulses from one or both ends allows the position of particles to be determined. The long axis of the detector is positioned horizontally along the focal surface.

The position-sensitive silicon detector (121, 122) has the advantage of commercial availability, and can measure position and energy (for particles of short range) or energy loss (for particles of intermediate range).

Its many disadvantages, however, make it unsuitable for general application in a spectrometer. The counters are small—at best 1-cm high × 5-cm long in the position-sensitive direction—and rather expensive. An array is needed to cover the full focal plane (123). The best units are made by ion implantation of boron into high-resistivity silicon to make a resistive front surface. One end of this surface is grounded; from the other end a signal is taken into a charge-sensitive preamplifier. The charge liberated by the passage of a particle divides inversely as the ratio of the resistances to the two ends of the counter. The preamplifier signal is therefore proportional to $\Delta E \cdot x/L$, where ΔE is the energy loss, L is the detector length, and x is the distance from the grounded end to the point of entry of the particle. The energy loss ΔE is obtained as a second signal from the back of the detector. Division of the two signals in an analog or digital divider gives the position x/L.

The position resolution is, at best, about 0.5–1% of the detector length (0.25–0.5 mm/5 cm) for particles that lose a few MeV in the depletion depth. In a detector 1.4-cm long, Laegsgaard, Martin & Gibson (121) measured a position resolution of 0.16 mm for stopping 5-MeV α particles. The system noise (30 keV) was equivalent to a position resolution of 0.08 mm. For particles that lose small amounts of energy, the resolution is limited by the signal/noise ratio so that even in the thickest available counters ($\sim$ 600 μm depletion layer), the resolution becomes unacceptable ($>$ 1 mm) for protons of $E > 50$ MeV. The theory of noise and resolution has been discussed by Owen & Awcock (124). The relatively small physical depth of silicon detectors (compared with that of gas-filled counters) makes their position resolution almost independent of the angle of entry of the particles.

The linearity of the silicon position-sensitive detector may be limited by the variation in rise time of the signals from the resistive surface as a function of position. The resistive surface and detector capacity act as a distributed RC line; so, the further the particle from the preamplifier end, the slower is the pulse rise time. The characteristic RC is approximately 0.1 μs, or about one tenth of the typical amplifier time constant, so that a flat-topped pulse should be used to reduce the dependence of amplitude on rise-time. In practice, the linearity is about 1%. The theory of the charge-division process has been discussed by Kalbitzer & Melzer (125), Doehring et al (126), and Kalbitzer & Stumpfi (127).

The position-sensitive gas proportional detector with charge-division readout is a device quite similar in principle to the silicon counter. The silicon counter (a solid ion chamber with a resistive surface) is replaced by a gas-filled proportional chamber with a resistive central anode wire (128, 129) having one end grounded or with a charge-sensitive preamplifier at each end as shown in Figure 28. The wire can be very thin (10-μm diameter) nichrome (130) or carbon-coated quartz (25–75 μm) (129, 131). The anode wire is stretched centrally between flat planes of thin aluminized plastic sheet, which form the counter cathode, as shown in Figure 28. As with the silicon detector, the position signal at one end measures $\Delta E \cdot x/L$ and must be divided by a signal proportional to ΔE obtained either from the cathode or from the sum of the signals from the two ends of the anode wire. Compared with the silicon detector, the proportional chambers have the advantage that they can be made as long as required. By using several horizontal wires placed vertically one above the other, the sensitive vertical dimension can also be made as large as required. Further, the large gas multiplications that are achievable—as high as 10^5 while retaining proportionality (132)—maintain an excellent signal-to-noise ratio even for particles of very low specific ionization.

For heavy ions of not too great a range, the ion chamber provides an attractive method for measurement of the total energy of stopped particles. In the detector developed by DeVries et al (105) it is combined with nichrome resistive-wire proportional counters in the configuration shown in Figure 29. Electrons drift through the two grids to the two nichrome wire proportional counters to give position signals by the charge division method. Using collimated α sources, the position resolution was about 1 mm. The use of two wires allows ρ and incident angle determination, and allows the counter to be used with normal incidence even

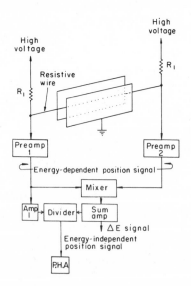

Figure 28   Position-sensitive proportional counter with charge division readout.

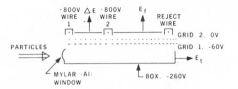

Figure 29   The detector system of Reference 105. Wires 1 and 2 measure particle position and hence the angle of entry. ΔE and E_f are measured on the two upper ion chamber plates and E_t on the box. The inverted U-shaped structures shield the ion chamber collectors from the proportional wire signals. The reject wire detects particles that do not stop in the ion chamber measuring E_f.

though the focal plane is inclined. The ΔE and E_f plates, operating in the ion chamber mode, provide energy loss and range information useful for charge and mass identification. As the electrons drift through the two grids, the total energy signal, E_t, is generated on the box and first grid (capacitively coupled to the box). With 30–60 MeV oxygen beams, resolutions of 3–4% and 4–5% are obtained for the ΔE and E_f plates, respectively, along with E_t resolutions of 1–2%.

In an alternative readout technique for the resistive-wire proportional chamber, Borkowski & Kopp (133, 134) showed that the anode wire can be used as a distributed RC line which slows down the rise-time of the pulse as it travels in both directions away from its point of origin. The highly resistive wire is, in practice, a carbon-coated quartz fiber,[2] diameter 25 μm and resistance 8000 Ω mm^{-1}. These fibers have excellent uniformity both in diameter and resistance and, having no tendency to curl, are easier to handle than metal wires of the same diameter. The rise-time method was first used in a spectrometer by Ford and co-workers (135).

Pulses are amplified at each end of the anode and differentiated to make bipolar pulses as shown in Figure 30. Crossover detectors generate pulses at the zero-crossing time, and the position of the particle is determined by measuring the time difference between these pulses, using a time-to-amplitude converter (TAC). The crossover pulse from one end passes through a fixed delay so that it always provides the stop signal for the TAC. A 25-μm anode fiber with a resistance of 8000 Ω mm^{-1} in a counter 1-cm deep gives a delay time of about 300 ns cm^{-1}. The energy loss can be obtained by adding suitably shaped pulses from each end of the anode.

Unlike charge division readout, the rise-time method has the advantage that the position signal, without additional processing, is independent of the amplitude of the event. The position resolution measured by charge

[2] Carl M. Zvanut Co., 14 Chetwynde Road, Paoli, Pennsylvania 19301.

division, where both the $\Delta E \cdot x/L$ and ΔE signals are of course amplitude-dependent, may depend on particle type when digital division is used. For example, if the pulse height given by α particles is adjusted just to saturate the ADC, the proton pulses—20 times smaller—are not accurately measured by the limited number of ADC channels. Position resolution is therefore lost (130). In both methods, the position resolution may also be limited by the signal-to-noise ratio. In the rise-time method, noise causes a jitter in the zero-crossing time of the bipolar pulse. If the bipolar pulses cross zero at a rate dV/dt and the (uncorrelated) rms noise on the signals is V_n, the time jitter in both the start and stop pulses to the TAC will be $V_n\, dt/dV$, and the position signal will be uncertain by $2\, k V_n\, dt/dV$ where k is the distance corresponding to unit time difference. Of course, dt/dV is larger for particles making small amplitude pulses, so the position resolution may be dependent on particle type if it is noise-limited. However, this is usually not the case. The position resolution and linearity of the rise-time method have been studied theoretically by Mathieson (136) who finds that both resolution and linearity degenerate rapidly near the ends of the resistive anode; particle detection should therefore be restricted to the central $\sim 75\%$ of the length. The time

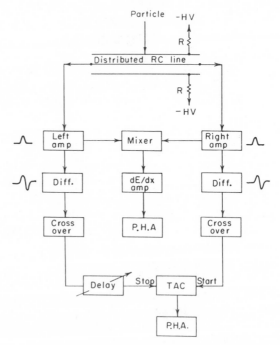

Figure 30   Position-sensitive proportional counter with rise-time readout.

constants of the differentiating network that produces the bipolar pulses should be about one tenth of the RC of the anode. Mathieson assumes that the ends of the resistive wire are terminated in charge-sensitive preamplifiers that essentially short-out any voltage changes at the ends. Since many of the chambers used in experiments do not operate in this mode, his conclusions are of limited validity.

One disadvantage of the rise-time readout method is the rather long resolving time (typically 10–15 μs). However, data rates are more likely to be limited by on-line computer speeds than by the chamber. It has been reported (137) that the position resolution of the rise-time readout can be as good as 0.16 mm at 3×10^5 counts sec^{-1}. Of course, the coincidence losses at this rate would be unacceptably high in most experiments. The position resolutions reported in spectrometer focal planes are typically 0.5 mm/20 cm [charge division (117)] and 0.6 mm 45 cm^{-1} [rise-time (107, 135, 138)]. Values as good as 0.15 mm are obtained in bench tests with collimated sources of radiation.

Neither charge-division nor rise-time readout gives a response that can be assumed to be exactly linear with particle position (101, 124). The rise-time method has been especially plagued by S-shaped response curves, but this problem can be cured by use of only the middle 75% of the anode length, by proper choice of the differentiating time constant (136), and by proper termination of the anode line (107). Even so, the response must be calibrated by, for example, moving a strong elastic peak to different positions by changing the magnetic field of the spectrometer. In a counter with several parallel wires, each wire will almost surely give a slightly different response: they must therefore all be calibrated separately and calibration data stored in an on-line computer.

In yet another readout system, one or both of the cathode planes of a proportional detector are replaced with a single or double helical delay line whose axis (see Figure 31a) is parallel with the anode wire or wires (139–141). The motion of the ion pairs near the anode induces a signal in several turns of the helical cathode. The pulse travels along the helix in both directions, delayed by about 30 ns cm^{-1} and the position is measured from the difference in arrival time of the pulse at the two ends.

For very long-range particles, the presence of helix wires in front of the anodes is unimportant, but it does cause trouble at lower particle energies (141). The configuration shown in Figure 31b is therefore preferable. In a typical helix (140), the wires are copper-clad aluminum, 75-μm diameter, 0.5-mm apart. The transparency of this helix to short-range particles is therefore only 85%; but no doubt somewhat more transparent helices could be built. When multiple detectors are required to measure ΔE or TOF, or to use multiple coincidences for the virtually essential

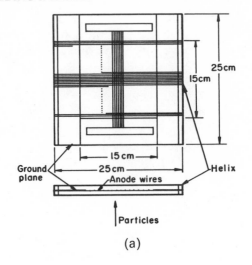

(a)

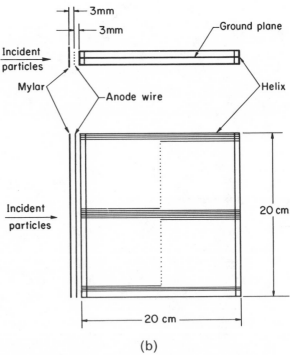

(b)

Figure 31 (*a*) Chamber with helical cathode wound around anode ground plane; (*b*) chamber with helical cathode wound around ground plane separate from the anode plane. Particles enter chamber without passing through the helix.

reduction of background (129, 130, 138), the helix detector would presumably have to be at the rear. Of course, the position-measuring device should be ahead of the other counters in order to avoid loss of position resolution from multiple scattering.

Although the delay-times in the helix—30 ns cm^{-1}—are about ten times shorter than in the rise-time system, pulses are reflected from the ends of the helix. In each reflection the pulse is inverted and attenuated about fivefold. After the second reflection, the pulse has its original polarity but is attenuated by a factor of 30. To prevent double-counting, discriminators are used between the amplifiers and the TAC; nevertheless, the pulse-storage time is only a factor of two or three shorter than in a rise-time system. Position resolution is excellent—0.33 mm FWHM has been reported in a counter 25-cm long (141).

The helix readout has the advantage that the position calibrations are identical for all the wires of a multiwire counter. The position calibration should be independent of time provided that the helix suffers no physical damage. Further, the position resolution appears to be independent of the angle of entry of the particles; this is discussed in more detail below.

Radeka (142) has made a theoretical analysis of the optimum performance to be expected from resistive wire charge-division chambers and delay-line readout systems. He showed that dispersion in the delay-line causes the timing and hence the position resolution to become signal-to-noise limited in long lines. In charge-division readout, the position resolution of an optimized system is determined only by the anode capacitance and not by its resistance. It should therefore be possible to use anodes of lower resistance than has been common, with consequent improvement in timing and energy resolution.

In the drift chamber (143) designed for the detection of 1-GeV protons (Figure 32), electrons liberated in the gas drift down a potential gradient

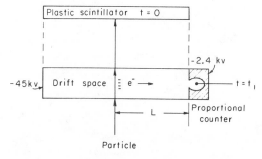

Figure 32   Drift counter (see Reference 143).

in a direction normal to the trajectory of the particle. At the end of the drift they are detected in a simple proportional chamber. A time-zero signal is obtained from a scintillator, and drift time is measured to give the position of the particle. Detectors up to 50 cm in length are used with electric field gradients of 800 V cm^{-1}, requiring a high voltage supply of at least 40 kV. The drift velocity W depends on the field gradient E, the nature of the gas and its pressure p, but W becomes independent of the gradient when the quantity E/p is equal to about unity (in units of V cm^{-1} and Torr). The preferred gas is pure methane, although neon-methane mixtures have been used. The limiting drift velocity is about 10 cm μs^{-1} in methane at a pressure of 1 atm and somewhat less in neon-methane mixtures.

The line of electrons along the particle track is broadened by diffusion during the drift, which makes a contribution σ to the resolution

$$\sigma = (2DL/W)^{1/2},$$ 38.

where L is the drift length and D is the electron diffusion coefficient in the chosen gas. In pure methane, σ was found to be 0.4 mm ($L = 12$ cm) and 1.3 mm ($L = 50$ cm). The position resolution in a long drift-chamber is therefore not quite as good as for other systems.

The proportional chamber at the end of the drift space (143) was found to measure the energy loss of 1-GeV protons with no apparent loss of resolution even after a 25-cm drift. However, the measured resolution must have been limited entirely by energy loss straggling so that the sensitivity to such effects as electron attachment to impurities would be quite low. It remains to be shown that good energy resolution can be obtained for particles for which energy loss straggling is small.

When particles cross the gas space of a proportional or drift chamber at a non-normal angle there may be a serious loss of position resolution. The readout systems correctly measure the center of ionization of the track; especially for tracks of low ionization density, however, statistical fluctuations can move the center of ionization with respect to the geometrical center. This problem has been analyzed by Miller and co-workers (117). As might be expected, the loss of resolution is reduced when the counter is made shallower (but then the ΔE resolution becomes worse). Alternatively, the energy loss can be increased and the fluctuations reduced by an increase in the gas pressure. Figure 33 shows the calculations of Miller et al for particles incident at 45°.

The volume of the sensitive zone around a proportional counter anode wire can be controlled at will by the addition of gases that capture electrons to form negative ions of low mobility, such as ethyl bromide or freon (112). Parts of the particle track that are more than the desired distance from the wire do not contribute to the pulse so that the effective depth of the counter is reduced. Although these gas mixtures have been used in multiwire proportional chambers for high-energy particles, they have not been used in the chambers described in this review. Of course it would be necessary to place the wires at a sufficiently close vertical spacing to avoid loss of events in dead spots. Further, good ΔE resolution would not be obtained; as we discuss below, however, it is usually advisable to use separate chambers for the position and ΔE measurements.

The helical delay line system is reported (141) to give a position resolution that is independent of angle of entry. This is because the readout system responds only to the fastest rising part of the cathode pulse, which comes from charges formed in that part of the track closest to the anode wire. It follows that particles far from an anode wire in the vertical direction, having no section of track close to a wire, will be detected with much reduced pulse height (and therefore perhaps poorer resolution). In the counter of Flynn and co-workers (141), the vertical anode wire

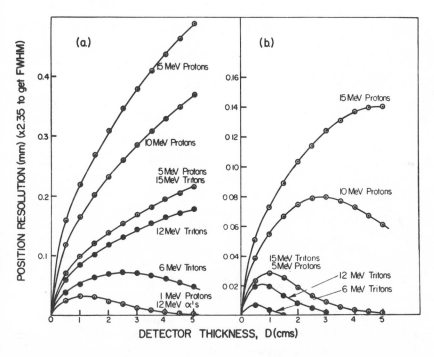

Figure 33   Position resolution as a function of detector thickness, (*a*) for 1 atm of argon, (*b*) 3 atm of argon (see Reference 117).

spacing was only 3 mm, whereas in the rise-time system the anode wires can be at least 30 mm apart without loss of position or ΔE resolution. The ΔE resolution of the helix system is about what would be expected for light particles where energy-loss straggling dominates, but there are no reported results for highly ionizing particles such as heavy ions.

Measurement of ΔE

The measurement of particle energy E and energy loss ΔE with semiconductor detectors in addition to the magnetic rigidity provides a powerful method of particle identification (144–146). Since the counters are small, $B\rho$ is determined approximately, even in the absence of a position measurement. Vorob'ev et al (146) used quadrupole lenses to define a small range of $B\rho$ (2–7%): they also measured TOF and E with a silicon detector.

Since

$$(B\rho)^2/E = 2A/q^2,\tag{39.}$$

measurement of $B\rho$ and E suffers from an A/q^2 ambiguity. For fully stripped ions $(q = Z)$, the ambiguity is in A/Z^2. Measurement of E and ΔE leaves an ambiguity in AZ^2, since approximately

$$E \cdot \Delta E \propto AZ^2.\tag{40.}$$

For *fully stripped* ions $(q = Z)$, equations 39 and 40 give

$$A^2 \propto (B\rho)^2 \cdot \Delta E\tag{41.}$$

and

$$Z^4 \propto E^2\Delta E/(B\rho)^2.\tag{42.}$$

In this case A and Z are completely determined by $B\rho$, E, and ΔE measurements. Both E and $B\rho$ are easy to measure with high resolution (at least over small areas and for particles of short range), but the resolution of ΔE measurements is usually limited by energy loss straggling. It is therefore fortunate that ΔE is proportional to A^2 (equation 41) and Z^4 (equation 42) so that the resolutions in the A and Z measurements are respectively twice and four times as good as they are in the ΔE determination. If $B\rho$ is not measured, equation 40 shows that ΔE depends only on the first and second powers of A and Z respectively.

Heckman et al (147) used a combination of magnetic analysis, dE/dx, and TOF measurements in a telescope of nine 3–5-mm thick silicon counters to identify the products from the interaction of 2.1 GeV A^{-1} heavy ions with various target nuclei. Identification was simplified because all products were fully stripped and (for $Z > 2$) they all had the same velocity—equal to that of the beam particles. Hence varying the field of the spectrometer produced a sharp transmission maximum for a particle at

$$B\rho \propto M\beta\gamma/Z, \qquad\qquad 43.$$

where $\gamma = (1 - \beta^2)^{-1/2}$ and $\beta\gamma$ was the same for all fragments. The various Z values were separated by the counter telescope with a Z resolution of ± 0.12 units.

With the possible exception of very low-energy heavy ions, particles do not stop in proportional counters, so that only energy loss is measured. It is important that the counters be designed so that the resolution in energy loss is limited only by the inevitable straggling rather than by instrumental problems. For small energy losses (e.g. with high-energy protons) the straggling is large (148), but for heavy ions, where the energy loss is large and energy loss straggling may be only a few percent, instrumental effects may limit the resolution.

It is important that the anode wires be smooth and uniform in diameter (149, 150). Experimentally (149), the gas multiplication M at a 20-μm wire varies with change of radius dr as

$$dM/M = dr/k, \qquad\qquad 44.$$

where $k \approx 1$ μm. Therefore a variation of M of $\pm 1\%$ requires that the wire radius uniformity be ± 0.01 μm, which is only $\pm 0.1\%$. This virtually impossible requirement can be relaxed in proportional counters in which position along the wire is measured, for, provided that the gas multiplication is sufficiently uniform over a length equal to the position resolution, a correction can be applied for slower diameter variations or for other effects by measuring the relative gas multiplication as a function of position. In counters with several wires the correction must be made for each position along each wire when the highest ΔE resolution is required.

Both energy loss and gas multiplication vary as a function of gas density, but in opposite directions. The resulting pulse height is approximately inversely proportional to the density. For 1% contributions to ΔE resolution, pressure must therefore be controlled to 1% and temperature to 3°C. In practice, pressures are controlled but temperatures usually are not, perhaps because they vary only slowly inside the spectrometer vacuum system.

The gas multiplication is, in addition, a complex function of the geometry of the counter (112, 151, 152). If several anode wires are used, they must be parallel, equally spaced from one another, and at equal distances from the cathode planes if equal gas multiplications are required. The mutual electrostatic repulsion of the anode wires causes them to move out of the plane in alternate directions (153, 154) by a distance Δy given by

$$\Delta y = 10^3 V^2 l^2 / TS \text{ mm}, \qquad\qquad 45.$$

where V is the potential of the wires (kV), l is the wire length (m), S is the distance between wires (cm), and T is the wire tension (dynes). Electrostatic stability and freedom from sustained vibrations is assured when the wire tension exceeds a critical value T_c,

$$T_c \geqq V^2 l^2 / 4\pi^2 a^2, \qquad\qquad 46.$$

where $2a$ is the depth of the counter (153). The effects of wire displacement, and other geometrical errors, were reviewed by Charpak (112). The maximum tension that 25-μm diameter quartz fibers can reliably sustain is about 12 g. The 20-μm gold-plated tungsten anode wires of a helix detector (140) are tensioned to 50 g.

In a detector with several wires, electrons drift towards the nearest anode wire, where the multiplication then occurs. The (negative) signal on that wire induces a smaller positive signal on adjacent wires. A negative signal may appear on two or more wires if the projection of the track on the plane of the wires crosses from one wire to another. The ΔE signal is therefore most reliably measured by adding the signals from all the wires. Corrections for different gas multiplications can be made by relative gain adjustments when each wire has a separate amplifier (141), by small adjustments (< 10 V) in the positive bias of each wire (155) or in an on-line computer that receives a logic signal to identify the wire on which a pulse occurred (107, 138).

Good ΔE resolution is especially important in the identification of heavy ions where the fractional difference between adjacent Z values is small. It has been found by several groups (129, 138, 156) that the requirements of good position and ΔE resolution cannot be optimized simultaneously in a single proportional counter. Good position resolution requires that the detector be shallow, especially when particle trajectories are non-normal, whereas deep detectors give greater energy loss and less energy-loss straggling. Good position resolution requires the greatest possible gas multiplication to obtain a very high signal-to-noise ratio. Good proportional behavior and energy resolution, on the other hand, are best obtained at low gas multiplications. The ideal system therefore consists of a shallow position-sensitive proportional counter in front of a deep ΔE counter. Harvey et al (138) use a 1 cm-deep resistive wire rise-time position detector and a 4 cm-deep ΔE counter with nickel wires. The ΔE resolution in each counter is close to the value calculated from energy-loss straggling, but it is twice as good in the deeper ΔE counter as in the position counter. Typical resolutions are 10% and 5% FWHM respectively for 100 MeV ^{16}O ions. The latter figure corresponds to a Z resolution of about $2\frac{1}{2}\%$ (since $\Delta E \propto Z^2$). Similar results have been obtained with the system at Argonne National Laboratory (156).

When the total charge density in the vicinity of an anode wire exceeds a certain value, the electric field is sufficiently perturbed that proportionality is lost (157). Since the total charge is proportional to the product of the energy loss and the gas multiplication M, it follows that M must be kept quite low for detection of heavy ions where the energy loss may be several MeV, especially when multiplication is confined to an extremely small volume as it will be in chambers at high gas pressures. In the Berkeley double counter (138), the gas multiplications in the position and ΔE chambers are only ~ 100 and 10 respectively for detection of 100 MeV ^{16}O ions. When the product of energy loss and M exceeds very roughly 200 MeV, peaks in the ΔE spectrum begin to develop "tails" on the high energy loss side when the gas pressure is $\frac{1}{3}$ atmosphere. This is of no consequence in a counter that is used only for a position measurement, but of course is highly undesirable for the ΔE counter.

In a chamber with horizontal wires spaced vertically one above another, guard wires must be used outside the upper and lower active anode wires; otherwise the electric fields and gas mutiplications of the outer wires may be considerably different from those of the inner wires. The maximum vertical spacing between wires consistent with good ΔE and position resolution is not well established. In one counter, the spacing is 30 mm so that particles up to 15 mm above or below a wire are detected. There appears to be no loss of position or ΔE resolution even in high-resolution heavy-ion measurements. In a 1-cm deep chamber with rise-time readout, the position resolution (for a collimated X-ray beam) was constant up to at least 2.2 cm from the anode (158).

Table 1 gives representative calculated values for energy loss and approximate energy-loss straggling of heavy ions in Ar gas (159). The straggling arises both from multiple scattering that changes the path length in the gas and from the statistical fluctuation in the charge state of the ion.

When ions of high Z and low velocity are detected, their energy loss is proportional to q_{eff}^2 rather than to Z^2, where q_{eff} is an effective ion charge $\leq Z$. The rate of energy loss $(dE/dx)_{A,Z,E/A}$ for an ion A, Z, and energy E per nucleon is related to the proton energy loss $(dE/dx)_{p,E}$ by

$$(dE/dx)_{A,Z,E/A} = q_{eff}^2 (dE/dx)_{p,E}. \qquad 47.$$

Independent of stopping medium, q_{eff} is given (empirically) by (56)

$$q_{eff} = Z\{1 - 1.032\exp[-v/(v_0 z^{0.69})]\} \qquad 48.$$

Table 1 Energy losses ΔE(MeV) and energy loss straggling (% of ΔE) in Ar gas (153). A fourfold increase in stopping power improves straggling by about a factor of 2. Almost identical results are obtained from Equation 49.

E(MeV)	^{11}B (ΔE)	^{11}B (%)	^{12}C (ΔE)	^{12}C (%)	^{16}O (ΔE)	^{16}O (%)	^{20}Ne (ΔE)	^{20}Ne (%)
1-cm atm								
50	2.19	5.80	3.29	4.63	6.45	3.15	9.90	2.56
100	1.30	9.81	2.01	7.61	4.29	4.74	7.34	3.46
150			1.48	10.30	3.20	6.37	5.73	4.44
200					2.59	7.86	4.68	5.44
250							4.00	6.37
4-cm atm								
50	9.15	2.78	14.20	2.15	28.50	1.42	46.70	1.09
100	5.27	4.83	8.21	3.72	18.00	2.26	31.40	1.62
150	4.06	6.29	5.86	5.22	13.10	3.11	23.90	2.13
200					10.40	3.93	19.20	2.65
250							16.20	3.16

for an ion of velocity v. The Bohr velocity v_0 equals 2.188×10^8 cm sec^{-1}. For the lighter heavy ions (e.g. ^{16}O), $q_{\text{eff}} \approx Z$ for energies above 5 MeV A^{-1}, (see Figure 2); but for heavier ion and lower velocities, separation by ΔE measurement becomes more difficult. For example, for $A = 90$, $E = -5$ MeV A^{-1},

$$\frac{(dE/dx)_{Z=40}}{(dE/dx)_{Z=39}} = \frac{(q_{\text{eff}}^2)_{Z=40}}{(q_{\text{eff}}^2)_{Z=39}} = 1.03,$$

whereas

$$Z^2(40)/Z^2(39) = 1.05.$$

Thus Z separation of these partially stripped ions requires a dE/dx resolution nearly twice as good as for fully stripped particles. For ^{24}Mg ions, q_{eff} rises from 9.6 at 50 MeV to 11.7 at 250 MeV. At the lower energy, the rate of energy loss is only 64% of what it would be for fully stripped ions of the same velocity.

Measurement of Particle Velocity

Particle velocity is measured by determining TOF over a known distance. The great advantage of making the measurement in a spectrometer is that the focusing permits a long flight path (typically several meters) with a much larger solid angle than would be possible in a simple evacuated pipe. The flight paths will be different for particles arriving at different points along the focal surface, but provided that position is measured, a correction can be applied to the TOF. There will be an additional dispersion in flight path for particles that leave the target at different angles and focus to a common radial point on the focal surface. A correction can be applied only if a particle's position is measured at two points. The amount of this dispersion decreases as the radial acceptance angle of the spectrometers is decreased: by operating at sufficiently small solid angles any required degree of isochronism can be obtained.

Since $B\rho \propto Av/q$, a two-dimensional plot of $B\rho$ vs TOF ($\propto 1/v$) shows a series of lines corresponding to discrete values of A/q. After correcting TOF for focal plane position, these lines become horizontal, as in Figure 34. Events corresponding to a given value of A/q can then be selected by a single-channel analyzer or digitally in a computer. Particles with different Z values but a common value of A/q can be separated by means of a ΔE measurement.

As discussed above, unambiguous identification is not obtained by measurement of $B\rho$, v, dE/dx. At a given value of $B\rho$ and v, two different species must satisfy

$$A_1/q_1 = A_2/q_2$$

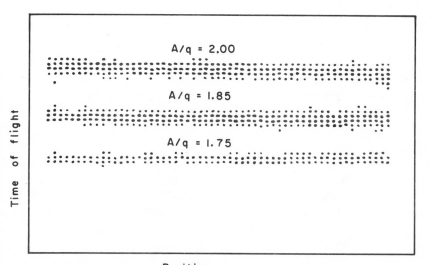

Figure 34 Plot of TOF vs position on detector. TOF has been corrected for differences in flight path.

and

$$E_1/E_2 = A_1/A_2.$$

If it is true that

$$dE/dx \propto AZ^2/E \propto Z^2/v^2,$$

then

$$(dE/dx)_1 = (dE/dx)_2$$

when

$$Z_1 = Z_2.$$

Hence two ions of the same element Z are indistinguishable when they have the same ratio of mass number to charge state A/q, for example $^{16}O(8+)$ and $^{14}O(7+)$. In most experiments there will be a range of $B\rho$ values where the two species do not overlap because of differences in Q value and reaction kinematics.

In a two-dimensional spectrum of TOF vs ΔE, species are separated both by A and Z (except for the ambiguities mentioned above). Figure 35 shows such a plot for the species obtained by bombarding ^{24}Mg with 86 MeV ^{11}B ions.

Figure 36 shows an unidentified particle position spectrum and the spectra of three ion species that were identified and separated by measuring $B\rho$, dE/dx, and TOF.

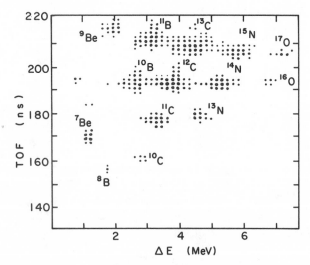

Figure 35   Plot of TOF vs ΔE showing separated particle groups.

Flight times for heavy ions in a spectrometer are typically 100–200 ns. As adjacent A values in typical heavy ion experiments ($A \approx 20$) differ by only 5%, the TOF resolution should be at least 1–2 ns.

When the beam is obtained from a cyclotron or other pulsed accelerator,

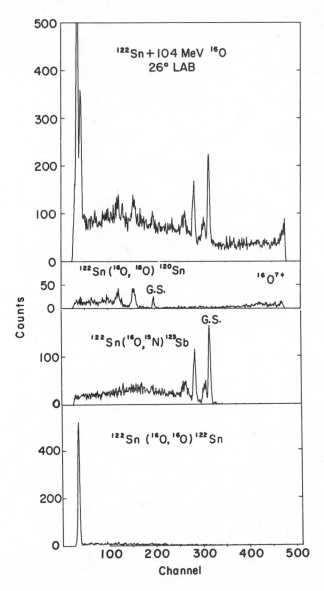

Figure 36   Total position spectrum with gated particle spectra below.

TOF *differences* can be measured by starting a TAC with the signal from a fast plastic scintillator in the focal plane and stopping it with a signal from the cyclotron oscillator (107, 160). The TOF resolution is then at best just the time width of the accelerator beam microstructure. With special precautions this has been reduced to 200 ps at the Michigan State University cyclotron (160), but in most cyclotrons the pulse width is about 5–10 ns, which is unacceptably long. Moreover, the time structure of the beam pulses in most machines is very sensitive to almost all the parameters of the cyclotron, such as the magnitude and shape of the magnetic field, the dee voltage and frequency, the ion source position, and so on. The time width of the microstructure is therefore not very stable, and the tuning conditions that give the sharpest time structure almost axiomatically produce less than the maximum amount of beam.

For most cyclotron work, or for experiments with dc beams, a time-zero detector between the target and the spectrometer entrance must be used. It should be sufficiently thin that energy-loss straggling and multiple scattering cause acceptably small losses of energy resolution. The effect of multiple scattering is minimized when the detector is placed as close as possible to the target. If it were actually *at* the target position (clearly impossible in practice), its effect would be exactly the same as the addition of the same material to the target. When the detector is placed downstream from the target, the change of direction due to multiple scattering makes the particle appear to have originated from a different part of the target. If the spectrometer is operated with a momentum dispersed beam rather than an analyzed (monoenergetic) beam, the multiple scattering causes a loss of the correlation between beam energy and its position on the target and this correlation is essential for the dispersion-matching in the spectrometer (106). Moreover, the change in angle of the trajectory causes a loss of the kinematic compensation. A particle emitted from the target at angle θ with respect to the beam, and having the kinematically correct energy E_θ will appear, after multiple scattering and energy loss in the time-zero detector, to have been emitted at $\theta \pm \Delta\theta$ with energy $E_\theta - \Delta E$. The new energy and the apparent angle are no longer related by the kinematics of the nuclear reaction. For heavy ions of energies used for nuclear reaction studies, the energy loss straggling effect would seem to dominate. The straggling δE is given approximately by

$$\delta E \,(\text{FWHM}) \,(\text{keV}) = 0.924 \, Z_1 (Z_2 t / A_2)^{1/2}, \qquad\qquad 49.$$

where Z_2, A_2 refer to the foil material of thickness t μg cm^{-2}. For a 50 μg cm^{-2} carbon foil, δE is approximately 40 keV for ^{16}O ions.

The multiple scattering half-angle $\delta\theta_{1/2}$ can be calculated from tables given by Meyer (161) or Sigmund & Winterbon (161). For 100 MeV ^{16}O and a 50 μg cm^{-2} carbon foil, $\delta\theta_{1/2}$ is about 0.3 mrad, but detailed calculations are required to estimate the effect of multiple scattering upon the energy resolution of a specific spectrometer system. Only about one third of the multiply scattered particles are contained in the cone of half-angle $\delta\theta_{1/2}$. For a given type of particle in a given foil, $\delta\theta_{1/2}$ is roughly proportional to q_{eff} and inversely proportional to the energy. Experimental values for the rms multiple scattering angles for ^{16}O and ^{32}S in C, BeO, and Al$_2$O$_3$ foils have been measured by Cline and co-workers (162). By extrapolating their results, a value of 1 mrad is obtained for $\langle\delta\theta^2\rangle^{1/2}$ for 100 MeV ^{16}O in a 50 μg cm^{-2} carbon foil. The experimental results of Cline et al agree with the following equation:

$$\theta_{rms} = 1.8 \left\{ \frac{3.922 \times 10^{-8} \, t Z_1(Z_1+1) q_{eff}^2}{A_1 E^2} \ln \left[\frac{208.1 t}{(Z_1^{2/3} + q_{eff}^{2/3})} \cdot \frac{Z_1+1}{A_1 Z_1} \right] \right\}^{1/2}, \quad 50.$$

where A_1, Z_1 refer to the scattering material, and q_{eff} to the particle whose energy E is in MeV. The thickness of the foil, t, is in μg cm^{-2}.

The time-zero detector will frequently be required to operate at an extremely high count rate, for regardless of what particles are detected at the focal surface, it will always be exposed to the elastically scattered beam particles. The best position for it is therefore immediately behind the entrance slits of the spectrometer. The detector should not be placed between magnetic elements when heavy ions are to be detected. The fields of the elements achieve proper focusing or deflection of particles of a fixed charge state, so that any ion that emerges from the detector with a charge state different from its value in the first element will behave incorrectly in subsequent magnetic elements. In spite of all these difficulties, the time-zero detector seems likely to become a standard feature in particle-identifying spectrometer systems, although at present it is rarely used.

The system of Vorob'ev et al (146) uses two quadrupole doublets to focus light nuclei formed in the thermal neutron fission of ^{235}U on to an energy-measuring silicon detector, thus selecting a range of $B\rho$ values 2–7% wide depending on the size of the detector. Time of flight is measured between the silicon detector and a thin (0.5 μm) aluminum foil placed at the focus of the first doublet. The foil is biased at -20 kV and secondary electrons are accelerated towards two scintillation-photomultiplier detectors. A coincidence between the two photomultiplier signals is used as the time-zero pulse. The time resolution of the system is 2 ns (FWHM). Combined with the 1% energy resolution, this gives a mass resolution of 2%.

The Berkeley spectrometer (107, 138) uses a thin NE III scintillator foil observed by two photomultipliers through a biconical light guide (Figure 37). For detection of heavy ions, foils of about 50 μg cm^{-2} prepared by the method of Muga and co-workers (163) give a detection efficiency close to 100%. It is unfortunate that the very great quenching of the light output (164) makes plastic scintillators basically unsuitable for detection of heavy ions: the high efficiency is obtained by setting the threshold for detection at a single photoelectron, and the count rate in the system is typically in the range of 10^5–10^6 counts sec^{-1}. The scintillator foil is extremely sensitive to electrons coming from the target so that an electrostatic deflector must be placed between the target and the foil. Time of arrival of a particle at the focal surface is measured with a 45×6 cm plastic scintillator placed behind the rise-time position-measuring proportional counter and the energy-loss counter. The intrinsic time resolution of the system is about 0.7 ns, but flight path length dispersion degrades it to 2 ns at a solid angle of 1 msr. The flight time for 100 MeV heavy ions is about 150 ns, so the time and mass resolution is typically better than 2%.

The use of a fourfold coincidence between the time-zero detector, two proportional counters and focal surface scintillator reduces the system background virtually to zero. In a study of the reaction ^{208}Pb$(^{20}$Ne,$\alpha)^{224}$Th, no α events were observed between 0 and 5 MeV of excitation in ^{224}Th. A single event would have corresponded to a cross section of 70 nb/sr.

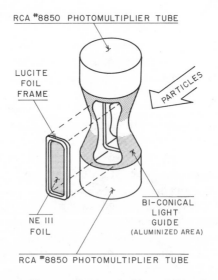

Figure 37 Schematic diagram of thin scintillator foil holder and light guide.

As already mentioned, the channel-plate secondary electron detector offers the possibility of simultaneously measuring the time and the position of the particle. With the apparatus shown in Figure 22b, Gabor, Homeyer & Kovar (165) obtained a position resolution of about 1.5 mm. The position-sensitive silicon counter that they used is not ideal for fast timing, but it could be replaced by an array of anodes to give both position and fast timing. The accelerating grid—transparency 99%— caused background problems in the position spectrum at the focal surface by scattering heavy ions. Muga (166) has shown that a low-resolution position measurement (~ 7.5 mm) can be obtained from a thin plastic scintillator through comparison of the light output from the two ends in a system similar to that of Figure 37. Clerc et al (167) used silicon counters to detect secondary electrons from the passage of heavy ions through thin carbon foils. They point out that the yield of electrons does not depend on foil thickness or uniformity, as all the electrons come from within a depth of only about 2.5 μg cm^{-2} from the foil surface. The yield obtained from two foils is somewhat greater than twice that from a single foil, presumably because a few high energy electrons from the front foil generate low energy secondaries in the second. In a multiple foil system, the electron yield varies with particle velocity and Z in the same way as dE/dx; with further improvements, the system therefore holds out the possibility of making at least low-resolution dE/dx measurements for particle identification.

A radial position measurement at the entrance of a spectrometer would be enormously valuable. If the radial width of the beam spot on the target is sufficiently small, a position measurement at a point downstream is equivalent to a determination of the angle at which a particle enters the spectrometer. Corrections can then be made for the angular dependence of the flight path length, for radial spectrometer aberrations and for the angular variation of the energy of nuclear reaction products. The spectrometer designer would be liberated from concern with isochronism, radial aberration control, and kinematic compensation with multipole elements or by focal plane displacement. Moreover, angular distributions of reaction products measured in a spectrometer with a large radial acceptance angle could be broken down into measurements over small angular increments. Unfortunately, many spectrometers are of the energy loss, dispersion matched type in which the dispersed beam on the target makes a spot that is too large to permit accurate determination of secondary particle angles by means of a single radial position measurement at a feasible distance from the target.

As an alternative, double or multiple position measurements can be made at or near the focal surface. This technique is especially useful for long range particles, and for spectrometers with a curved focal surface (143). For heavy ion detection, careful attention will have to be paid to the design in order to minimize the inevitable loss of angular information produced by multiple scattering in the first position counter. When the distance between the position measuring planes is short, the angular resolution may be limited by the intrinsic position resolution. At large separations, multiple scattering dominates. There is therefore an optimum separation, but unfortunately it must depend upon the particular ion that is being detected. This is because the multiple scattering in a given detector configuration is a function of particle Z, A, and energy, as shown in equations 48 and 50. Multiple scattering will be reduced when the counter gas contains only low-Z elements. Helium would be excellent but cannot be contained by mylar windows. $N_2 + CO_2$ works well as a proportional counter gas. Pentane combines the advantages of low multiple scattering with a molecular weight of 72, which makes it possible to obtain high stopping power (and dE/dx resolution) at a low pressure. It has been used at pressures of 1–2 Torr in MWPCs for the detection of 1-MeV-per-nucleon heavy nuclei recoiling from heavy ion reactions (168).

With increasing interest in the study of heavy ion reactions, the designer of spectrometers and particle-identifying instruments will be presented with many fascinating new problems. Fortunately, the field has never contained so many potentially fruitful new ideas, nor has the rate of advance ever been so high.

Literature Cited

1. Price, P. B., Fleischer, R. L. 1971. *Ann. Rev. Nucl. Sci.* 21:295
2. Livingston, M. S., Bethe, H. A. 1937. *Rev. Mod. Phys.* 9:261
3. Hudson, G. M. 1970. *Nucl. Instr. Methods* 77:197
4. Bichsel, H. 1970. *Nucl. Instr. Methods* 78:277
5. Northcliffe, L. C. 1963. *Ann. Rev. Nucl. Sci.* 13:67
6. Northcliffe, L. C. 1970. *Nucl. Data Tables A* 7:233
7. Stokes, R. G., Northrop, J. A., Boyer, K. 1958. *Rev. Sci. Instr.* 29:61
8. Griffiths, R. J. et al 1962. *Nucl. Instr. Methods* 15:309
9. Sachs, M. W., Chasman, C., Bromley, D. A. 1966. *Nucl. Instr. Methods* 41:213
10. Mark, S. K., More, R. B. 1966. *Nucl. Instr. Methods* 44:93
11. Broadhurst, J. H., Pyle, G. J. B. 1967. *Nucl. Instr. Methods* 48:117
12. Gupta, S. K. 1971. *Nucl. Instr. Methods* 92:33
13. Goulding, F. S., Landis, D. A., Cerny, J., Pehl, R. H. 1964. *Nucl. Instr. Methods* 31:1
14. Cerny, J., et al 1966. *Nucl. Instr. Methods* 45:337
15. Alderson, P. R., Bearpark, K. 1968. *Nucl. Instr. Methods* 62:217
16. Harmanci, A. D. 1970. *Nucl. Instr. Methods* 86:109
17. Fischer, P. S., Scott, D. K. 1967. *Nucl. Instr. Methods* 49:301
18. Armstrong, D. W. et al 1969. *Nucl. Instr. Methods* 70:69

19. Skyrme, D. J. 1967. *Nucl. Instr. Methods* 57:61
20. Butler, G. W., Poskanzer, A. M., Landis, D. A. 1970. *Nucl. Instr. Methods* 89:189
21. Chaminade, R., Faivre, J.-C., Fain, J. 1967. *Nucl. Instr. Methods* 49:217
22. Ohkawa, S., Husimi, K. 1974. *Nucl. Instr. Methods* 116:61
23. Bird, B., Ollerhead, R. W. 1968. *Nucl. Instr. Methods* 71:231
24. Chulick, E. T., Natowitz, J. B., Schnatterly, C. 1973. *Nucl. Instr. Methods* 109:171
25. Radeka, V. 1964. *IEEE Trans. Nucl. Sci.* NS-11:302
26. Sluiters, J. E., Klein, S. S. 1974. *Nucl. Instr. Methods* 120:305
27. England, J. B. A. 1973. *Nucl. Instr. Methods* 106:45
28. Muga, M. L. 1971. *Nucl. Instr. Methods* 95:349
29. Anderson, C. E., Bromley, D. A., Sachs, M. 1961. *Nucl. Instr. Methods* 13:238
30. Gamp, A. et al 1974. *Nucl. Instr. Methods* 120:281
31. Goulding, F. S., Stone, Y. 1970. *Science* 170:280
32. Ewan, G. T. 1968. *Progress in Nuclear Techniques and Instrumentation*, ed. F. J. M. Farley, 3:69. Amsterdam: North-Holland
33. Dabbs, J. W. T., Walter, F. J., eds. 1961. *Nat. Acad. Sci. Publ.* Vol. 871
34. *Nucl. Instr. Methods,* 1966. Vol. 43
35. Brown, W. L. et al, eds. 1969. *Nat. Acad. Sci. Publ.* Vol. 1593
36. Gibson, W. M. et al 1965. *Alpha, Beta and Gamma Ray Spectroscopy* ed. K. Siegbahn, 1:343. Amsterdam: North-Holland
37. Bertolini, G., Coche, A., eds. 1968. *Semiconductor Detectors.* New York: Wiley
38. Dearnaley, G., Northrop, D. C. 1966. *Semiconductor Counters for Nuclear Radiations.* New York: Wiley. 2d ed.
39. Bertolini, G., Cappellini, F., Restelli, G. 1973. *Nucl. Instr. Methods* 12:219
40. Meek, R. L., Gibson, W. M., Brown, R. H. 1971. *Nucl. Instr. Methods* 94:435
41. Ponpon, J. P., Siffert, P. 1973. *Nucl. Instr. Methods* 112:465
42. Inskeep, C., Elad, E., Sareen, R. A. 1974. *IEEE Trans. Nucl. Sci.* NS-21:379
43. Forcinal, G., Siffert, P., Coche, A. 1968. *IEEE Trans. Nucl. Sci.* NS-15:275
44. Hansen, N. J. 1971. *Nucl. Instr. Methods* 96:378
45. Wilkins, B. D. et al 1971. *Nucl. Instr. Methods* 92:381
46. Wegner, H. E. 1965. *Nucleonics* 23:52-56
47. Dearnaley, G. 1964. *IEEE Trans. Nucl. Sci.* NS-11:249
48. Gibson, W. M. 1966. *IEEE Trans. Nucl. Sci.* NS-13:162
49. Fairstein, E., Hahn, J. 1965. *Nucleonics* 23:56
50. Goulding, F. S. 1966. *Nucl. Instr. Methods* 43:1
51. Dabbs, J. W. T. See Ref. 31, pp. 375–742
52. Van Roosebroeck, W. 1965. *Phys. Rev. A* 139:1702
53. Klein, C. A. 1966. *J. Phys. Soc. Jpn.* 21:307
54. Eberhart, J. E. 1970. *Nucl. Instr. Methods* 80:291
55. Northcliffe, L. C. 1960. *Phys. Rev.* 120:1744
56. Betz, H. D. 1972. *Rev. Mod. Phys.* 44:466
57. Landau, W. 1944. *J. Phys. USSR* 8:201
58. Symon, K. 1958. PhD thesis. Harvard Univ., Cambridge, Mass.; also see Rossi, B. 1952. *High Energy Particles.* New York: Prentice Hall
59. Vavilov, P. V. 1957. *Zh. ETP* 32:920. Transl. 1957 in *JETP* 5:749
60. Maccabee, H. D., Raju, M. J., Tobias, C. A. 1966. *IEEE Trans. Nucl. Sci.* NS-13:76
61. Gooding, T. J., Eisberg, R. M. 1957. *Phys. Rev.* 105:357
62. Selzer, S. M., Berger, M. J. 1964. *Nat. Acad. Sci. Publ.* 1133:187
63. Bohr, N. 1915. *Phil. Mag.* 30:581
64. Lindhard, J., Nielson, V. 1962. *Phys. Lett.* 2:209
65. Haines, E. L., Whitehead, A. B. 1966. *Rev. Sci. Instr.* 37:190
66. Steinberg, E. P. et al 1972. *Nucl. Instr. Methods* 99:309

67. Finch, E. C. 1973. *Nucl. Instr. Methods* 113:29 (see also p. 41)
68. Kassirov, S. A. et al 1974. *Nucl. Instr. Methods* 119:301
69. Goulding, F. S. Landis, D. A., Cerny, J., Pehl, R. H. 1966. *IEEE Trans. Nucl. Sci.* NS-13:514
70. Greiner, D. E. 1972. *Nucl. Instr. Methods* 103:308
71. Jackson, H. G. 1973. *IEEE Trans. Nucl. Sci.* NS-20:3
72. Gedke, D. A., McDonald, W. J. 1967. *Nucl. Instr. Methods* 55:377
73. Gedke, D. A., McDonald, W. J. 1968. *Nucl. Instr. Methods* 58:253
74. Chase, R. L. 1968. *Rev. Sci. Instr.* 39:1318
75. Maier, M. R., Sperr, P. 1970. *Nucl. Instr. Methods* 87:13
76. Maier, M. R., Landis, D. A. 1974. *Nucl. Instr. Methods* 117:245
77. Jaffe, L. 1962. *Ann. Rev. Nucl. Sci.* 12:153
78. Muga, M. L., Burnsed, D. J., Steeger, W. E., Taylor, H. E. 1970. *Nucl. Instr. Methods* 83:135
79. Gelbke, C. K. et al 1971. *Nucl. Instr. Methods* 95:397
80. Cormier, T. M. et al 1974. *Nucl. Instr. Methods* 119:145
81. Schneider, W. F. W., Kohlmeyer, B., Bock, R. 1970. *Nucl. Instr. Methods* 87:253
82. Stern, W. E., Leachman, R. B. 1950. *Rev. Sci. Instr.* 27:1049
83. Milton, J. D., Frazer, J. S. 1962. *Can. J. Phys.* 40:1626
84. Grachev, V. T. et al 1968. *Bull. Acad. Sci. USSR* Ser. 32:653
85. Deitz, E. et al 1971. *Nucl. Instr. Methods* 97:581
86. Pfeffer, W., Kohlmeyer, B., Schneider, W. F. W. 1973. *Nucl. Instr. Methods* 107:121
87. Gabor, G. Lawrence Berkeley Laboratory, Berkeley, California. Unpublished report
88. Parkinson, E. R., Bodansky, D. 1965. *Nucl. Instr. Methods* 35:347
89. Pleyer, H. et al 1971. *Nucl. Instr. Methods* 96:263
90. Blignaut, E. et al 1967. *Nucl. Instr. Methods* 51:102
91. Emerson, S. T. et al 1967. *Nucl. Instr. Methods* 52:229
92. Zeidman, B., Henning, W., Kovar, D. G. 1974. *Nucl. Instr. Methods* 118:361
93. Mak, H. B., Jenson, H. B., Barnes, C. A. 1973. *Nucl. Instr. Methods* 109:529
94. Watson, B. A., Chang, C. C., Tabor, S. L. 1971. *Particles Nucl.* 2:376
95. Henning, W., Zeidman, B., Kovar, D. Argonne National Laboratory, Chicago, Illinois. Unpublished report
96. Poskanzer, A. M., Cosper, S. W., Hyde, E. K., Cerny, J. C. 1966. *Phys. Rev. Lett.* 17:1271
97. Poskanzer, A. M. et al 1968. *Phys. Rev. Lett.* 27B:414
98. Davids, C. N., Laumer, H., Austin, S. M. 1969. *Phys. Rev. Lett.* 22:1388
99. Davids, C. N., Laumer, H., Austin, S. M. *Production of Light Elements Li, Be and B by Proton Spallation of* ^{12}C, Center for Nuclear Studies, University of Texas, Austin, Texas. Unpublished
100. Raisbeck, G. M. et al. Princeton University, Princeton, New Jersey. Unpublished report
101. Klapish, R., *Special Experimental Techniques in the Study of Nuclei Far From Stability,* Centre de Spectrometric Nucleaire et de Spectrometric de Masse du CNRS, Orsay, France. Unpublished report
102. Laumer, H., Austin, S. M., Panggabean, L. M., Davids, C. N. 1973. *Phys. Rev. C* 8:483
103. Bowman, J. D., Poskanzer, A. M., Korteling, R. G., Butler, G. W. 1973. *Lawrence Berkeley Lab. Rep. LBL-1967;* Bowman, J. D., Poskanzer, A. M., Korteling, R. G., Butler, G. W. 1973. *Phys. Rev. Lett.* 31:614
104. Becchetti, F. D. 1975. Personal communication
105. DeVries, R., Fulbright, H., Gunn, G., Shapira, D., Toke, J., Bennett, C. 1974. Personal communication
106. Hendrie, D. L. 1974. In *Nuclear Spectroscopy and Reactions,* ed. J. Cerny, Vol. A. New York: Academic
107. Harvey, B. G. et al 1972. *Nucl.*

Instr. Methods 104:21
108. Wiedner, C. A., Goldschmidt, M., Rieck, D., Enge, H. A., Kowalski, S. B. 1972. *Nucl. Instr. Methods* 105:205
109. Matsuda, K., Nonaka, I., Omata, K., Yagi, K., Koike, M. 1967. *Nucl. Instr. Methods* 53:82
110. Cohen, B. L., Rubin, A. G. 1958. *Nucl. Instr. Methods* 111:1568
111. Kobayashi,T., Takayanagi, S. 1967. *Nucl. Instr. Methods* 53:77
112. Charpak, G. 1970. *Ann. Rev. Nucl. Sci.* 20:195 and references therein
113. Becker, H., Kalbitzer, S., Rieck, D., Wiedner, C. A. 1971. *Nucl. Instr. Methods* 95:525
114. Saudinos, J., Vallois, G., Laspalles, C. 1967. *Nucl. Instr. Methods* 46:229; Specht, H. J. 1968. *Bull. Am. Phys. Soc.* 13:1363
115. Hardacre, A. 1967. *Nucl. Instr. Methods* 52:309
116. Fulbright, H. W., Robbins, J. A. 1969. *Nucl. Instr. Methods* 71:237
117. Miller, G. L., Williams, N., Senator, A., Stensgaard, R., Fischer, J. 1971. *Nucl. Instr. Methods* 91:389
118. Grove, R., Lee, K., Perez-Mendez, V., Sperinde, J. 1970. *Nucl. Instr. Methods* 89:257
119. Grove, R., Ko, I., Leskovar, B., Perez-Mendez, V. 1972. *Nucl. Instr. Methods* 99:381
120. Grove, R., Perez-Mendez, V., Sperinde, J. 1973. *Nucl. Instr. Methods* 106:407
121. Laegsgaard, E., Martin, F. W., Gibson, W. M. 1968. *Nucl. Instr. Methods* 60:24
122. Jolly, R. K., Trentelman, G. E., Kashy, E. 1970. *Nucl. Instr. Methods* 87:325
123. Bock, R., Duhm, H. H., Melzer, W., Pühlhofer, F., Stadler, B. 1966. *Nucl. Instr. Methods* 41:190
124. Owen, R. B., Awcock, M. L. 1968. *IEEE Trans. Nucl. Sci.* NS-15:290
125. Kalbitzer, S., Melzer, W. 1967. *Nucl. Instr. Methods* 56:301
126. Doehring, A., Kalbitzer, S., Melzer, W. 1968. *Nucl. Instr. Methods* 59:40
127. Kalbitzer, S., Stumpfi, W. 1970. *Nucl. Instr. Methods* 77:300

128. Kuhlmann, W. R., Lauterjung, K. H., Schummer, B., Sistermich, K. 1966. *Nucl. Instr. Methods* 40:118
129. Williams, M. E., Kruse, T., Bayer, D., Williams, N., Savin, W. 1972. *Nucl. Instr. Methods* 102:201
130. Fulbright, H. W., Markham, R. G., Lanford, W. A. 1973. *Nucl. Instr. Methods* 108:125
131. Williams, N., Kruse, T. H., Williams, M. E., Fenton, J. A. 1971. *Nucl. Instr. Methods* 93:13
132. Fuzesy, R. Z., Jaros, J., Kaufman, L., Marriner, J., Parker, S., Perez-Mendez, V., Redner, S. 1972. *Nucl. Instr. Methods* 100:267
133. Borkowski, C. J., Kopp, M. K. 1968. *Rev. Sci. Instr.* 39:1515
134. Borkowski, C. J., Kopp, M. K. 1970. *IEEE Trans. Nucl. Sci.* NS-17:340
135. Ford, J. L. C., Stelson, P. H., Robinson, R. L. 1972. *Nucl. Instr. Methods* 98:199
136. Mathieson, E. 1971. *Nucl. Instr. Methods* 97:171
137. Gabriel, A., Dupont, Y. Institute des Sciences Nucleaires, University Grenoble, France. Unpublished report
138. Harvey, B. G., Homeyer, H., Mahoney, J. A. 1974. *Nucl. Instr. Methods* 118:311
139. Lee, D. M., Sobottka, S. E., Thiessen, H. A. 1972. *Nucl. Instr. Methods* 104:179
140. Lee, D. M., Sobottka, S. E., Thiessen, H. A. 1973. *Nucl. Instr. Methods* 109:421; 1974. *Nucl. Instr. Methods* 120:153
141. Flynn, E. R. et al 1973. *Nucl. Instr. Methods* 111:67
142. Radeka, V. 1974. *IEEE Trans. Nucl. Sci.* NS-21:51
143. Saudinos, J., Duchazeaubeneix, J. C., Laspalles, C., Chaminade, R. 1973. *Nucl. Instr. Methods* 111:77
144. Artukh, A. G. et al 1970. *Nucl. Instr. Methods* 83:72
145. Jackmart, J. C., Liu, M., Mazloum, F., Riou, M. 1966. *Proc. Int. Conf. Heavy Ion Physics,* Dubna, USSR
146. Vorob'ev, A. A. et al 1969. *At. Energ.* 27:31; 1969. *Sov. At. Energy* 27:713

147. Heckman, H. H., Greiner, D. E., Lindstrom, P. J., Bieser, F. S. 1972. *Phys. Rev. Lett.* 28:926
148. *Studies in Penetration of Charged Particles in Matter,* 1964. *NAS-NRC Nucl. Sci. Ser. Rep. No. 39* and references therein
149. Peterson, V. Z., Yount, D. 1971. *Nucl. Instr. Methods* 97:181
150. Charles, M. W., Cooke, B. A. 1968. *Nucl. Instr. Methods* 61:31
151. Waligorski, M. P. R. 1973. *Nucl. Instr. Methods* 109:43
152. Tomitani, T. 1972. *Nucl. Instr. Methods* 100:179
153. Schilly, P. et al 1971. *Nucl. Instr. Methods* 91:221
154. Walenta, A. H. 1973. *Nucl. Instr. Methods* 111:467
155. Parker, S. et al 1971. *Nucl. Instr. Methods* 97:181
156. Greenwood, L. R. et al 1972. *Rep. PHY-1972b,* Argonne Nat. Lab., Phys. Div.; Kovar, D. G. Personal communication
157. Hanna, G. C., Kirkwood, D. H. W., Pontecorvo, B. 1949. *Phys. Rev.* 75:985
158. Mahoney, J. 1974. Personal communication
159. Maples, C. C. 1974. Personal communication
160. Benenson, W., Kashy, E., Proctor, I. D., Preedom, B. M. 1973. *Phys. Lett. B* 43:117
161. Meyer, L. 1971. *Phys. Stat. Solidi B* 44:253; Sigmund, P., Winterbon, K. B. 1974. *Nucl. Instr. Methods* 119:541
162. Cline, C. K., Pierce, T. E., Purser, K. H., Blann, M. 1969. *Phys. Rev.* 180:450
163. Muga, M. L., Burnsed, D. J., Steeger, W. E. 1972. *Nucl. Instr. Methods* 104:605
164. Muga, M. L., Griffith, G. L., Schmitt, H. W., Taylor, H. E. 1973. *Nucl. Instr. Methods* 111:581
165. Gabor, G., Homeyer, H., Kovar, D. G. 1973. Personal communication
166. Muga, M. L. 1972. *Nucl. Instr. Methods* 105:61
167. Clerc, H.-G., Gerhardt, H. J., Richter, L., Schmidt, K. H. 1974. *Nucl. Instr. Methods* 113:325
168. Ghiorso, A. 1974. Personal communication

KAONIC AND OTHER EXOTIC ATOMS

×5562

Ryoichi Seki
California State University, Northridge, California 91324

Clyde E. Wiegand
Lawrence Berkeley Laboratory, University of California, Berkeley, California 94720

CONTENTS

1 INTRODUCTION

In this article we attempt to tell what is known about kaonic, sigmonic, and antiprotonic atoms. These are atoms in which an electron has been replaced by a K$^-$-meson, Σ^--hyperon, or antiproton. Loosely speaking, the term "mesonic atom" has been applied to any atom that contains a negative particle other than a regular electron. But according to present nomenclature only pionic and kaonic atoms are truly mesonic atoms. Other exotic systems could include Ξ^- (cascade), Ω (1), and anti-deuteron but these have not yet been observed. Sometimes pionic, kaonic, sigmonic, and antiprotonic atoms are called hadronic atoms to signify that they contain a strongly interacting particle. We confine most of our remarks to kaonic, sigmonic, and antiprotonic systems. Review articles on muonic atoms by Wu & Wilets (2) and pionic atoms by Backenstoss (3) have appeared in the *Annual Review of Nuclear Science*.

A complete description of mesonic atoms involves a gamut of disciplines in physics ranging from atomic and solid state through nuclear to strong interactions of elementary particles. Atomic or solid state physics is involved at the moment a meson stops in matter; the decision must be made as to which nucleus will become the host and into which Bohr orbit and angular momentum state will the meson be captured. From capture to disappearance in a reaction with a nucleon, hadronic atoms must obey the rules of atomic physics in which Auger emission and electromagnetic radiation are the dominant processes. Nuclear physics enters through the distribution of nuclear matter that the mesons encounter near the nuclear surface. Finally particle physics comes into play when annihilation in a reaction with a nucleon occurs.

Among the exotic atoms, kaonic atoms are the most studied. However, even the exact history of a kaonic atom has not yet been determined. Recently indications were found that mesons were captured into atomic angular momentum states that have a strong dependence on the atomic number Z, as discussed in a later section.

Kaonic atoms have been studied since 1958 when stacks of photographic emulsion were exposed to beams of K$^-$-mesons at the Berkeley Bevatron (4). Evidence of the formation of kaonic atoms came from the emission of Auger electrons of energies > 10 keV. Auger electron spectra were consistent with the cascade of kaons down to low-lying Bohr radii in Ag and Br atoms. From some of the early work with emulsions it was concluded that heavy nuclei might carry neutron-rich surfaces as suggested by Johnson & Teller and by Wilets (5). The question of the neutron halos has been discussed pro and con. Some recent experiments appear to support the often criticized claim that neutron-rich atmospheres are present on nuclear surfaces. This subject is discussed later. Early attempts were made to observe X-ray emission from kaonic atoms, but the experiments were not successful mainly because of the poor resolution of the detectors and the low intensities of the beams. Investigations of kaonic atoms got underway in earnest with the invention and application of semiconductor detectors. In 1966 prominent X-ray lines of the light elements Li, Be, B, and C were recorded by Wiegand & Mack (6).

For the study of kaonic and sigmonic atoms, the experimenter must have access to an accelerator that produces protons of at least 5 GeV; for antiprotons even higher energy is required. Experiments are thus limited to a few laboratories in the world: Lawrence Berkeley, Argonne National, Brookhaven National, European Organization for Nuclear Research (Geneva), Rutherford High Energy Laboratory, Dubna (USSR) and Serpukhov (USSR).

Negative kaon production occurs mainly by the reaction

$$p + N \rightarrow K^- + K^+ + 2N.$$

The primary target is usually heavy metal. Newly generated kaons and pions enter a secondary beam; the channel in which the beam is found is really a particle spectrometer that transports the kaons from the production target to a secondary target, in which kaonic atoms will be formed. Kaons are relatively scarce particles. Many thousands of pions are produced for each negative kaon, and most of the pions must be removed by the beam spectrometer. Typical beams can supply a few hundred stopped kaons per second. However, several hundred pions may pass through the targets without harmful consequences.

Kaon beams should be constructed as physically short as practicable; otherwise, kaons are unnecessarily lost by decay during their flight along the beam path. Typical beams are around 15 m long and operate at particle momenta from 450 to 800 MeV/c. Figure 1 shows the arrangement used by Wiegand & Godfrey (7). At 500 MeV/c, 3% of the kaons survive a beam length of 13 m and require an absorber of about 50 g cm^{-2} of graphite to bring them to rest. Figure 2 illustrates a series of scintillation and Cerenkov counters used in the experiment of (7) to signal the arrival and stoppage of kaons. Targets were 9 cm in diameter and 2 g cm^{-2} thick. At CERN thicker targets were used because the main objective of the

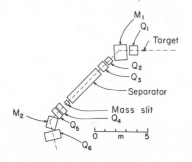

Figure 1 Plan of kaon beam showing arrangement of quadrupoles Q, bending magnets M, and electrostatic separator (Berkeley) (7).

experiments was to observe the highest energy X-rays of a given species, and in this case target self-absorption was not a problem.

The heart of any system to study kaonic atoms is the X-ray spectrometer. At present the most practical way to measure mesonic X-rays is by semiconductor detectors. We do not discuss in detail the theory and operation of these devices, but refer the reader to a recent article in the *Annual Review of Nuclear Science* by Goulding & Jaklevic (8) in which pertinent information and references are given. Some remarks, however, are in order. Semiconductor detectors have several advantages over gas proportional counters and scintillation counters of the NaI(Tl) variety. A disadvantage at present is their relatively small size. All X-ray detectors convert photons to electrons that in turn ionize the sensitive medium whether it be in a gaseous or solid state. Si and Ge require the expenditure of about 3 eV and noble gases require about 25 eV to make an ion pair. Scintillators take more than 100 eV per effective unit charge.

Measurement accuracy can be expressed as the full width at half-maximum (FWHM) of the distribution of energy output of the detector system when the input energy is constant.

$$\text{FWHM} = 2.35(E\varepsilon F)^{1/2},$$

where E is the energy deposited in the detector, ε the energy expended per ion pair, and F is a factor that depends upon the nature of the ionization process. F is about 0.5 for gases and 0.1 for Ge. Thus the resolution is significantly better for semiconductors. For energies above about 20 keV, ionization statistics dominate the ultimate resolution. At lower energies,

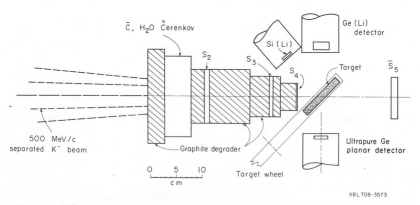

XBL 708-3573

Figure 2   Diagram of the arrangement of the beam counters, target wheel, and semiconductor detectors. The targets were approximately 65 cm from the exit of quadrupole Q_6 (Berkeley) (7).

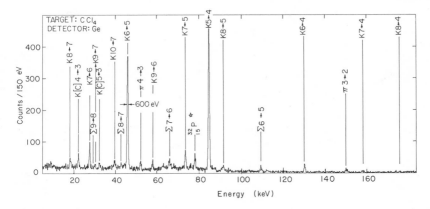

Figure 3   Kaonic X-ray spectra of CCl_4. The line at 78 keV is attributed to nuclear γ rays from the first excited level of $^{32}_{15}P$ (7).

electronic noise, equivalent to about 100 eV FWHM, becomes important. Background noise comes mostly from leakage currents across the surfaces of the detectors and from the first transistor in the amplifier. To measure the number of charges per X ray, amplifiers are connected to give pulse amplitudes proportional to the amount of charge collected, and the amplified pulses are then shaped to give the best discrimination of signal to noise.

The signal processing systems used at Berkeley were those of Landis, Goulding & Jaklevic (9). They contained a branch of circuitry that inspected the time distribution of detector pulses and rejected those that occurred closer together than the resolution period of the apparatus. After the pulses were formed to a near optimum shape, they passed through a linear gate. For the gate to open, these conditions were required: 1. that a kaon-stop signal and a timing pulse from the detector be in coincidence (± 0.5 μsec), and 2. that there be no time-wise interfering pulses. Amplitudes of signals that passed the linear gate were digitized and stored on magnetic tape. Later the laboratory's central computer system was used to tabulate the data and generate graphs of the X-ray spectra of which Figure 3 is an example.

2   BASIC CHARACTERISTICS OF EXOTIC ATOMS

Before discussing the detailed physics of exotic atoms, we present in this section some basic characteristics of the atoms that are relevant to the following sections. Since information about the exotic atoms comes mostly from X-ray spectroscopy of the atoms, we make also a few pertinent remarks on the spectroscopy.

The exotic atoms, namely kaonic, sigmonic, and antiprotonic atoms, differ from pionic atoms in the following essential points: the negatively charged heavy hadrons interact with nuclei more strongly than pions do, and these hadrons create more complicated final states as a consequence of nuclear absorptions.

The exotic atoms have been investigated mainly when the orbiting handron is closer to the nucleus than are the atom's ground state electrons. The system is therefore hydrogenic, with the electron cloud contributing an almost negligible screening effect, and the fundamental equations of the hydrogen atom are applicable: In terms of "Bohr radii" of the exotic atoms, $a_0 = \hbar^2/e^2\mu$, the mean values of the radii and their inverses are given by

$$\langle r \rangle = a_0[3n^2 - l(l+1)]/2Z$$

and

$$\langle r^{-1} \rangle^{-1} = a_0 n^2/Z,$$

where μ is the reduced mass of the hadron in the atom and Z is the atomic number. The rms value of the speed of the hadron is

$$\langle v^2 \rangle^{1/2}/c = \alpha Z/n, \quad \text{where} \quad \alpha = e^2/\hbar c.$$

The above quantities are evaluated nonrelativistically. Energy levels of the atoms are determined from the Dirac energy for point nuclei,

$$E_{n,j} = \mu c^2 \left[1 + (\alpha Z)^2 / \{n - j - \tfrac{1}{2} + [(j + \tfrac{1}{2})^2 - (\alpha Z)^2]^{1/2}\}^2\right]^{-1/2}$$
$$= \mu c^2 - \tfrac{1}{2}(\alpha Z/n)^2 \mu c^2 - \tfrac{1}{2}[n/(j + \tfrac{1}{2}) - \tfrac{3}{4}](\alpha Z/n)^4 \mu c^2 + \cdots, \qquad 1.$$

where $j = l \pm \tfrac{1}{2}$ and the second term is the familiar Bohr atom energy. For spin zero hadrons, a substitution $j = l$ reduces the expression to the Klein-Gordon energy.

Thus the large masses of the hadrons decrease the radii of the exotic atoms and increase the binding energies by a factor of more than 950 times those of ordinary atoms. On the other hand the "speeds" of the orbiting hadrons remain the same.

Strong nuclear absorption affects the X-ray lines in an important way; it sets a boundary on the lowest angular momentum state in which an exotic atom can exist. Hadrons with higher angular momentum have a lower probability to be found within the nucleus than do those with lower l. This causes the low-lying n, l transitions to be absent from all but the lightest elements. The probability of finding the hadron at the distance between r and $r + dr$ from the nuclear center behaves roughly as

$$|\Psi \, (\text{hadron})|^2 r^2 \, dr \sim [(2l+1)!]^{-1}(2Zr/na_0)^{2l+2}(Z/n^2 a_0) \, dr \qquad\qquad 2.$$

for circular orbits ($l = n - 1$), and with the additional factor $n^{2l+1}/(2l+1)!$ for $n \gg l$ (l fixed). Here we are neglecting perturbation of the hadronic wave function due to the strong interaction. The overlap of the hadronic density with the nucleus is a crude estimate of the region where nuclear absorption occurs. For large l it is peaked in the neighborhood of the nuclear surface. (See Figure 4.)

High rates of nuclear absorption have an important effect on the last observable X-ray lines of some hadronic atoms. The lines are broadened and shifted in energy from that predicted by the Dirac (or Klein-Gordon) energy plus other electromagnetic energies. Line broadening is a direct measurement of the rate of nuclear absorption. This effect is discussed in Section 3.2.

As seen in equation 1 and the similar equation for Klein-Gordon energy, the X-ray spectra of kaonic atoms will not exhibit fine structure splitting (the spin of kaons is zero), whereas sigmonic and antiprotonic lines will be doublets (analogous to those of the alkali elements) because

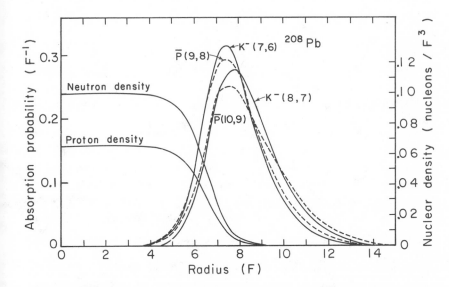

Figure 4   An illustrated example of the regions of kaonic and anti-protonic nuclear absorption in ^{208}Pb. The major absorptions from these (n, l) atomic states were found to occur as a result of cascade calculations (132). The absorption probabilities were computed numerically (Section 4.2) and the total probability from each state was normalized to 100%. The proton and neutron distributions are taken to be proportional to the same Wood-Saxon form ρ with the half-density radius 6.5 F and with the skin thickness 2.35 F.

their spins are $\frac{1}{2}$. These splittings are altered because of the anomalous magnetic moments of the hadrons. The effects have been observed in sigmonic and antiprotonic atoms (Section 3.3).

In addition to the more intense principal lines ($\Delta n = -1$), X rays have been observed from transitions of $\Delta n = -2$ with about 10% of the intensity of $\Delta n = -1$. Progressively weaker lines have been seen for transitions up to $\Delta n = -4$. (See Figure 3.)

For spin 0 hadrons, energies $E_{n,l}$ are almost equal for constant n, and the degeneracy for transitions between various l states has not been resolved experimentally. However, for high Z atoms $\Delta E = (n, l = n-2 \rightarrow n-1, l = n-3) - (n, l = n-1 \rightarrow n-1, l = n-2)$ can be significant. For example: $\Delta E = 0.5$ keV for $U, n = 10$.

3   EXPERIMENTAL FINDINGS

In this section we review experimental findings in the field of exotic atoms. Quite unexpectedly it was discovered that X-ray-intensity data exhibited an intricate competition between atomic processes (capture and cascade) and the strong nuclear interaction. Energy level shifts and line widths that are direct manifestations of the strong interactions were found as anticipated.

Exotic atoms serve as tools to learn more about the properties of negative hadrons (K^-, $\bar{p}$, Σ^-). In fact the best knowledge of some of these properties came from studies of exotic atoms (Section 3.3). Phenomena related to nuclear electric quadrupole moments are treated in Section 3.4. Results of the study of hadron-nucleon reaction products are reviewed in Section 3.5.

3.1   X-Ray Intensities

In this section we discuss the absolute X-ray-intensity data. The relative intensities, which have been utilized to deduce the small widths of the next to last observed levels, are treated in Section 3.2.

3.1.1   KAONIC ATOMS   Wiegand & Godfrey (7) have reported on an extensive survey of the absolute X-ray intensities in kaonic atoms of natural elements and pure isotopes ranging from $Z = 2$ through $Z = 92$. Except for ^{4}He (10) the data supersede that previously obtained by Wiegand and his collaborator (6, 11). Figure 5 summarizes the results of most of the $\Delta n = -1$ transitions. Here we see the mechanism of "kaonic atoms in action." The effect of the strong nuclear absorption is reflected in the rather sudden disappearance of X-ray lines for lower-n transitions. Disappearance in small Z for higher transitions is probably

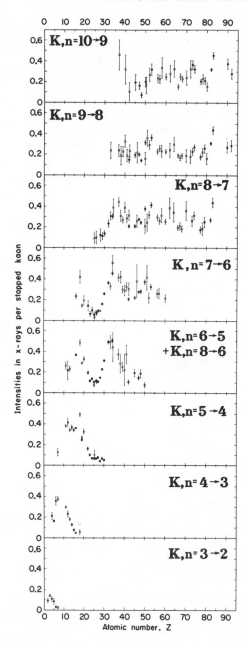

Figure 5 Intensity versus Z of observed principal (Δn = − 1) kaonic X-ray lines (7).

present as a result of competition with Auger emission. The intensities observed are at most 0.5 X rays per stopped kaon and often much less, indicating that nuclear absorption is significantly active before the kaons reach observable transitions. Other striking features of the data are the variations or dips as a function of Z. Figure 4 suggests that the maxima occur near closed atomic electron shells. The origin of these deviations is unknown. A less conspicuous valley was observed in pionic X-ray intensities (12). Related kinds of variations were reported in intensity ratios of $(n = 3 \to 1)/(n = 2 \to 1)$ transitions in muonic atoms (13) and also in intensities of muonic metal atoms in oxide (14). In the data of (7) no significant differences were observed among the isotopes of the same element, although it might have been expected that n, l states where nuclear absorption was strong would show more absorption for added neutrons. Apparently additional neutrons caused very little change in nuclear radii.

To illustrate the state of the art of X-ray spectroscopy of exotic atoms, Figure 3 shows a spectrum of kaonic Cl (target: CCl_4). Several sigmonic lines and a nuclear γ ray are also visible.

3.1.2 SIGMONIC ATOMS Σ^--hyperons (negative sigmons) are produced as a result of K^--nucleus absorption. In experiments where kaons are stopped, most of the sigmons that emerge from target nuclei remain within the targets and either decay in flight or are captured into sigmonic atoms. It has been reported that about 10% of the kaons stopped in photographic emulsion resulted in the ejection of Σ^- (15) and about 14% made Σ^- in CF_3Br (heavy liquid bubble chamber) (16). A theoretical calculation by Zieminska predicts about 12% to 3% roughly monotonously decreasing for an increase of Z (17). The predominant reaction is $K^- + N \to \Sigma^- + \pi^{+,0}$ where N stands for proton or neutron. Wiegand & Godfrey have measured the absolute intensities of some X-ray lines of sigmonic atoms (7), as shown in Table 1. X-ray intensities observed are less than 4% per stopped kaon.

3.1.3 ANTIPROTONIC ATOMS   No absolute intensity data for X-ray lines of antiprotonic atoms are available to date.

3.2 Level Shifts and Widths

Shifts and widths of the energy levels in exotic atoms are a direct consequence of the hadron-nucleus strong interaction. The shift in energy ε is defined as the total calculated electromagnetic energy minus the experimentally measured energy. The high nuclear absorption rate $1/\tau$ is also manifested by line broadening, $\Gamma = \hbar/\tau$ where Γ is the line width.

Table 1 Measured absolute intensities of Σ^- hyperonic X-ray lines per stopped kaon (7)

	Transition	I X Rays per K Stop	Error in Relative I	Error in Absolute I		Transition	I X Rays per K Stop	Error in Relative I	Error in Absolute I
^{6}Li	$4 \to 3$	0.011	0.25	0.004	CCl_4	$8 \to 7$	0.008	0.50	0.004
^{7}Li	$4 \to 3$	0.009	0.24	0.003		$7 \to 6^a$	0.023	0.11	0.004
^{12}C	$4 \to 3$	0.012	0.39	0.006		$6 \to 5^b$	0.021	0.07	0.004
^{14}N	$5 \to 4$	0.021	0.35	0.008	^{40}Ar	$7 \to 6^a$	0.034	0.26	0.010
$H_2{}^{16}$O	$5 \to 4$	0.005	0.40	0.002		$6 \to 5$	0.029	0.26	0.009
^{23}Na	$7 \to 6^a$	0.019	0.69	0.013	K	$8 \to 7$	0.015	0.37	0.006
	$6 \to 5$	0.019	0.39	0.008		$7 \to 6^a$	0.015	0.47	0.007
	$5 \to 4$	0.016	0.23	0.004		$6 \to 5$	0.007	0.47	0.003
^{27}Al	$6 \to 5$	0.021	0.36	0.008	Fe	$7 \to 6$	0.023	0.52	0.013
S	$7 \to 6^a$	0.021	0.10	0.004	^{59}Co	$7 \to 6^a$	0.022	0.42	0.010
	$6 \to 5$	0.011	0.11	0.002	Ge	$9 \to 8$	0.040	0.28	0.013
						$8 \to 7$	0.037	0.23	0.011
					Se	$8 \to 7$	0.042	0.21	0.011

[a] Includes K, $n = 10 \to 6$ which is estimated to contribute about 0.02 I_{max}, where I_{max} is the most intense kaonic line of the target.

[b] Possibly includes a contribution from ^{19}F* nuclear γ rays.

In practice ε and Γ are directly observed only in transitions to the lowest n levels that undergo strong but not complere nuclear absorption. An example of kaonic X-ray spectra in which ε and Γ are directly observed is shown in Figure 6.

The technique of extracting the absolute energies and widths of the X-ray lines from experimental measurements is given in the review article on pionic atoms by Backenstoss (3). The same technique was used in analyses of X-ray lines of exotic atoms.

In addition to the Klein-Gordon or Dirac energy for the point Coulomb interaction, the total electromagnetic energy includes effects due to the following: vacuum polarization, finite charge distribution of the nucleus, shielding by atomic electrons, nuclear and hadron polarization, and the center-of-mass motion of the nucleus. Among these, vacuum polarization contributes most appreciably. For the purpose of energy shift determinations, vacuum polarization can be evaluated by first order perturbation theory in the lowest order of $\alpha(\alpha Z)$ with better accuracy than that allowed by the present errors in the transition energy measurement or the errors reflected from kaon mass (discussed later). If desired, higher

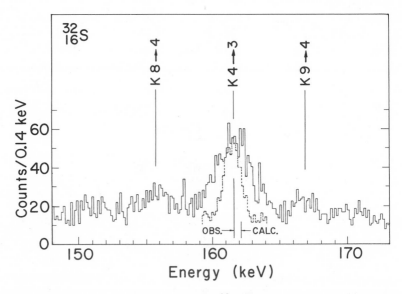

Figure 6   A part of the kaonic spectrum of ^{32}S, showing the line broadening and energy shift of the $n = 4 \rightarrow 3$ transition. The dotted spectral line under the broadened $n = 4 \rightarrow 3$ peak is the $n = 6 \rightarrow 4$ line that actually appeared at 130 keV. The width of the $n = 6 \rightarrow 4$ line represents the instrumental resolution, but the peak height is not up to scale (7).

order corrections of vacuum polarization can be evaluated by a known technique (18).

In contrast to pionic atoms, the effect of finite charge distribution is small because of the large angular momentum and high n-values involved in exotic atoms and is computed safely by use of first order perturbation theory. The rest of the effects are normally negligible.

The largest uncertainty in the evaluation of energy shifts in kaonic atoms comes from the uncertainty in the kaon mass. The error enters through the Klein-Gordon equation where energies are directly proportional to the reduced kaon mass. As described in Section 3.3, only recently has the kaon mass been determined to 8×10^{-5} accuracy (19) so as to yield lesser uncertainties in the total electromagnetic energies than most of the errors reported in X-ray energy measurements. At the same time we expect that experimental errors will decrease in forthcoming X-ray energy measurements. Considering this rapid improvement in the field, we list the shifts and widths in kaonic atoms as they have been

Table 2   Energy level shifts and widths of K^- atoms

Transitions		E_{em}(keV)	ε(keV)	Γ(keV)	Ref.	Γ_{up}(eV)	Ref.
^{10}B	3–2	43.576	0.208 ± 0.035	0.81 ± 0.10	a	—	—
^{11}B	3–2	43.776	0.167 ± 0.035	0.70 ± 0.08	a	—	—
^{12}C	3–2	63.317	0.59 ± 0.08	1.73 ± 0.15	a	0.98 ± 0.19	b
Al	4–3	106.45	0.13 ± 0.05	0.49 ± 0.16	e	—	—
Si	4–3	123.51	0.24 ± 0.05	0.81 ± 0.12	e	—	—
^{31}P	4–3	142.354	0.33 ± 0.08	1.44 ± 0.12	a	1.94 ± 0.33	b
S	4–3	162.141	0.57 ± 0.18	2.50 ± 0.32	c	—	—
S	4–3	162.141	0.33 ± 0.15	2.23 ± 0.20	c	—	—
S	4–3	162.107	0.55 ± 0.06	2.33 ± 0.20	a	3.25 ± 0.41	b
Cl	4–3	183.351	1.08 ± 0.22	2.79 ± 0.25	c	—	—
Cl	4–3	183.306	0.77 ± 0.40	3.8 ± 1.0	a	5.69 ± 1.50	b
Cl	4–3	182.23	0.94 ± 0.40	3.92 ± 0.99	d	—	—
Ni	5–4	231.67	0.18 ± 0.07	0.59 ± 0.21	e	6.0 ± 2.3	e
Cu	5–4	248.74	0.24 ± 0.22	1.65 ± 0.72	e	7.1 ± 3.8	e
Pb	8–7	426.27	—	0.37 ± 0.15	f	4.1 ± 2.0	f
U	8–7	538.86	0.26 ± 0.40	1.50 ± 0.75	f	45.5 ± 24.0	f

a $m_k = 493.73$ MeV (20).
b (21, 32).
c $m_k = 493.84$ MeV (7).
d $m_k = 493.87$ MeV (22).
e $m_k = 493.715$ MeV (23).
f $m_k = 493.715$ MeV (24).

reported. Each experimental group has used different kaon masses that were judged to be best at the time of publication. Table 2 includes this information.

X-ray energy measurements of antiprotonic S atoms have been made to an accuracy of 3×10^{-4} (25). The $\bar{p}$ mass determination was reported with a little less accuracy (26). However, when the theorem of the con- servation of parity, charge, and time reversal invariance (PCT) is assumed to set the antiprotonic mass equal to the protonic mass [accuracy 3×10^{-6} (27)] the problem of the uncertainty of the $\bar{p}$ mass does not occur. At present the shifts and widths data of antiprotonic atoms are scarce and generally crude, as shown in Table 3.

No direct shift or width measurement has been made in sigmonic atoms.

Before continuing the discussion on line widths, we insert a few words on what is meant by "last observable transition" or "last observed X-ray line." From the experimentalist's viewpoint, it means the lowest n assigned to identified X-ray lines, but it might not be the lowest n to which all hadrons arrived without being absorbed. The lowest observed intensity of an X-ray line may depend upon how much effort was put into detecting the lines. Also, broadening tends to cause the lines to blend into the background. For example, in Figure 5, $n = 7 \to 6$ transi- tions are not shown for $Z > 60$, but undoubtedly $7 \to 6$ X rays would be observed for higher Z if more accelerator running time had been used

Table 3 Energy level shifts and widths of $\bar{p}$ atoms[a]

Transition		E_{em}(keV)	ε(eV)	Γ(eV)	Ref.	Γ_{up}(eV)	Ref.
[4]He	$3 \to 2$	—	—	191 ± 170	(24)	—	—
[6]Li	$4 \to 3$	—	—	≥ 2	(24)	—	—
[14]N	$4 \to 3$	55.824	39 ± 51	173 ± 34	(28)	—	—
[16]O	$4 \to 3$	73.562	60 ± 72	648 ± 150	(28)	—	—
P	$5 \to 4$	—	—	—	—	$1.14^{+0.25}_{-0.19}$	(21)
S	$5 \to 4$	140.50	80 ± 40	310 ± 180	(25)	$3.04^{+0.70}_{-0.51}$	(21)
Cl	$5 \to 4$	—	—	—	—	$8.00^{+2.22}_{-1.48}$	(21)
K	$5 \to 4$	—	—	—	—	$26.8^{+7.0}_{-4.7}$	(21)
Sn	$8 \to 7$	—	—	—	—	$3.07^{+1.79}_{-1.28}$	(21)
I	$8 \to 7$	—	—	—	—	$9.93^{+7.68}_{-4.05}$	(21)
Pr	$8 \to 7$	—	—	—	—	$24.7^{+56.8}_{-13.6}$	(21)

[a] The proton mass is used for $\bar{p}$ mass (see text).

and/or detector efficiency had been greater. On the other hand, in kaonic Cl where the $n = 4 \rightarrow 3$ line is about 3 keV wide, we expect vanishingly few kaons to have arrived at $n = 2$. Therefore, we should use caution in applying the term "last observed." Likewise, high n transitions may have failed to appear because of target self-absorption, and not necessarily because of X-ray emission being overwhelmed by Auger processes. Investigations using thin targets (less absorption, less background) await the availability of "kaon factories."

For upper levels of last observed transitions the nuclear absorption rate is often comparable to the X-ray transition rate, yet its magnitude is too small to be observed by line broadening. The widths of such levels can be indirectly determined primarily from the relative intensities of the last and next-to-last observed X-ray lines. The determination of the upper level widths $\Gamma_{abs}(n+1, l+1)$ is based on a simple observation that the nuclear absorption reduces the X-ray intensities of the last observed lines that consist almost entirely of $(n+1, l+1) \rightarrow (n, l)$ transitions between two circular states. Only orbits of maximum angular momentum are involved because of the strongly favored radiative transitions of $\Delta l = -1$ and because of the large (two orders of magnitude or more) nuclear absorption width for "parallel" $(n+1, l) \rightarrow (n, l-1)$ transitions (see Figure 7). The amount of the reduction is expressed as

$$I_{out}/\sum I_{in} = \Gamma_{rad}/(\Gamma_{rad} + \Gamma_{abs}(n+1, l+1)), \qquad\qquad 3.$$

where Γ_{rad} is the X-ray width of the $(n+1, l+1) \rightarrow (n, l)$ transition, and is computed by a standard dipole approximation (29) with a correction factor for the center-of-mass motion of the nucleus (30, 31).[1] The Auger width of this transition is normally negligible. I_{out} is the X-ray intensity of the $(n+1, l+1) \rightarrow (n, l)$ transition, while $\sum I_{in}$ is the sum of all X-ray and Auger transition intensities coming to the $(n+1, l+1)$ state.

I_{out} and $\sum I_{in}$ are obtained from the X-ray spectrum after subtracting the background, correcting for detection efficiency and target self-absorption. Detector efficiency and target attenuation are well-measured factors. For weak lines the largest error comes from the statistics in the numbers of X rays and background. Also determination of the background is endangered by unexpected lines from such sources as nuclear γ rays. Furthermore, an accurate value of $\sum I_{in}$ has to rely on a cascade calculation because X-ray transitions with $|\Delta n|$ larger than about 4 or 5

---

[1] In the field of mesonic and hadronic atoms, this large factor (e.g. about 2 for $\bar{p}$ atoms) had been curiously neglected for a decade since the original paper by Fried & Martin (30). Most publications on exotic atoms prior to the summer of 1973 that deal with de-excitation rates must be corrected to include this factor.

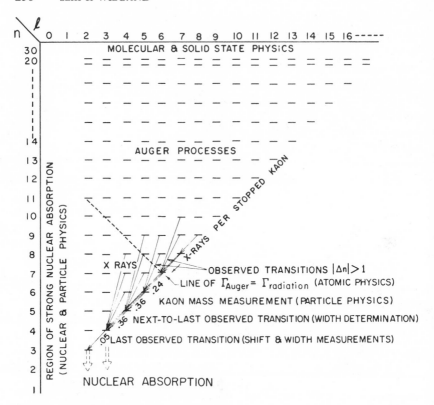

Figure 7 Schematic diagram of kaonic $^{32}_{16}$S atoms showing some of the features of the cascade and nuclear absorption. The cascade was started at $n = 30$ where the angular momentum distribution of kaons was assumed to be proportional to $2l + 1$ and $l_{\max}$ equaled $n - 1$. Regions of applicability of various disciplines in physics are indicated.

and the Auger transitions are not observed. The calculated corrections are reported to amount to about 10% in kaonic atoms of light nuclei (32). We note that the correction has to be treated with caution because it inherits the uncertainties associated with cascade calculations (see Section 4.1).

The values of $\Gamma_{abs}(n+1, l+1)$ thus determined are listed as Γ_{up} in Tables 2–4.

3.3   Particle Properties Derived from Exotic Atoms

Precise measurements of the radiative transition energies provide information on some particle properties that are less accurately known by other means. In order as much as possible to avoid uncertainties arising from the evaluation of nuclear effects, strong interaction, and nuclear polarization, the measurements are done on high-n transitions that are far from the nucleus. At the same time the transitions must not be too high, because then the effect of electron screening will become important. Corrections for shielding and finite nuclear size are usually so small as to be almost negligible. They are evaluated by standard procedures. However, the choice of higher transitions causes a complication. The high n levels are almost, but not quite, degenerate in energy. Some of the steps in de-excitation involve noncircular orbits and thereby introduce X rays of slightly different energy than those of circular orbits. Here we must rely on cascade calculations with their inherent uncertainties in order to determine the amount of mixing-in of "parallel" transitions as well as Auger transitions.

3.3.1 MASS The Klein-Gordon (or Dirac) energies for the point Coulomb interaction are directly proportional to the reduced mass of the atoms. The transitions chosen are almost entirely dominated by these energies. Correction for vacuum polarization amounts to several tenths of a percent but can be computed to the desired accuracy. Therefore, the accuracy of the hadron mass determination from exotic atoms is practically the same as the accuracy of the energy measurement of the chosen X-ray lines.

3.3.1.1 K^- Currently the K^- mass as determined by kaonic atom measurements is considerably more accurate than that determined by other means. It is

$$m_K c^2 = 493.691 \pm 0.040 \text{ MeV},$$

claimed by Backenstoss et al (19) from studies of kaonic Au and Ba atoms. The determination from emulsion tracks remains at ± 0.17 MeV (27). In these kaonic atoms the transition energies are determined to an

Table 4 Widths of Σ^- atoms (21)

Transitions		Γ_{up} (eV)
C	4–3	0.031 ± 0.012
Ca	6–5	0.40 ± 0.22
Ti	6–5	0.66 ± 0.43
Ba	9–8	1.68 ± 3.60

average error $\Delta E/E$ of about 10^{-4}. The old mass was corrected essentially by finding Δm in

$$E_{\text{exp}} = (1 + \Delta m/m_0)E_{\text{calc}},$$

where $m_0 = 493.750$ MeV.

The final error (8×10^{-5}) was obtained by taking a weighted average of the measurements. Note that the world average is (27)

$$m_K c^2 = 493.707 \pm 0.037 \text{ MeV}.$$

3.3.1.2 $\bar{p}$ When the first observation of antiprotonic atoms was made in 1970 (26) the equality of the proton and antiproton masses was confirmed within ± 0.5 MeV. This verification of the charge independence-parity-time invariance (CPT) theorem to an error of $\pm 5 \times 10^{-4}$ could be improved within the present status of the art.

3.3.1.3 Σ^- Owing to the low intensities of sigmonic X-rays, a precise energy measurement of the lines leading to an improvement in the Σ^- mass is difficult. The measured energies are in agreement with the Σ^- mass obtained by other means. The world average is (27)

$$m_{\Sigma^-} c^2 = 1197.35 \pm 0.06 \text{ MeV}.$$

3.3.2 MAGNETIC MOMENT   The fine structure splitting in exotic atoms of spin-$\frac{1}{2}$ negative baryons is given by (29)

$$\Delta E = \tfrac{1}{2}(1 + 2g_1)\mu c^2 (\alpha Z/n)^4 n/[l(l+1)] \qquad \qquad 4.$$

between $j = l + \frac{1}{2}$ and $l - \frac{1}{2}$ states of the same l. Here g_1 is the anomalous part of the magnetic moment in the unit of the baryon magneton and μ is the reduced mass of the atom. The magnetic moment of a negative baryon, s, of mass m_s is written as $\mu_s = -(1 + g_1)$ in units of the baryon magneton, $eh/2m_s c$. When $g_1 > -\frac{1}{2}$, the binding energy of $j = l + \frac{1}{2}$ is always less than that of $l - \frac{1}{2}$.

Equation 4 is obtained by first order perturbation theory in the Pauli approximation. The restriction $(v/c)^2 \approx (\alpha Z/n)^2 \ll 1$ exists, where v is the hadron speed in the atom. States of exotic atoms between which radiative transitions are observable satisfy this condition.

Exotic atoms convenient for measurement of fine structure splitting are large Z nuclei where Z/n would not be too small to be observed even for relatively higher-n transitions. A precise measurement of the fine structure splitting in these atoms is difficult, especially for sigmonic atoms. In $\bar{p}$ atoms the doublet lines are resolvable, whereas in sigmonic atoms the instrumental resolution is larger than the amount of the splitting. Contamination by noncircular transitions is also a problem. Furthermore, the sigmonic X-ray lines are found only in kaonic X-ray spectra where the kaonic lines are more than an order of magnitude more intense. In order to unfold the fine structure splitting, one has to perform a subtle computer fitting that is influenced directly by results of a cascade calculation. Moreover, the results of such analyses have an ambiguity caused by the fact that the measurements of ΔE themselves do not determine the sign of $\pm(1+2g_1)$. In principle, the ambiguity could be removed by an intensity difference in two major transitions between the fine-structure doublets (33–36) when statistical population of the levels is assumed (29). However, the difference is so small that the analyses of data (33–36) have yielded two statistically acceptable values of the magnetic moments that are of the opposite sign with slight preference for the negative value. The results are presently reported to be (33, 34 respectively)

$$\mu_{\bar{p}} = -2.819 \pm 0.056 \text{ N magnetons}$$

and   5.

$$\mu_{\Sigma^-} = -1.89 \pm 0.47 \ \Sigma \text{ magnetons.}$$

The errors assigned are defined to increase χ^2 by one. For $\bar{p}$, $\chi^2 = 43$ for 57 degrees of freedom; for Σ^-, $\chi^2 = 172$ for 144 degrees of freedom. The other minima that yield *positive* magnetic moments occur with $\chi^2 = 50$ ($\bar{p}$) and 173 (Σ^-). Each χ^2 curve is almost symmetric with two deep minima. Although the small errors are assigned, one has to be aware of the complexity of the analyses and the ambiguity in the results; especially for Σ^-. The previous value of $\mu_{\bar{p}}$ was -2.83 ± 0.10 N magnetons reported in 1972 (35), and that of μ_{Σ^-} was between 1.0 and $-2.0 \ \Sigma$ magnetons reported in 1973 (36).

When the PCT theorem is assumed, the magnetic moment of $\bar{p}$ is the negative of the proton $(2.7928456 \pm 0.0000011)$ N magnetons (27) in agreement with the $\bar{p}$ atoms result, equation 5. SU(3) theory predicts that the magnetic moment of Σ^- has a value equal to the negative difference between the proton and neutron magnetic moments $\mu_{\Sigma^-} = -(\mu_p - \mu_n) = -0.88 \ \Sigma$ magnetons. According to the equal-mass quark model, $\mu_{\Sigma^-} = -\frac{1}{3}\mu_p = -0.93 \ \Sigma$ magnetons (37). Equation 5 disagrees with these values.

3.3.3 POLARIZABILITY The strong electric field created between the nucleus and the hadron in an exotic atom may induce a polarization in the hadron structure (38–40) in addition to the nuclear polarization. The energy shift resulting from these induced polarizations is estimated to be, in the adiabatic treatment (39),

$$(\Delta E)_{\text{pol}} = -\tfrac{1}{2}e^2(p_N + Z^2 p)\langle r^{-4} \rangle,$$ 6.

where p_N and p are the nuclear and hadron polarizabilities respectively.

Equation 6 shows a large enhancement factor Z^2 on p that may make a direct observation of p possible.

When the precise measurement of the kaon mass was made (19), such an attempt was undertaken. There was a difference of a few tens of eV between the measured and calculated transition energies, which was attributed to $(\Delta E)_{pol}$. By use of p_N estimated from photonuclear cross sections (39), an upper limit of p defined in Equation 6 was found to be 0.020 F^3 at 90% confidence (19).

The experiment was quite difficult, and it may not be justifiable to place significance on the value comparable to the experimental error. However, to observe the hadron structure directly is an exciting possibility.

3.4   Atomic Phenomena Directly Related to Nuclear Properties

3.4.1   NUCLEAR QUADRUPOLE MOMENT   An effect of hyperfine structure splitting due to the electric quadrupole moment of a nucleus has been seen in kaonic atoms. When X-ray lines of $n = 8 \rightarrow 7$ transitions in ^{167}Er and ^{170}Er were compared, a clear broadening in the ^{167}Er line in contrast to the one of ^{170}Er was observed (7). Note that ^{167}Er has spin 7/2 and quadrupole moment of 2.83×10^{-24} cm^2, whereas ^{170}Er has spin zero. The components are not resolved because of instrumental resolution.

3.4.2   DYNAMICAL QUADRUPOLE $(E2)$ MIXING   When a massive negative particle cascades down in an atom to reach the vicinity of the nucleus, it can couple with the nucleons in the nucleus and cause combinations of nuclear excitations and nonradiative transitions in which the angular momentum of the total system, nucleus + hadron, is conserved. That is, the wave function of the total system becomes an admixture of the various atomic states and nuclear excited states, and the energy of the total system is perturbed from that of no admixture. The strength of the admixture is proportional to the ratio of the coupling strength and the energy difference between the mixed states. Thus occurs the nuclear polarization, and its effect is normally small. Deformed nuclei with large quadrupole moments, however, can induce a strong coupling involving a large admixture of angular momentum states of $\Delta j = 2$, especially when the energy difference between nuclear excited states and atomic transitions is small. This phenomenon is called dynamical quadrupole (or $E2$) mixing. In muonic atoms $E2$ mixing shows up prominently in some states involving fine structure doublets of small n and l, e.g. $2P_{1/2}$ and $2P_{3/2}$ states (2).

In exotic atoms the phenomenon is less prominent because the strong interaction prevents hadrons from reaching small angular momentum states. Recently an observation of dynamical quadrupole mixing has been reported in kaonic uranium atoms (41). The mixing was identified by the fact that the transition energies were perturbed by the amounts predicted from a detailed theoretical calculation (42).

The admixtured states in exotic atoms may involve atomic states of small angular momentum that, as a result of nuclear absorption, cannot be seen. Consequently, X-ray intensities of transitions from mixed states may be reduced drastically when the admixture is large. Recently Leon (43) proposed to utilize this method to gain additional understanding of the K^--nucleus strong interaction by predicting X-ray intensity reductions in several specific transitions.

Late in 1974 the effects of $E2$ nuclear resonance were seen in pionic (44) and kaonic (45) atoms. The Berkeley group (45) obtained kaonic X-ray spectra of ^{98}Mo and ^{95}Mo. The intensity ratio $I(n = 6 \to 5)/I(n = 7 \to 6)$ in ^{98}Mo was predicted to be attenuated from 0.93 (without mixing) to 0.18 (with mixing). The experimental ratio was found to be 0.35 ± 0.2. A control target ^{95}Mo also showed attenuation, but the reason for this is not understood. The importance of these observations is that a new method has been found to obtain information on previously inaccessible states in hadronic atoms, and thereby to verify calculations of hadron-nucleus strong interactions.

3.5 Products from Nuclear Absorption of K^-, Σ^-, and $\bar{p}$

3.5.1 PARTICLES OBSERVED IN NUCLEAR EMULSIONS AND BUBBLE CHAMBERS Since the late 1950s there have been extensive investigations on particles produced by K^--nuclear absorption in photographic emulsions designed to show particle tracks (4, 46–53) and in bubble chambers filled with mixtures of Freon (CF_3Br) and propane (C_3H_8) (54, 55), pure Freon (15, 56), neon (57, 58), helium (59, 60), and deuterium (61). Recently a hydrogen bubble chamber was used to detect charged pions produced by the nuclear absorption of negative kaons (62) and antiprotons (63) in several solid targets. An investigation was also made specifically on products of Σ^--nuclear absorption in emulsion in which K^- were stopped (64). Results of these investigations would be of importance in learning the distribution of nuclear matter if the reaction products could be assigned to specific hadron-nucleon interactions that occurred in the nucleus. Unfortunately there are complications. In many cases the reaction products undergo secondary reactions before they escape from the nuclei under consideration. We mention the major secondary reactions and estimates of their occurrences not that they may be used as a basis for new analyses, but simply to indicate their general features: (a) Σ to Λ conversion $(\Sigma^- + N \to \Lambda + N)$ occurs for about 0.5 of the Σ^- produced, (b) Λ trapping (hypernucleus formation) occurs 0.2 to 0.35 per Λ produced, (c) pion absorption occurs for 0.08 to 0.15 per pion produced, (d) Σ and Λ decays occur with known partial decay modes.

Besides the estimate of the secondary reaction occurrences, there is a question of bias and detectability of the particles in the experimental techniques. For these reasons some reports of the primary process occurrences published in the past have been questioned because of later findings. See, for example, (53, 65, 66). Under these circumstances we show in Table 5 the primary process occurrences obtained from Freon (16) and Freon + propane (55) bubble chamber studies as these two relatively recent and independent results agree quite well.

Burhop (65, 66) made careful studies of K^--emulsion and K^--bubble chamber data to review the values of the quantities

$$R_{pn} \equiv [\mathcal{N}(\Sigma^+\pi^-) + \mathcal{N}(\Sigma^-\pi^+)]/\mathcal{N}(\Sigma^-\pi^0)$$

and

$$R_{+-} \equiv \mathcal{N}(\Sigma^+\pi^-)/\mathcal{N}(\Sigma^-\pi^+),$$

Table 5 Primary processes and their occurrences in K^- nuclear absorption

Primary Processes	Occurrences[a]	
	$CF_3Br + C_3H_8$[b]	CF_3Br[c]
$K^- + p \to \Sigma^- + \pi^+$	10.3 ± 1.7	11.1 ± 0.5
$\to \Sigma^+ + \pi^-$	14.4 ± 2.3	11.3 ± 1.4
$\to \Sigma^0 + \pi^0$	11.8 ± 1.4	9.6 ± 1.5
$\to \Lambda + \pi^0$	13.3 ± 1.1	11.8 ± 1.2
	(11.8 ± 1.0)	
$K^- + n \to \Sigma^- + \pi^0$	1.2 ± 0.4	3.2 ± 0.5
$\to \Sigma^0 + \pi^-$	1.2 ± 0.4	3.2 ± 0.5
$\to \Lambda + \pi^-$	26.6 ± 2.1	23.6 ± 2.4
	(23.6 ± 1.9)	
Multinucleon absorption (no pion)	21.3 ± 2.5 (25.7 ± 3.1)	[d]

[a] The total K^- absorption events = 100.

[b] The π absorption rates 0.1 (with Σ) and 0.2 (with Λ) per π are assumed except the numbers in parentheses for which 0.1 (with Σ or Λ) is assumed (55).

[c] Reference 16.

[d] Not given.

Table 6 Significant quantities of primary processes in K^- nuclear absorption (see text on the definitions of the quantities)

| | Bubble Chamber | | | | CF$_3$Br | Emulsion | |
	H	D	He	Ne		Light nuclei (C, N, O)	Heavy nuclei (Ag, Br)
R_{pn} [a]	2.3, 4.0	10.5±4	9.0±4	28 ±12	18 ±8	21 ±4	4.25±0.7
R_{+-} [a]	0.43	1.2±0.3	3.5±1	1.8±0.15	1.5±0.3	1.75±0.15	1.5 ±0.3
Multinucleon [b] (non-pion)		~1.2 [c]	16.5±2.6 [d]	22.6 [e]	21±3 26±3 [f]	17±5 [g]	16±5 [g]

[a] Reference 66.
[b] Out of the total K^--absorption events (=100).
[c] Reference 68, not corrected for primary multinucleon processes. The corrections are expected to be small.
[d] Reference 60.
[e] Reference 58.
[f] Reference 55. See footnote b in Table 5.
[g] Reference 51.

where the $\mathcal{N}$ represents an occurrence of the primary processes labeled by their final states. Because of charge conservation, emission of $\Sigma^+\pi^-$ or $\Sigma^-\pi^+$ implies that the reaction was $K^- + p$. In Table 6 we list the values taken from (66). Note that the sum of these three reactions (emission of charged Σ) represents only a fraction of the total K^- absorption— about 25% in CF_3Br, as seen in Table 5.

These quantities are important since they depend critically on the K^--nucleon interactions in the nucleus. For example, the ratio of R_{pn} for light to heavy nuclei is about 5. The large ratio suggests that neutron absorption is either stronger or more frequent than proton absorption. If the latter, the nuclear surface would be rich in neutrons, as originally suggested by Johnson & Teller (5). After a controversy extending over many years this problem has been found to be subject to a frustrating complication and is not yet settled. We discuss it further in Section 4.3.

The recent measurement of charged pions emitted from kaon absorption in targets placed inside a hydrogen bubble chamber augments the previous K^--nuclear emulsion and heavy liquid bubble chamber data. The rates of total K^- absorption occurrences by neutrons and protons are determined by combining all of this information. Here also the deductions are subject to the question of reliability in sorting out the primary absorptions. The ratios of proton to neutron absorption were found to be 0.63 for C, 1.08 for Ti, 4.96 for Ta, and 4.01 for Pb (62).

Charged pions resulting from antiproton absorption were also measured in a similar way in a hydrogen bubble chamber (63); from these data the ratios of total $\bar{p}$ absorption in various nuclei by neutrons and protons were deduced. It has been shown in a simple model, however, that the energetic pions produced by $\bar{p}$ absorption undergo appreciable charge exchange reactions within the nuclei (67). In view of this finding, the charged pion data for $\bar{p}$ absorption have to be reanalyzed.

We turn now to kaon absorption by more than one nucleon. Rates of two-nucleon reactions of the type $K^- + N + N \rightarrow N + (a\ hyperon)$ have been obtained for some nuclei. These absorptions do not accompany pions and produce distinctly higher energy nucleons than those from one-nucleon absorptions. The rates are listed in Table 6 as reported in the literature. We discuss the problem of two-nucleon absorptions in Section 4.3.

3.5.2 NUCLEAR γ RAYS   Interest in nuclear γ rays induced by stopped negative kaons is concerned with the formation of excited nuclei. High resolution semiconductor detectors give the energies of the lines with accuracy sufficient to identify unambiguously the excited nuclear states responsible for the radiation. In the energy range less than 200 keV,

28 nuclear γ rays have been identified throughout the periodic table (7). Between 750 and 1500 keV, about 25 nuclear γ rays emitted from kaons stopped in Ni and Cu have been identified (69). Unfortunately one cannot tell how many of these γ rays result from the primary reactions— that is, how many come from the nuclei of the atoms in which K^- are captured—and how many result from secondary reactions, in which the particles produced by the primary reactions are absorbed in other target nuclei. To resolve this problem it is necessary to study γ-ray production versus target thickness. This has been done only for ^{55}Mn using pions (70) and kaons (7). It was found that the intensities per stopped meson depended linearly on target thickness. In this particular case, secondary particles were responsible for the γ rays. Some speculations have been made concerning the processes of the excited state formation, specifically, γ-ray emission after formation of hypernuclei (7) and the removal of one, two, or three α particles (69). These types of excitation are of interest because large momentum transfer components of the nuclear wave functions are involved. In particular, recent $\pi^\pm$-nucleus reaction experiments reported results that tend to confirm multiparticle removal (71–73); but the interpretation of another experiment seems to disagree (74). Further effort to produce concrete evidence is needed.

In order to identify the K^--neutron and K^--proton absorptions specifically, some nuclear γ-ray lines have been theoretically predicted from kaons stopped in ^{16}O and ^{208}Pb (75). These γ-ray lines have not yet been observed.

4   THEORETICAL UNDERSTANDING

The idea that kaonic atoms would be a useful means for investigating the nuclear surface was originally proposed by Jones in 1958 (76) and subsequently advanced strongly by Wilkinson (77, 78). In 1959 Wilkinson (77) pointed out that nuclear emulsion data of kaon capture products could be interpreted as the formation of α-particle clusters in the nuclear surface.

The experimental and theoretical investigations that followed have revealed unexpected complications. The first is that the K^--proton scattering is so highly energy dependent as to cause a severe difficulty for constructing the K^- nucleus optical potential in relation to the nuclear density distribution. The multichannel nature of the K^--nucleon interactions further deepens this difficulty. At present no reliable optical potential exists that agrees with all experimental data and at the same time stands on overall firm theoretical ground. The second complication comes from the fact that less than half of the kaons captured in target atoms were observed in the X-ray intensity experiments (7). An understanding of the atomic cascade scheme thus becomes essential if we are to determine from which atomic states the kaon is absorbed by the nucleus. Unfortunately the cascade scheme depends directly on how the kaon is captured in the atom. We thus face a highly complicated problem in atomic physics. At present this problem is very poorly understood. We present a brief account of previous theoretical investigations, including what has been learned from them. As less work has been done on the other exotic atoms, we primarily discuss kaonic atoms.

4.1   Atomic Capture and Cascade

The atomic capture processes for kaons are believed to be described in the same way as for pions and muons except that muons do not undergo strong interaction. The theoretical understanding of these complicated atomic processes is not an easy task. This is particularly true in the case of the capture process. Currently, however, some advances are being made in the study of this field. For the kaonic cascade calculation the same methods have been used as those for the muonic and pionic cascade processes. Some short reviews of the latter are available (3, 79, 80) and we refer the reader to these articles for further details.

4.1.1   ATOMIC CAPTURE   When a kaon is captured in an atom, perhaps primarily via an Auger process, the kaon velocity is smaller than the electron velocities in the atom. Most of the quantum mechanical calculations that have been done are based on Born approximation (81–84) and for this reason do not seem to give reliable results. Seemingly the most systematic and reliable current account of the capture process is given by a semiclassical model of Fermi & Teller (85–89). This model is constructed in the adiabatic limit. The Auger process is described as a continuous loss of the meson's energy and angular momentum by dissipation of the zero temperature electron gas to states above the Fermi level. Radiative processes can be accounted for by the classical radiation formula of an accelerated charge.

Based on this model, lengthy computer calculations have been performed by Leon & Seki (88) for a completely isolated single atom, and subsequently by Haff et al (89) for an agglomeration of neutral atoms. The former authors found that, as a consequence of the atomic capture, the population distribution $P(l)$ of kaons just inside the electronic ground state orbit is strongly suppressed for the circular and nearly circular kaonic states. A cascade calculation making use of this $P(l)$ gave crude agreement with the kaonic X-ray intensity data, but did not reproduce the Z-dependence structure. In a more realistic model of the target, the latter authors found that the energy distribution of the muon captured by an atom in agglomeration is concentrated in a small range so as to yield an almost statistical distribution of $P(l)$. This $P(l)$ is expected, however, to yield about twice as large kaonic X-ray intensities (88) as are found by experiment. On the other hand, during the atomic capture process in congregated atoms, the nuclear absorption of kaons is expected to take place when the perigee of the kaon orbits is close to the nuclei in the target. It is not known how many kaons are lost by this mechanism. The loss due to kaon decay was found experimentally to be insignificant (7).

In these calculations the meson is assumed to experience the Thomas-Fermi potential or a similar potential. Even if the known Hartree-Fock type potentials are adopted instead, these approaches may not sufficiently include the dynamical atomic structure. Meson capture involves the outer region of the atom whose dynamical response to the meson is an unexplored domain of atomic physics. The exotic atoms may turn out to be a useful means of studying the atomic surface. It has in fact been recognized that the Z-dependence structure in the kaonic X-ray intensities follows the same pattern as the statistic atomic size for metals (90). Beyond speculation, no detailed study has yet been made along this line.

The rate of the primary capture process (Auger) depends critically on how many electrons are available during the capture. Although in metals electron vacancies would be refilled instantaneously, electron depletion is a completely open question involving further areas of molecular and solid state physics.

The Fermi-Teller model is based on a simplification intended to replace the probability distribution of possibly discrete meson energy losses by a continuous energy loss. The attractive illustration of a definite trajectory of the meson is certainly a distortion. We do not know how much distortion is introduction in $P(l)$. Clearly, further work is needed in this area.

4.1.2 ATOMIC CASCADE A cascade calculation describes successive de-excitations of a kaon in a hydrogenic atom. Cascades are usually started inside the shell of the electronic atom's ground state where $n_K < 30$ ($n_K \approx (m_K/m_e)^{1/2}$). Because no confident calculation of the capture process is available, one has to assume a form of the initial population distribution $P(l)$. It is often assumed to be the convenient form

$$P(l) \propto (2l+1)\exp(al), \qquad\qquad 7.$$

where a is an adjustable parameter, $a = 0$ corresponding to the statistical distribution. The value of a is typically less than 0.2, tends to be positive, and varies from atom to atom (32, 91–93).

The cascade calculation then follows the fate of the kaon that descends via Auger and radiative (X-ray) transitions. The transition rates are safely computed by first order perturbation theory. Ferrell's formula (94),

$$\Gamma_{\text{Auger}}(\omega)/\Gamma_{\text{rad}}(\omega) = (3c^4/8\pi)\sigma_{\text{photo}}(\omega)/(Z-1)^2,$$

can also be used for the transition energy, ω, where experimental data of the photoelectric cross section for the neutral $Z-1$ atom, $\sigma_{\text{photo}}(\omega)$, are available. The Auger transitions, whose rates are proportional to $\omega^{-1/2}$, dominate for large n while radiative transitions, proportional to ω^3, dominate for small n, and both transitions tend to favor $\Delta l = -1$. As with the capture process, the Auger transition rates cannot be accurately obtained unless the electron depletion problem is well understood. This seems to be the major uncertainty in the present status of the cascade calculation (92, 95).

The nuclear absorption rates are computed in practice by use of phenomenologically introduced optical potentials. As discussed in Section 4.2, these potentials are not based on a firm theoretical foundation, but seem to agree well with the experimental data. Since the basic cascade scheme is not sensitive to minor variations in the nuclear absorption rates, a phenomenological form of the potentials would not give substantially distorted results. By use of such a potential it is found (88) that nuclear absorption dominates in small angular momentum states throughout all n's, and that only a fraction of kaons are absorbed from the last-observed circular state and often overwhelmingly more from the next-to-last state.

In Figure 7 we show a result of a cascade calculation that illustrates how kaons cascade down and where they are absorbed. We see a prominent competition between atomic and nuclear effects.

4.2   Strong Interactions with Nuclei

As discussed in the previous section, strong interactions play an important role in the atomic capture and cascade, but no reliable means is yet known for isolating the effects of the strong interaction. On the other hand, the energy-level shifts and widths of the last observed circular states are believed to be independent of these atomic processes. They are extensively used for the study of the strong interactions. The widths of the next-to-last observed circular states are also used. As discussed in Section 3.2, these widths are determined indirectly from the relative intensity data. The determination depends on atomic processes, and care must be taken in this respect.

The formation of hadronic hydrogen-like atoms via the attractive Coulomb interaction prepares the hadrons to interact with the nuclei in a way fundamentally different from the strong interactions in scattering (1). The hadrons interact with nuclei at definite kinetic energies of less than a few MeV (2). The hadrons interact with nuclei in definite angular momentum states of non-zero values, except with the proton for which the $l = 0$ interaction is expected.[2]

The observed widths and shifts reflect the strength of the strong interactions in the same way as the phase shifts do in scattering. Hadrons in non-zero angular momentum states of exotic atoms overlap the nuclei mainly in their surface region (equation 2 and Figure 4). This fact has encouraged the idea that these shifts and widths would also reflect the nuclear surface distribution. In order to proceed with this idea one has to establish a reliable means of relating the strong interaction potential to the nuclear density.

4.2.1   K⁻-NUCLEUS OPTICAL POTENTIAL

4.2.1.1 Phenomenological potentials   The simplest means of constructing the K^--nucleus optical potential is to assume the K^--(ith) nucleon interaction $v_i(K^--N)$ to be point-like, described in the form of a pseudo-potential,

$$v_i(K^--N) \approx -4\pi(\hbar^2/2\mu_K)f_i(K^--N)\delta^3(\mathbf{r}_i-\mathbf{r}_K),\qquad\qquad 8.$$

so that in Born approximation the K^--nucleon amplitude is exact at the energy and momentum of interest. In equation 8 μ_K is the reduced mass of the kaon and nucleon system, $\mu_K = m_K m_N/(m_K+m_N)$ and $\mathbf{r}_i$ and $\mathbf{r}_K$ are the coordinates of the ith nucleon and the kaon respectively. Once $v_i(K^--N)$ is set, the K^--nucleus potential V is obtained as

$$V = \sum_i \langle\psi|v_i|\psi\rangle$$

with neglect of nuclear excitations, ψ is the ground-state nuclear wave function and $\langle\ \rangle$ denotes integrations over $\mathbf{r}_i$'s.

For the sake of simplification one makes a further approximation, $f_i(K^--N) = -a$, where a is the K^--(ith) nucleon scattering length in free space. The negative sign is a nuclear physics convention, while most of the literature in elementary particles adopts the opposite sign. Nuclear physics convention always gives $Ima \leqq 0$ and, when the potential is weak, this convention yields the same sign of the real parts of the potential and its scattering length. The approximation, $f(K^--N) = -a$, involves neglect of all nuclear effects on $f(K^--N)$ such as on- and off-shell energy, momentum dependence, and nuclear correlation. The K^--nucleus optical potential is then obtained in the form

$$V = 4\pi(\hbar^2/2\mu)(m_K/\mu_K)(Za_p\rho_p(r)+Na_n\rho_n(r)),\qquad\qquad 9a.$$

---

[2] Mesonic atoms are "natural angular momentum selectors," as depicted by T. E. O. Ericson.

where $a_p(a_n)$ and $\rho_p(\rho_n)$ are the K^--proton (neutron) scattering lengths and the proton (neutron) density distribution normalized to unity, respectively; μ is the reduced mass of the kaon-nucleus system. The kinematic factor m_K/μ_K arises due to a shift of the center of mass from K^--nucleon to the K^--nucleus system.

After the various approximations above, the nuclear potential of equation 9a is now proportional to $\rho_n(r)$ and $\rho_p(r)$, and its form is convenient for investigating them. Nevertheless, before pursuing the investigation, one has to insure that the potential form of equation 9a is reliable, at least in the surface region where the potential is expected to be most sensitive to the kaonic atom data. In practice, the potential has generally been applied in a form

$$V = 4\pi(\hbar^2/2\mu)(m_K/\mu_K)A\bar{a}\rho(r), \qquad \qquad 9b.$$

where $\bar{a} = (Za_p + Na_n)/A$ and $A = Z + N$. The further assumption leading to equation 9b from equation 9a is $\rho_n(r) = \rho_p(r) \equiv \rho(r)$; the Wood-Saxon and harmonic-well forms have been commonly adopted as $\rho(r)$. These would not, of course, be a rigorous description of the nuclear density distributions, but have been used for the sake of a technical simplification.

When the known K^--nucleon scattering lengths (Table 7) were applied in equation 9b, the shifts and widths computed were found to disagree with the experimental data, yielding less than about half of the values of the observed widths (96–99). A remedy was then taken to keep the form of equation 9b and to treat $\bar{a}$ as an "effective scattering length in nuclei," the exact value of which was to be determined phenomenologically. It has been found that a single value of $\bar{a}$ produces kaonic atom data of various transitions very well (100, 101). Figure 8 exemplifies the overall good description of the experimental data by a phenomenological potential. The most recent value of $\bar{a}$, fit to 1972 CERN data (20), is (98)

$$a = -0.44 \pm 0.04 - i(0.83 \pm 0.07)F, \qquad \qquad 10.$$

which corresponds to a square well K^--nucleus potential of $(-42 - i79)$ MeV depth when the radius is taken to be $1.25\ A^{1/3}\ F$. More complicated versions of equation 9 were examined, but it was found that adding $\nabla\rho \cdot \nabla$ terms (the K^--nucleon p-wave interaction) (20), a $\rho^2(r)$ term (two-nucleon absorption) (20), or a $(N-Z)/A$ dependence term in $\bar{a}$ (101) does not alter the good agreement and is not required at least for the present experimental accuracy. The nuclear density dependence on $\bar{a}$ was also examined: for given shift and width data the explicit value of $\bar{a}$ is rather sensitive to the parameters in $\rho(r)$ (102, 103).

The conclusion that the kaonic atom data provide no evidence to revoke $\rho_n = \rho_p$ is tempting, but must be avoided until the theoretical

Table 7 K$^-$-nucleon scattering lengths (a and $\bar{a}$), and the mass (M) and width (Γ) of $\Lambda(1405)$ obtained by multichannel analyses.

		Martin-Sakitt[a]	Kim[b]
$a(I = 0)$	(F)	$1.66 \pm 0.02 - i(0.69 \pm 0.02)$	$1.65 \pm 0.04 - i(0.73 \pm 0.02)$
$a(I = 1)$	(F)	$0.09 \pm 0.03 - i(0.54 \pm 0.02)$	$0.13 \pm 0.03 - i(0.51 \pm 0.03)$
$\bar{a}$	(F)	$0.48 \pm 0.03 - i(0.58 \pm 0.02)$	$0.51 \pm 0.03 - i(0.57 \pm 0.03)$
$M + i\Gamma/2$	(MeV)	$1416 \pm 4 + i(29 \pm 6)/2$	$1403 \pm 3 + i(50 \pm 5)/2$

[a] Reference 113, 134.
[b] Reference 112, 134.

origin of $\bar{a}$ is well understood in terms of the K$^-$-nucleus interaction in the nucleus. We discuss this problem below in this section, and more in detail in the following section. The good agreement found so far must also be examined critically through more accurate data since some indication of alarming disagreements has recently been reported (21, 23). In connection with this matter, there exists an interesting region of the kaonic transitions, $n = 6 \rightarrow 5$ and $7 \rightarrow 6$, which often involve deformed nuclei (Figure 8). Dynamic effects of the deformed nuclei would appear in the X-ray spectra as additional shifts and widths of the energy levels (104, 105), admixtures of level splitting (105), and $E2$ dynamics (Section 3.4). A systematic measurement of these transitions has not been made, but its careful analysis may turn out to be an important means of examining the validity of the phenomenological potentials and other potentials discussed below and in the following section.

Comparing equation 10 and Table 7, one observes that the phenomenological $\bar{a}$ has a negative real part (an attractive K$^-$-nucleus interaction by our convention), whereas the known K$^-$-nucleon scattering lengths have positive real parts (yielding a repulsive K$^-$-nucleus interaction via equation 9b). The confusing sign change originates from the fact that the K$^-$-nucleon interaction is so strong and absorptive that the potential and its scattering length no longer have the same sign of the real parts (in our convention). This fact is neglected in the point-like or zero-range approximation (equation 8). In fact, the K$^-$-nucleon potential of a reasonable range is found to be attractive and strongly absorptive (98, 106–108). The strong absorption then causes the scattering length to be positive by overcoming the attractive real part of the interaction. In the case of the K$^-$-proton interaction, this effect is greatly amplified because "the kaon and proton spend more time together" (106) via K$^- + \mathrm{p} \rightarrow \Lambda(1405) \rightarrow \Sigma + \pi(100\%)$.

The underlying physics discussed here has been exemplified by direct

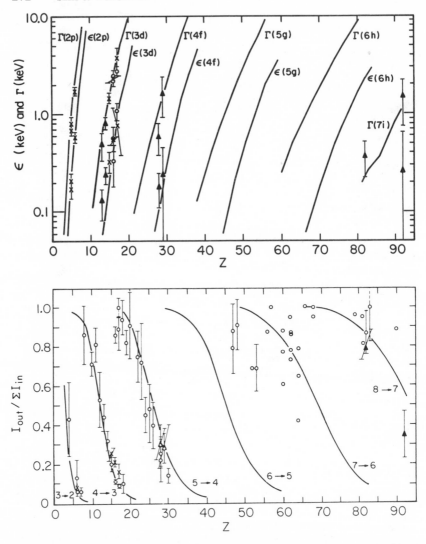

Figure 8   Kaonic energy level shifts (ε), broadening (Γ), and $I_{out}/\Sigma I_{in}$ via equation 3 computed from a phenomenological potential. The "effective scattering length in nuclei" $\bar{a}$ is $-0.60 - i0.71$ F, by fitting only to 1970 CERN ^{32}S data (133) and a 1971 Cl data (22). The experimental data shown are from the Berkeley group (○), the CERN group (✗), and the Argonne group (▲). Most of the experimental $I_{out}/\Sigma I_{in}$ data are not corrected by cascade calculations (Section 3.2) and the experimental errors are not shown when they are greater than 20%. No data are shown for $6 \to 5$ $I_{out}/\Sigma I_{in}$ because the transitions $6 \to 5$ and $8 \to 6$ have nearly the same energies, which made intensity measurements difficult.

applications of finite range K^--nucleon potentials. K^--nucleus optical potentials were constructed either by "smearing the K^--nucleon potentials over the nucleus" (98, 106, 107) or "folding in the nuclear density distribution" (106, 108). Both methods have been shown to yield potentials similar to the phenomenological ones. As seen in the above discussion, however, it is clear that a substantial K^--nucleon dynamics is at work in the nucleus. In order to obtain a K^--nucleus optical potential reliable enough to extract a detailed nuclear density distribution, one has to establish its microscopic foundation starting from the detailed K^--nucleon interactions in the nucleus. We discuss theoretical efforts on this in the following section.

Before closing this section, we make a pertinent remark. In computing the shifts and widths of two observed states from an optical potential, one must avoid first order perturbation theory. In 1971 it was found independently by Krell (96) and Seki (97) that perturbation theory for a repulsive, complex potential, such as with the known K^--nucleon scattering lengths, yields substantially smaller widths than the values obtained by numerical integration of the Klein-Gordon equation. In some situations perturbation theory gives only one third of the values obtained by a numerical integration. In the case of an attractive, complex potential, such as with a phenomenologically fitted $\bar{a}$, the attractive real part of the potential enhances the widths so as to bring about a fair agreement for the widths by the two computational methods. This is an accidental agreement depending critically on the magnitude of both real and imaginary parts, exemplified by the fact that the shifts obtained by two methods have opposite signs. In principle, the smallness of the shifts and widths is a necessary, but not a sufficient, condition for first order perturbation theory to be applicable, as previously pointed out in the case of the pionic atoms (109). One must use care in interpreting results of analyses in the literature, particularly prior to 1971. But this is a numerical problem, rather than one of physics.

4.2.1.2 K^--nucleon interactions in nuclei and theoretical K^--nucleus optical potentials   The K^--nucleon interactions are as follows:

$$K^- + N \rightarrow K^- + N, \quad \bar{K}^0 + N$$

$$\rightarrow \Sigma + \pi$$

$$\rightarrow \Lambda^0 + \pi.$$

The prominent feature in the K^--proton interaction is formation of Λ (1405) about 27 MeV below the K^--proton threshold (zero kinetic energy). Λ (1405) has isospin-0 and decays only to $\Sigma + \pi$. The Particle

Data Group (27) gives the mass and width of Λ (1405) as 1405 ± 5 and 40 ± 10 MeV, respectively. The K⁻-nucleon scattering amplitudes below the threshold are obtained up to now by extrapolations based on multi-channel, effective-range-type methods (110, 111).

Figure 9 and Table 7 summarize the results most frequently quoted (112, 113). In Figure 9 we observe the highly energy-dependent K⁻-proton scattering amplitude that differs recognizably in the two analyses despite the very close values of the scattering lengths shown in the table. Because of Λ (1405) the K⁻ proton amplitude is not so well determined that it can be relied on explicitly. The K⁻-neutron amplitude is expected to be a much less "pathological case." Recently Chao et al (114)

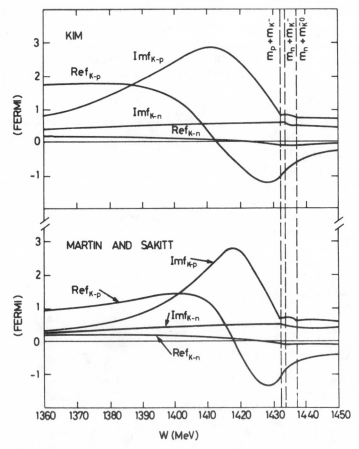

Figure 9   K⁻-P and K⁻-n scattering amplitudes based on multichannel effective-range-type calculations (99, 115).

showed that $\Sigma \pi$ production data give severe constraints on the mass and width of Λ (1405). Further studies are required to determine with confidence the K^--nucleon amplitudes below threshold.

In order to incorporate the strong energy dependence in the K^--p scattering amplitude, Bardeen & Torigoe (99, 115) proposed using an effective amplitude in equation 8. The effective $f_i(K-N)$ was obtained by weighting $f_i(K-N)$ with the probability of finding the K^--nucleon reaction taking place in the nucleus at various K^--nucleon center-of-mass energies. Most of the K^--nucleon interactions are between about -10 and -40 MeV in the K^--nucleon center-of-mass kinetic energy system, peaking around -20 MeV. 20 MeV is close to the nucleon binding energy in nuclear matter plus the kaon binding energy in the atom. The averaged $f_i(K-N)$ for the proton turns out to have the sign of the real part opposite to that at threshold. That is, the K^--nucleus optical potential thus constructed is attractive and proportional to the nuclear density distribution. This potential gave a reasonable agreement with 1972 CERN data (20). Ideas similar to the above modification of $f_i(K-N)$ have been also discussed by Bethe & Siemens (116), Burhop (66), Bloom, Johnson & Teller (117), and recently Chattarji & Ghosh (118).

However, the K^--nucleon interactions in the nucleus take place in the domain of the off-energy shell caused by nuclear binding. Furthermore, the presence of the other nucleons near the interacting kaon-nucleon pair may induce dynamical effects via the Pauli exclusion principle and nuclear correlations. Wycech (119) performed a detailed calculation of the K^--nucleon scattering amplitudes in the nucleus by examining their dependence on the position, momentum, and energy of the kaon in the nuclear Fermi gas. Since the scattering amplitudes obey multichannel Lippmann-Schwinger equations that involve the K^--nucleus potential itself, the scattering amplitudes and in turn the K^--nucleus potential are calculated in a self-consistent way. The scattering amplitudes are approximated so as to yield a local K^--nucleus potential, but not proportional, to the nuclear distribution. The shifts and widths calculated with this potential give a fair agreement with the experimental data (120). Two-nucleon correlation and the Pauli exclusion principle have been shown to have a prominent effect on $\Lambda(1405)$ in the nucleus. In the region where the nuclear density is greater than about $20 \sim 30\%$ of the central value, the width is found to be broadened to about 70 MeV and the position of the bound state to be shifted to about -10 or -15 MeV.

The effects of the nonlocality and off-energy shell on the K^--nucleon interactions have been explicitly examined and found to be appreciable by Alberg, Henley & Wilets (121), and Alberg (122). The off-energy-shell K^--nucleon scattering amplitudes are obtained from separable K^--nucleon potentials that describe the multichannel K^--nucleon scatterings in free space. The K^--nucleus potential is then constructed from these scattering amplitudes via a multiple scattering theory without nuclear correlations. The resulting potential is nonlocal, so different localizations are examined. None of the localizations is found to follow the nuclear distribution, but agreement with shift and width data turns out to be fair for these localized potentials.

The K^--nucleon interactions in the nucleus thus seem to be highly sensitive to all nuclear effects considered. One feature appears to be clear, however: the behavior of the K^--proton interaction dominates the K^--nuclear interaction. In order to learn the detailed nuclear distribution, then, it seems that we require more reliable descriptions of the K^--nucleon interactions and how they are influenced by nuclear effects. It is not yet entirely clear whether the good overall agreement between the data and the phenomenological potentials described in the previous section is accidental, due to the choices of the nuclear density distribution, or a consequence of some important physics.

In closing this section, we take note of a peculiar contradiction of the above works. Revai (123) performed a three-body calculation for a system consisting of the kaon, a nucleon, and the rest of the nucleus. He found that the effect of $\Lambda(1405)$ due to the $\Sigma\pi$ channel yields little variation in the K^--proton amplitude as a function of the energy below threshold. Despite recent work on the K^--deuteron three-body calculation (124), this puzzle seems to remain unsolved.

4.2.2 $\bar{p}$- AND Σ^--NUCLEUS OPTICAL POTENTIAL   Because reliable information is not available on the $\bar{p}$ and Σ^--nucleon scattering near threshold, little is known about the $\bar{p}$ and Σ^--nucleus optical potential. The effective scattering length $\bar{a}$ in the phenomenological potential (equation 9) has been determined by fitting the available shift and width data in $\bar{p}$ atoms. It is (28)

$$\bar{a} = -2.9^{+1.0}_{-1.4} - i(1.5^{+0.6}_{-1.2}) \text{ F}.$$

This is to be compared with the $\bar{p}$-nucleon scattering length (125)

$$\bar{a} = +0.88 - i0.81 \quad \text{or} \quad +0.91 - i0.69 \text{ F},$$

which is computed from a semitheoretical $\bar{p}$-nucleon optical potential that agrees reasonably well with the $\bar{p}$-nucleon scattering data below 300 MeV. Again we seem to have the sign change in Re $\bar{a}$ that appeared in the K^--nucleus optical potentials (Section 4.2.1). The $\bar{p}$-

nucleus potential constructed by the method of "folding in the nuclear distribution" yields the correct sign change (126). The $\bar{p}$-nucleon scattering amplitudes in nuclei should, however, be carefully examined in order to understand the $\bar{p}$-nucleus interaction. For this purpose more reliable information would be needed on the $\bar{p}$-nucleon scattering amplitudes in free space.

No information is available on the Σ^--nucleus optical potential. Based on poorly known Σ^--nucleon scattering lengths (127), an educated guess on $Im\,\bar{a}$ is within a few F.

4.3   Two- and One-Nucleon Absorption Products

According to Table 6 in Section 3.5, kaonic two-nucleon absorption occurs in deuterium for only about 1% of the total absorption, whereas in ^{4}He and other nuclei it occurs with a probability of about 20%. Wilkinson (78) proposed this experimental observation as possible evidence of α-particle cluster formation in the nuclear surface. However, detailed analyses by Wycech (128) and Aslam & Rook (129) showed that the α-cluster formation is not necessary to explain the large two-nucleon absorption rates in nuclei other than deuterium. The mechanism considered by these authors is that the virtual $\Lambda(1405)$ produced by proton aborption interacts with a second nucleon in a finite range via a virtual pion that is a decay product of $\Lambda(1405)$. Thus Σ and Λ and one nucleon are produced. It is argued that this mechanism occurs even in that part of the nuclear surface region in which two nucleons are more tightly bound than deuterium. This region corresponds to nuclear densities greater than roughly 10% of the nuclear density at the center.

Nevertheless, there is a question of how much K^- absorption takes place in the nuclear surface region. As discussed in Section 4.1, answering this question requires a better understanding of the atomic capture and cascade processes. The same problem exists in understanding the various rates that result from one-nucleon absorption, such as R_{pn} and R_{+-}, and the occurrence ratio of neutron and proton absorption. In Section 3.5 we mentioned the question of reliability in deducing the values of these quantities from the observed data. Besides this question, an uncertainty in understanding the atomic schemes worsens the possibility of verifying the theoretical K^--nucleus optical potentials by comparing their predictions with these quantities. Several attempts (65, 66, 115–117, 119, 130) have been made to combine these quantities in relation to one-nucleon absorption and the K^--nucleon interactions in the nucleus. Efforts (131, 132) have also been made to explicitly include the atomic cascade scheme. Yet the complexity of the problem seems to prevent us from obtaining confident, detailed information.

5 CONCLUDING REMARKS

We have seen that exotic atoms are complex phenomena, rich in physics that remains to be completely understood. Here three stages of physics—atomic, nuclear, and elementary particle physics—are interrelated, and better understanding of all three disciplines is needed. This is particularly true of the early hope of learning about the nuclear surface structure. Regrettably, we are far from reaching that goal. Whether or not this complicated mixture of disciplines will turn out to be a fertile source of physics, must be answered in the future.

Clearly, theoretical and experimental effort should continue. "Kaon factories," whose construction should be studied seriously, could foster major experimental advances. Nuclear physics involving strangeness could be one of the interesting fields to be pursued.

ACKNOWLEDGMENTS

We wish to express thanks to our colleagues in the field of the exotic atoms for various discussions of the material presented. Part of the work was done when one of us (R.S.) stayed at Lawrence Berkeley Laboratory and Rutherford High Energy Laboratory. He is grateful to Professor O. Chaimberlain and Professor C. Batty for their hospitality.

Literature Cited

1. Sternheimer, R. M., Goldhaber, M. 1973. *Phys. Rev. A* 8:2207
2. Wu, C. S., Wilets, L. 1969. *Ann. Rev. Nucl. Sci.* 19:527
3. Backenstoss, G. 1970. *Ann. Rev. Nucl. Sci.* 20:467
4. European K⁻ Collaboration. 1959. *Nuovo Cimento* 13:690
5. Johnson, M. H., Teller, E. 1954. *Phys. Rev.* 93:357; Wilets, L. 1956. *Phys. Rev.* 101:1805
6. Wiegand, C. E., Mack, D. A. 1967. *Phys. Rev. Lett.* 18:685
7. Wiegand, C. E., Godfrey, G. L. 1974. *Phys. Rev. A* 9:2282; 1974. *Lawrence Berkeley Lab. Rep.* 3080
8. Goulding, F. S., Jaklevic, J. M. 1973. *Ann. Rev. Nucl. Sci.* 23:45
9. Landis, D. A., Goulding, F. S., Jaklevic, J. M. 1970. *Nucl. Inst. Methods* 87:211
10. Wiegand, C. E., Pehl, R. H. 1971. *Phys. Rev. Lett.* 27:1410
11. Wiegand, C. E. 1969. *Phys. Rev. Lett.* 22:1235
12. Kunselman, A. R. 1969. *Lawrence Berkeley Lab. Rep.* UCRL-18654
13. Quitmann, D. et al 1964. *Nucl. Phys.* 51:609
14. Zinov, V. G., Konin, A. D., Mukhin, A. I. 1965. *Yad. Fiz.* 2:859; *Sov. J. Nucl. Phys.* 2:613
15. European K⁻ Collaboration. 1959. *Nuovo Cimento* 14:1959
16. Schorochoff, G. 1969. *Univ. Libre Bruxelles Int. Phys. Bull.* No. 40
17. Zieminska, D. 1971. *Phys. Lett. B.* 37:403
18. Blomqvist, J. 1972. *Nucl. Phys. B* 48:95
19. Backenstoss, G. et al 1973. *Phys. Lett. B* 43:431

20. Backenstoss G. et al 1972. *Phys. Lett. B* 38:181
21. Koch, H. 1974. *Proc. Int. Conf. High Energy Phys. Nucl. Struct., 5th,* 225
22. Kunselman, A. R. 1971. *Phys. Lett. B* 34:485
23. Barnes, P. D. et al 1974. *Nucl. Phys. A* 231:477
24. Miller, J. 1974. *An experimental investigation of strong interaction effects in kaonic and antiprotonic atoms.* PhD thesis. Carnegie-Mellon Univ., Pittsburgh, Pa.
25. Backenstoss, G. et al 1972. *Phys. Lett. B* 41:552
26. Bamberger, A. et al 1970. *Phys. Lett. B* 33:233
27. Particle Data Group. 1974. *Phys. Lett. B* 50:1
28. Barnes, P. D. et al 1972. *Phys. Rev. Lett.* 29:1132
29. Bethe, H. A., Salpeter, E. E. 1957. *Quantum Mechanics of One- and Two-Electron Atoms.* Berlin: Springer-Verlag
30. Fried, Z., Martin, A. D. 1963. *Nuovo Cimento* 29:574
31. Leon, M., Seki, R. 1973. *Phys. Lett. B* 46:1973
32. Backenstoss, G. et al 1974. *Nucl. Phys. B* 73:189
33. Roberts, B. L. 1974. *Measurements of magnetic moments of antiproton and* Σ^- *hyperon.* PhD thesis. College of William and Mary, Williamsburg, Va.
34. Roberts, B. L. 1974. *Phys. Rev. Lett.* 32:1265
35. Fox, J. D. et al 1972. *Phys. Rev. Lett.* 29:193
36. Fox, J. D. et al 1973. *Phys. Rev. Lett.* 31:1084
37. Kokkedee, J. J. J. 1969. *The Quark Model.* New York: Benjamin
38. Iachello, F., Lande, A. 1971. *Phys. Lett. B* 35:205
39. Ericson, T. E. O., Hufner, J. 1972. *Phys. Lett. B* 40:459; 1972. *Nucl. Phys. B* 47:205
40. Schroder, U. E. 1972. *Acta Phys. Austr.* 36:248
41. Cheng, S. C. et al 1974. *Bull. Am. Phys. Soc.* 19:599
42. Ara, G., Chen, M. Y. 1974. *Bull. Am. Phys. Soc.* 19:599
43. Leon, M. 1974. *Phys. Lett. B* 50:425
44. Bradbury, J. N., Leon, M., Daniel, H., Reidy, J. J. 1975. *Phys. Rev. Lett.* 34:303
45. Godfrey, G. L., Lum, G. K., Wiegand, C. E. 1975. *Lawrence Berkeley Lab. Rep.* No. 3616
46. European K⁻ Collaboration. 1959. *Nuovo Cimento* 14:315
47. Evans, D. et al 1961. *Proc. Roy. Soc. London A* 262:73
48. Jones, B. D. et al 1961. *Nuovo Cimento* 19:1077
49. Eisenberg, Y., Friedmann, M., Alexander, G., Kessler, D. 1961. *Nuovo Cimento* 22:1
50. Cester, R. et al 1961. *Nuovo Cimento* 22:1069
51. Condo, G. T., Hill, R. D. 1963. *Phys. Rev.* 129:388
52. Davis, D. H. et al 1967. *Nucl. Phys. B* 1:434
53. Lovell, S. P., Schorochoff, G. 1968. *Nucl. Phys. B* 5:381
54. Knight, W. L. et al 1964. *Nuovo Cimento* 32:598
55. Davis, H. et al 1968. *Nuovo Cimento A* 53:313
56. Csejthey-Barth, M., Schorochoff, G., Van Binst, P. 1969. *Nucl. Phys. B* 14:316
57. Cohn, H. O. et al 1968. *Phys. Lett B* 27:527
58. Moulder, J. W. et al 1971. *Nucl. Phys. B* 35:332
59. Sudarshan, E. C. G., Tinlet, J. H., Melissinos, A. C., eds. 1960. *Proc. Int. Conf. High Energy Phys. Rochester,* 419, 423, 426. Rochester, NY: Univ. Rochester
60. Katz, P. A. et al 1970. *Phys. Rev. D* 1:1267
61. Dahl, O. et al 1963. *Nuovo Cimento* 27:343
62. Bugg, W. M. et al 1973. *Nucl. Phys. B* 64:29
63. Bugg, W. M. et al 1973. *Phys. Rev. Lett.* 31:475
64. Andersen, B., Skjeggestad, O., Davis, D. H. 1963. *Phys. Rev.* 132:2281
65. Burhop, E. H. S. 1967. *Nucl. Phys. B* 1:438
66. Burhop, E. H. S. 1972. *Nucl. Phys. B* 44:445

67. Gerace, W. J., Sternheim, M. M., Walker, J. F. 1974. *Phys. Rev. Lett.* 33:508
68. Dahl, O. I. et al 1961. *Phys. Rev. Lett.* 6:142
69. Barnes, P. D. et al 1972. *Phys. Rev. Lett.* 29:230
70. Riddle, R. A. J. et al 1973. *Int. Conf. High Energy Phys. Nucl. Struct., 5th, Contrib. Pap.* 147
71. Jackson, H. E. et al 1973. *Phys. Rev. Lett.* 31:1353
72. Lind, V. G. et al 1974. *Phys. Rev. Lett.* 32:479
73. Ashery, D. et al 1974. *Phys. Rev. Lett.* 32:943
74. Ullrich, H. et al 1974. *Phys. Rev. Lett.* 33:433
75. Bloom, S. D., Weiss, M. S., Shakin, C. M. *Phys. Rev. C* 5:238
76. Jones, P. B. 1958. *Phil. Mag.* 3:33
77. Wilkinson, D. H. 1959. *Phil. Mag.* 4:215
78. Wilkinson, D. H. 1960. *Proc. Kingston Conf. Nucl. Struct.,* 24; 1961. *Proc. Rutherford Jubilee Int. Conf.,* 339; 1968. *Proc. Int. Conf. Nucl. Struct., Suppl. J. Phys. Soc., Japan,* 469; 1967. *Comments Nucl. Particle Phys.* 1:36, 80, 112
79. Burhop, E. H. S. 1969. *High Energy Phys.* 3:110
80. Kim, Y. N. 1971. *Mesic Atoms and Nuclear Structure.* Amsterdam: North-Holland
81. de Borde, A. H. 1954. *Proc. Roy. Soc. London* 67:57
82. Mann, R. A., Rose, M. E. 1961. *Phys. Rev.* 121:293
83. Martin, A. D. 1963. *Nuovo Cimento* 27:1359
84. Au-Yang, M. Y., Cohen, M. L. 1963. *Phys. Rev.* 174:468
85. Fermi, E., Teller, E. 1947. *Phys. Rev.* 72:399
86. Rook, J. R. 1970. *Nucl. Phys. B* 20:14
87. Aslam, K. 1970. *Theoretical study of the decay of K-mesic atoms.* PhD thesis. Univ. Oxford, Oxford, England
88. Leon, M., Seki, R. 1974. *Phys. Rev. Lett.* 32:132; 1973. *Los Alamos Sci. Lab. Rep.* LA-UR-73-943
89. Haff, R. K., Vogel, P., Winther, A. 1974. *Cal. Tech. Rep.* BAP-3; 1975. CALT-63-214
90. Condo, G. T. 1974. *Phys. Rev. Lett.* 33:126
91. Eisenberg, Y., Kessler, D. 1963. *Phys. Rev.* 130:2352; 1961. *Phys. Rev.* 123:1472
92. Sapp, W. W. Jr. 1970. *Pionic X-ray yields and level broadening in low-Z atoms.* PhD thesis. College of William and Mary, Williamsburg, Va.
93. Atarashi, M., Narumi, H. 1971. *Progr. Theor. Phys.* 45:1779
94. Ferrell, R. A. 1960. *Phys. Rev. Lett.* 4:425
95. Suzuki, A. 1967. *Phys. Rev. Lett.* 19:1005
96. Krell, M. 1971. *Phys. Rev. Lett.* 26:584
97. Seki, R. 1971. *Bull. Am. Phys. Soc.* 16:80; 1972. *Phys. Rev. C* 5:1196
98. Koch, J. H., Sternheim, M. M. 1972. *Phys. Rev. Lett.* 28:1061
99. Bardeen, W. A., Torigoe, E. W. 1972. *Phys. Lett. B* 38:135
100. Kunselman, A. R., Seki, R. 1973. *Phys. Rev. C* 8:2492
101. Seki, R. 1972. *Phys. Rev. Lett.* 29:240
102. Ericson, T. E. O., Scheck, F. 1970. *Nucl. Phys. B* 19:450.
103. Seki, R., Kunselman, A. R. 1973. *Phys. Rev. C* 7:1260
104. Haff, P. K., Eisenberg, J. M. 1970. *Phys. Lett. B* 33:133
105. Scheck, F. 1972. *Nucl. Phys. B* 42:573
106. Uretsky, J. L. 1967. *High Energy Phys. Nucl. Struct.,* 395; 1966. *Phys. Rev.* 147:906
107. von Hippel, F., Douglas, J. H. 1966. *Phys. Rev.* 146:1042
108. Deloff, A., Law, J. 1974. *Phys. Rev. C* 10:1688
109. Seki, R., Cromer, A. H. 1967. *Phys. Rev.* 156:93
110. Dalitz, R. H., Tuan, S. F. 1960. *Ann. Phys.* 10:307
111. Ross, M. H., Shaw, G. L. 1961. *Ann. Phys.* 13:147
112. Kim, J. K. 1967. *Phys. Rev. Lett.* 19:1074
113. Martin, B. R., Sakitt, M. 1969. *Phys. Rev.* 183:1345

114. Chao, Y. A. et al 1973. *Nucl Phys.* *B* 56:46
115. Bardeen, W. A., Torigoe, E. W. 1971. *Phys. Rev. C* 3:1785
116. Bethe, H. A., Siemens, P. J. 1970. *Nucl. Phys. B* 21:589
117. Bloom, S. D., Johnson, M. H., Teller, E. 1969. *Phys. Rev. Lett.* 23: 28; 1972. *Magic Without Magic: John Archibald Wheeler,* p. 89. San Francisco: Freeman
118. Chattarji, D., Ghosh, P. 1973. *Phys. Rev. C* 8:2115
119. Wycech, S. 1971. *Nucl. Phys. B* 28: 541; 1974. *Proc. Int. Conf. High Energy Phys. Nucl. Struct., 5th,* 239
120. Rook, J. R., Wycech, S. 1972. *Phys. Lett. B* 39:469
121. Alberg, M., Henley, E. M., Wilets, L. 1973. *Phys. Rev. Lett.* 30:255; 1972. *Comments in Nucl. Particle Phys.* 5:1; 1975. Preprint. Univ. Washington, Seattle, Wash.
122. Alberg, M. A. 1974. *K-nucleus optical potential for kaonic atoms.* PhD thesis. Univ. Washington, Seattle, Wash.
123. Revai, J. 1970. *Phys. Lett. B* 33: 587
124. Myhrer, F. 1973. *Phys. Lett. B* 45: 96
125. Bryan, R. A., Phillips, R. J. N. 1968. *Nucl. Phys. B* 5:201
126. Deloff, A., Law, J. 1974. *Phys. Rev.* 10:2657
127. Alexander, G., Gell, Y., Steiner, I. 1972. *Phys. Rev. D* 6:2405
128. Wycech, S. 1967. *Acta Phys. Pol.* 32:161
129. Aslam, K., Rook, J. R. 1970. *Nucl. Phys. B* 20:159
130. Aslam, K., Rook, J. R. 1970. *Nucl. Phys. B* 20:397
131. Burhop, E. H. S., Davis, D. H., Sacton, J., Schorochoff, G. 1969. *Nucl. Phys. A* 132:625
132. Leon, M., Seki, R. 1974. *Nucl. Phys. B* 74:68; 1974. *Phys. Lett. B* 48: 173
133. Backenstoss, G. et al 1970. *Phys. Lett. B* 32:399
134. Ebel, G. et al 1970. *Nucl. Phys. B* 17:1

RADIOMETRIC CHRONOLOGY OF THE EARLY SOLAR SYSTEM

×5563

George W. Wetherill
Department of Terrestrial Magnetism, Carnegie Institution of Washington, Washington DC 20015

CONTENTS

1 INTRODUCTION

This article reviews the present status of our knowledge of major events in the earliest history of the solar system, primarily as learned from dating methods using radioisotopes.

From the provincial point of view of a resident of the solar system, the time scale of the universe can be divided into two major subdivisions: presolar and solar. The former includes all time prior to the separation from the interstellar medium of the dust and gas that has evolved into the sun, planets, and other bodies in the solar system. Thus presolar time includes events such as the initiation of the mutual recession of the galaxies, as evidenced by their spectral red shift, the nucleosynthesis of the light elements associated with that primordial event (1), the formation of our galaxy, and the formation, evolution, and in some cases the terminal phases of stars formed before the sun. The synthesis of the elements (at least those with $Z \geqq 6$) which comprise the matter of the solar system occurred in the interior of presolar stars. They were returned to the interstellar medium by processes of mass loss from stars, primarily supernova explosions. Considerable information regarding the presolar time scale is provided by radiometric chronology, as reviewed by Schramm (2). The present review is confined to part of the subsequent solar portion of history, and simply makes use of conclusions regarding presolar history when necessary.

Much of the clarification achieved in this area during the past 25 years is a consequence of the clear separation of events during various stages in the history of the universe. Processes associated with the "Big Bang," $\sim 10^{10}$ yr ago, are distinguished from the synthesis of the elements that were to form the solar system in the interior of ordinary presolar stars (3, 4). The synthesis of the elements in turn is distinguished from the formation of the sun and solar system ~ 4.6 b.y. ago, fairly late in galactic history. This has permitted a "demythologizing" of early solar system history by disentangling it from philosophical questions concerning the concept of time and from the need to consider unfamiliar or extremely high density and high temperature states of matter. Although there is still very much to be understood regarding the origin of the solar system, the problem is now well defined as an adjunct of normal processes of star formation, such as are occurring at present in regions of the galaxy such as the Orion Nebula (5, 6).

2 METHODS USED FOR RADIOMETRIC CHRONOLOGY

This review primarily considers results rather than techniques. To make these results more understandable, however, we give a brief treatment of the various approaches to radiometric dating of solar system events. The various radiometric methods have one thing in common: they all measure a radioactive decay interval.

If at some time t_1 the quantity of a radioisotope was P_1, whereas at some previous time t_0 this quantity was P_0, the interval between t_1 and t_0 is given by the law of radioactive decay as

$$(t_1 - t_0) = (1/\lambda)\ln P_0/P_1. \qquad\qquad 1.$$

Many methods of radiometric chronology are included in this class, e.g. ^{14}C, uranium fission-track, ^{87}Rb-^{87}Sr, and ^{129}I-^{129}Xe dating, as is discussed more fully later. The fundamental thing they have in common is that the time measurement is based on radioactive disintegration and the time scale is proportional to the value of a radioactive decay constant.

2.1 Formation Intervals

There are a number of radioisotopes with half-lives in the range 10^6–10^8 years. These isotopes are formed in galactic nucleosynthesis and therefore were present along with the stable isotopes at the time of solar system formation. However, they are sufficiently short-lived to have decayed to extremely low, usually undetectable, concentrations during the 4.6 b.y. of earth and solar system history. Such radioisotopes are referred to as "extinct activities." Their former presence is indicated by an anomaly in the isotopic abundance of their stable daughter isotope.

The first and most important of these extinct activities, 17 m.y., ^{129}I, was discovered by Reynolds in 1960 (7), as evidenced by a marked excess of its stable daughter ^{129}Xe in the mass spectrum of Xe extracted from the stone meteorite Richardton, and subsequently measured in many other meteorites (8) and in lunar materials (9). This excess was caused by the in situ decay of ^{129}I in the sample, the quantity of ordinary ^{129}Xe being sufficiently small to permit a pronounced enhancement to occur.

The time interval Δ between the isolation of the solar system material from galactic nucleosynthesis and the time at which the parent body of the meteorite had cooled sufficiently to retain Xe is given by

$$(^{129}Xe^*/^{127}I) = (^{129}I/^{127}I)_0 \exp(-\lambda_{129}\Delta), \qquad\qquad 2.$$

where $^{129}Xe^*$ is the excess Xe attributed to the decay of ^{129}I, ^{127}I is the concentration of the ordinary nonradiogenic stable isotope of iodine, $(^{129}I/^{127}I)_0$ is the ratio of the abundance of the iodine isotopes at the time of isolation from galactic nucleosynthesis, and λ_{129} is the decay constant of ^{129}I.

Solving equation 4 for Δ, we find that

$$\Delta = (1/\lambda_{129})\ln[(^{129}I/^{127}I)_0/(^{129}Xe^*/^{127}I)]. \qquad\qquad 3.$$

The use of the ratio of $^{129}I_0$ to $^{127}I_0$ instead of the absolute concentration of $^{129}I_0$ permits the calculation to be independent of the very large chemical fractionations that occurred in the condensation, accretion, and internal evolution of bodies in the solar system. Knowledge of the ratio

$(^{129}I/^{127}I)_0$ is both provided and limited by our understanding of the production rates of these isotopes in nucleosynthesis, and by the time dependence of the rate of galactic nucleosynthesis. A particularly simple case is that of steady state nucleosynthesis (10),

$$(^{129}I/^{127}I)_0 = (P_{129}/P_{127})\lambda_{129} T = 5 \pm 2 \times 10^{-3}, \qquad \qquad 4.$$

where (P_{129}/P_{127}) is the ratio of the production rates and T is the age of the galaxy. More complex and possibly more realistic models have also been discussed (10, 11), in which the ratio $(^{129}I/^{127}I)_0$ differs from the steady state value; for example it is larger by factors of 10–100 as a consequence of a "late spike" of nucleosynthesis immediately prior to the formation of the solar system. The accuracy of the absolute value of the formation interval depends on the actual initial $(^{129}I/^{127}I)_0$ ratio and hence on details of galactic nucleosynthesis. It should be noted, however, that the difference between the formation interval of two bodies is not dependent upon the initial ratio of the iodine isotopes. Furthermore, the logarithmic dependence of Δ on the $^{129}Xe/^{127}I$ ratio results in Δ being insensitive to experimental errors in this measured ratio.

The actual experimental determination of formation intervals is considerably more complex than may be apparent from the foregoing discussion. This involves conversion of the ^{127}I to ^{128}Xe (stable) by neutron irradiation, stepwise heating of the sample to distinguish between ^{129}Xe generated by the in situ decay of ^{129}I and trapped radiogenic ^{129}Xe, and numerous corrections, such as that for spallation Xe produced by galactic cosmic rays. These important details may be found elsewhere (12).

Another extinct radioactivity of considerable importance is 82 m.y. ^{244}Pu. The effect of this nuclide was first identified by Rowe & Kuroda (13) by observation in meteorites of abundance anomalies in ^{136}Xe, ^{134}Xe, and ^{132}Xe, which they attributed to the spontaneous fission branch of the decay of this extinct transuranic isotope. This interpretation was completely confirmed in 1971 by Alexander et al (14) who showed that the xenon anomaly observed in meteorites was identical to the mass spectrum of xenon generated by the decay of ^{244}Pu produced in a nuclear reactor. ^{244}Pu may be used to determine formation intervals in a manner similar to that described above for ^{129}I, with the important exception that there is no stable isotope of Pu that can be used analogously to ^{127}I in the denominator of the left-hand side of equation 2. It is conventional to use the geochemically similar isotopes ^{238}U or ^{232}Th for this purpose. There is both experimental and theoretical evidence, however, that Pu, U, and Th are fractionated by uncertain factors in geochemical processes (15). This uncertainty is a practical difficulty in the interpretation of ^{244}Pu-Xe formation intervals.

It is no coincidence that xenon isotopes are the stable daughter products of both of these extinct radioactivities that have proven to be important. There are a number of other radioactive isotopes of similar half-life and original abundance, but with a stable daughter element much more abundant than the rare gas Xe. Observation of an excess of the daughter isotope requires reduction of the daughter/parent element ratio relative to average solar system composition prior to the decay of the extinct radioisotope. Therefore when the daughter is not a depleted element it will be much more difficult to identify the associated isotopic anomaly in the mass spectrum of the daughter element. Some effort has gone into searches for such anomalies [e.g. those produced by 15 m.y. ^{205}Pb (16, 17), 0.7 m.y. ^{26}Al (18), and 100 m.y. ^{146}Sm (19, 20)] but no clear effects have yet been observed, although ~ 100 m.y. ^{92}Nb has been found on earth (21). ^{244}Pu formation intervals have also been determined by identification and counting of charged particle fission tracks in meteoritic material (22, 23). This approach is nearly analogous to the use of fissiogenic Xe to determine the initial ^{244}Pu/^{238}U ratio of the meteorite. However, the end of the fission track formation interval is established by the meteorite parent body falling below a temperature sufficient to anneal the tracks. It is probable that this temperature is well below that required to retain xenon. In this way it is likely that these measurements can provide additional information regarding the cooling times of the meteorite parent bodies (24).

2.2 Formation and Metamorphism Ages

The extinct radioactivities discussed above were so short-lived that the increase in abundance of their daughter isotopes has been negligible subsequent to the first few hundred million years of solar system history. In contrast there are other radioisotopes with half-lives > 500 m.y. that still exist in significant quantities. For the purpose of age measurement, the most important of these are 50 b.y. ^{87}Rb, 1.3 b.y. ^{40}K, 0.7 b.y. ^{235}U, 4.5 b.y. ^{238}U, and 13 b.y. ^{232}Th. The long half-lives of these radioisotopes make it possible to measure decay intervals beginning at various times in the past and ending at the present. The relevant equation is still equation 1, where now P_1 is the still measurable present quantity of the parent radioisotope, P_0 is the quantity at some time in the past, and the decay interval $(t_1 - t_0)$ is the time that has elapsed since the quantity of the radioisotope was equal to P_0. Although P_1 is measurable, there is usually no direct way of measuring P_0, as there is no universal quantity of say, uranium, in the object (such as a meteorite or a rock) to be dated. However, if there has been no gain or loss of parent and daughter, except by decay of parent into daughter, the quantity P_0 can be determined indirectly by measuring the quantity of the stable daughter that has accumulated,

$$P_0 = P + D - D_0, \qquad\qquad 5.$$

where P and D are the present quantities of the parent and daughter isotopes in the object to be dated, and D_0 is the quantity of daughter at the beginning of the decay interval to be determined. It may seem that the problem of identifying the original conditions has merely been shifted from the parent element to the daughter element, but this is not the case, provided the daughter element has at least one other stable isotope that is not the daughter of a long-lived parent. Dividing equation 5 by the quantity of this other isotope (D_x), writing P_0 in terms of P by the equation of radioactive decay, and rearranging terms as follows,

$$(D/D_x) - (P/D_x)\{\exp[\lambda(t_1 - t_0)] - 1\} - (D/D_x)_0 = 0, \qquad 6.$$

we may identify D_x with its original value $(D_x)_0$, as it is neither radioactive nor the daughter of a naturally occurring radioisotope. All of the quantities in the first two terms of equation 6 are measurable, with the exception of the desired decay interval $(t_1 - t_0)$. The initial ratio $(D/D_x)_0$ is still unknown. The most straightforward way of determining this is to identify two objects with different values of P/D_x that had the same daughter isotope composition at time t_0. Measurement of D/D_x and P/D_x for both of these "cogenetic" or "coeval" objects then provides two equations which can be solved simultaneously for the unknown decay interval $(t_1 - t_0)$ and the unknown initial daughter element ratio $(D/D_x)_0$. The plausibility of the two objects actually fulfilling these conditions can be based on other information. For example, two igneous rocks that on geological evidence have both crystallized in a short period of time from a well-mixed melt could be expected to have the same age $(t_1 - t_0)$ and the same initial ratios $(D/D_x)_0$, even though the absolute concentrations of P and D are very different. Similarly two minerals in one of these igneous rocks, or two minerals in a thoroughly metamorphosed rock, would be expected to fulfill these conditions. In both cases the "age" $(t_1 - t_0)$ refers to the time since the mutual initial ratio was equal to $(D/D_x)_0$. In the case of the igneous rocks this would be the time of crystallization from the melt; for the metamorphosed rock it would be the time of metamorphism.

In practice it is usual to overconstrain equation 6 by measuring more than two such cogenetic objects. This is done to provide some check on the assumption that there has been no migration of parent or daughter elements in or out of the object during the decay interval, and to check on the validity of the arguments that were used to establish whether the objects were cogenetic or coeval. Least squares fitting is used for choosing the values of $(t_1 - t_0)$ and $(D/D_x)_0$ which best fit the system of simultaneous equations of form 6, and the standard deviation of the fit is some measure of the validity of the above assumptions.

It is also useful to note that if the experimental points are plotted on a diagram with D/D_x on the ordinate and P/D_x on the abscissa, cogenetic objects will yield points lying on a straight line of slope $m = \exp[\lambda(t_1 - t_0) - 1]$ and ordinate intercept $= (D/D_x)_0$. Fitting the straight line on such diagrams is equivalent to the least squares procedure mentioned above; the graphical presentation, however, permits a visual appraisal of the goodness of fit. This approach is most commonly associated with ^{87}Rb-^{87}Sr age determination, leading to the so-called "isochron" or "strontium evolution" diagrams, an example of which is given in Figure 1. This approach has also been applied to all the other

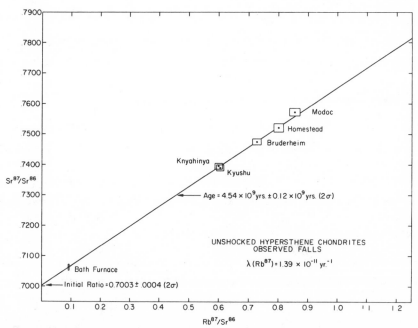

Figure 1 ^{87}Rb-^{87}Sr evolution diagram for six hypersthene chondrite meteorites (93). The data can be interpreted as showing that 4.54 b.y. ago all of these rocks had the same ^{87}Sr/^{86}Sr ratio of 0.7003.

parent-daughter decay schemes in common use for the measurement of geological ages or decay intervals.

The fundamental equation 6 is basically applicable to all the parent-daughter pairs used, but must be appropriately interpreted so as to take into account the intermediate daughter products of the ^{238}U, ^{235}U, and ^{232}Th series and the branching decay of a single parent into more than one stable daughter in the case of ^{40}K-^{40}Ar and spontaneous fission of ^{238}U. More complete details concerning these dating methods and the experimental techniques employed may be found in (25, 26) and in the references cited in connection with the results to be discussed subsequently.

2.3   Reservoir Ages

The reservoir age approach to radiometric chronology is closely related to the principles discussed above, but also differs sufficiently to require special discussion. A slight transformation of equation 6 for the case of ^{87}Rb-^{87}Sr decay is

$$(^{87}Sr/^{86}Sr)_{t_1} = (^{87}Rb/^{86}Sr)_{t_0}\{1 - \exp[\lambda(t_0 - t_1)]\} + (^{87}Sr/^{86}Sr)_0$$

$$\cong (^{87}Rb/^{86}Sr)_{t_0}\lambda(t_1 - t_0) + (^{87}Sr/^{86}Sr)_0 \qquad\qquad 7.$$

when $\lambda(t_1 - t_0) \ll 1$.

This states that in a well-mixed system or "reservoir" with a Rb/Sr ratio that is fixed (except for radioactive decay) the $^{87}Sr/^{86}Sr$ ratio will grow from its value of $(^{87}Sr/^{86}Sr)_0$ at time $t_1 = t_0$ to a higher value at a subsequent time t_1. If this system is sampled at various times, i.e. different values of t_1, values of $^{87}Sr/^{86}Sr$ characteristic of each t_1 will be obtained. The growth of $^{87}Sr/^{86}Sr$ in such a system can then be used as a measure of time.

An example of this is the use of $^{87}Sr/^{86}Sr$ ratios to measure the time at which the material of various objects in the solar system separated from the primitive solar nebula of dust and gas, hypothesized to represent a well-mixed reservoir of the kind discussed above. This is possible because some meteorites and/or their minerals contain such a low abundance of Rb relative to Sr that it is possible to correct for their internal generation of radiogenic ^{87}Sr during the past 4.6 b.y. This permits calculation of the value of $^{87}Sr/^{86}Sr$ at the time the material of which they consist was separated from the solar nebula. Thus the initial $^{87}Sr/^{86}Sr$ ratio of Rb-poor inclusions in the Allende carbonaceous chondrite has been found to be the lowest $^{87}Sr/^{86}Sr$ ratio found in the solar system: 0.69877 (27). Slightly higher initial values are found in the differentiated achondrite Angra dos Reis (0.69884) (28), and still higher values for the basaltic achondrites (0.69898) (29). If one assumes a solar nebula Rb/Sr ratio of 0.25 inferred to be the mean value for the solar system (30) based on meteoritic data, equation 7 shows that the $^{87}Sr/^{86}Sr$ ratio increases by 0.00010 per 10^7 years. Therefore the entire span in time represented by the three values of $^{87}Sr/^{86}Sr$ listed above is about 20 m.y. This result shows that if the assumptions involved are correct, these measurements provide very high resolution in the dating of events in the early solar system. Although relative ages measured in this way are obviously very model dependent, they also establish some key facts concerning the early solar system, when used together with other data. A good example of this use of initial $^{87}Sr/^{86}Sr$ ratios, including a discussion of the problems involved in their interpretation, has been given by Gray et al (27).

Another application of "reservoir ages" is the use of the isotopic composition of lead in uranium-poor terrestrial minerals to infer the time of crystallization of the mineral. This makes use of the fact that the $^{206}Pb/^{238}U$, $^{207}Pb/^{235}U$, $^{206}Pb/^{207}Pb$, and $^{208}Pb/^{232}Th$ ratios of the earth have been increasing throughout geological time as a consequence of the decay of U and Th. It involves the assumption that the source of the lead represents in some approximation the sampling of a well-mixed reservoir having average terrestrial U/Pb and Th/Pb ratios. While this assumption is frequently valid, the isotopic composition of the lead shows that in many cases this assumption is grossly incorrect. Lead samples of this kind are termed "anomalous." Even in such cases, it is sometimes possible to draw valid inferences concerning the history of the rocks being considered. There is an extensive literature on this subject that cannot be reviewed here. See for example Doe (31) and Russell & Farquhar (32).

2.4   Model Ages

"Model ages" is a somewhat ambiguous term in that it has been used in various senses by different authors. In general it refers to calculated ages that are more model dependent than the formation ages discussed previously. An example is a ^{87}Rb-^{87}Sr age in which the initial $^{87}Sr/^{86}Sr$ ratio is assumed rather than measured. The term is also frequently applied to the lead reservoir ages discussed above. In some cases there is an implication that the age can probably be literally interpreted as the time of formation of a rock or mineral, provided one holds some reservations about this because of the model dependence of this interpretation. In other cases this implication is not intended. Examples of this are the approximately 4.5 b.y. model ^{87}Rb-^{87}Sr ages of the lunar basalts, calculated from equation 7 using the whole rock $^{87}Sr/^{86}Sr$ and $^{87}Rb/^{86}Sr$

measurements, and an assumed initial Sr ratio, usually the very primitive initial ratio of the basaltic achondrites, $^{87}Sr/^{86}Sr = 0.69898$ (29). In these cases it is clearly recognized by the authors that the age has nothing to do with the time of crystallization of the rock, but does provide some information concerning the Rb/Sr fractionation that occurred at the time of its formation. Probably the only general rule that can be applied in the use of model ages is to understand clearly in each case exactly what data is put into the calculation and how alternative assumptions or

Table 1   Solar system time scale

			Start	Finish	Duration (yr)
			\(10^9 yr\)		
		Formation of Galaxy			
Presolar			12 ± 2	4.70 ± 0.1	$\sim 5 \times 10^9$
	Formational	Nebular	4.70	4.55	$1.5 \pm 0.5 \times 10^8$
		Condensation of solar nebula			
		Accretional	4.55	4.54–4.45	10^3 to 10^8
		Early heavy bombardment			
		Early	4.54–4.45	3.85	0.7×10^9
		Late heavy bombardment			
		Oldest known earth rocks			
Solar	Planetary	Middle	3.85	2.85	1.0×10^9
		(End of major lunar volcanism, extensive terrestrial volcanism, and metamorphism)			
		Late	2.85	present	2.85×10^9
		~ 2.0 Sudbury, Vredefort impacts			
		~ 1.0 Copernicus impact?			
		~ 0.6 Last breakup of typical iron meteorite Oldest well-identified animal fossils			
		0.2 Separation of North America and Europe			
		~ 0.02 Last breakup of typical stone meteorites Present Apollo asteroids placed in Earth-crossing orbit			
		~ 0.002 Evolution of man			
		~ 0.000001 Encke's comet in short-period orbit Formation of Taurid meteor stream			

additional information will affect the interpretation of the number obtained from equation 7 or its equivalent for other parent-daughter pairs.

3 TIME SCALE FOR THE SOLAR SYSTEM

Geologists have divided geological time into major subdivisions such as the Archean, Proterozoic, and Phanerozoic, which in turn are further subdivided. We wish to make use of a similar time scale to which events in the history of the solar system can be referred. This could be done independently of particular hypotheses regarding solar system history by arbitrarily assigning these subdivisions to absolute age intervals. In some cases, this is done here. In other cases, where generally accepted events in the history of the solar system can be identified as bounding these subdivisions, it is more useful to employ measured data to assign absolute ages and time intervals to these boundaries. A schematic representation of such a time scale is given in Table 1. This review is confined to the formational period. The time scale is given to help place this period into its more general context of solar system history.

Radioactive decay constants used in the calculation of ages in this table and elsewhere in this article are listed in Table 2. The references in Table 2 refer in some cases to experimental papers and in more complex cases, too lengthy to review here, to reviews of the data in which reasons for the choice are given. Use of the particular decay constants shown in Table 2 does not constitute "endorsement" of them by this writer as being preferred to slightly different values that have also been published.

In much of the following discussion, meteoritic data and to some extent

Table 2 Radioactive decay constants

Isotope	Branch	$\lambda(\text{yr}^{-1})$	Ref.
^{40}K	λ_e (to Ar)	5.85×10^{-11}	33, 34
	λ_β (to ^{40}Ca)	4.72×10^{-10}	33, 34
^{87}Rb		1.39×10^{-11}	33
^{129}I		4.03×10^{-8}	35
^{232}Th		4.95×10^{-11}	36
^{235}U		9.85×10^{-10}	37
^{238}U	λ_α	1.551×10^{-10}	37
	$\lambda_{\text{S.F.}}$	7.03×10^{-17}	38
^{244}Pu	λ_α	8.47×10^{-9}	39
	$\lambda_{\text{S.F.}}$	1.06×10^{-11}	39

lunar data are used. A general introduction to meteorites, their classification, age, and origin is given by Wasson (40). An earlier review of both meteoritic and lunar age measurements is that of Burnett (41); data on inert gases in meteorites have been reviewed by Bogard (42). Lunar rocks have been reviewed by Wood (43), lunar geology by El-Baz (44), and the lunar interior, including its thermal evolution, by Toksöz (45).

The use of specialized meteoritic nomenclature is held to a minimum, but to some extent is unavoidable. Explanation of a few terms may help the reader who is unacquainted with this field.

3.1 Meteorites

Meteorites are extraterrestrial objects moving in heliocentric orbits that intersect the orbit of the earth and possess sufficient mechanical strength to survive passage through the atmosphere. Atmospheric alteration is confined to a thin surface layer and the interior of the meteorite is in the same physical and chemical state as it was in interplanetary space. The source of meteorites are not clearly understood, but meteorites are certainly derived from minor bodies in our solar system, probably asteroids, comets, or both. The unknown source is referred to as "the parent body" without intention to limit the number or variety of these parent bodies. Current understanding of the sources of meteorities has been reviewed elsewhere (46).

Meteorites may be classified into two broad types: undifferentiated and differentiated. The undifferentiated meteorites are also called *chondrites,* because most of them contain ~1 mm diameter spherical objects, called *chondrules,* which consist of silicate minerals. They are described as undifferentiated because their relative concentrations of rock-forming nonvolatile elements (such as Si, Mg, Fe, Ca, Al, Na, and trace elements with similar geochemical properties) are very similar to the relative abundance of these elements in the sun and in average solar system material. They are greatly depleted, relative to the sun, in extremely volatile elements, such as hydrogen, helium, other inert gases, and nitrogen. They vary in their depletion in more moderately volatile elements such as carbon, oxygen, and sulfur. The *carbonaceous chondrites,* particularly those designated as type I, are the least depleted in these volatile elements. Most chondrites have undergone various degrees of thermal metamorphism and therefore are in chemical equilibrium. Other chondrites, including the carbonaceous chondrites, are unequilibrated.

There is a considerable variety of differentiated meteorites, of quite different chemical composition. Some consist principally of Ni-Fe alloy (iron meteorites) whereas others consist primarily of silicate minerals (achondrites). Still others contain similar quantities of both silicate and iron (e.g. pallasites and mesosiderites). Most achondrites, particularly those rich in calcium, resemble in many ways terrestrial and lunar igneous rocks, although in most cases the meteoritic material has been fragmented (brecciated) and reconstituted since the crystallization of the igneous material. The most numerous objects of this kind are the *basaltic achondrites.*

It can be seen in the subsequent discussion that the formation of most meteorite parent bodies took place near the very beginning of the solar era (Table 1), and it is through the study of meteorites that most has been learned regarding the earliest history of the solar system. Presolar time is beyond the scope of the present article, and is reviewed elsewhere (2, 10, 11). The value of $12 \pm 2 \times 10^9$ yr for the age of the galaxy is obtained from Fowler (10).

3.2   Nebular Period

The nebular period, earliest period in solar system history, begins with the isolation from galactic nucleosynthesis of the dust and gas that are to become the solar nebula, and ultimately the sun, planets, and other bodies that comprise the solar system. It ends with the condensation of chemical compounds from the solar nebula and the beginning of their aggregation and accretion into larger bodies. Although much serious and even profound attention has been given to this problem, events during this period are not generally understood, as may be familiar to those who have read various discussions of this subject. At most one person on this planet has a correct understanding of these events.

At this stage the origin of the planetary system is intimately associated with the origin of the sun and stars, reviewed by Larson (5). As an indication of the lack of agreement among various workers in this area (47), there is no consensus as to whether the sun formed before the onset of accretion, which then required 10^7–10^8 years (e.g. Safranov 48), whether accretion on a time scale of 10^3 yr occurred first (Cameron 49), or even whether solar systems involve more than one "sun" in their earliest stages (Larson 50).

Because of this situation it is impossible to give a firm theoretical value for even the order of magnitude of the length of the nebular period. However, there is a general feeling that it should be comparable to the free-fall time of the initial cloud, i.e. $\sim 10^6$ years. This is much less than the experimental value of ~ 100 m.y. discussed below, and has given rise to an interesting speculation regarding the interpretation of this larger value (51). This is that the collapse of the interstellar cloud, which after further fragmentation ultimately became the solar nebula, was initiated by the shock wave experienced by the gas and dust upon entering a spiral arm of the galaxy. This is to be expected in accordance with the density wave theory of galactic structure (52, 53). As it is thought that most supernovae occur in spiral arms, this material would have been largely free of admixture with newly synthesized material during the $\sim 10^8$ years required for it to pass between successive spiral arms.

In contrast to the theoretical problems sketched above, there is clear experimental evidence regarding the $\sim 10^8$-year interval between the isolation of solar system material from galactic nucleosynthesis and the onset of accretion. This evidence is provided by calculation of a formation interval (Section 2.1) for the solar system from measurements of the Xe mass spectrum produced by the decay of ^{129}I and ^{244}Pu formed in galactic (r-process) nucleosynthesis. By simultaneous use of both of these short-lived isotopes, it is possible to largely eliminate the dependence of the calculated formation interval on the age of the galaxy and the nucleosynthesis rate during the early portion of galactic history. This is because the short lifetimes of both these radioisotopes precludes a significant contributrion from nucleosynthesis much before the beginning of the solar era. The relative change in their abundance, during the formational period, as a consequence of their different half-lives, then permits a calculation of the length of this period. Although this simultaneous use of ^{244}Pu and ^{129}I eliminates most of the dependence of the formation interval Δ on early nucleosynthesis, the result is still sensitive to whether or not the rate of galactic nucleosynthesis was slowly or rapidly varying near the end of presolar time. Schramm & Wasserburg (11) give expressions for two extreme cases, smoothly varying (approximately steady state) nucleosynthesis and sudden nucleosynthesis at the beginning of the formation interval,

$$\Delta_{129,244} \cong (\lambda_{129} - \lambda_{244})^{-1} \left\{ \ln \left[\frac{\lambda_{244}}{\lambda_{129}} \frac{P_{129}}{P_{127}} \left(\frac{^{127}I}{^{129}I} \right)_m \middle/ \frac{P_{244}}{P_{232}} \left(\frac{^{232}Th}{^{244}Pu} \right)_m \right] \right.$$
$$\left. + \ln \left(1 - \tau\lambda_{232} \right)^{-1} \right\} \qquad 8.$$

(smooth nucleosynthesis)

and

$$\Delta_{129,244} \cong (\lambda_{129} - \lambda_{244})^{-1} \left[\frac{P_{129}}{P_{127}} \left(\frac{^{127}I}{^{129}I} \right)_m \middle/ \frac{P_{244}}{P_{232}} \left(\frac{^{232}Th}{^{244}Pu} \right)_m \right] \qquad 9.$$

(sudden last-minute nucleosynthesis),

where the ratios $(^{127}I/^{129}I)_m$ and $(^{232}Th/^{244}Pu)_m$ refer to the values of the ratios of these isotopes at the end of the formation interval, as determined by the Xe mass spectrum measured in a meteorite, and the concentration of the elements I and Th in the meteorite. Dependence on the age of the galaxy and early galactic nucleosynthesis arises only because of the fact that, unlike ^{127}I, ^{232}Th is not a stable isotope, but has a half-life comparable to the age of the galaxy. This gives rise to the second logarithmic term in equation 8, where $\tau \cong 3 \times 10^9$ years is the "mean age of the elements" measured back from the beginning of the formation interval. [For a more complete discussion of this quantity, see (11).] Inclusion of this term is not very consequential, as it makes only about a 4% contribution to the result. A schematic representation of the time intervals appearing in these equations is given in Figure 2.

The data necessary to calculate Δ from (8) and (9) has been obtained for two chondrites [St. Severin (54–56) and Allende (57)] and the achondrite Peña Blanca Spring (12). The principal problem in the use of these data is the probability that there has been geochemical fractionation of Pu from Th and U in the formation of the meteoritic material (15). Uranium is greatly enriched over its mean solar system value in the phosphate mineral whitlockite in St. Severin (54) and somewhat enriched in the Allende inclusions (57). This is the probable explanation of the

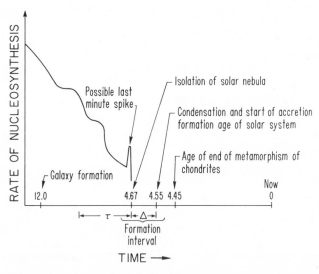

Figure 2   Illustration of terms used in discussion of formation intervals. The solid curve represents a hypothetical rate of galactic nucleosynthesis up to 4.67 b.y. ago, at which time the solar nebula was isolated from the galaxy. The time τ is the "mean age of the elements" (11), and Δ is the formation interval of the solar system, i.e. the duration of the Nebular period.

result that the $^{244}Pu/^{238}U$ value inferred from the Xe data for these different meteorites varies over a factor of six, even though the $^{129}I/^{127}I$ ratios are nearly the same. It might be thought that the data obtained on the bulk sample of St. Severin (55) would be free of this difficulty, since chondrites are relatively unfractionated with respect to solar or average solar system material, at least for refractory elements such as Pu and U. However this is not entirely the case. The measurement carried out on the bulk sample of St. Severin involved neutron irradiation of the sample to produce ^{235}U fission xenon. Stepwise heating of the sample then permits determination of the $^{244}Pu/^{238}U$ ratio by measurement of the xenon mass spectrum alone, obviating a separate determination of the uranium. Correlated release of xenon produced by the decay of ^{244}Pu and ^{235}U also serves to discriminate between fissiogenic xenon produced by the in situ decay of ^{244}Pu and parentless fissiogenic xenon simply trapped in the meteoritic mineral lattices at the time of their formation. The fact that this correlated release occurs only in the high temperature fractions of the step-heating experiment means that the calculated $^{244}Pu/^{238}U$ ratio is based not on the entire bulk sample but on the lattice sites that release xenon at high temperature. It is quite possible that Pu/U fractionation exists for these mineralogically unidentified high temperature sites, even though plutonium and uranium may be unfractionated in the meteorite as a whole. Nevertheless most authors (2, 10, 58) have adopted the result of this bulk measurement $(^{244}Pu/^{238}U)_m = 0.0154 \pm 0.0014$, as revised in (56). Use of this value, and substitution of the production ratios, $^{238}U/^{232}Th$, and decay constants given by Schramm (2) in equations 8 and 9, give values of the formation interval Δ of 122 m.y. for the case of a smooth nucleosynthesis rate at the beginning of the formation interval and 164 m.y. for the case of all the xenon isotope anomaly being produced by a sudden last-minute "spike" in the rate of r-process nucleosynthesis. When the experimental uncertainties in the quantities used in equations 8 and 9 are included, the range in the formation interval extends from 75 to 255 m.y.

Some of the motivation for the introduction of the late spike may be disappearing. One of the reasons for which a late spike of r-process nucleosynthesis was introduced was in association with the hypothesis (59) that the collapse of the solar nebula was triggered by the shock wave generated by a nearby supernova. An alternative way of providing similar shocks by galactic density waves has already been mentioned. In addition Reeves (60) has argued on dynamical grounds against significant addition of newly synthesized material even if a last minute supernova did occur. Although this is an area of "knowledge" in which dogmatism is inappropriate, a "best value" for the formation interval could reasonably be taken to be about 125 m.y.

The foregoing calculation of the formation interval of the solar system was based on simultaneous use of data based on the decay of ^{244}Pu and ^{129}I. At the cost of making the calculation more model dependent, a formation interval can be calculated from ^{129}I data alone by use of equation 4 or its equivalent for nonuniform models of galactic nucleosynthesis. These calculations also lead to formation intervals of ~ 100 m.y. using plausible values of the relevant parameters without the necessity of invoking a late spike.

It may be understood from the foregoing discussion that the absolute value of the formation interval is somewhat uncertain, and also of considerable relevance to galactic astrophysics. In contrast, relative ^{129}I-^{129}Xe formation intervals can be measured with much greater accuracy and make a major contribution to our understanding of later stages of star and solar system formation.

3.3   Accretional Period

3.3.1   RELATIVE FORMATION INTERVALS

Relative formation intervals can be calculated by taking the difference between two equations of the form of equation 3,

$$\Delta_1 - \Delta_2 = (1/\lambda_{129}) \ln \left[(^{129}Xe^*/^{127}I)_2 / (^{129}Xe^*/^{127}I)_1 \right].$$ 10.

This difference is independent of the ratio $(^{129}I/^{127}I)_0$ at the beginning of solar system formation. Early measurements of these differences yielded considerable scatter. In 1967, however, Hohenberg et al (61) showed that when various experimental difficulties are eliminated the range of the difference in formation intervals between different chondrites is only a few million years—a phenomenon that these authors named the "sharp isochronism." This phenomenon was later shown to include not only the relatively undifferentiated chondritic meteorites, but differentiated objects such as enstatite achondrites and even silicate inclusions from an iron meteorite. The chondrites participating in the sharp isochronism include primitive material such as magnetite (Fe_3O_4) from the Orgueil type 1 and Murchison type 2 carbonaceous chondrites (62) and the Ca-Al rich inclusions of the Allende type 3 carbonaceous chondrite (57). Highly metamorphosed chondrites are also included. Figure 3 is an updated version of the figure given by Podosek (12). The relative ages are arbitrarily referred to the chondrite Bjurböle. On this scale the primitive Orgueil, Murchison, and Allende material have ages of -5.7, -5.5, and -2.4 m.y., whereas metamorphosed chondrites such as Karoonda appear as early

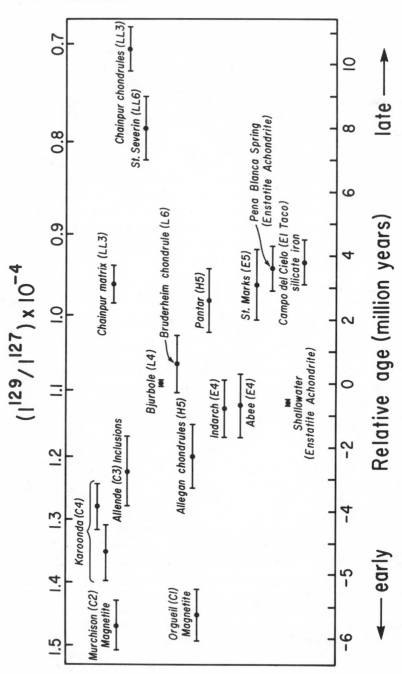

Figure 3 Relative formation intervals of meteorites exhibiting the "sharp isochronism." Figure is updated version of that of Podosek (12) and includes data from (57) and (62) as well.

as −3.9 m.y. and silicate inclusions in the iron metorite Campo del Cielo (El Taco) measure +3.8 m.y. One metamorphosed chondrite (St. Severin) appears as late as 7.5 m.y.; two enstatite achondrites (Peña Blanca Spring and Shallowater) fall in the middle of the range. Together the members of this very diverse group of meteorites have formation intervals all within ∼15 m.y. of one another, in spite of their very different chemical history. If these formation intervals are interpreted in the most straightforward (or possibly naive) way, these results mean that for each object at least some of their mineral grains were separated from the solar nebula at nearly the same time, and that major losses of neither xenon nor iodine have occurred since that time. Insofar as these objects underwent metamorphism, melting and subsequent cooling, or other processes of chemical fractionation, all of these events were confined to these few million years. This point of view has been taken by some authors (62). However, as is not unusual in this field, an apparently straightforward interpretation introduces serious difficulties, discussed in the next section.

An alternative is that these various meteorites were formed and subsequently were heated, metamorphosed, melted, or chemically fractionated, as appropriate. Then at nearly the same time they were all cooled to a sufficiently low temperature to retain xenon. In this case the "sharp isochronism" dates this time of cooling. This alternative is not very attractive in view of the unmetamorphosed state of the low temperature mineral assemblage in Orgueil and the preservation within the Allende inclusions of oxygen and magnesium isotope anomalies attributed to presolar events (63, 64). These meteorites would have to have formed at the time of this cooling. This hypothesis is also hard to reconcile with the results discussed in the next section.

Another alternative interpretation is that all of these objects actually did separate from the solar nebula at nearly the same time, but that their metamorphism and other chemical differentiation occurred at a later time. Under this interpretation the record of ^{129}I-^{129}Xe formation intervals must be very resistant to these later processes and is able to "see through" these events (27, 58). At first this may seem surprising because of the volatile nature of both iodine and xenon. Since the age is based on the correlated high temperature (> 1000°C) release of xenon, however, it could be that the sites in which this iodine and xenon are found are very retentive. These sites have not yet been identified in a mineralogical sense.

3.3.2 INITIAL STRONTIUM ISOTOPE RATIOS   The introduction of techniques for the measurement of high precision (error ≲ 5 × 10^{-5}) ^{87}Sr/^{86}Sr ratios by Papanastassiou & Wasserburg (29) has permitted application of an additional method for achieving high resolution of events during the accretional period. This is the use of $^{87}Sr/^{86}Sr$ solar nebula reservoir ages, as discussed in Section 2.3. Application of this approach makes use of the fact that the Rb/Sr ratio in the solar nebula as inferred from relatively unfractionated chondritic material is high (~ 0.25), and that early in the history of the solar system planetary bodies, rocks, and minerals formed with Rb/Sr ratios lower by factors of 10–1000. Thus the record of $^{87}Sr^{86}Sr$ growth at a rate of $\sim 0.0001/10^7$ yr in the previous high Rb/Sr environment was "frozen in" these materials and now can be measured. The first application of this to high resolution of early solar system events, was the calculation (29) that the factor of ~ 100 reduction of the Rb/Sr ratio of the parent material of the various basaltic achondrite meteorites, from that presumably of the solar nebula, was simultaneous to within about 4 m.y. The strontium isotope composition of the material at that time was 0.69898 ± 0.00005. The augite achondrite Angra dos Reis was found to have the significantly lower value 0.69884 ± 0.00004 (28). Measurements of similar low values for this ratio in the inclusions of the carbonaceous chondrite Allende (27, 65, 66) strongly suggested another sharp isochronism in the time of separation of material from the solar nebula. Some measured low values of $^{87}Sr/^{86}Sr$ are shown in Table 3. In each case the authors cited assigned an age value to the rock by use of some other technique, which then permits correction for the increase of $^{87}Sr/^{86}Sr$ during the period since its formation. These corrected values are given in the fourth column of Table 3 as the initial $^{87}Sr/^{86}Sr$ ratio. Although in some cases the rocks themselves formed well after the formation of the solar system, the material of the planetary body of which they are a part must have had an equal or lower value of this ratio at the time of separation from the solar nebula. The difference between these minimum values and the Allende value is used to calculate maximum Δt values for the time of separation of this material from the solar nebula, assuming the solar nebula to increase its $^{87}Sr/^{86}Sr$ ratio at the typical chondritic rate of $0.0001/10^7$ yr. If the Rb/Sr ratio derived from stellar spectra (70, 71) had been used instead, the values of Δt would be smaller by a factor of about 2. If it is assumed that the ratio found in the Allende inclusions is that characteristic of the solar nebula at the time this material was separated, the measurements on these inclusions connect the ^{129}I-^{129}Xe time scale discussed in the previous section to that based on initial Sr isotope ratios. The spread in values of $(\Delta t)_{max}$ is similar in both cases, at least for those samples discussed so far.

Where do the metamorphosed chondrites fit into this picture? Based on the ^{129}I-^{129}Xe results described, it might be expected that similar low initial ratios would be found for these objects. Instead, Rb-Sr

Table 3 Some $^{87}Sr/^{86}Sr$ ratios in solar system material

Sample	$^{87}Sr/^{86}Sr$ Measured	$^{87}Rb/^{86}Sr$	$^{87}Sr/^{86}Sr$ Initial	Δt Formation (Maximum) (m.y.)	Reference
Allende (chondrite) Rb-poor inclusions	0.69877	0.00014	0.69876	$\equiv 0$	27
Juvinas (basaltic achondrite)	0.69938	0.00644	0.69896	20	29
Moore County (basaltic achondrite)	0.69898	0.00014	0.69891	15	27
Lunar rock 60015 average of plagioclase samples	0.69896	0.002	0.69884	8	67
Lunar rock 60025	0.69896	0.00028	0.69894	18	68
~3400 m.y. old terrestrial rock	0.70107	0.012	0.70048	172	69

measurements on phases separated from the metamorphosed chondrite
Guareña (72) showed that the $^{87}Sr/^{86}Sr$ ratio of this meteorite at the time
of metamorphism was 0.69995, ~ 100 m.y. later than Allende on the time
scale used in Figure 3. Measurements of initial $^{87}Sr/^{86}Sr$ of two other
metamorphosed chondrites, Bath Furnace and Ensisheim (66), with
markedly lower Rb/Sr ratios than those typical of chondrites, indicated
values of Δt of 81 and 36 m.y. respectively, assuming this excess initial
$^{87}Sr/^{86}Sr$ had been generated in a chondritic environment. Even if solar
values were used, these time differences do not fit well with the ^{129}I-^{129}Xe
sharp isochronism of *both* metamorphosed and unmetamorphosed
chondrites. As discussed in the previous section, this result can be understood if the ^{129}I-^{129}Xe ages are thought to "see through" these metamorphic effects which permitted isotopic equilibration of strontium as
much as 100 m.y. later. Hohenberg & Reynolds (73) have presented
some laboratory data that supports this idea. Although this interpretation
provides the simplest reconciliation of the data, it is not well established,
and the integrated picture it provides should be regarded as tentative
rather than final.

3.3.3 ^{244}PU-XE FORMATION INTERVALS In principle, absolute and
relative formation intervals can also be obtained from the spontaneous
fission of ^{244}Pu. The clearer results of such measurements are given in
Table 4. These are given in terms of the $^{244}Pu/^{238}U$ ratio at the time
of xenon retention. Other results are discussed by Reynolds (78). These
measurements are difficult because of uncertainties in the contribution of
spallation and trapped xenon (79) and because of the presence of heavy

Table 4 Fissiogenic (^{244}Pu) Xe in meteorites

Meteorite	Type	$(^{244}Pu/^{238}U)_M$	Ref.
Allende	Chondrite-C3 (Ca-rich inclusions)	0.087	57[a]
St. Severin	Chondrite-LL6	0.0154	55, 56[a]
	(Whitlockite)	0.033	54
Peña Blanca Spring	Enstatite Achondrite	0.026	56[a]
Angra dos Reis	Augite Achondrite	0.006	77
Pasamonte	Basaltic Achondrite	0.007	8, 74, 75
Petersburg	Basaltic Achondrite	0.0045	56[a]
Kapoeta	Basaltic Achondrite	0.011	76
Lafayette	Diopside Achondrite	<0.0007	56[a]

[a] Neutron activation determination to identify fissiogenic xenon generated in situ.

isotope anomalies in many samples of chondritic xenon (80). Insofar as comparisons are possible, ^{244}Pu formation intervals are in agreement with those based on ^{129}I decay. Although the 82 m.y. half-life and geochemical fractionation of ^{244}Pu precludes the high time resolution of the ^{129}I method, its longer half-life permits effects to be found in rocks formed well after the formation period. It may be expected that future ^{244}Pu-Xe measurements will make an important contribution to our understanding of events during the early planetary period.

Charged particle fission tracks, far in excess of those expected from uranium fission, are also found in meteorites (81) and are attributed to the fission of ^{244}Pu. In several cases this identification is confirmed by observation of Xe produced by ^{244}Pu fission in the same material, St. Severin whitlockite (23, 54) and Allende inclusions (81, 57). There is no doubt that at least in those cases the attribution of the tracks to ^{244}Pu fission is correct. However, the agreement is qualitative, rather than quantitative, because in both these cases there is a marked deficiency of tracks relative to the number expected from the xenon data. The explanation of this discrepancy is not known, but could represent loss of tracks because of some fading or annealing effect.

3.3.4 EXCEPTIONS TO THE SHARP ISOCHRONISM   Before leaving the discussion of the sharp isochronism, it should be pointed out there are cases where the ^{129}Xe/^{127}I ratio does not fit into the grouping shown in Figure 3, even though this discussion may have led to that expectation. No case is yet known of a basaltic achondrite fitting into this group. The only meteorite of this class for which high quality ^{129}I-^{129}Xe data has been reported (Petersburg) has a formation interval relative to Bjurböle of > 27 m.y. Measurements on a Ca-rich achondrite of another type (Lafayette), give a formation interval of > 66 m.y. In a later section we discuss the measurement of the ages of formation and metamorphism of these Ca-rich achondrites. In some cases they are not primitive objects like the chondrites but are in some ways more analogous to terrestrial and lunar rocks in that they were formed within a planetary body at a time long after the formation of the solar system. The fact that the basaltic achondrites conform to the sharp isochronism indicated by the initial ^{87}Sr/^{86}Sr ratios thus would be interpreted as showing that the material of the parent planets of the basaltic achondrites, probably asteroids, condensed and accreted from the solar nebula during the accretional period. Like the moon and earth, at least some of these parent planets continued to form new rocks by internal igneous processes and impact metamorphism long after that time.

It is difficult to reconcile the formation interval of the earth with the sharp isochronism of the meteorites. Comparison of the isotopic composition of xenon in the earth's atmosphere (82) with that of presumably solar wind xenon in the very xenon-rich achondrite Novo Urei (83) and with xenon in the solar wind itself (84) suggests that the earth's xenon is very similar to solar xenon except for a mass-dependent fractionation which, relative to the solar wind, depleted terrestrial xenon in the lighter isotopes with respect to the heavier isotopes. In addition, terrestrial xenon has a rather clear excess ^{129}Xe of about 6% and possibly some enrichment of the fissiogenic heavy isotopes. The difference between the formation interval for the earth and that of the meteorites participating in the sharp isochronism can be calculated subject to several alternative sets of assumptions. The simplest assumption is that all of the earth's radiogenic ^{129}Xe is represented by the atmospheric anomaly, and that the iodine now present in the crust and the ocean is a measure of the ^{127}I associated with the extinct ^{129}I which generated the ^{129}Xe. Using a crustal ^{127}I abundance of 0.5 μg g^{-1} (85) and a seawater iodine concentration of 0.05 μg g^{-1} (86), the mass of iodine in the crust and ocean is calculated to be $\sim 1.0 \times 10^{19}$ g. This can be compared with the quantity of ^{129}Xe in the atmosphere, $\sim 5 \times 10^{14}$ g of which 6% or 3×10^{13} g is excess ^{129}Xe attributable to the decay of ^{129}I. The ^{129}Xe*/^{127}I ratio is found to be 3×10^{-6}. Use of this result in equation 12 leads to a formation interval of 113 m.y. relative to Bjurböle. Taken at face value this would seem to indicate that the earth did not participate in the sharp isochronism. This long formation interval is at the very upper end of the 10^3–10^8 range given by various theories. A possible explanation is that the required additional ^{129}Xe is still buried within the earth. This alone will not suffice, as the crust must be at least largely outgassed, and if xenon is trapped in the mantle of the earth, some iodine is probably there also. An excess enrichment of xenon over iodine in the mantle is required. There is direct evidence that there is actually radiogenic ^{129}Xe in the earth's interior that has never been mixed with atmospheric xenon. Measurements of the isotopic composition of xenon in CO_2 wells by Butler et al (87) and by Boulos & Manuel (88) exhibit an excess at mass 129. However, it is hard to know the quantity of ^{127}I with which to associate this ^{129}Xe.

Some attempt at this may be made by use of the fissiogenic xenon anomaly also reported in well gas by these authors. The isotopic composition of this fission Xe appears to be intermediate between that produced by spontaneous fission of ^{244}Pu (14) and ^{238}U (38). A model can be considered in which the xenon fission anomalies consist of two components:

1.   that from a primordial source containing a mixture of fissiogenic

xenon from ^{244}Pu and ^{238}U in accordance with the ^{244}Pu/^{238}U ratio at the time the earth first retained xenon;

2. a younger component containing only fissiogenic xenon from ^{238}U, analogous to the excess ^{40}Ar and ^{4}He found in well gas primarily generated by young crustal rocks.

In this model the ^{129}Xe anomaly is found only in the primordial component.

The contribution of ^{244}Pu to fissiogenic ^{132}Xe in the primordial component will be

$$^{132}\text{Xe}_{244} = (^{244}\text{Pu}/^{238}\text{U})_M {}^{238}\text{U}_1 \exp(\lambda_u T)(\lambda_{SF}/\lambda_\alpha) Y_{\text{Pu}}^{132}, \qquad 11.$$

where the subscript M refers to ratio at the time the earth retained xenon, and T is the value of that time measured back from the present. $U_1 \exp(\lambda_u T)$ is the uranium content of the primordial source, and the last three factors on the right describe the fraction of ^{244}Pu disintegration that leads to ^{132}Xe production. The ratio of radiogenic ^{129}Xe to ^{127}I will be

$$(^{129}\text{Xe*}/^{127}\text{I}) = (^{129}\text{I}/^{127}\text{I})_M = (^{129}\text{Xe*}/^{132}\text{Xe}_{244})(^{132}\text{Xe}_{244}/^{127}\text{I}). \qquad 12.$$

Substitution of ^{132}Xe$_{244}$ from equation 11 into equation 12 then gives

$$\left(\frac{^{129}\text{I}}{^{127}\text{I}}\right)_M = \left(\frac{^{129}\text{Xe*}}{^{132}\text{Xe}_{244}}\right)\left(\frac{^{244}\text{Pu}}{^{238}\text{U}}\right)_M \left(\frac{^{238}\text{U}}{^{127}\text{I}}\right)_1 \exp(\lambda_u T)\left(\frac{\lambda_{SF}}{\lambda_\alpha}\right) Y_{\text{Pu}}^{132}. \qquad 13.$$

The isotopic composition of the measured xenon fission anomaly in the well gas may be used to determine the ratio of fission produced by ^{244}Pu to that produced by ^{238}U. Calling this quantity R, we find

$$\left(\frac{^{129}\text{I}}{^{127}\text{I}}\right)_M = \left(\frac{R+1}{R}\right)\left(\frac{^{129}\text{Xe*}}{^{132}\text{Xe}_f}\right)\left(\frac{^{244}\text{Pu}}{^{238}\text{U}}\right)_M \left(\frac{^{238}\text{U}}{^{127}\text{I}}\right)_1 \frac{\lambda_{SF}}{\lambda_\alpha} Y_{\text{Pu}}^{132} \exp(\lambda_u T), \qquad 14.$$

where now ^{129}Xe*/^{132}Xe$_f$ is the ratio of the ^{129}Xe excess to the total ^{132}Xe fission excess. Using equation 10, we find the formation interval relative to Bjurböle to be

$$\left(\frac{^{129}\text{I}}{^{127}\text{I}}\right)_M \Big/ \left(\frac{^{244}\text{Pu}}{^{238}\text{U}}\right)_M = \left(\frac{R+1}{R}\right)\left(\frac{^{129}\text{Xe}}{^{132}\text{Xe}_f}\right)\left(\frac{^{238}\text{U}}{^{127}\text{I}}\right)_1 \left(\frac{\lambda_{SF}}{\lambda_\alpha}\right) Y_{\text{Pu}}^{232} \exp(\lambda_u T)$$

$$= \left[\left(\frac{^{129}\text{I}}{^{127}\text{I}}\right)_B \Big/ \left(\frac{^{244}\text{Pu}}{^{238}\text{U}}\right)_B\right] \exp\left[-\Delta(\lambda_{129} - \lambda_{244})\right], \qquad 15.$$

where the subscript B refers to the values of the quantities at the time the Bjurböle chondrite retained xenon. This equation can be solved for

Δ_1 by substituting $R \sim 1$ and $(^{129}\text{Xe}/^{132}\text{Xe}_f) \sim 2$ from the well gas data (87, 88), the molar ratio $(^{238}\text{U}/^{127}\text{I}) = 0.5$ from (85), the λ's from Table 2, the fission yield $= 0.053$ from (14), and the previously used values of 1.1×10^{-4} and 0.015 for the extinct radioactivity abundances at the time of the sharp isochronism. This calculation gives a result of $+127$ m.y., in agreement with the value of $+113$ m.y. found for the ^{129}I-^{129}Xe formation interval using atmospheric data. While it is true that there is considerable uncertainty in much of the input data to equation 15, this calculation shows that a closer look at the presence of the ^{129}Xe anomaly in the well gas gives no support for the idea that the formation interval of the earth is actually short.

There is a third way to calculate the formation interval of the earth: use of equation 15 for atmospheric xenon. This cannot actually be done as yet. While the ^{129}Xe excess in the atmosphere is clear, it is hard to tell if there is a fission anomaly as well, because of uncertainties in the mass fractionation correction to be applied to the solar wind data. An atmospheric $(^{132}\text{Xe}_f/^{129}\text{Xe}^*)$ ratio of one fourth is entirely compatible with the data that would lead to the same formation interval found above.

The trapped excess ^{129}Xe and fission Xe from ^{244}Pu found in lunar breccias (9) formally give a formation interval of 140 m.y. if one uses a lunar I/U ratio a factor of 10 less than the terrestrial value to take into account the depletion of the moon in volatile elements relative to uranium. The interpretation of this result is difficult because of a lack of understanding of the mechanisms whereby the radiogenic gases were transferred from the place of origin and implanted in the grains of the lunar breccias.

3.3.5   FORMATION AGES OF CHONDRITIC METEORITES   The formation intervals discussed in Section 3.2 date the time of formation of the meteorites relative to the time at which the matter of the solar nebula was isolated from galactic nucleosynthesis. The ^{129}I-^{129}Xe relative formation intervals and the initial $^{87}\text{Sr}/^{86}\text{Sr}$ ratios dated the time of accretion of various bodies relative to one another and to the "sharp isochronism." The time in the past at which the sharp isochronism occurred remains to be dated, in order that this fundamental event in solar system history may be placed on the same absolute time scale used to date lesser events, such as the formation times of ancient rocks on the earth.

The long-lived radioisotope methods described in Section 2.2 are used for this purpose. The simplest ages to interpret are $^{87}\text{Rb}/^{86}\text{Sr}$ ages based on measurements of $^{87}\text{Sr}/^{86}\text{Sr}$ and $^{87}\text{Rb}/^{86}\text{Sr}$ on different minerals or mixtures of minerals separated from metamorphosed chondrites. If the strontium isotope ratios in the minerals were homogenized at the time of metamorphism, and the rubidium and strontium unaltered by subsequent chemical processes, one would expect that plotting the results of these analyses on a Rb-Sr evolution diagram (Figure 4) would yield a straight line. The intercept on the ordinate would indicate the value of $(^{87}Sr/^{86}Sr)$ in the meteorite at the time of metamorphism, and the time in the past at which this metamorphism occurred could be calculated from the slope (m) of the line through the points:

$$T = \frac{1}{\lambda_{Rb}} \ln(m+1).$$
16.

These measurements have been reported for four chondrites. The results

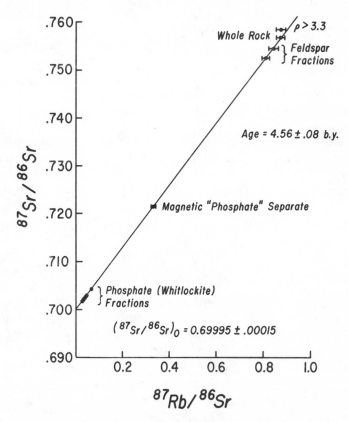

Figure 4   Internal mineral isochron for the chondrite Guareña; data from (72). Note that the initial $^{87}Sr/^{86}Sr$ ratio is elevated over the primordial value of 0.6988 as a consequence of metamorphism of the meteorite.

Table 5 ^{87}Rb-^{87}Sr ages of chondrites, based on mineral density separates from a single meteorite

Meteorite	Type	Age	$(^{87}Sr/^{86}Sr)_0$	Ref.
Guareña	H6	4.56 ± 0.08	0.69995 ± 0.00015	72
Indarch	E4	4.56 ± 0.08	0.7005 ± 0.0017	89
Olivenza	LL5	4.63 ± 0.16	0.6994 ± 0.0017	90
Krähenberg	LL5	4.70 ± 0.03	0.6989 ± 0.0010	91

of these measurements are given in Table 5. The isochron for one of these is shown in Figure 4. These ages all agree rather well with one another and all lie in the range 4.56–4.70 b.y. The Krähenberg result appears to be somewhat higher than the others. This could represent a real difference, but also is likely to be caused by slight differences in the concentration of Rb and Sr standard solutions used in different laboratories. The time of metamorphism of Guareña can be placed on the formation interval time scale by use of its initial $(^{87}Sr/^{86}Sr)$ ratio at the time of metamorphism, as discussed in Section 3.2.2. Probably the best $^{87}Rb/^{86}Sr$ ratio to use in correcting Guareña's initial $^{87}Sr/^{86}Sr$ ratio back to that of Allende is the $^{87}Rb/^{86}Sr$ ratio of Guareña itself (0.87). This gives a time of metamorphism of +95 m.y. and an age of 4.65 b.y. for the sharp isochronism referred to Bjurböle as before. It may be argued that there is little point in combining these numbers considering that the quoted experimental error on the age of Guareña is 80 m.y., comparable to the interval calculated between the sharp isochronism and the time of metamorphism. This is a valid objection, but it it also felt that it is at least instructive to press these data somewhat beyond their limits so as to define a problem that can be more definitely resolved with more accurate future data.

In principle it might have been possible to "see through" the metamorphism of the chondrites by Rb-Sr measurements on whole meteorite samples rather than on separated minerals from a single meteorite. This would be the case if at some time all chondrites had a common initial $^{87}Sr/^{86}Sr$ ratio (e.g. that of Allende) and the Sr of the whole rocks had not been chemically distributed by the metamorphism, even though strontium isotopic equilibrium had been established between the mineral grains. Such a phenomenon is common in terrestrial rocks. The results of a number of whole rock chondrite measurements made in a single laboratory are given in Figure 5, and the calculated age is seen to be nearly the same as that given by Guareña. The reason for this is that Sr apparently did equilibrate on a macrodimensional scale subsequent to the sharp isochronism. As a consequence the points with the lowest $^{87}Sr/^{86}Sr$

ratios in Figure 5, e.g. Bath Furnace and Ensisheim, have received radiogenic Sr from more Rb-rich material, leading to the elevated initial $^{87}Sr/^{86}Sr$ ratios, discussed in Section 3.3.2. A similar interpretation is possible for the high initial $^{87}Sr/^{86}Sr$ ratio (0.6988) reported on the meta-

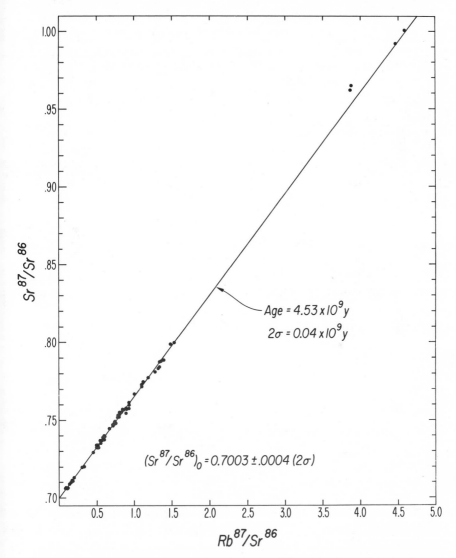

Figure 5   Rb-Sr evolution diagram for whole chondrites. The slope of the isochron and the initial $^{87}Sr/^{86}Sr$ ratio are nearly the same as those found for internal mineral isochrons of chondrites; data from (89, 92–95).

morphosed St. Severin chondrite (29). If the entire spread of samples in Figure 5 is assumed to represent a good sample of chondritic material, then the average point can be used to calculate the time since these samples had the Allende ratio. [This type of calculation would be termed a "model age" (Section 2.4).] The result of such a calculation is 4.67 b.y. in agreement with the result found by adding the metamorphic age of Guareña to its relative formation interval. The "best value" of the age of the sharp isochronism on the Rb-Sr time scale appears to be ~ 4.65 b.y. It is of interest to compare this with the results found by other methods.

There are no completely analogous U-Pb data on separated minerals. In spite of diligent consideration of severe problems of laboratory contamination at the extremely low Pb and U concentration involved, ^{238}U-^{206}Pb and ^{235}U-^{207}Pb ages on chondrites usually disagree. This indicates either that the meteorites have failed to behave as closed systems, or that analytical problems still exist. There is a way to at least partially get around these problems. Equation 6 for the ^{238}U-^{206}Pb decay and with $t_1 = 0$ (the present) is:

$$^{206}\text{Pb}/^{204}\text{Pb} = (^{238}\text{U}/\text{Pb}^{204})[\exp(\lambda_{238} t) - 1] + (^{206}\text{Pb}/^{204}\text{Pb})_0, \qquad 17.$$

where the symbol t represents the decay interval (>0), indicated by $(t_1 - t_0)$ in equation 6; t is also the time measured back into the past that the $^{206}\text{Pb}/^{204}\text{Pb}$ ratio was equal to $(^{206}\text{Pb}/^{204}\text{Pb})$. A similar expression can be written for the ^{235}U-^{207}Pb decay. If we rearrange terms and divide these two expressions we obtain:

$$\left[\frac{207_{\text{Pb}}}{204_{\text{Pb}}} - \left(\frac{207_{\text{Pb}}}{204_{\text{Pb}}} \right)_0 \right] \Big/ \left[\frac{206_{\text{Pb}}}{204_{\text{Pb}}} - \left(\frac{206_{\text{Pb}}}{204_{\text{Pb}}} \right)_0 \right]$$

$$= (^{235}\text{U}/^{238}\text{U}) \frac{\exp(\lambda_{235} t) - 1}{\exp(\lambda_{238} t) - 1} \qquad 18.$$

Equation 18 represents a straight line on a plot of $^{207}\text{Pb}/^{204}\text{Pb}$ vs $^{206}\text{Pb}/^{204}\text{Pb}$ which passes through the points corresponding to the material measured and the initial ratio and has a slope given by the right-hand side of the equation. Two or more samples that have the same age and initial ratio will plot on the same line, and the slope can thereby be measured without knowledge of the initial ratio. This line is again called an "isochron." All of the quantities appearing in the slope are known constants with the exception of the age t. This age can therefore be calculated from the slope.

Use of equation 17 still requires that the samples studied be closed systems with respect to uranium and lead, but the age calculated from the ratio (equation 18) is insensitive to gain or loss of lead or uranium provided that this occurred at a recent time. This results from the fact that a finite amount of time is required to produce sufficient radiogenic lead to effect a change in the average isotopic composition of the sample. Obtaining a good linear fit to a set of experimental points is evidence either that the systems have been closed or that failure of this assumption occurred only in recent times.

This procedure has been carried out by Tatsumoto et al (96) and Chen & Tilton (97) for various fragments from the primitive unmetamorphosed chondrite Allende. In both cases an excellent linear array was obtained from which ages of 4.55 and 4.56 b.y. were calculated. Gray et al (27) found evidence that Allende was not internally closed to Rb and Sr. The uranium-lead data also show evidence for open system behavior, but only in recent times, thereby not affecting these ages.

The unmetamorphosed and unequilibrated nature of Allende, its primitive $^{87}Sr/^{86}Sr$ ratio, and its oxygen (63) and magnesium (64) isotopic anomalies, probably of presolar origin, all indicate that this age should refer to the time of the sharp isochronism.

The isochron determined by the $^{207}Pb/^{204}Pb$ and $^{206}Pb/^{204}Pb$ ratios does not uniquely define the initial lead isotope composition, as equation 18 requires only that the isochron pass through the initial ratio. However, Tilton (98) has determined the lead isotopic composition of another unmetamorphosed chondrite, Mezö-Madaras, which has a sufficiently low U/Pb ratio to permit extrapolation to its initial ratio. This value $(^{206}Pb/^{204}Pb) = 9.310$, $(^{207}Pb/^{204}Pb) = 10.296$ represents the best determination of the initial lead isotopic composition of the solar system at the stage of condensation. The Allende isochron passes through this point, which may be assumed to represent Allende's initial lead isotopic composition as well. It is remarkable that recent measurements of the isotopic composition of lead separated from FeS inclusions in a highly differentiated iron meteorite, Canyon Diablo, give exactly the same result (98, 99). This shows that negligible growth of radiogenic lead occurred during the time required to form the parent body of this iron meteorite and to differentiate and cool it sufficiently to form the FeS inclusions.

A number of U-Pb determinations have been reported recently on whole rock samples of chondrites, principally metamorphosed meteorites (98–101). With the exception of a few points, some of which have suffered terrestrial alteration, all of these data plot well on the Allende isochron (Figure 6). They therefore yield nearly the same age as Allende. Although the initial ratio is not uniquely determined, it is unlikely that two different initial ratios lying on this same isochron were present at the same time in the early solar system. Therefore it seems that if the metamorphism of these chondrites persisted as late as that of Guareña and Bath Furnace,

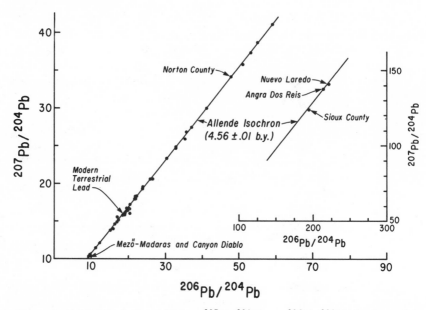

Figure 6 Solid line is the 4.56 b.y. $^{207}Pb/^{204}Pb$ vs $^{206}Pb/^{204}Pb$ isochron for Allende (96, 97). Unlabeled points represent various chondritic data (98–101). Inset figure at right is a continuation of the figure to include the much more radiogenic Pb of the basaltic achondrites.

the metamorphism of the chondrites did not involve extensive migration of lead on the scale separating different meteorites in their parent bodies over this entire period of time. A corollary of this is that the $\sim$ fiftyfold increase of the U/Pb ratio of the metamorphosed chondrites relative to average solar system material did not occur at a time distinguishably later than the time at which Allende was formed. This could be a consequence of the temperature being high at the beginning of the metamorphic period and declining on a ~ 100 m.y. time scale.

The time of the sharp isochronism is therefore given as 4.65 b.y. by the Rb-Sr measurements and as 4.55 b.y. by U-Pb measurements. Although the difference between these two results could reflect analytical errors, its consistency even when comparing work done in different laboratories suggests that it is real. It is quite possible that the difference simply reflects an incorrect choice of the ^{87}Rb decay constant, which has an experimental uncertainty of at least 5%. The uranium decay constants are much easier to measure, have recently been redetermined by Jaffey et al (37), and may be considered to be accurate to $\sim 1\%$. If the uranium decay constants generally in use prior to this redetermination were used, the discrepancy would be reduced to only 0.02 b.y., i.e. would essentially vanish. Although one cannot be at all conclusive about this, comparison of Rb-Sr and U-Pb ages is probably best done by reduction of the Rb-Sr ages by 2%.

The formation ages of meteorites have also been extensively studied by ^{40}K-^{40}Ar decay. Much of the earlier work by this method suffered from problems of high and variable argon loss, together with the fact that most of the measurements in the literature assume an average K concentration in the meteorite rather than a measured value. This older work has been summarized by Zähringer (102).

In 1966 Merrihue & Turner (103) proposed a technique for ^{40}K-^{40}Ar dating that circumvents many of these problems and that has now essentially supplanted the older bulk sample approach. This method is a neutron activation method analogous to that used for ^{129}Xe and fission Xe measurements (Sections 2.1 and 3.2). ^{39}K is converted to ^{39}Ar in a nuclear reactor by $^{39}K(n, p)^{39}Ar$. The sample is then heated in vacuum to successively higher temperatures and the $^{39}Ar/^{40}Ar$ ratio of the argon released at each temperature step is measured with a mass spectrometer of high sensitivity and low background. Comparison with a flux monitor of known K and Ar content permits conversion of $^{39}Ar/^{40}Ar$ ratios to $^{39}K/^{40}Ar$ and thence to $^{40}K/^{40}Ar$ ratios by use of the known K^{40}/K^{39} ratio. A ^{40}K-^{40}Ar age can thereby be calculated for each temperature step. Many meteorites yield a constant age at all or a number of temperature steps (Figure 7). Such a result is termed a "plateau." The corresponding age is taken to be the age of a significant event, e.g. formation or outgassing associated with metamorphism. The appearance of a plateau is analogous to Rb-Sr measurements on separated mineral or density fractions from a single meteorite. In most cases, however, the

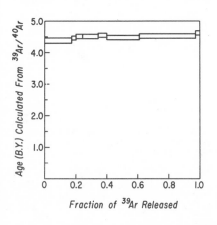

Figure 7 ^{39}Ar-^{40}Ar plateau indicating the 4.50 b.y. age of the St. Severin chondrite; data from (105, 107).

actual mineral primarily contributing to the various temperature steps in a plateau has not been identified. A plateau consisting of only the high temperature steps is commonly interpreted as either a formation time or the time of a very thorough metamorphic outgassing. A low temperature plateau with higher ages at higher temperature is usually interpreted as the time of metamorphism with incomplete loss of ^{40}Ar.

This procedure has the advantages that a K concentration is actually measured for each calculated age, and that this K is at least nearly that associated with the lattice sites in which the ^{40}Ar resides for each temperature step. The measurements at various steps that form a plateau provide a method of detecting evidence for invalid ages resulting from the migration of K and Ar.

The method is not without difficulties, however. Corrections must be made for interfering nuclear reactions caused by the neutron irradiation, and problems arising from nuclear recoil effects in the production of ^{39}Ar properly taken into account. Corrections for nonradiogenic ^{40}Ar caused by cosmic ray spallation, trapped argon of extraterrestrial origin, and atmospheric contamination must also be made (104–106).

^{39}Ar-^{40}Ar ages of chondrites that have given well-defined plateaus and ages >4.0 b.y. are given in Table 6. Ages measured in the same laboratory

Table 6   Chondrites with ^{39}Ar-^{40}Ar ages >4.0 b.y.

Meteorite	Type[a]	Age	Ref.
Marion	L6	4.40	108
Bjurböle	L4	4.60	108
Mt. Browne	H6	4.58 ± 0.05	109
Olivenza	LL5	4.55 ± 0.05	109
Queen's Mercy	H6	4.55 ± 0.05	109
Sutton	H	4.6	110
Wellington	H	4.5	110
Karoonda	C4	4.55	105
Allegan	H5	4.59	105
St. Severin	LL6	4.50	107
Guareña	H6	4.50	111
Ochansk	H4	4.56 ± 0.04	
Barwell:	L6		G. Turner, personal
matrix		4.53 ± 0.04	communication, 1975
chondrules		4.55 ± 0.04	

[a] These designations are those given by the Wood-Van Schmus classification of chondrites (112). Types designated by the numerals 5 and 6 are highly metamorphosed, whereas those designated as 4 are less, but still significantly metamorphosed.

can be compared using the errors given by the authors, usually $\sim 1\%$. Between laboratories an additional uncertainty of $\sim 1\%$ arises because of use of different monitors. When possible the data of Table 6 have been recalculated using the interlaboratory cross-calibration of monitors reported by Alexander & Davis (107) in order to make the results as comparable as possible.

The ages in Table 6 agree very well at 4.55 ± 0.05 b.y. and show no tendency to scatter toward the lower end of the minimum age selected for tabulation. All of these chondrites are metamorphosed. If it is assumed that the light inert gas Ar is lost during the metamorphism which re-equilibrated the chondritic strontium isotopes, then the formation time and the sharp isochronism should be ~ 100 m.y. earlier, based on the Guareña Sr data. Subject to the reservation expressed earlier that these differences are very near the resolution of these techniques using long-lived isotopes, we see again about a 100 m.y. difference with the U-Pb time scale. As before, an adjustment of this magnitude is permissible considering the experimental errors in the ^{40}K decay constants. However, to actually make this change in reported ages on the basis of this evidence is probably premature.

In addition to these ages that group at 4.55 b.y. there are a similar number of results extending to as low as 0.03 b.y., with a large number between 0.3 and 0.6 b.y. (108–110). These younger ages are generally interpreted as the result of relatively recent impact shock metamorphism. It is likely that the younger ages record events in the parent bodies during the late planetary period. As discussed elsewhere (93) the absolute age of these events and the resolution of these events from one another is of considerable importance to the problem of identifying the parent bodies in the present solar system. Late events of this kind are beyond the scope of the present article, and will be reviewed in a subsequent article on the planetary period.

3.3.6 FORMATION AND METAMORPHISM AGES OF DIFFERENTIATED METEOR-ITES   Most of our understanding regarding the chemistry and time scale of the early solar system has come from the study of chondritic meteorites. Although these objects have been in some cases metamorphosed and brecciated, to a considerable extent the primordial record is preserved.

From a chemical and mineralogical point of view the differentiated meteorites represent much less primordial material. Many consist of igneous rock or fragments of igneous rocks. There is no a priori reason why these igneous rocks had to form early in solar system history. In fact it would be easier to understand late melting in a large (radius $\gtrsim 200$ km) asteroidal body, after long-lived radioactive elements had billions of years to heat its interior. However, we have already seen that some iron meteorites and enstatite achondrites conform to the sharp isochronism, and therefore the mineralogy has been sufficiently stable to preserve ^{129}Xe for the entire last 4.55 b.y. It appears that in some sense at least most differentiated meteorites still preserve some record of the formational period.

Papanastassiou & Wasserburg (29) and Gray, Papanastassiou & Wasserburg (27) have reported a 4.47 ± 0.24 b.y. whole rock isochron for the subclass of basaltic achondrites known as eucrites. Although most of these meteorites have been fragmented and reconstituted as breccias, this has occurred to a much lesser extent than in the other subclass of basaltic achondrites, the howardites, which more than anything else resemble lunar surface breccias. This age shows that either these rocks or their sources have not experienced much Rb/Sr fractionation since this early time, even though mineral crystallization and Sr isotopic equilibration may have occurred at later times. In this regard this result suggests comparison with the lunar mare basalts. Although these were crystallized as recently as 3.2–3.8 b.y. ago, data on whole rock samples show that major Rb/Sr fractionation of their source has not occurred since the moon was formed (123).

Inspection of the measured formation ages of differentiated meteorites (Tables 7–9) shows that this comparison is not always valid. The basaltic achondrite Juvinas has, within small experimental error, a Rb-Sr mineral age as old as any in the solar system. Lugmair (20) has reported a ^{147}Sm-^{143}Nd age of 4.56 ± 0.08 b.y. for Juvinas. Although this is the first age obtained by this method, the data appear to be of high quality, and the agreement with the Rb-Sr result is excellent. Another basaltic achondrite, Ibitira, could be as old. In agreement with the same or similar meteorites belonging to the sharp isochronism (Figure 3), the old ages of the enstatite achondrites and irons are not surprising.

To some extent the analogy with the late-forming rocks of the moon is also valid. Well-defined young (i.e. 1.3–3.8 b.y.) ages of igneous crystallization or metamorphism are well established in differentiated meteorites (124–127). Data on differentiated meteorites, however, also strongly indicate that heat sources adequate to effect differentiation were present during the formational period. Demonstration of a satisfactory way to achieve this is a major unsolved problem, even though interesting suggestions have been made regarding its possible solution (128, 129).

These results also tell much about the cooling of these rocks. The fact that radiogenic argon and Sr isotope disequilibrium has been preserved for ∼4.5 b.y. shows that cooling rates of ≳ 5°/10⁶ yr occurred at the position of these rocks in their parent bodies. These rates imply that

Table 7 Differentiated meteorites with ^{87}Rb-^{87}Sr mineral isochron ages >4.0 b.y.

Meteorite	Type	Age	Initial ^{87}Sr/^{86}Sr	Ref.
Juvinas	basaltic achondrite	4.60 ± 0.07	0.69898 ± 0.00005	114
Ibitira	basaltic achondrite	4.48 ± 0.12	0.69893 ± 0.00007 (C. Allegre, personal communication, 1975)	
Norton County	enstatite achondrite	4.7 ± 0.1	0.700 ± 0.002	113
Weekeroo Station	iron (silicate inclusion)	4.37 ± 0.13	0.703 ± 0.003	115, 116
Colomera	iron (silicate inclusion)	4.61 ± 0.04	0.69940 ± 0.00004	117
Campo del Cielo (El Taco)	iron (silicate inclusion)	4.7 ± 0.1	0.700 ± 0.001	118

Table 8 Differentiated meteorites with ^{39}Ar-^{40}Ar ages >4.0 b.y.

Meteorite	Type	Age	Ref.
Peña Blanca Spring	enstatite achondrite	4.5±0.1	104
Bishopville	enstatite achondrite	4.5±0.1	104
Petersburg	basaltic achondrite	4.40±0.03	119
Pasamonte	basaltic achondrite	4.10±0.03	119
Campo del Cielo (El Taco)	iron	4.5	105
Mundrabilla	iron olivine:	4.62±0.02	120
	plagioclase:	4.58±0.02	

Table 9 Uranium-lead whole rock ages of achondrites (based on ^{207}Pb/^{206}Pb ratio)

Meteorite	Type	Age	Ref.
Norton County	enstatite achondrite	4.550±0.003	100
Angra dos Reis	augite achondrite	4.555±0.005	99
Sioux County	basaltic achondrite	4.526±0.010	99
Nuevo Laredo	basaltic achondrite	4.529±0.005	99
		4.49	121
		4.53±0.02	122
Pasamonte	basaltic achondrite	4.53±0.02	122

either the parent bodies were fairly small (radius $\sim$50 km) or the rocks were near the surface of a larger body. These results do not preclude the possibility that the differentiation occurred in somewhat larger bodies (e.g. 500 km in radius) and that cooling followed fragmentation of these bodies during an "early heavy bombardment" initiated by the formation of Jupiter (130, 131).

3.3.7 FORMATION AGE OF THE EARTH AND THE MOON   Where do the planets fit into this story? The only large bodies in the solar system from which the samples necessary to study this question have been obtained are the earth and the moon. It is very plausible that at least the other terrestrial planets were formed at nearly the same time. The time of formation of the giant planets is more obscure, although there is a prejudice that Jupiter and Saturn formed very early.

The extensive external bombardment and internal evolution of the earth and moon constitute an obstacle to the straightforward approach that was successful in establishing the formation age of the meteorite parent bodies.

It has already been shown (Section 3.3.4) that insofar as there is evidence regarding the xenon formation interval of the earth, this planet does not conform to the sharp isochronism, but records a formation interval ~ 100 m.y. longer. However, the very low initial $^{87}Sr/^{86}Sr$ ratio of some lunar rocks (Table 3) does place the time of separation of the lunar material within the sharp isochronism. The extrapolation required to draw a similar conclusion for the earth is much larger. Linear extrapolation of the $^{87}Sr/^{86}Sr$ ratio (0.7025) found in modern terrestrial oceanic volcanic rocks (134) through the ~ 2700–3250 m.y. values in Table 3 leads to an initial ratio of ~ 0.6993 for the earth. If, as is generally believed, terrestrial chemical evolution was more rapid early in its history, this ratio would be lower. Therefore the material of the earth probably also separated from the solar nebula at the time of separation of the meteoritic material.

However, it is entirely conceivable that $\sim 10^8$ years was required for the earth and moon to accrete from smaller objects that were possibly similar in some ways to the meteorite parent bodies (Safranov 48). If so, this would greatly reduce the role of gravitational energy in the early heating and differentiation of the earth and moon. In fact, evidence for this early differentiation may be the best argument against such a long accretion time. However, because the differentiation of the small meteorite parent bodies and the metamorphism of chondrites also require early non-gravitational heat sources of an unknown nature, it is felt premature to draw this conclusion too strongly. Evidence has been presented for lunar rocks 4.5–4.6 b.y. in age (132, 133). However, the uncertainties in these ages are too great to attempt to use them in resolving events on a ~ 100 m.y. time scale. The oldest known terrestrial rocks are too young, ~ 3.7 b.y. (135), to say anything significant about the age of the earth.

The classical method for measuring the age of the earth is that of Patterson (136) and Houtermans (137). Lead isotope measurements from young terrestrial rocks are combined with primordial lead of Canyon Diablo or Mezö-Madaras composition in equation 18. Solving this equation for t then gives the time required for terrestrial lead to evolve to its present isotopic composition, assuming the source of the basalts did not undergo U/Pb fractionation during earth history. If the isotopic composition of lead (138) from a modern fumarole ($^{206}Pb/^{204}Pb = 18.83$, $^{207}Pb/^{204}Pb = 15.72$) is used with Mezö-Madaras lead (Section 3.3.5) in equation 18, an age of 4433 m.y. is calculated. There is, of course, considerable choice in the value used for modern lead, but reasonable choices yield ages in the range 4350–4500 m.y. These are all significantly less than that given by the Allende isochron (Figure 6) of 4550 ± 10 m.y.

In view of the complexity of terrestrial geochemical history the assumption of a uniform U-Pb ratio used in this calculation might seem too absurd to take seriously. There is strong evidence, however, that this assumption is approximately correct. Equation 17 may be used to calculate the isotopic composition of lead generated in a uniform earth during the interval between the time the earth's lead had the primordial value and any subsequent time. This isotopic composition can then be compared with isotopic measurements of lead of known age. Use of a compilation of these data (139) shows that assumption of a uniform 4433 m.y. earth not only yields the isotope ratio assumed for modern lead, but also yields that measured for ore and rock leads as far back in time ($\sim$3700 m.y.) as data are available. A similar approach using lunar leads generated between the time of formation of the moon and $\sim$3.85 b.y. ago gave a result of 4.42 b.y. for the "age of the moon" (i.e. the time since the lunar material had the primordial isotope ratio) (140). This could, for example, have come about as a result of the lunar material condensing and accreting into planetesimals with the primordial lead isotope ratios and the very low $^{238}U/^{204}Pb$ ratios of the carbonaceous chondrites (30). Formation of the moon from these planetesimals at 4.42 b.y., followed by internal differentiation to the lunar $^{238}U/^{204}Pb$ ratio of $\sim$300 (140), would then produce the observed result.

This discussion is not presented to argue that the formation of the earth and moon was actually delayed by $\sim$100 m.y. The point to be made is that a simple calculation of the age of the earth and the moon that fits all the lead isotope data leads to this result, rather than to $\sim$4.55 b.y. planets. Early formation of the earth and moon requires a modification of this history sufficiently subtle not to violate the significant constraints imposed by the lead isotope data. We must "slow down" the growth of lead in the earth so as to make an earth that is actually 4.55 b.y. old look as though it is only $\sim$4.43 b.y. old. This requires a low U/Pb or equivalently a low $^{238}U/^{204}Pb$ ratio early in earth history, because the formally calculated age of the earth is very sensitive to growth of lead at the time when the abundance of rather short-lived ^{235}U was high. However, it is not plausible to assume, as suggested above, that the very low $^{238}U/^{204}Pb$ of the carbonaceous chondrites prevailed long after the separation of terrestrial or lunar material from the solar nebula. The Sr isotopic composition reflects an early depletion in the fairly volatile element Rb relative to the refractory element Sr, and a similar depletion of Pb relative to U should have occurred. This is incorporated in the model used to calculate the earth's Pb evolution shown in Figure 8. A twelvefold increase in the $^{238}U/^{204}Pb$ ratio is assumed to occur at 4.55 b.y. and a further threefold increase at 4.45 b.y., as a consequence of early terrestrial internal differentiation. These modifications will preserve the agreement between calculated and observed lead isotopic composition over most of earth history. However, in order to obtain agreement at ages $\lesssim 1100$ m.y., a minor (11%) increase in the $^{238}U/^{204}Pb$ ratio of the source of terrestrial surface lead was assumed to occur 2800 m.y. ago. Of course this solution is not unique, nor must the increases in $^{238}U/^{204}Pb$ ratios be episodic. Averaging the increases over finite time intervals would serve as well. In some average way, however, an *early* increase in the $^{238}U/^{204}Pb$ ratio is required. Similar considerations apply to the moon. The same model with a tenfold increase over the carbonaceous chondrite value of $^{238}U/^{204}Pb$ at 4.55 b.y. and a subsequent twentyfold increase at 4.45 b.y. (in order to achieve the observed higher lunar $^{238}U/^{204}Pb$ ratio) will satisfy the lunar data. Similar results are obtained using a continuous model for lunar differentiation (140).

Therefore the lead isotope data in no way argue against a 4.55 b.y. old earth or moon, provided the necessary ~ 100 m.y. interval of differentiation proves to be geochemically reasonable. The question of the age of the earth and moon must be settled in some other way. Plausible arguments can be given for formation of both earth and moon on a

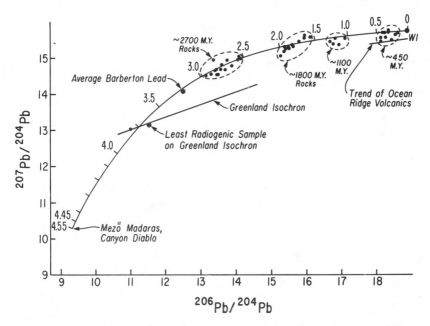

Figure 8   Evolution of the isotopic composition of terrestrial lead. Measured data (135, 138, 139) are fitted to a model in which the source of the lead undergoes a twelve-fold increase of the $^{238}U/^{204}Pb$ ratio over its primordial value of 0.25 at 4.55 b.y. The ratio again increases to 8.96 at 4.45 b.y. and undergoes a further 11% increase at 2800 m.y.

similar time scale, even though very different theories for the origin of the moon are assumed. Therefore a good age for the moon would tell much about the age of the earth and the time scale for its early differentiation. Perhaps the most definite way to accomplish this would be through ^{129}I-^{129}Xe data on ~ 4.5 b.y. lunar rocks, if suitable well-preserved samples can be found.

4 SUMMARY AND A TENTATIVE SYNTHESIS

We have discussed plausible, but not necessarily correct, hypotheses that permit defining a consistent picture for events in the formational era of the solar system. Experimental and theoretical testing of these hypotheses could support such a synthesis, or destroy it, and thereby demonstrate the need for alternative syntheses.

The principal features of such a synthesis are as follows:

1. The nebular period had a duration of 100–150 m.y., following which time condensation of solid matter took place.

2. Planetesimals up to ~ 100 km in radius accreted rapidly ($\sim 10^4$ yr) (48, 141) as a consequence of gravitational instability of solid matter in the central plane of the nebula.

3. The interiors of at least some of these planetesimals were at temperatures of 500–1600°C within $\lesssim 10^7$ years, resulting in metamorphism and/or igneous differentiation.

4. Some of these planetesimals cooled quickly ($\sim 10^7$ to 10^8 yr) either because of their original small size or because of disruption during an early heavy bombardment associated with the formation of Jupiter and Saturn.

5. Most of the planetesimals and their debris accreted further to form planets and large satellites on a very uncertain time scale of 10^4 to 10^8 years. The importance of gravitational energy of accretion as a heat source is critically dependent on the length of this time scale.

6. The larger surviving planetesimals and planetary objects combined to evolve internally in the subsequent planetary era.

Literature Cited

1. Reeves, H. 1974. *Ann. Rev. Astron. Astrophys.* 12:437–69
2. Schramm, D. N. 1974. *Ann. Rev. Astron. Astron. Astrophys.* 12:383–406
3. Burbidge, E. M., Burbidge, G. R., Fowler, W. A., Hoyle, F. 1957. *Rev. Mod. Phys.* 29:457–650
4. Arnett, W. D., ed. 1965. *Nucleosynthesis, Proc. Conf. Jan. 25–26, NASA Goddard Inst. Space Stud.* New York: Gordon & Breach. 273 pp.
5. Larson, R. B. 1973. *Ann. Rev. Astron. Astrophys.* 11:219–37
6. Heiles, C. 1971. *Ann. Rev. Astron.*

Astrophys. 9:293–322
7. Reynolds, J. H. 1960. *Phys. Rev. Lett.* 4:8–10
8. Reynolds, J. H. 1967. *Ann. Rev. Nucl. Sci.* 17:253–316
9. Behrmann, C. J., Drozd, R. J., Hohenberg, C. M. 1973. *Earth Planet. Sci. Lett.* 17:446–55
10. Fowler, W. A. 1972. *Cosmology, Fusion, and Other Matters,* ed. F. Reines, 67–123. Boulder, Colo: Colorado Univ. Press
11. Schramm, D. N., Wasserburg, G. J. 1970. *Astrophys. J.* 162:57–69
12. Podosek, F. A. 1970. *Geochim. Cosmochim. Acta* 34:341–65
13. Rowe, M. W., Kuroda, P. K. 1965. *J. Geophys. Res.* 70:709–14
14. Alexander, E. C., Lewis, R. S., Reynolds, J. H., Michel, M. C. 1971. *Science* 172:887–40
15. Crozaz, G. 1974. *Earth Planet. Sci. Lett.* 23:164–69
16. Anders, E., Stevens, C. M. 1960. *J. Geophys. Res.* 65:3043–47
17. Kohman, T. P., Huey, J. M. 1972. *Earth Planet. Sci. Lett.* 16:401–12
18. Schramm, D. N., Tera, F. Wasserburg, G. J. 1970. *Earth Planet. Sci. Lett.* 10:44–59
19. Notsu, K., Mabuchi, H., Yoshioka, O., Matsuda, J., Ozima, M. 1973. *Earth Planet. Sci. Lett.* 19:29–36
20. Lugmair, G. W. 1974. *Meteoritics* 9:369 (Abstr.)
21. Apt, K. E., Knight, J. D., Camp, D. C., Perkins, R. W. 1974. *Geochim. Cosmochim. Acta* 38:1485–88
22. Shirck, J., Hoppe, M., Maurette, M., Walker, R. 1969. *Meteorite Research,* ed. P. Millman, 41–50. Dordrecht: Reidel. 940 pp.
23. Cantelaube, Y., Maurette, M., Pellas, P. 1967. *Radioactive Dating and Methods of Low Level Counting,* ed. P. M. Millman, 215–29. Vienna: 940 pp.
24. Pellas, P., Störzer, D. 1974. *Meteoritics* 9:388–90 (Abstr.)
25. Harper, C. T., ed. 1973. *Geochronology.* Stroudsburg, Pa: Dowden, Hutchison & Ross. 469 pp.
26. Faure, G., Powell, J. L. 1972. *Strontium Isotope Geology.* Berlin: Springer-Verlag. 188 pp.

27. Gray, C. M., Papanastassiou, D. A., Wasserburg, G. J. 1973. *Icarus* 20:213–39
28. Papanastassiou, D. A. 1970. *The determination of small time differences in the formation of planetary objects.* PhD thesis. Calif. Inst. Technol., Pasadena, Calif.
29. Papanastassiou, D. A., Wasserburg, G. J. 1969. *Earth Planet. Sci. Lett.* 5:361–76
30. Mason, B. 1971. *Handbook of Elemental Abundances in Meteorites.* New York: Gordon & Breach. 555 pp.
31. Doe, B. 1970. *Lead isotopes.* Berlin: Springer-Verlag. 137 pp.
32. Russell, R. D., Farquhar, R. M. 1960. *Lead Isotopes in Geology.* New York: Interscience. 243 pp.
33. Wetherill, G. W. 1966. *Handbook of Physical Constants,* ed. S. Clark. *Geol. Soc. Am. Mem.* 97:513–19
34. Beckinsale, R. D., Gale, N. H. 1969. *Earth Planet. Sci. Lett.* 6:289–94
35. Katcoff, S., Schaeffer, O. A., Hastings, J. M. 1951. *Phys. Rev.* 82:688–90
36. Le Roux, L. J., Glendenin, L. E. 1963. *Proc. Nat. Conf. Nucl. Energy,* 83–94. S. Africa: Pretoria
37. Jaffey, A. H., Flynn, K. F., Glendenin, L. E., Bentley, W. C., Easling, A. M. 1971. *Phys. Rev. C* 4:1889–1906
38. Roberts, J. M., Gold, R., Armani, R. J. 1968. *Phys. Rev.* 174:1482–84
39. Fields, P. R. et al 1966. *Nature* 212:131–37
40. Wasson, J. T. 1974. *Meteorites.* New York: Springer. 316 pp.
41. Burnett, D. S. 1971. *EOS* 52: IUGG 435–40
42. Bogard, D. D. 1971. *EOS* 52: IUGG 429–35
43. Wood, J. A. 1975. *Proc. Sov. Am. Conf. Cosmochem. Moon & Planets,* Moscow, 1974, ed. J. Pomeroy. In press
44. El-Baz, F. 1974. *Ann. Rev. Astron. Astrophys.* 12:135–65
45. Toksöz, M. N. 1974. *Ann. Rev. Earth Planet. Sci.* 2:151–77
46. Wetherill, G. W. 1974. *Ann. Rev. Earth Planet. Sci.* 2:303–31
47. Reeves, H., ed. 1972. *On the origin*

326   WETHERILL

of the solar system. Paris: Cen. Nat. Rech. Sci. 383 pp.
48. Safranov, V. S. 1972. *Evolution of the Protoplanetary Cloud & Formation of the Earth & Planets.* Jerusalem: Isr. Program Sci. Transl. 206 pp.
49. Cameron, A. G. W. 1973. *Icarus* 18:407–50
50. Larson, R. B. 1972. *Mon. Notic. Roy. Astron. Soc.* 156:437–58
51. Reeves, H. 1972. In *On the Origin of the Solar System,* ed. H. Reeves, 376–79. Paris: CNRS
52. Lin, C. C., Shu, F. H. 1964. *Astrophys. J.* 140:646–55
53. Shu, F. H., Milione, V., Gebel, W. 1972. *Astrophys. J.* 173:557–92
54. Wasserburg, G. J., Huneke, J. C., Burnett, D. S. 1969. *J. Geophys. Res.* 74:4221–32
55. Podosek, F. A. 1970. *Earth Planet. Sci. Lett.* 8:183–87
56. Podosek, F. A. 1972. *Geochim. Cosmochim. Acta* 36:755–72
57. Podosek, F. A., Lewis, R. S. 1972. *Earth Planet. Sci. Lett.* 15:101–9
58. Reynolds, J. H. *Proc. Sov.-Am. Conf. Cosmochem. Moon & Planets,* Moscow, June 1974. In press
59. Cameron, A. G. W. 1962. *Icarus* 1:13–69
60. Reeves, H. 1972. *Astron. Astrophys.* 19:215–23
61. Hohenberg, C. M., Podosek, F. A., Reynolds, J. H. 1967. *Science* 156:202–6
62. Lewis, R. S., Anders, E. 1975. *Proc. Nat. Acad. Sci. USA* 72:268–73
63. Clayton, R. N., Grossman, L., Mayeda, T. K. 1973. *Science* 182:485–88
64. Lee, T., Papanastassiou, D. A. 1974. *Geophys. Res. Lett.* 1:225–28
65. Nyquist, L. E., Hubbard, N. J., Gast, P. W., Bansal, B. N., Wiesmann, H. 1973. *Proc. 4th Lunar Sci. Conf., Geochim. Cosmochim. Acta:* Suppl. 4, 1823–46
66. Wetherill, G. W., Mark, R., Lee-Hu, S. 1973. *Science* 182:2781–83
67. Tatsumoto, M., Nunes, P. D., Unruh, D. H. *Proc. Sov.-Am. Conf. Cosmochem. Moon & Planets,* Moscow, 1974. In press
68. Papanastassiou, D. A., Wasserburg,

G. J. 1972. *Earth Planet. Sci. Lett.* 17:52–62
69. Jahn, B.-M., Shih, C.-Y. 1974. *Geochim. Cosmochim. Acta* 38:873–85
70. Hauge, O. 1972. *Solar Phys.* 26:263–75
71. Hauge, O. 1972. *Solar Phys.* 26:276–82
72. Wasserburg, G. J., Papanastassiou, D. A., Sanz, H. G. 1969. *Earth Planet. Sci. Lett.* 7:33–43
73. Hohenberg, C., Reynolds, J. 1969. *J. Geophys. Res.* 74:6679–83
74. Eberhardt, P., Geiss, J. 1966. *Earth Planet. Sci. Lett.* 1:99–101
75. Hohenberg, C. M., Munk, M. N., Reynolds, J. H. 1967. *J. Geophys. Res.* 72:3139–77
76. Rowe, M. W. 1970. *Geochim. Cosmochim. Acta* 34:1019–28
77. Hohenberg, C. M. 1970. *Geochim. Cosmochim. Acta* 34:185–91
78. Reynolds, J. H. 1968. *Nature* 218:1024–28
79. Podosek, F. A., Huneke, J. C. 1971. *Earth Planet. Sci. Lett.* 12:73–82
80. Pepin, R. O. 1968. *Origin and Distribution of the Elements,* ed. L. H. Ahrens, 379–86. Oxford: Pergamon. 1178 pp.
81. Shirck, J. 1974. *Earth Planet. Sci. Lett.* 23:308–12
82. Nier, A. O. 1950. *Phys. Rev.* 79:450–54
83. Marti, K. 1967. *Earth Planet. Sci. Lett.* 3:243–48
84. Eberhardt, P. et al 1972. *Proc. 3rd Lunar Sci. Conf., Geochim. Cosmochim. Acta:* Suppl. 3, 1821–56
85. Becker, V., Bennett, J. H., Manuel, O. K. 1968. *Earth Planet. Sci. Lett.* 4:357–62
86. Goldberg, E. D. 1965. *Chemical Oceanography,* ed. J. P. Riley, G. Skirrow, 1:163–96. London: Academic. 712 pp.
87. Butler, W. A., Jeffery, P. M., Reynolds, J. H., Wasserburg, G. J. 1963. *J. Geophys. Res.* 68:3283–91
88. Boulos, M. S., Manuel, O. K. 1971. *Science* 174:1334–36
89. Gopalan, K., Wetherill, G W. 1970. *J. Geophys. Res.* 75:3457–67
90. Sanz, H. G., Wasserburg, G. J. 1969. *Earth Planet. Sci. Lett.* 6:335–45

91. Kempe, W., Müller, O. 1969. *Meteorite Res.*, ed. P. M. Millman, 418–28. Dordrecht, Holland: Reidel. 940 pp.
92. Gopalan, K., Wetherill, G. W. 1968. *J. Geophys. Res.* 73 : 7133–36
93. Gopalan, K., Wetherill, G. W. 1971. *J. Geophys. Res.* 76 : 8484–92
94. Gopalan, K., Wetherill, G. W. 1969. *J. Geophys. Res.* 74 : 4349–58
95. Kaushal, S. K., Wetherill, G. W. 1969. *J. Geophys. Res.* 74 : 2717–26
96. Tatsumoto, M., Unruh, D. M., Desborough, G. A. 1973. *Meteoritics* 8 : 446 (Abstr.)
97. Chen, J. H., Tilton, G. R. 1974. *Meteoritics* 9 : 325–26 (Abstr.)
98. Tilton, G. R. 1973. *Earth Planet. Sci. Lett.* 19 : 321–29
99. Tatsumoto, M., Knight, R. J., Allegre, C. J. 1973. *Science* 180 : 1279–83
100. Huey, J. M., Kohman, T. P. 1973. *J. Geophys. Res.* 78 : 3227–44
101. Gale, N. H., Arden, J., Hutchison, R. 1972. *Nature Phys. Sci.* 240 : 56–57
102. Zähringer, J. 1968. *Geochim. Cosmochim. Acta* 32 : 209–37
103. Merrihue, C., Turner, G. 1966. *J. Geophys. Res.* 71 : 2852–57
104. Turner, G. 1971. *Earth Planet. Sci. Lett.* 10 : 227–34
105. Podosek, F. A. 1971. *Geochim. Cosmochim. Acta* 35 : 157–73
106. Turner, G., Cadogan, P. H. 1974. *Proc. 5th Lunar Sci. Conf., Geochim. Cosmochim. Acta* : Suppl. 5, 1601–15
107. Alexander, E. C. Jr., Davis, P. K. 1974. *Geochim. Cosmochim. Acta* 38 : 911–28
108. Turner, G. 1969. *Meteorite Research,* ed. P. M. Millman, 407–17. Dordrecht, Holland: Reidel. 940 pp.
109. Turner, G., Cadogan, P. H. 1973. *Meteoritics* 8 : 477–88 (Abstr.)
110. Bogard, D. D., Wright, R. J., Husain, L. 1974. *EOS* 55 : 334 (Abstr.)
111. Podosek, F. A., Huneke, J. C. 1973. *Meteoritics* 8 : 64 (Abstr.)
112. Van Schmus, W. R., Wood, J. A. 1967. *Geochim. Cosmochim. Acta* 31 : 747–65
113. Bogard, D. D., Burnett, D. S.,
114. Allegre, C. J., Birck, J. L., Fourcade, S., Semet, M. P. 1975. *Science* 187 : 436–38
115. Wasserburg, G. J., Burnett, D. S. Frondel, C. 1965. *Science* 150 : 1814–18
116. Burnett, D. S., Wasserburg, G. J. 1967b. *Earth Planet. Sci. Lett.* 2 : 397–408
117. Sanz, H. G., Burnett, D. S., Wasserburg, G. J. 1970. *Geochim. Cosmochim. Acta* 34 : 1227–39
118. Wasserburg, G. J., Burnett, D. S. 1969. *Meteorite Research,* ed. P. M. Millman, 467–79. Dordrecht, Holland: Reidel. 940 pp.
119. Podosek, F. A., Huneke, J. C. 1973. *Geochim. Cosmochim. Acta* 37 : 667–84
120. Kirsten, T., Horn, P. 1973. *Jahresbericht,* p. 195. Heidelberg: Max Planck Inst. Kernphys.
121. Patterson, C. C. 1955. *Geochim. Cosmochim. Acta* 7 : 151–53
122. Silver, L. T., Duke, M. B. 1971. *EOS* 52 : 269 (Abstr.)
123. Albee, A. L., Gancarz, A. J. *Proc. Sov.-Am. Conf. Cosmochem. Moon & Planets,* Moscow, 1974. In press
124. Burnett, D. S., Wasserburg, G. J. 1967a. *Earth Planet. Sci. Lett.* 2 : 137–47
125. Podosek, F. A. 1973. *Earth Planet. Sci. Lett.* 19 : 135–44
126. Papanastassiou, D. A., Rajan, R. S., Huneke, J. C., Wasserburg, G. J. 1974. *Lunar Sci. V* 583–85. (Abstracts of papers submitted to 5th Lunar Sci. Conf., Houston, Lunar Sci. Inst.)
127. Papanastassiou, D. A., Wasserburg, G. J. 1974. *Geophys. Res. Lett.* 1 : 23–6
128. Fish, R. A., Goles, G. G., Anders, E. 1960. *Astrophys. J.* 132 : 243–58
129. Sonett, C. P., Colburn, D. S., Schwartz, K., Keil, L. 1970. *Astrophys. Space Sci.* 7 : 446–88
130. Safranov, V.. S. 1972. *On the origin of the Solar System,* ed. H. Reeves, 89–113. Paris: CNRS. 383 pp.
131. Wetherill, G. W. 1972. *Tectono-*

physics 13:31–45
132. Papanastassiou, D. A., Wasserburg, G. J. 1975. *Lunar Science VI* (Abstracts of papers submitted to the 6th Lunar Sci. Conf., Mar. 17–21, 1975, Houston, Lunar Sci. Inst.)
133. Jessburger, E. K., Huneke, J. C., Podosek, F. A., Wasserburg, G. J. 1974. *Proc. 5th Lunar Sci. Conf., Geochim. Cosmochim. Acta*: Suppl. 5, 1419–49
134. Hart, S. R. 1971. *Phil. Trans. Roy. Soc. London A* 268: 573–87
135. Moorbath, S., O'Nions, R. K., Pankhurst, R. J., Gale, N. H., McGregor, V. R. 1972. *Nature Phys. Sci.* 240: 78–82

136. Patterson, C. C. 1956. *Geochim. Cosmochim. Acta* 10: 230–37
137. Houtermans, F. G. 1953. *Nuovo Cimento* 10: 1623–33
138. Ostie, R. G., Russell, R. D., Stanton, R. I. 1969. *Can. J. Earth Sci.* 4: 245–69
139. Sinha, A. K., Tilton, G. R. 1973. *Geochim. Cosmochim. Acta* 37: 1823–49
140. Tera, F., Wasserburg, G. J. 1974. *Proc. 5th Lunar Sci. Conf., Geochim. Cosmochim. Acta*: Suppl. 5, 1571–99
141. Goldreich, P., Ward, W. R. 1973. *Astrophys. J.* 183: 1051–61

ELECTRON SCATTERING AND NUCLEAR STRUCTURE[1]

✺5564

T. W. Donnelly[2] and J. D. Walecka

Institute of Theoretical Physics, Department of Physics, Stanford University, Stanford, California 94305

CONTENTS

1 INTRODUCTION AND GENERAL THEORY

There are several reasons why electron scattering is a very powerful tool for studying nuclear structure. First, the interaction is known, as the electron interacts electromagnetically with the local current density $J_\lambda(\mathbf{x})$ in the nuclear target. Since this interaction is relatively weak, of order

[1] Research supported by the National Science Foundation, grant MPS 073-08916.
[2] Alfred P. Sloan Foundation Fellow.

$\alpha \approx 1/137$ the fine structure constant, one can make measurements without greatly disturbing the structure of the target. For example, in light nuclei only a single scattering need be considered. In constrast, in hadronic scattering of strongly interacting particles, it is difficult to separate the reaction mechanism from the target structure. Of course, these same comments apply to real photon processes, but a second advantage of electrons is that for a fixed nuclear excitation energy ω, one can vary the three-momentum $q = |\mathbf{q}|$ transferred by the scattered electron to the nucleus and thus map out the Fourier transforms of the transition matrix elements of the current density $J_\lambda(\mathbf{x})$. This means that not only the charge density, but also the nuclear convection and magnetization current densities may be studied: by inverting the transforms, one has a microscope for locating the spatial distribution of the densities themselves. To illustrate these points, consider the elastic scattering of high energy electrons by an extended charge distribution as a problem of Fraunhofer diffraction in classical optics (see Figure 1.1). The phase difference of an arbitrary path relative to the central ray can be written $\Delta\phi = (2\pi/\lambda)[\hat{\mathbf{k}} \cdot \mathbf{x} - \hat{\mathbf{k}}' \cdot \mathbf{x}] = (\mathbf{k} - \mathbf{k}') \cdot \mathbf{x} \equiv -\mathbf{q} \cdot \mathbf{x}$ where $\mathbf{q}$ is the three-momentum transfer. The amount of charge scattering from the volume element $d\mathbf{x}$ will be proportional to $\alpha f(\mathbf{q})\rho(\mathbf{x}) \, d\mathbf{x}$ where $f(\mathbf{q})$ is an elementary scattering amplitude and $\rho(\mathbf{x})$ is the charge density. Superposing the scattered waves, the resulting diffraction pattern will be proportional to

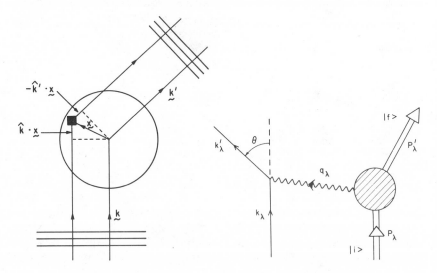

Figure 1.1   High-energy electron scattering as a problem of Fraunhofer diffraction.

Figure 1.2   Electron scattering in the one-photon-exchange approximation.

$$I(\mathbf{k}, \mathbf{k'}) \propto \alpha^2 |f(\mathbf{q})|^2 \left| \int \exp\left[-i\mathbf{q} \cdot \mathbf{x} \right] \rho(\mathbf{x}) \, d\mathbf{x} \right|^2 \qquad 1.1$$

and measures the Fourier transform of the charge distribution as maintained above. Clearly the larger the value of q, the finer the details of the charge density that can be probed. The quantum mechanical structure of the target can be included by replacing $\rho(\mathbf{x})$ by the appropriate expectation value or transition density. A similar analysis holds for the scattering from the nuclear current.

A third feature of electron scattering is that by varying ω, the energy transferred to the target, we obtain an excitation profile of the nucleus. At low momentum transfer, this is similar to the excitation functions seen in photonuclear reactions and is dominated by electric dipole transitions. However, at high q, transitions that require high angular momentum may take place and it becomes possible to investigate high spin states. Furthermore, because the interaction of the electron with the nuclear spin-magnetization current is enhanced at high q and large scattering angle, it is possible to examine states of a magnetic character, including high spin magnetic excitations. Electron scattering provides the only means for directly studying these nuclear properties.

In recent years, some progress has been made on interrelating the semileptonic electromagnetic (electron scattering) and weak (β-decay, μ-capture, ν-reactions, and ν-scattering) interactions with nuclei. As a fourth feature of electron scattering, we show how information obtained from this process can be used to eliminate many of the uncertainties in the nuclear many-body problem and provide a relatively reliable nuclear laboratory with which to study the less well known weak interactions.

Several recent review articles (1–10) on electron scattering from nuclei, as well as review talks at conferences (11–20) appear in the literature.[3] There are also some recent books on this subject (21–24). The main point of departure for the present work is the review article of deForest & Walecka in 1966 (1). Since that time, considerable progress has been made in this field. With the advent of a new generation of electron accelerators, such as the machines at MIT and Saclay, one can look forward to exciting new developments. Within the scope of a review article like this, only a representative selection of topics can be treated in detail; however, we attempt to summarize the current status of the field, emphasizing those aspects that appear to us to demonstrate the power and uniqueness of electron scattering as a tool for studying nuclear structure. Our emphasis is on the theoretical understanding of nuclear structure and the ideas about the nature of electromagnetic and weak interactions with nuclei that can be obtained using electron scattering. We do not attempt to

---

[3] See also the papers by Beck in (15) and Lührs, J. W. Negele, H. Theissen, D. G. Ravenhall, and J. W. Lightbody Jr. in (19).

provide a summary of all existing experimental data on the subject, but rather use only enough data to illustrate the theoretical ideas involved. We attempt to keep the present article reasonably self-contained; however, the reader is referred to (1) for further technical background and to the book by Überall (23), which also contains an extensive bibliography.

1.1   Qualitative Discussion of Nuclear Response Surfaces

In this article, we discuss electron scattering within the framework of the one-photon-exchange approximation as illustrated in Figure 1.2 (higher-order processes are discussed in Section 5.1). Here a plane-wave electron is scattered through an angle θ from an initial state of four-momentum $k_\lambda(\mathbf{k}, i\varepsilon)$ to a final state $k'_\lambda = (\mathbf{k}', i\varepsilon')$.[4] In this process, a single virtual photon with four-momentum $q_\lambda = k'_\lambda - k_\lambda = (\mathbf{q}, -i\omega)$ is exchanged with the nuclear target. We consider energies high enough so that the electron mass can be neglected. Then

$$q_\lambda^2 = 4\varepsilon\varepsilon' \sin^2 \theta/2,$$

$$\mathbf{q}^2 = q_\lambda^2 + \omega^2,$$ 1.2

$$\omega = \varepsilon - \varepsilon'.$$

In the scattering process, the nucleus makes a transition from an initial state $|i\rangle$ with four-momentum P_λ to a final state $|f\rangle$ with $P'_\lambda = P_\lambda - q_\lambda$. In the nuclear rest system, the state vectors are assumed to have total angular momentum, parity, and isospin, $J^\pi T$, as good quantum numbers.

If the target is unpolarized and unobserved, then the double-differential cross section for the process (e, e') in the one-photon-exchange approximation takes the form

$$\frac{d^2\sigma}{d\Omega\, d\varepsilon'} = \frac{4\pi\sigma_M}{M_T} \left\{ (q_\lambda^2/q^2)^2 S_L(q, \omega) + \left[\frac{1}{2}(q_\lambda^2/q^2) + \tan^2 \frac{\theta}{2} \right] S_T(q, \omega) \right\},$$ 1.3

where σ_M is the Mott cross section for scattering an electron from a point charge

$$\sigma_M = \left[\alpha \cos \frac{\theta}{2} \middle/ \left(2\varepsilon \sin^2 \frac{\theta}{2} \right) \right]^2$$ 1.4

and M_T is the initial target mass.

The functions $S_L(q, \omega)$ and $S_T(q, \omega)$ are the longitudinal and transverse dynamic structure functions, or nuclear response surfaces,[5] and contain

[4] We use units where $\hbar = c = 1$ and $e^2/4\pi = \alpha \approx 1/137$. Four-vectors are denoted $v_\lambda = (\mathbf{v}, iv_0)$ and $a \cdot b = a_\lambda b_\lambda = \mathbf{a} \cdot \mathbf{b} - a_0 b_0$ where repeated indices are summed. We denote the magnitude of a three-vector by $v = |\mathbf{v}|$.

[5] A general discussion of linear response in many-body systems is contained in (25).

all the information on the distribution of the nuclear electromagnetic current density. They are related to the response surfaces $W_{1,2}(q_\lambda^2, \omega)$ commonly used in high energy particle physics by (1)

$$4\pi S_T(q, \omega) = 2W_1(q_\lambda^2, \omega)$$

$$4\pi[(q_\lambda^2/q^2)^2 S_L(q, \omega) + \tfrac{1}{2}(q_\lambda^2/q^2)S_T(q, \omega)] = W_2(q_\lambda^2, \omega). \qquad 1.5$$

Since the structure functions depend only on q and ω but not on θ, they may be separated either by making a plot of the cross section against $\tan^2 \theta/2$ at fixed (q, ω)—a "Rosenbluth plot"—or by working at $\theta = 180°$ where only the transverse surface contributes.

Before proceeding to a detailed derivation and discussion of this result we present a qualitative survey of the nuclear structure information contained in these response surfaces. Schematic representations of the longitudinal and transverse surfaces are given in Figures 1.3 and 1.4 respectively. Only a few of the main features are included for clarity. In electron scattering, only the region of space-like momentum transfer can be probed

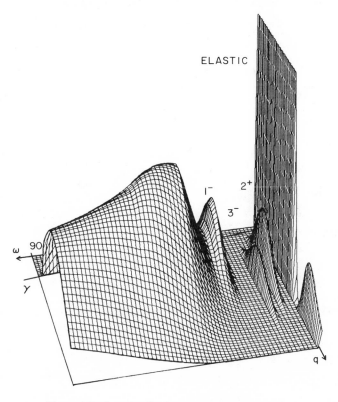

Figure 1.3   Longitudinal response surface $S_L(q, \omega)$.

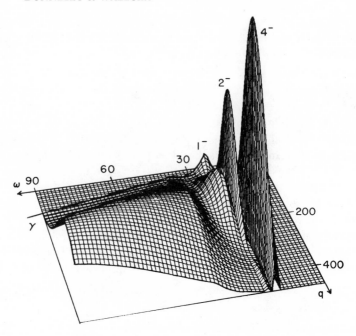

Figure 1.4   Transverse response surface $S_T(q, \omega)$.

$(q_\lambda^2 \geq 0$ or $q \geq \omega)$ and so the boundary $q = \omega$, which corresponds to real photons, is indicated in the figures (labeled γ). At this boundary, one measures the real photoabsorption cross section at a photon energy ω through the relation (26)

$$\sigma_\gamma = [(2\pi)^3 \alpha/\omega M_T] S_T(q = \omega). \qquad 1.6$$

The three-momentum transfer region that has so far been explored in electron scattering from nuclei runs from $q = \omega$ to $q \approx 600$ MeV. In the present article we are concerned with excitation energies between $\omega \approx 0$ (elastic scattering) and $\omega \approx m_\pi$ beyond which pion and isobar electroproduction are important. Let us briefly consider the various parts of this kinematic region.

For $\omega = 0$ (neglecting recoil) we have elastic scattering where the nucleus is left in its ground state, $|f\rangle = |i\rangle$. The longitudinal structure function (Figure 1.3) is dominated by the monopole moment of the nuclear ground-state charge distribution

$$(4\pi/M_T) S_L(q, \omega) = |F_{el}(q)|^2 \, \delta(\omega)$$

$$F_{el}(q) = \int [\sin(qx)/(qx)] \rho_{00}(\mathbf{x}) \, d\mathbf{x}. \qquad 1.7$$

Thus a measurement of

$$\int_{\text{elastic peak}} (4\pi/M_T)S_L(q, \omega)\, d\omega = |F_{\text{el}}(q)|^2 \qquad 1.8$$

maps out the Fourier transform of the spherical average of the nuclear charge distribution. Higher even-multipole moments of the charge distribution may also contribute if $J_i \geq 1$. Elastic charge scattering has been reviewed in detail by Hofstadter (27). Our best knowledge of the size and shape of the nuclear charge distribution comes from elastic electron scattering. Recent work on this topic has concentrated on seeing the higher Fourier components, or shell structure, on the nuclear charge density. Also as the quality and quantity of elastic scattering data has increased, a great deal of effort has gone into so-called "model-independent" analysis of the data in an attempt to extract features of the charge distribution that are independent of any explicit functional form initially assumed for the charge density. We briefly review these developments in Section 2.1.

If $J_i \geq \frac{1}{2}$, then the nuclear ground state may possess odd magnetic multipole moments and a similar analysis of the elastic contribution to $S_T(q, \omega)$ maps out the Fourier transform of the nuclear ground state current distribution. The dominant multipole at low q is the magnetic dipole, but if J_i is large enough, magnetic octupole and higher magnetic moments can come into play as q is increased. In recent years considerable work has been done on this problem, and we discuss elastic magnetic scattering in some detail in Section 2.2.

For $\omega > 0$ we have *inelastic* electron scattering. The precise nature of the response surfaces depends, of course, on the excited state structure of the nucleus in question, but typically there will be sharp peaks as a function of ω corresponding to discrete bound states below the particle emission threshold. Several of these are indicated in Figures 1.3 and 1.4. The first excited state illustrated here is assumed to be excited through the electric quadrupole operator. For instance, in even-even nuclei, the most common sequence of levels is 0^+-2^+ where the very collective 2^+ state undergoes a strong E2 γ decay to the ground state. Measuring the q dependence of S_L and S_T in this case leads to the Fourier transform of the nuclear transition charge and current densities. In many cases a very collective 3^- state may be seen as well at low excitation energies and Figure 1.3 shows such a feature. We discuss some examples of such transitions for both vibrational and rotational excitations in Section 3.1. Since the excited states of heavy deformed nuclei are closely spaced, very high resolution is needed in electron scattering to track the inelastic form factors out to the interesting and informative region of high momentum transfer. It is only with the new generation of electron accelerators that a systematic study of deformed nuclei will be possible.

Similarly, a systematic study of the much weaker single-particle transitions in nuclei remains to be carried out.

The most prominent feature of photoexcitation in nuclei is the giant electric dipole resonance that occurs systematically in nuclei as a rather broad peak above particle-emission threshold. So too, in electron scattering at low momentum transfers the giant dipole resonance dominates the inelastic response surfaces. As q is increased, however, other modes of excitation may be as important or more important than the giant dipole. We see, for example, that the spin-flip dipole resonance becomes more important than the usual giant dipole as q increases and accounts for the shift of strength toward higher excitation energies ω with increasing q (see Figure 1.4). In electroexcitation, in contrast to photoexcitation, multipolarities higher than one also play a crucial role. For example, Figures 1.3 and 1.4 illustrate the appearance of the strongly excited $J^{\pi}T = 2^{-}1$ and $4^{-}1$ particle-hole states that occur in nuclei such as ^{12}C and ^{16}O. The $2^{-}1$ state shown here corresponds to a magnetic quadrupole vibration of the nucleus and, being magnetic in character, shows up only in the transverse response function. The same is true for the excitation of the $4^{-}1$ state, a magnetic hexadecupole vibration. The cross section for exciting such a mode at low q, for instance in photoexcitation where $q = \omega$, is very small indeed. However, as shown in Figure 1.4, the structure function may rise to a significant value at medium-to-large values of momentum transfer. The excitation of such high spin states via the isovector magnetic moment of the nucleon has been a topic of interest in recent years and we discuss electron excitation of collective particle-hole states (including the giant dipole resonance) in detail in Section 3.2.

In contrast to all the features considered thus far, which occur as peaks in the response function centered on a fixed value of ω, there is also a "quasi-elastic peak" corresponding to nucleon emission, which occurs as a curving hill in the q-ω plane. If the nucleus were a collection of noninteracting nucleons at rest, this feature would appear at $\omega = q^2/2M$ where M is the nucleon mass, as this is the energy transferred to a nucleon at rest if it is given a momentum q. Because of the interaction between the nucleons in the nucleus, the ridge if shifted to approximately $\omega = q^2/2M + \bar{\varepsilon}$ where $\bar{\varepsilon}$ is the average separation energy for the nucleus in question. Furthermore, the nucleons are not at rest in a nucleus, but for a Fermi gas have velocities up to k_F/M. Thus this feature, which would otherwise be a sharp peak, is spread out by the Fermi motion of the target particles and has a width (versus ω) characterized by k_F. In Section 3.3, we discuss the process of quasi-elastic electron scattering within the framework of some simple nuclear models.[6]

[6] Coincidence measurements of the process (e, e'N) are discussed in the article by Dieperink and deForest in this volume, pp. 1–26. See also (3).

In Section 4 we briefly review the relationship between electromagnetic interactions and semileptonic weak interactions, stressing the extra knowledge that electron scattering brings to bear on the latter. Finally, in Section 5 we address a few of the special topics of interest in intermediate-energy physics, which relate to the main body of this article.

1.2   Multipole Analysis and the Nuclear Current Operator

We proceed to the general theory of electron scattering (1) in the one-photon-exchange approximation where the nuclear target makes a transition between the states $|J_i^{\pi_i}\rangle$ and $|J_f^{\pi_f}\rangle$. This can include elastic scattering. Here we assume that we have a well-localized target that presents a well-localized transition density to scatter the electron. (Thus we assume the target is heavy and neglect recoil in the transition matrix elements. We include recoil in the density of final states.) The electron thus feels the perturbation

$$\mathscr{H}^{(1)}(\mathbf{x}, t) = -ej_\mu(\mathbf{x})A_\mu^{\text{ext}}(\mathbf{x}, t), \qquad\qquad 1.9$$

where $j_\mu(\mathbf{x}) = i\bar{\psi}(\mathbf{x})\gamma_\mu\psi(\mathbf{x})$ is the electron current.[7] The external vector potential scattering the electron can be related to the nuclear transition current density through Maxwell's equations

$$\Box A_\mu^{\text{ext}}(\mathbf{x}, t) = -e_p\langle f|\hat{J}_\mu(\mathbf{x}, t)|i\rangle = -e_p\langle f|\hat{J}_\mu(\mathbf{x})|i\rangle \exp\left[-i(E-E')t\right],$$
$$1.10$$

where $|i\rangle$ and $|f\rangle$ are exact (Heisenberg) nuclear target states with energies E and E'. All we assume about the target at this point is that there is a local current density operator[8] $\hat{J}_\mu(\mathbf{x}) = [\hat{\mathbf{J}}(\mathbf{x}), i\hat{\rho}(\mathbf{x})]$. This can include exchange currents coming from the flow of charged mesons between the nucleons, for example. We also give the operator the Heisenberg time dependence $\hat{J}_\mu(\mathbf{x}, t) = \exp(i\hat{H}t)\hat{J}_\mu(\mathbf{x})\exp(-i\hat{H}t)$; e_p is the proton charge. Going to the interaction representation for the electron, we obtained the

---

[7] The electron field is given by

$$\psi(\mathbf{x}) = \frac{1}{(\Omega)^{1/2}}\sum_{\mathbf{k}\lambda}\left[a_{\mathbf{k}\lambda}u_\lambda(\mathbf{k})\exp(i\mathbf{k}\cdot\mathbf{x}) + b_{\mathbf{k}\lambda}^\dagger v_\lambda(-\mathbf{k})\exp(-i\mathbf{k}\cdot\mathbf{x})\right],$$

where Ω is the normalization volume, u and v are Dirac spinors normalized to $u_\lambda^\dagger(\mathbf{k})u_{\lambda'}(\mathbf{k}) = \delta_{\lambda\lambda'}$ and $a^\dagger$ and $b^\dagger$ are creation operators for electrons and positrons. The γ matrices are hermitian and satisfy $\gamma_\mu\gamma_\nu + \gamma_\nu\gamma_\mu = 2\delta_{\mu\nu}$. We use $\not{k} \equiv k_\lambda\gamma_\lambda$. We also make use of the explicit representation

$$\gamma = \begin{pmatrix} 0 & -i\boldsymbol{\sigma} \\ i\boldsymbol{\sigma} & 0 \end{pmatrix}, \qquad \gamma_4 = \begin{pmatrix} 1 & 0 \\ 0 & -1 \end{pmatrix}.$$

[8] We use a circumflex over a symbol, e.g. $\hat{J}$, to denote an operator in the nuclear Hilbert space.

scattering operator $\hat{S} = -i \int \mathscr{H}_I^{(1)}(\mathbf{x}, t)\, d^4x$ where the electron fields now carry the free-field time dependence. The required scattering matrix element between electron states $|\mathbf{k}, \sigma\rangle$ and $|\mathbf{k}', \sigma'\rangle$ is thus given by

$$\langle \mathbf{k}', \sigma' | \hat{S} | \mathbf{k}, \sigma \rangle = -(2\pi/\Omega)\delta(E' + \varepsilon' - E - \varepsilon)(ee_p/q_\lambda^2)\bar{u}_{\sigma'}(\mathbf{k}')\gamma_\mu u_\sigma(\mathbf{k})J_\mu(\mathbf{q}),$$

1.11

where

$$J_\mu(\mathbf{q}) = \int \exp(-i\mathbf{q}\cdot\mathbf{x})\langle f|\hat{J}_\mu(\mathbf{x})|i\rangle\, d\mathbf{x}.$$

1.12

(Note that this expression follows directly from the usual Feynman rules for the one-photon exchange graph shown in Figure 1.2.) The cross section now follows from the scattering matrix in familiar fashion (the incident flux is just $1/\Omega$)

$$d\sigma = 2\pi\delta(E' + \varepsilon' - E - \varepsilon)\frac{1}{(1/\Omega)}\left|\frac{4\pi\alpha}{\Omega q_\lambda^2}\bar{u}_{\sigma'}(\mathbf{k}')\gamma_\mu u_\sigma(\mathbf{k})J_\mu(\mathbf{q})\right|^2 \frac{\Omega\, d\mathbf{k}'}{(2\pi)^3}.$$

1.13

If this expression is summed over final, and averaged over initial, electron spins it takes the form

$$d\sigma = \frac{1}{(2\pi)^2}\frac{1}{2\varepsilon}\delta(E' + \varepsilon' - E - \varepsilon)\left(\frac{4\pi\alpha}{q_\lambda^2}\right)^2\frac{d\mathbf{k}'}{2\varepsilon'}\eta_{\mu\nu}J_\mu(\mathbf{q})J_\nu(\mathbf{q})^*,$$

1.14

where $J_\mu^* = (\mathbf{J}^*, +iJ_0^*)$ and

$$\eta_{\mu\nu} \equiv 4\varepsilon\varepsilon'\tfrac{1}{2}\sum_{\sigma'}\sum_{\sigma}\bar{u}_\sigma(\mathbf{k})\gamma_\nu u_{\sigma'}(\mathbf{k}')\bar{u}_{\sigma'}(\mathbf{k}')\gamma_\mu u_\sigma(\mathbf{k}) = \tfrac{1}{2}Tr\gamma_\mu \not{k}\gamma_\nu \not{k}'.$$

1.15

Here we have assumed $m_e = 0$. It is convenient to make a change of variables $Q_\lambda \equiv \tfrac{1}{2}(k' + k)_\lambda$, $q_\lambda = (k' - k)_\lambda$ and use nuclear current conservation $\partial\hat{J}_\mu/\partial x_\mu = 0$. By taking matrix elements of this expression and using the Heisenberg time dependence, current conservation can be recast in the form $q_\mu J_\mu(\mathbf{q}) = \mathbf{q}\cdot\mathbf{J}(\mathbf{q}) - q_0 J_0(\mathbf{q}) = 0$. Thus, in equation 1.15, $\eta_{\mu\nu}$ can be written

$$\eta_{\mu\nu} \doteq 2[2Q_\mu Q_\nu + \tfrac{1}{2}q_\lambda^2\delta_{\mu\nu}]$$

1.16

and the cross section then takes the simple form

$$d\sigma = \frac{4\alpha^2}{q_\lambda^4}\frac{1}{\varepsilon}\delta(E' + \varepsilon' - E - \varepsilon)\frac{d\mathbf{k}'}{2\varepsilon'}[2(Q_\mu J_\mu)(Q_\nu J_\nu^*) + \tfrac{1}{2}q_\lambda^2 J_\mu J_\mu^*],$$

1.17

where J_μ is given by equation 1.12.

In order to utilize the nuclear selection rules involving angular momentum and parity, we make a multipole analysis of the electromagnetic current operator. Let us start with the charge density where we need

$$\rho(\mathbf{q}) = \int \exp(-i\mathbf{q}\cdot\mathbf{x})\langle J_f M_f|\hat{\rho}(\mathbf{x})|J_i M_i\rangle\, d\mathbf{x}. \qquad 1.18$$

(We suppress the other nuclear quantum numbers for clarity.) The familiar plane-wave expansion

$$\exp(-i\mathbf{q}\cdot\mathbf{x}) = 4\pi \sum_{JM} (-i)^J j_J(qx) Y_{JM}(\Omega_x) Y_{JM}^*(\Omega_q) \qquad 1.19$$

can now be introduced and we may define charge density (or "Coulomb") multipoles:

$$\hat{M}_{JM}(q) \equiv \int j_J(qx) Y_{JM}(\Omega_x)\hat{\rho}(\mathbf{x})\, d\mathbf{x}. \qquad 1.20$$

These are irreducible tensor operators of rank J in the nuclear Hilbert space, and the Wigner-Eckart theorem may be used to extract the dependence of the nuclear matrix element on magnetic quantum numbers

$$\langle J_f M_f|\hat{M}_{JM}(q)|J_i M_i\rangle = (-)^{J_f-M_f}\begin{pmatrix} J_f & J & J_i \\ -M_f & M & M_i \end{pmatrix}\langle J_f\|\hat{M}_J(q)\|J_i\rangle. \qquad 1.21$$

A sum over final, and average over initial, nuclear orientations may now be performed according to

$$(2J_i+1)^{-1}\sum_{M_i}\sum_{M_f}|\rho(\mathbf{q})|^2 = 4\pi(2J_i+1)^{-1}\sum_J |\langle J_f\|\hat{M}_J(q)\|J_i\rangle|^2. \qquad 1.22$$

The treatment of the matrix elements of the current $\mathbf{J}$ is a little more complicated. We introduce an orthogonal basis of unit vectors with respect to $\mathbf{q}$, $\mathbf{e}_{q\pm1} \equiv \mp[1/(2)^{1/2}](\mathbf{e}_x \pm i\mathbf{e}_y)$, $\mathbf{e}_{q0} = \mathbf{e}_z = \mathbf{q}/q$, and express $\mathbf{J}(\mathbf{q})$ in this basis $\mathbf{J}(\mathbf{q}) = \sum_\lambda \mathbf{e}_{q\lambda}^\dagger(\mathbf{e}_{q\lambda}\cdot\mathbf{J}) \equiv \sum_\lambda J_\lambda \mathbf{e}_{q\lambda}^\dagger$. From Blatt & Weisskopf (28) we have the expansions [see (1)]

$$\mathbf{e}_{q\lambda}\exp(-i\mathbf{q}\cdot\mathbf{x}) = (i/q)\sum_{J=0}^{\infty}[4\pi(2J+1)]^{1/2}(-i)^J\nabla[j_J(qx)Y_{J0}(\Omega_x)]$$

$$\text{for}\quad \lambda = 0,$$

$$= -\sum_{J=1}^{\infty}[2\pi(2J+1)]^{1/2}(-i)^J[\lambda j_J(qx)\mathscr{Y}_{JJ1}^\lambda - (1/q)\nabla\times(j_J(qx)\mathscr{Y}_{JJ1}^\lambda)]$$

$$\text{for}\quad \lambda = \pm 1 \qquad 1.23$$

in terms of the vector spherical harmonics defined by

$$\mathscr{Y}_{JJ1}^M(\Omega_x) \equiv \sum_{mm'}(Jm1m'|J1JM)Y_{Jm}\mathbf{e}_{m'}. \qquad 1.24$$

[For the properties of these vector spherical harmonics, see Edmonds (29) and (1).] We define the longitudinal and transverse multipoles of the current by

$$\hat{L}_{JM}(q) \equiv (i/q) \int d\mathbf{x} \{\nabla[j_J(qx) Y_{JM}(\Omega_x)]\} \cdot \hat{\mathbf{J}}(\mathbf{x})$$

$$\hat{T}_{JM}^{el}(q) \equiv (1/q) \int d\mathbf{x} \{\nabla \times [j_J(qx) \mathscr{Y}_{JJ1}^M(\Omega_x)]\} \cdot \hat{\mathbf{J}}(\mathbf{x}) \qquad 1.25$$

$$\hat{T}_{JM}^{mag}(q) \equiv \int d\mathbf{x}[j_J(qx) \mathscr{Y}_{JJ1}^M(\Omega_x)] \cdot \hat{\mathbf{J}}(\mathbf{x}).$$

These are again irreducible tensor operators of rank J in the nuclear Hilbert space. The continuity equation on the nuclear current $\mathbf{e}_{q0} \cdot \mathbf{J}(\mathbf{q}) = (q_0/q)\rho(\mathbf{q})$ can be used to eliminate the longitudinal multipoles in favor of the charge multipoles in electron scattering. Only the *transverse* multipoles of the current are therefore required and for $\lambda = \pm 1$,

$$\hat{J}_\lambda(\mathbf{q}) = -(2\pi)^{1/2} \sum_{J=1}^\infty (-i)^J (2J+1)^{1/2} [-\hat{T}_{J\lambda}^{el}(q) + \lambda \hat{T}_{J\lambda}^{mag}(q)], \qquad 1.26$$

where use has been made of equations 1.12, 1.23, and 1.25. These are the same operators that govern the emission and absorption of real photons, only in that case $q = \omega$. Using the Wigner-Eckart theorem to again extract the dependence of the matrix elements on magnetic quantum numbers, and noting that for unoriented nuclei it is most convenient to quantize the nucleus along the direction $\mathbf{q}/q$, we find for $\lambda = \pm 1$,

$$\sum_{M_i M_f} J_\lambda(q)\rho(q)^* = 0 \qquad 1.27$$

and

$$\sum_{M_i} \sum_{M_f} J_\lambda(q) J_{\lambda'}(q)^* = \delta_{\lambda\lambda'} 2\pi \sum_{J=1}^\infty \{|\langle J_f \| \hat{T}_J^{el}(q) \| J_i \rangle|^2 + |\langle J_f \| \hat{T}_J^{mag}(q) \| J_i \rangle|^2\}.$$

$$1.28$$

We have used the fact that $\hat{T}_J^{el}$ and $\hat{T}_J^{mag}$ have different parity, and therefore one or the other matrix element in the brackets vanishes. We may now put everything together and after some algebra we get

$$\frac{d\sigma}{d\Omega} = 4\pi\sigma_M \left[1 + \left(2\varepsilon \sin^2 \frac{\theta}{2}\right)\Big/ M_T\right]^{-1} F^2$$

$$F^2 \equiv (q_\mu^2/q^2)F_L^2 + \left(\frac{1}{2}(q_\mu^2/q^2) + \tan^2 \frac{\theta}{2}\right)F_T^2$$

$$F_L^2 = (2J_i + 1)^{-1} \sum_{J=0}^\infty |\langle J_f \| \hat{M}_J(q) \| J_i \rangle|^2 \qquad 1.29$$

$$F_T^2 = (2J_i + 1)^{-1} \sum_{J=1}^\infty \{|\langle J_f \| \hat{T}_J^{el}(q) \| J_i \rangle|^2 + |\langle J_f \| \hat{T}_J^{mag}(q) \| J_i \rangle|^2\},$$

where σ_M is defined in equation 1.4. We have allowed the nucleus to recoil in computing the density of final states and have made the approximation $E'/M_T = 1 + q_\lambda^2/2M_T^2 \approx 1$ in this factor.

In the same manner the transition rate for real photon emission is readily calculated to be

$$\omega_\gamma = (8\pi\alpha)K(2J_i+1)^{-1} \sum_{J=1}^{\infty} \{|\langle J_f \| \hat{T}_J^{\text{el}}(K) \| J_i \rangle|^2 + |\langle J_f \| \hat{T}_J^{\text{mag}}(K) \| J_i \rangle|^2\},$$

1.30

where $K = \omega$ is the photon wave number.

Similarly, the photoabsorption cross section integrated over a resonance line is given by

$$\int_{\text{abs. line}} \sigma_\gamma(\omega) \, d\omega = (2\pi)^3(\alpha/K)(2J_i+1)^{-1} \sum_{J=1}^{\infty} \{|\langle J_f \| \hat{T}_J^{\text{el}}(K) \| J_i \rangle|^2$$
$$+ |\langle J_f \| \hat{T}_J^{\text{mag}}(K) \| J_i \rangle|^2\}.$$

1.31

Let us summarize the main features of these results:

1. Because of current conservation, only the matrix elements of the transverse current and charge density are independent. The multipole analysis of the transverse interaction is the same as in real photon processes; however, for a given energy transfer ω in electron scattering, we can vary the three-momentum transfer q, which enters through the argument of the spherical Bessel functions appearing in the multipole operators. The only restriction in electron scattering is $q \geqq \omega$.
2. The selection rules on the multipole operators are the same as for real photons. Thus the parity of $\hat{M}_{JM}$ and $\hat{T}_{JM}^{\text{el}}$ is $(-)^J$ and of $\hat{T}_{JM}^{\text{mag}}$ is $(-)^{J+1}$.
3. The Coulomb and transverse multipoles cannot interfere if only the electron is observed, and their relative contribution can be separated by the methods previously discussed.
4. The transverse multipoles start with $J = 1$ since a transverse photon carries unit helicity; in contrast there is a $J = 0$ Coulomb multipole.
5. The long-wavelength behavior of the multipole operators follows from general considerations [see Blatt & Weisskopf (28) and in (1)]. By increasing q, we can escape the long-wavelength limit and bring out interesting features of high multipolarity or of a magnetic character.
6. There is an additional long-wavelength relation $(q \to 0)$ between the Coulomb and transverse electric multipoles, which follows from current conservation

$$\langle f | \hat{T}_{JM}^{\text{el}}(q) | i \rangle \to [(E - E')/q][(J+1)/J]^{1/2} \langle f | \hat{M}_{JM}(q) | i \rangle.$$

1.32

These results are all quite general.

To test our understanding of nuclear structure, we must eventually relate the nuclear current operator to the properties of the nucleon. The usual prescription of nuclear physics is to take the matrix elements of the interaction with *free* nucleons in plane wave states $\langle \mathbf{p}'\sigma'\rho'|J_\mu(0)|\mathbf{p}\sigma\rho\rangle$ (here σ and ρ refer to the third components of spin and isospin) and write the nuclear current-density operator at the origin in second quantization as

$$\hat{J}_\mu(0) = \sum_{\mathbf{p}'\sigma'\rho'} \sum_{\mathbf{p}\sigma\rho} c^\dagger_{\mathbf{p}'\sigma'\rho'}\langle \mathbf{p}'\sigma'\rho'|J_\mu(0)|\mathbf{p}\sigma\rho\rangle c_{\mathbf{p}\sigma\rho}. \qquad 1.33$$

The equivalent first-quantized densities $J_\mu(\mathbf{x}) = \sum_{i=1}^{A}[J_\mu(i)\delta(\mathbf{x}-\mathbf{x}_i)]$ can now be constructed in standard fashion (25). Equation 1.33 assumes the energy relation $E_p = (p^2 + M^2)^{1/2}$ appropriate to free nucleons. It also assumes that the nuclear coordinates provide a complete description of the nuclear many-body system. Thus it neglects meson exchange currents coming from the flow of mesons *between* nucleons in the nucleus. Although there are many indications that exchange current contributions in the nucleus are generally rather small, this is one of the most interesting open areas of nuclear physics and we return to this question in Section 5.2.

The general form of the single-nucleon matrix element of the current in equation 1.33 is

$$\langle \mathbf{p}'\sigma'\rho'|J_\mu(0)|\mathbf{p}\sigma\rho\rangle = (i/\Omega)\bar{u}_{\sigma'}(\mathbf{p}')\eta^\dagger_{\rho'}[F_1\gamma_\mu + F_2\sigma_{\mu\nu}q_\nu]\eta_\rho u_\sigma(\mathbf{p}). \qquad 1.34$$

Here $q_\lambda = p_\lambda - p'_\lambda$ and $\sigma_{\mu\nu} \equiv (1/2i)[\gamma_\mu, \gamma_\nu]$. The Dirac charge and anamolous magnetic moment form factors are functions of the four-momentum transfer q_λ^2. The isospin structure is given by $F_{1,2} = \frac{1}{2}(F^s_{1,2} + \tau_3 F^v_{1,2})$. Thus $F^s_{1,2} = F^p_{1,2} + F^n_{1,2}$ and $F^v_{1,2} = F^p_{1,2} - F^n_{1,2}$. Note the following values: $F^s_1(0) = F^v_1(0) = 1$, and also $2MF^s_2(0) = \lambda_p + \lambda_n = -0.120$ and $2MF^v_2(0) = \lambda_p - \lambda_n = +3.706$. For the discrete nuclear transition of interest in the present work, the interesting $\mathbf{q}^2$ region is determined by the *nuclear* form factor and the motion of the target nucleons can be treated nonrelativistically. We therefore make a reduction of the single-nucleon amplitude through order $p^2/M^2 \approx (v/c)^2_{\text{nucleon}}$. To do this, the four-component Dirac spinors of the nucleon are written

$$u_\sigma(\mathbf{p}) = \left(\frac{E_p + m}{2E_p}\right)^{1/2}\left(\begin{array}{c}\chi_\sigma \\ [\boldsymbol{\sigma}\cdot\mathbf{p}/(E_p+M)] \ \chi_\sigma\end{array}\right), \qquad 1.35$$

where χ_σ is a two-component Pauli spinor $[\chi_\uparrow = \binom{1}{0}, \chi_\downarrow = \binom{0}{1}]$ and the components of $\boldsymbol{\sigma}$ are the Pauli matrices. This expression can be substituted into equation 1.34 and the resulting matrix element evaluated to order $1/M^2$ using an explicit representation of the γ matrices.[7] A little algebra leads to the expression (1)

$$\langle \mathbf{p}'\sigma'\rho'|J_\mu(0)|\mathbf{p}\sigma\rho\rangle = (1/\Omega)\eta_{\rho'}^\dagger\chi_{\sigma'}^\dagger[\mathcal{M}_\mu]\chi_\sigma\eta_\rho,$$

$$\mathcal{M} = F_1[(\mathbf{p}+\mathbf{p}')/2M]-(F_1+2MF_2)[i\boldsymbol{\sigma}\times\mathbf{q}/2M],$$

$$\mathcal{M}_0 = F_1-[q^2/8M^2-i\mathbf{q}\cdot(\boldsymbol{\sigma}\times\mathbf{p})/4M^2](F_1+2MF_2),$$

1.36

where $\mathcal{M}_\mu \equiv (\mathcal{M}, i\mathcal{M}_0)$. The isospin structure has been noted above. In arriving at these results it is assumed that F_2 is already of $0(1/M)$, and in addition that q_0, which for a free nucleon is $q_0 = (p^2-p'^2)/2M = \mathbf{p}\cdot\mathbf{q}/M - q^2/2M$ is also of $0(1/M)$. As a consequence of this assumption q_0 does not appear in equation 1.36. There are no terms of $0(1/M^2)$ in $\langle J\rangle$, and the equivalent nuclear current needed in equation 1.33 is given by the familiar expression $\hat{\mathbf{J}}(\mathbf{x}) = \hat{\mathbf{J}}_c(\mathbf{x})+\nabla\times\hat{\boldsymbol{\mu}}(\mathbf{x})$, where $\hat{\mathbf{J}}_c$ is the convection current and $\hat{\boldsymbol{\mu}}$ is the intrinsic spin magnetization. The charge density, on the other hand, contains two corrections of $0(1/M^2)$. These are the well-known Darwin-Foldy and spin-orbit terms [see Bertozzi et al (30) for a discussion of the role played by these terms in a precise analysis of charge scattering from nuclei at high q]. We assume $\mathcal{M}_0 \cong F_1$ in our subsequent discussion.

Apart from an overall single-nucleon form factor $f_{sn}(q^2) \cong (1+[q/855\text{ MeV}]^2)^{-1}$ which multiplies all the transition matrix elements and takes into account the finite size of the nucleon (we employ this shape as a representation of the data adequate for our purposes), the nuclear densities that then follow from equation 1.36 and the prescription of equations 1.33 are, in first quantization,

$$\rho(\mathbf{x}) = \sum_{j=1}^{A} e(j)\delta(\mathbf{x}-\mathbf{x}_j)$$

$$\mathbf{J}_c(\mathbf{x}) = \sum_{j=1}^{A} e(j)[\nabla(j)\delta(\mathbf{x}-\mathbf{x}_j)]_{\text{sym}}/iM$$

1.37

$$\boldsymbol{\mu}(\mathbf{x}) = \sum_{j=1}^{A} \mu(j)[\boldsymbol{\sigma}(j)\delta(\mathbf{x}-\mathbf{x}_j)]/2M,$$

where $[AB]_{\text{sym}} \equiv \frac{1}{2}\{A, B\}$ and $e(j) = \frac{1}{2}(1+\tau_3(j))$ and $\mu(j) = \mu_p\frac{1}{2}(1+\tau_3(j))+\mu_n\frac{1}{2}(1-\tau_3(j))$. Here $\mu_p = 2.793$ and $\mu_n = -1.913$ are the magnetic moments of the proton and neutron. It is these operators with which we are concerned when we discuss electron scattering between discrete nuclear states.

1.3 General Relations on Nuclear Matrix Elements of One-Body Operators

We can make some useful statements about the nuclear matrix elements of the multipoles of the current (equations 1.20, 1.25) if the current is a one-body operator as defined through equation 1.33. In second quantization, the multipole operators then take the form

$$\hat{T}_{JM_J;TM_T}(q) = \sum_{\alpha,\beta} c_\alpha^\dagger \langle \alpha | T_{JM_J;TM_T}(q) | \beta \rangle c_\beta, \qquad 1.38$$

where $\alpha \equiv \{a, m_\alpha\}$ denotes a complete set of single-particle quantum numbers. If matrix elements are now taken between exact nuclear states, it follows that

$$\langle f | \hat{T}_{JM_J;TM_T}(q) | i \rangle = \sum_{\alpha,\beta} \langle \alpha | T_{JM_J;TM_T}(q) | \beta \rangle \psi_{\alpha\beta}^{(fi)}, \qquad 1.39$$

which expresses the matrix element of an arbitrary multipole operator as a linear combination of single-particle matrix elements multiplied by simple numerical coefficients

$$\psi_{\alpha\beta}^{(fi)} \equiv \langle f | c_\alpha^\dagger c_\beta | i \rangle. \qquad 1.40$$

The basis of equation (1.39) can be reduced with respect to both angular momentum and isospin (1, 25)

$$\langle f; J_f^\pi T_f \vdots T_{J;T}(q) \vdots i; J_i^\pi T_i \rangle = \sum_{a,b} \langle a \vdots T_{J;T}(q) \vdots b \rangle \psi_{J;T}^{(fi)}(ab). \qquad 1.41$$

No matter how complicated the nuclear states, once the set of numerical coefficients $\psi_{J;T}^{(fi)}(ab)$ has been determined, the exact transition matrix element of any multipole operator can be computed as a linear combination of single-particle matrix elements.

Some general properties of the multipoles of the current can be used to simplify these results.

1. Assuming hermiticity, isospin invariance, and time reversal: $\mathcal{T}\hat{J}_\lambda(\mathbf{x})\mathcal{T}^{-1} = -\hat{J}_\lambda(\mathbf{x})$, $\mathcal{T}|JM_J\rangle = (-)^{J+M_J}|J-M_J\rangle$, it follows that the matrix elements may be turned around in the following manner:

$$\langle i \vdots \hat{T}_{J;T} \vdots f \rangle = (-)^{J_i - J_f + J + \eta}(-)^{T_i - T_f}\langle f \vdots \hat{T}_{J;T} \vdots i \rangle \qquad 1.42$$

where $\eta = 0$ for charge multipoles and $\eta = 1$ for current multipoles.

2. Equation 1.42 also holds for the single-particle matrix elements if the complete set of single-particle wave functions consists of bound states and standing waves in the continuum. In this case, time reversal implies that the coefficients $\psi_{J;T}^{(fi)}(ab)$ are real.

3. Combining equations 1.41 and 1.42, we may simplify the sum over single-particle matrix elements to

$$\langle f \vdots \hat{T}_{J;T} \vdots i \rangle = \sum_{a \geq b} \langle a \vdots T_{J;T} \vdots b \rangle \chi_{J;T}^{(fi)}(ab)_\eta, \qquad 1.43$$

where

$$\chi_{J;T}^{(fi)}(ab)_\eta = (1 + \delta_{ab})^{-1}[\psi \mathcal{Y}_{J;T}^{(fi)}(ab) + (-)^{j_a - j_b + J + \eta} \psi \mathcal{Y}_{J;T}^{(fi)}(ba)]. \qquad 1.44$$

4. For elastic scattering where the states $|f\rangle$ and $|i\rangle$ are identical, equation 1.42 implies that $J + \eta$ *must be even*. In combination with the additional parity selection rules, we arrive at the quite general result that only the even Coulomb and odd transverse magnetic multipoles contribute to elastic electron scattering.

2. ELASTIC SCATTERING

We begin our study of the response surfaces shown in Figures 1.3–1.4 with a discussion of elastic electron scattering where, neglecting recoil, $\omega = 0$ and $|i\rangle = |f\rangle$ with $J_i = J_f \equiv J_0$. We see from the preceding section (equation 1.42) that $J + \eta$ must be even and so only even Coulomb multipoles C0, C2,... and odd magnetic multipoles M1, M3,... can contribute. Of course, conservation of angular momentum requires that $J \leq 2J_0$. In the following two sections we discuss elastic charge scattering and elastic magnetic scattering. In the former case only the basic ideas are presented with a few representative results and references, as this very wide subject has recently been reviewed (31) and will be treated in detail in a book to appear in the near future (R. Barrett and D. Jackson, to be published by Oxford). In the latter case, however, we attempt to be as complete as possible.

2.1 *Charge Scattering*

The elastic charge form factor is given (see Section 1.2) by

$$F_L^2(q) = (2J_0 + 1)^{-1} \sum_{J \text{ even}} |\langle J_0 \| \hat{M}_J(q) \| J_0 \rangle|^2. \qquad 2.1$$

It is sometimes convenient to define a form factor with a different normalization: $F_{ch}^2(q) \equiv F_L^2(q) 4\pi / Z^2$, where now $F_{ch}^2(0) = 1$.

Consider first the case $J_0 = 0$ where only the monopole term C0 is allowed. In Figure 2.1 the elastic form factor for ^{16}O from (32) is shown out to large values of momentum transfer q. If a form factor were known at all q with no error bars and if the first Born approximation were strictly valid, then the form factor could be Fourier transformed to obtain the ground-state charge distribution $\rho_{00}(\mathbf{x})$ as discussed in Section 1.1:

$$F_{el}(q) = ZF_{ch}(q)|_{J_0 = 0} = \int \sin(qx)/(qx) \rho_{00}(\mathbf{x}) \, d\mathbf{x}. \qquad 2.2$$

However, these conditions are not met in practice. One possible avenue of approach is to assume a model for the nuclear charge density. This may be a phenomenological distribution such as the familiar Fermi

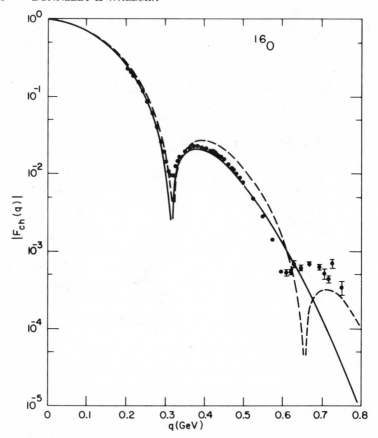

Figure 2.1 Elastic charge form factor for ^{16}O showing results using harmonic oscillator (*solid curve*) and Woods-Saxon (*dashed curve*) wave functions (32).

function (1) or may be the density obtained from a microscopic model of the nucleus. An example of the latter frequently applied to elastic electron scattering is the independent particle model with Hartree-Fock (HF) single-particle wave functions:[9]

[9] If the A-particle wave function is constructed from antisymmetrized products of single-particle wave functions in the independent-particle shell model, then the center-of-mass motion of the system is not treated correctly. Wave functions describing different internal states of the nucleus should be constructed from the $3(A-1)$ independent internal coordinates of the system. This problem can be examined within the framework of harmonic oscillator wave functions and the result is that the actual form factor $F(q)$ involving true internal states of the target is related to the form factor $F_{SM}(q)$ computed in the independent-particle shell

$$\rho_{00}(\mathbf{r}) = \sum_\alpha n_\alpha |\phi_\alpha(\mathbf{r})|^2, \qquad\qquad 2.3$$

where α represents a complete set of single-particle quantum numbers for the basis wave functions $\phi_\alpha(\mathbf{r})$ and where n_α specifies the proton occupancy of the αth single-particle state. Perhaps the simplest basis to use to approximate the HF wave functions is a set of spherical harmonic oscillator wave functions. If, for example, ^{16}O is taken to have the $1s_{1/2}$, $1p_{3/2}$ and $1p_{1/2}$ single-particle proton states completely filled and all others completely empty then the solid curve in Figure 2.1 is obtained (32). Here the oscillator parameter $b = 1.77$fm has been chosen to give a good fit to the data at low to medium values of q. At large q, however, higher Fourier components in the density are being probed and a more realistic model of the density must be used that allows for more spatial structure than is possible with a small number of harmonic oscillator wave functions. The dashed curve in Figure 2.1 shows the results of using Woods-Saxon wave functions where the Woods-Saxon potential parameters are chosen to give the correct proton separation energies in ^{16}O. This is in no way a best fit to the data, but only serves to illustrate the effects of using more realistic wave functions. The high momentum transfer region where the data are hardest to fit is also the hardest to describe theoretically, because, in addition to these finite-well effects, there may be many-body effects such as long-range and short-range correlations, dispersion effects (see Section 5.1) and relativistic effects. The overall conclusion, however, must be that the relatively simple model single-particle wave functions gives a rather good representation of $\rho_{00}(\mathbf{r})$ and that we may have reason to believe that the same basis may be applied to the other processes discussed in this work.

An alternative to this approach of using models to represent the density is to try to extract information directly from the data in a "model-independent" fashion. We conclude this subsection with a very brief discussion of a few of these approaches; the reader is directed to (31) and the coming book by Barrett & Jackson for a comprehensive treatment of this subject. The problem reduces to one of having insufficient information to obtain the Fourier transform of the experimental form factor: the data are only known to a finite maximum value of q and have finite error bars. Perhaps the least model-dependent way of handling the data employs integral quantities involving the density such as the partial moment function (33, 34) defined by $T(Q) \equiv \int_0^Q R(Q')\,dQ'$, where Q is the

---

model by [see (1)] $F(q) = \exp[(bq/2)^2(1/A)]F_{SM}(q)$. We employ such a factor in this and all subsequent discussions. Because of the $1/A$ dependence in the exponential, this correction is only important for light nuclei.

fraction of the charge between radii 0 and R. Also used are the moments
(34, 35) $M(K) \equiv [\int_0^\infty \rho_{00}(\mathbf{r}) r^K \, d\mathbf{r}/Z]^{1/K}$ which include the familiar rms
charge radius when $K = 2$. It is shown in (33–35) that $T(Q)$ and $M(K)$ for
K integral and between -2 and $+2$ may be obtained in a model-
independent manner from elastic charge scattering data.

These are integrals over the density and it is sometimes desirable to
have information on the density itself, albeit with some model input.
There have been several approaches to this problem; we mention only
two, the works of Friar & Negele (36) and Sick (37). Friar and Negele
have proposed an iterative procedure for determining $\rho_{00}(\mathbf{r})$ from experimental (e, e') data and muonic X-ray transition energies. In their approach
the density is separated into two terms, $\rho_{00} = \rho^0 + \delta\rho$ where some physical

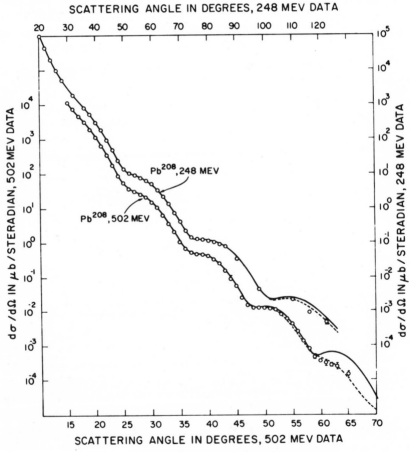

Figure 2.2   Elastic scattering from ^{208}Pb (38).

input enters into the choice for ρ^0: (*a*) they do not expect spatial density fluctuations with wavelength $\lambda < \pi/k_F$ and so restrict the form of the model density, and (*b*) the decay of the wave function in the surface of the nucleus is governed by the single-particle energies and the Coulomb potential and is thus highly constrained in that region. The additional term $\delta\rho$ contains L free parameters describing spatial fluctuations with $\lambda > \pi/k_F$ and is treated as a small perturbation. Perturbation theory for the Dirac equation enables them to use the wave functions determined by ρ^0 (see Section 5.1) to obtain values for the L parameters in $\delta\rho$, which simultaneously improves the agreement with L different combinations of experimental cross sections and bound state energies. This procedure is then iterated using the sum of the nth estimate of ρ^0 and $\delta\rho$ as the $n+1$st estimate of ρ^0 and so on. For ρ^0 they make two choices: the two-parameter Fermi function and the distribution obtained using density-dependent Hartree-Fock wave functions. For $\delta\rho$ they use a Fourier sine series keeping the first L terms. They apply their analysis to ^{208}Pb. In Figure 2.2 we show some of the Stanford data (38) taken at high momentum transfer. Friar and Negele have analyzed such data to obtain the density shown in Figure 2.3 [see (36) for a discussion of the

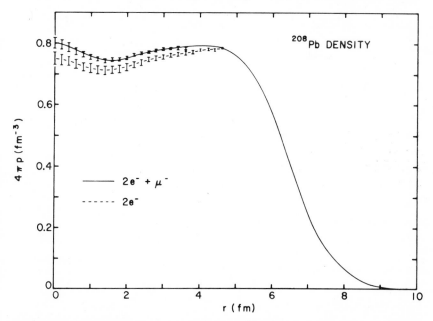

Figure 2.3   Ground-state charge density of ^{208}Pb (36) using (e, e′) data alone (*dashed curve*) and including muonic X-ray data (*solid curve*). The statistical error envelope is indicated.

difference between the results using the (e, e') data alone and using these data plus the muonic X-ray data]. The error estimates reflect the high information content and power of electron scattering.

Sick (37) has an alternative approach to the same problem. He assumes that no structures in $\rho_{00}(\mathbf{r})$ having a width smaller than some width Γ (chosen on physical grounds) are to be admitted. To introduce this condition and in order to decouple the densities at different radii as much as possible he writes the density as a sum of many individual components all Gaussian in shape and having the width Γ. A set of Gaussians is distributed over radii R_i, $i = 1, N$ and the amplitudes of the Gaussians, A_i, $i = 1, N$ are chosen to yield a best fit to the experimental data. Many sets are then tried and all sets with good χ^2 are retained as possible densities; their superposition defining a band for the possible values of $\rho_{00}(\mathbf{r})$. In (39) results are presented for elastic charge scattering from ^{12}C and ^{32}S, along with a complete development of the ideas summarized here [see also (31, 40)].

2.2   Magnetic Scattering

Just as elastic charge scattering provides the best picture of the nuclear charge distribution, elastic magnetic electron scattering out to high momentum transfer brings out fine details of the nuclear ground-state current and magnetization distributions.

Because the magnetic moments of the nuclear ground state are generally determined by the valence nucleons, we are in a position to examine directly the properties of these valence particles. In contrast to charge scattering, neutrons also contribute to the nuclear current through their intrinsic spin magnetization (equation 1.37); thus we also have the capability of examining the distribution of valence neutrons through elastic magnetic scattering.

Consider elastic magnetic scattering from nuclei with $J_i = J_f = J_0 \neq 0$. The quantity measured is the transverse form factor

$$F_T^2(q) = (2J_0 + 1)^{-1} \sum_{\text{odd } J} |\langle J_0 \| \hat{T}_J^{\text{mag}}(q) \| J_0 \rangle|^2, \qquad 2.4$$

where we have made use of the selection rule that only the odd transverse magnetic multipoles can contribute to elastic electron scattering (see Section 1.3). The contribution of this term to the cross section in equation 1.29 can be separated by a straight line plot of F^2 against $\tan^2(\theta/2)$ at fixed q^2, or by scattering electrons through $180°$, in which case only F_T^2 contributes to the cross section. In the long-wavelength limit $q \to 0$, the contributions of the transverse multipoles can be ordered just as they are for real photons [see (1)]. The dominant contribution will be $M1$, and the matrix elements of this operator are proportional to the magnetic moment of the nucleus

$$F_T^2 \underset{q \to 0}{\to} [(J_0+1)/J_0](q/2M)^2\mu^2/6\pi, \qquad\qquad 2.5$$

where μ is the nuclear magnetic moment in units of nuclear magnetons. At higher momentum transfers, the transverse cross section is modified by the form factor of the magnetic moment and the contribution of the higher magnetic multipoles (41–49). These contributions cannot be separated in any model-independent way; however, calculations indicate (see below) that different multipoles dominate the magnetic scattering at different q^2. By mapping out the transverse elastic cross section, either by 180° scattering (50), or by making Rosenbluth plots at large angles where the transverse term is enhanced by a factor $[\frac{1}{2}+\tan^2(\theta/2)]$, we can in fact see *all the multipoles that can contribute*.

Low q^2 experiments studying the magnetic-dipole form factor have been carried out using 180° scattering (51–57). Magnetic scattering from ^{3}He and ^{3}H where only the magnetic-dipole term contributes, as $J_0 = 1/2^+$, has also been examined out to large q^2 (58–60). In p-shell nuclei with $J_0 = 3/2^-$, calculations indicate that the magnetic-dipole form factor will have a diffraction minimum at higher q^2 (see below), and in this region the entire transverse cross section will be due to magnetic octupole scattering. Making use of this observation, Rand, Frosch & Yearian (53) have mapped out M3 form factors using 180° scattering for several light p-shell nuclei. These experiments have been interpreted by Griffy & Yu (42) who observed that quite independently of the nuclear coupling scheme in the p-shell (L-S, intermediate, or j-j), a simple scaling of the theoretical result gave an excellent fit to the data and allowed an essentially model-independent determination of the static M3 moment. A simple understanding and substantial generalization of this result is given in (46).

The study of F_T^2 at high q^2 has been discussed by Li et al (45) who observed that since the monopole charge form factor of a nucleus falls off very rapidly with q^2, and the highest magnetic multipoles do not peak until large values of q^2, the contribution of the *highest* magnetic multipoles can be obtained relatively easily with high q^2, large-angle experiments, and Rosenbluth plots. In this way, the M5 distribution in ^{27}Al($5/2^+$) has been examined (see Figure 2.4) and a very preliminary point on the M9 distribution in ^{209}Bi($9/2^-$) has been obtained from a Rosenbluth plot of $(F_{209_{Bi}})^2/(F_{208_{Pb}})^2$.

With a nuclear target, the current can be decomposed into a convection-current part and magnetization current part as indicated in Section 1.2.

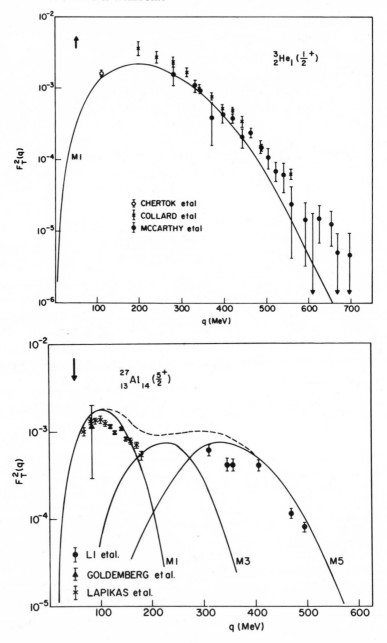

Figure 2.4 (*above, opposite, and following page*) Elastic magnetic form factors
(46, 51–59).

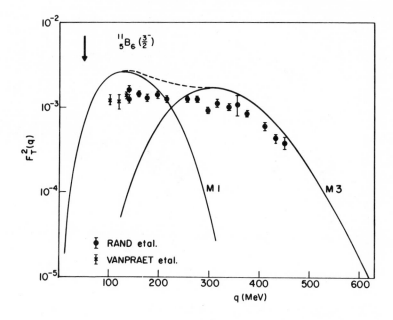

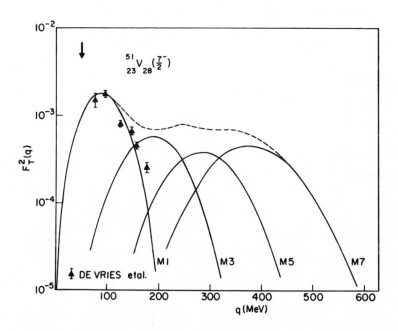

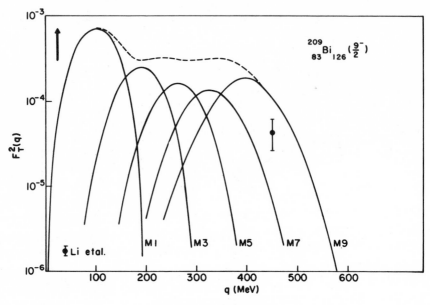

$$\textit{Figure 2.4} \quad (\textit{continued})$$

If meson-exchange currents are neglected, then these contributions can be expressed in terms of single-particle operators according to equations 1.37. Inserting equations 1.37 into equation 1.25, we can write the one-body transverse magnetic multipole operators after a little algebra as

$$-i\hat{T}_{JM}^{\text{mag}}(q) = q \int d\mathbf{r}\{[J(J+1)]^{-1/2}[(1/q)\nabla j_J(qr)Y_{JM}(\Omega_r)] \cdot \hat{\mathbf{l}}(\mathbf{r})/M$$
$$+ [(1/iq)\nabla \times j_J(qr)\,\mathscr{Y}_{JJ1}^M(\Omega_r)] \cdot \hat{\boldsymbol{\mu}}(\mathbf{r})\}, \quad 2.6$$

where the angular-momentum density operator is defined by

$$\mathbf{l}(\mathbf{r}) \equiv \sum_{j=1}^{A} e(j)\delta(\mathbf{r}-\mathbf{r}_j)[\mathbf{r}(j) \times (1/i)\nabla(j)]. \quad 2.7$$

We are allowed to write this operator in either an unsymmetrized or symmetrized form when it is used in equation 2.6 because $(\mathbf{r} \times \nabla) \cdot \nabla j_J(qr)Y_{JM}(\Omega_r) = 0$. The long-wavelength reduction of equation 2.6 (1) takes the familiar form

$$-i\hat{T}_{JM}^{\text{mag}}(q) \xrightarrow[q\to0]{} \frac{q^J}{(2J+1)!!}\left(\frac{J+1}{J}\right)^{1/2} \int d\mathbf{r}[\nabla r^J Y_{JM}(\Omega_r)]$$
$$\cdot \left[\hat{\boldsymbol{\mu}}(\mathbf{r}) + \frac{1}{J+1}\frac{1}{M}\hat{\mathbf{l}}(\mathbf{r})\right]. \quad 2.8$$

In particular, the transverse magnetic dipole operator in this limit $q \to 0$ becomes $-i\hat{T}_{1M}^{mag}(q) \to [q/(6\pi)^{1/2}]\hat{\mu}_{1M}$ where $\hat{\mu}_{1M}$ is the magnetic dipole moment operator for the target. This leads immediately to equation 2.5. The single-particle reduced matrix elements of the terms appearing in equation 2.6 are given in (46) and will not be repeated here.

In Section 1.3 we made some general observations on the *nuclear* matrix elements of the multipole operators. We have already indicated the experimental simplifications that occur when studying the *highest* magnetic multipoles that can contribute. There are also some theoretical simplifications that occur for these highest multipoles:

1. If $j = l + \frac{1}{2}$ (stretched configuration) and $J = 2j$, *only the intrinsic magnetization* contributes to the magnetic multipole. This contrasts with the situation at low q where the M1, which has contributions from both the orbital and intrinsic magnetization (unless $l = 0$), dominates the cross section.

2. Consider the highest $-j$ states within an oscillator shell and $J = 2j$. In order to have additional single-particle matrix elements contributing to the sum in equations 1.41 and 1.43 it is necessary to admix 2p-2h states into the ground state where the 2p states have $j_1 + j_2 \geq J_{max}$. Such states are far removed in energy (at least $2\hbar\omega$ to get states of the same parity) and thus we expect to be able to describe these highest multipoles by simply *scaling* the single-particle values with a coefficient $\chi_{J_{max}}(aa)$. For example, for M3 moments in the p-shell, the only contributing single-particle matrix elements will be $\langle 1p_{3/2} \| T_3^{mag} \| 1p_{3/2} \rangle$ and the next set of single-particle matrix elements that can contribute are $\langle 1d_{5/2,3/2} \| T_3^{mag} \| 1d_{5/2,3/2} \rangle$. Thus we have a simple understanding of the results of Griffy & Yu (42) mentioned previously, and we see that this will be a quite general result.

The situation with the highest moment contrasts with that of the lowest moment, the M1. In this case, many different single-particle matrix elements with different q^2 dependences can contribute to the sums in equations 1.41 and 1.43. For example, if we are discussing the M1 moment in the p-shell, all three matrix elements $\langle 1p_{3/2} \| T_1^{mag} \| 1p_{3/2} \rangle$, $\langle 1p_{3/2} \| T_1^{mag} \| 1p_{1/2} \rangle$ and $\langle 1p_{1/2} \| T_1^{mag} \| 1p_{1/2} \rangle$ can contribute.

3. There is an additional simplification that occurs for all the magnetic multipoles if the odd group of nucleons contributing the magnetic properties is composed of neutrons, for then $e(i) = 0$ and again, only the intrinsic magnetization contributes.

4. In the case of harmonic oscillator wave functions, the reduced matrix elements of the convection and magnetization current densities can always be written $p(y)\exp(-y)$ where $y = (\frac{1}{2}bq)^2$ and $p(y)$ is a polynomial in y. An important observation is that the degree of the polynomial for the

magnetization current case is one higher than the convection current case. So at high momentum transfer (*viz.* $y \gg 1$) the magnetization current contributions dominate for all multipoles.

In Figure 2.4 we show the transverse form factors (equation 2.4) for a series of odd-A nuclei characterized by the fact that they have the highest j possible in an oscillator shell and hence the maximum number of magnetic multipoles. The theoretical curves have been computed by Donnelly & Walecka (46) from appropriate single-particle matrix elements using harmonic oscillator wave functions with an oscillator parameter determined by fitting elastic monopole charge scattering. The difference obtained by using single-particle wave functions generated by a Woods-Saxon nuclear potential shows up only at large q^2; this is illustrated for the case of ^{27}Al in Figure 2.5. The heavy arrows in Figure 2.4 indicate how the theoretical M1 contribution should be scaled to fit the *experimental* magnetiç moment (compare equation 2.5). The results for ^{3}He $(1/2^+)$ (58, 59) clearly map out the M1 distribution here and this simple

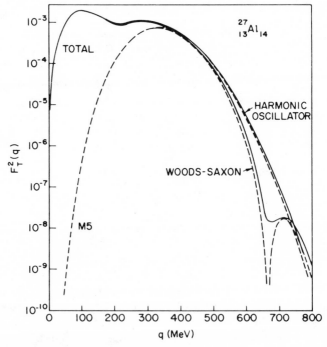

Figure 2.5 Same as Figure 2.4 for ^{27}Al using harmonic oscillator and Woods-Saxon wave functions. The M5 contributions (*dashed curves*) and the sums of the M1, 3, 5 contributions (*solid curves*) are shown (46).

model does very well on the shape. For ^{11}B $(3/2^-)$ the configuration assignment is $(1p_{3/2})_p^{-1}$. Comparison between theory and experiment (52, 53) clearly shows the presence of magnetic octupole scattering and a simple scaling of the M1 and M3 contributions gives an excellent fit to the data (42). This is one of the very few cases where the whole magnetic cross section has been explored. For ^{27}Al $(5/2^+)$ the configuration assignment is $(1d_{5/2})_p^{-1}$. Note the clear demonstration of M5 scattering in this case, and a simple scaling of the theoretical curve will again fit the data [(45), (51), and G. C. Li, private communication]. For ^{51}V $(7/2^-)$ an assumed configuration $(1f_{7/2})_p$ is also shown. The low-q points (55) map out the M1 distribution; the higher multipoles, up to M7 here, remain to be explored. Finally, Figure 2.4 shows the calculated transverse form factor for ^{209}Bi $(9/2^-)$ which is assumed to be $(1h_{9/2})_p$. The experimental point from (45) must be viewed as very preliminary, and was obtained from forward-angle experiments, originally designed to enhance charge scattering [in this connection, see (61)]. This figure clearly illustrates how the magnetic scattering holds up at high q^2, and that there is a whole world of magnetization structure waiting to be explored. This is also illustrated in Figure 2.5.

To illustrate the information being obtained here, we note that in the stretched case $j = l+\frac{1}{2}$ we have $\langle j, m = j|\boldsymbol{\mu}(\mathbf{x})|j, m = j\rangle = \mu(\mathbf{x})\mathbf{e}_z$ where $\mathbf{e}_z$ is a unit vector in the z-direction; the orbital contribution is similar to this (46). Figure 2.6 shows the surface of $\frac{1}{2}$-maximum density for the distribution of intrinsic magnetization $\mu(\mathbf{x})$ in ^{51}V. Thus the valence particle is traversing a circular orbit with angular momentum in the z-

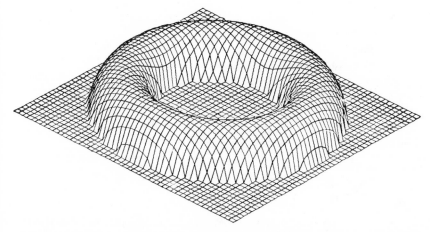

Figure 2.6 Surface of half-maximum density for the ^{51}V intrinsic magnetization distribution $\mu(\mathbf{x})$ [(46) see text].

direction and some spatial distribution, and its intrinsic magnetic moment serves to label the distribution (for neutrons as well as protons). If we measure just the static magnetic moment of the nucleus, we measure only the volume integral of this distribution. But by mapping out all the multipoles that can contribute to the transverse form factor, we can determine that the nucleus actually forms a tiny current loop confined to the x-y plane, with an accompanying intrinsic magnetization. By studying the transverse form factors at all q^2, we can actually plot out the detailed spatial distribution of this current and magnetization. (Note that as j gets very large, the orbit approaches the well localized classical circular orbit.) Many more experiments of this type will be done on the new generation of electron accelerators, and we can look forward with great anticipation to the information on the nucleus that will be obtained through this means.

3   INELASTIC SCATTERING

In this section we discuss inelastic electron scattering within the framework of three classes of nuclear states, employing nuclear models designed to exhibit the main properties of each class: low-lying collective states, states near or below particle threshold that involve primarily the excitation of a particle-hole pair, and excitation into the continuum. The distinction between these classes is not clearly defined; for example, the giant electric dipole resonance may be treated as a linear combination of a few particle-hole states or as a collective state as in the Goldhaber-Teller model. Or, for example, a given particle-hole state may involve an unbound nonresonant particle wave function and appear in the third class; or it may involve an unbound but resonant particle wave function and appear in the second class. Furthermore, the specific models discussed here by no means exhaust the possibilities in each of the three classes, but are meant to illustrate the kind of information that can be extracted from inelastic electron scattering data.

3.1   Collective Vibrations and Rotations

The low-lying nuclear states that show up most strongly in inelastic electron scattering are collective excitations of the nucleus where a finite fraction of the total charge Z of the nucleus can contribute to the transition charge density (1) (see Figure 1.3). These states are correspondingly excited most strongly through the Coulomb interaction. The two types of collective nuclear motion that are most commonly observed are vibrations and rotations. In the vibrations, the whole nucleus oscillates about an equilibrium configuration (which may be either spherical or deformed); in the rotations, an intrinsically deformed nucleus can be set into rotational motion just as in molecular spectroscopy. The theory of collective nuclear motions is well developed [see, for example Bohr & Mottelson (62)] and present work in nuclear theory centers on relating the collective excitations back to the basic nucleon-nucleon interaction between nucleons in the nucleus. For example, the Random-Phase (RPA) and Tamm-Dancoff (TDA) approximations [see (25)], two approximation schemes for describing the collective vibrations in terms of linear combinations of particle-hole states (these linear combinations are very complicated for the strongly collective vibrations), are described in the next section.

Inelastic electron scattering provides an extremely valuable tool for studying the collective excitations, for it can be used to determine the *detailed spatial distribution* of the transition charge density. Thus in contrast to the situation where just one static or transition moment of the charge distribution is determined (for example, through a measurement of a γ-decay width where the wavelength of the emitted photon is very large compared to the size of the nucleus) there is now a whole *function* that must be understood. To examine this point in a little more detail, we observe that the orthogonality of the 3-j symbols can be used to invert equation 1.21 and we can write

$$\langle J_f \| \hat{M}_J(q) \| J_i \rangle = \int_0^\infty j_J(qx) \rho_{\text{tr}}(x) x^2 \, dx, \qquad 3.1$$

where

$$\rho_{\text{tr}}(x) \equiv \int d\Omega_x \, Y_{JM}(\Omega_x) \left[(2J+1) \sum_{M_i} \sum_{M_f} (-)^{J_f - M_f} \begin{pmatrix} J_f & J & J_i \\ -M_f & M & M_i \end{pmatrix} \right.$$
$$\left. \times \langle J_f M_f | \hat{\rho}(\mathbf{x}) | J_i M_i \rangle \right] \qquad 3.2$$

is the radial transition charge density. As the left-hand-side of equation 3.1 is independent of M, equation 3.2 must also be independent of M. A convenient choice is $M = 0$. If one examines the excitations of even-even nuclei with $J_i = 0$, then equation 3.2 can be simplified to

$$\rho_{\text{tr}}(x) = (2J+1)^{1/2} \int d\Omega_x \, Y_{J0}(\Omega_x) \langle J0 | \hat{\rho}(\mathbf{x}) | 00 \rangle. \qquad 3.3$$

To date, of the various possible nuclear excitations, the nuclear vibrations have been most thoroughly studied by inelastic electron scattering, as they are fairly well separated in energy (spacings of the order of 1 MeV) and the energy resolution requirements in the scattering experiments are not too stringent. Figure 3.1 shows an example of the present experimental

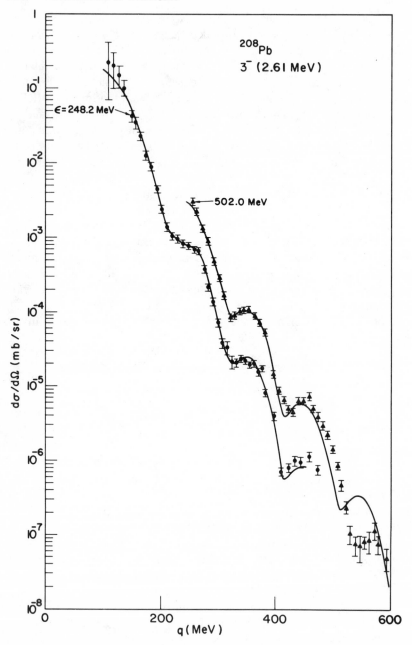

Figure 3.1 Inelastic cross sections for the first-excited state of ^{208}Pb (63).

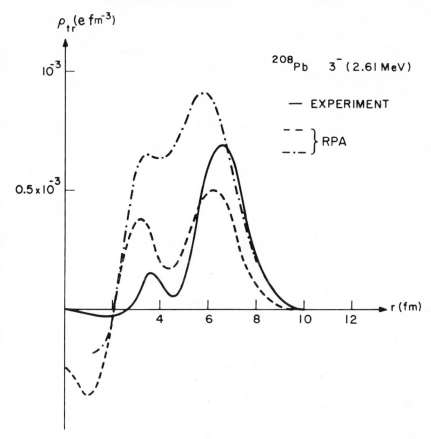

Figure 3.2 Transition charge density for the first-excited state of ^{208}Pb (63), normalized to fit the measured $B(E3) \equiv (7/4\pi)\left|\int r^3 \rho_{tr}(r)\,dr\right|^2$. The broken curves are RPA results (64, 65).

capabilities in this area. This is work of Heisenberg & Sick (63) on the excitation of the first excited state of ^{208}Pb, a 3^- oscillation at 2.61 MeV. The transition charge density is shown in Figure 3.2 and is compared with the RPA calculations of Gillet, Green & Sanderson (64) and Blomqvist (65).[10] Note that these high-q experiments show ripples on the inelastic densities, at least qualitatively similar to those expected within

[10] These experiments were analyzed by using the distorted-wave Born approximation where the initial and final electron wave functions are distorted by the static, spherically symmetric Coulomb field of the nucleus, but still only one *inelastic* scattering interacting is retained (see Section 5.1).

the framework of particle-hole models. Some recent representative work on electron excitation of nuclear vibrations is contained in (66–77), and representative work on the theoretical interpretation in (78–90). Monopole oscillations are discussed in (91–96). Reference (24) also contains a good survey of recent work on vibrations.

In contrast to the vibrations, the levels in a nuclear rotational band

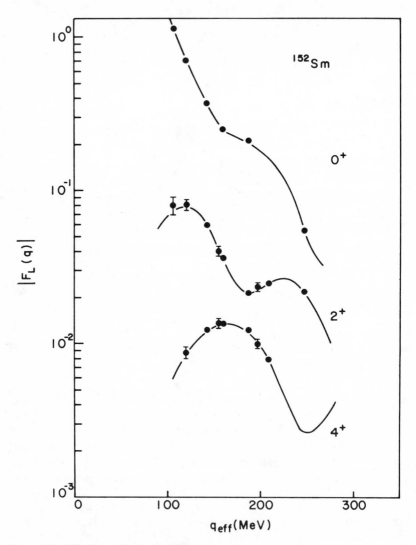

Figure 3.3   Elastic and inelastic form factors for the ground-state rotational band of ^{152}Sm (97). Here $q_{\text{eff}} \equiv q(1 + 4Z\alpha/3\langle r^2 \rangle^{1/2}\varepsilon)$.

are very closely spaced (on the order of hundreds of keV). Very high resolution is required in order to study these inelastic transitions at high momentum transfer, where the full power of electron scattering as a microscope for spatially locating the transition densities comes into play. Figure 3.3 shows recent work of Bertozzi et al (97) on elastic and inelastic scattering to the 2^+ (0.122 MeV) and 4^+ (0.367 MeV) rotational states of ^{152}Sm. These experiments were done on the NBS electron linac with relatively low incident electron energies between 50 and 105 MeV. In the picture of nuclear rotation developed by Bohr & Mottelson (62), the excitations correspond to the different rotational states of an intrinsically deformed nucleus. For an even nucleus, a general result on the transition matrix elements for such an intrinsically deformed object states that

$$\langle J \| \hat{M}_J(q) \| 0 \rangle = M_{J0}^{int}(q), \qquad\qquad 3.4$$

where $M_{J0}^{int}(q)$ is the corresponding multipole moment computed in the intrinsic coordinate frame. Bertozzi et al parametrize their results[10] by assuming a distorted Fermi charge distribution of the following form

$$\rho(r, \theta)^{int} = \rho_0 \left(1 + \exp\left[(r - R(\theta))/t\right]\right)^{-1}$$
$$R(\theta) = C\left[1 + \beta_2 Y_{20}(\theta) + \beta_4 Y_{40}(\theta) + \beta_6 Y_{60}(\theta) + \cdots\right]. \qquad 3.5$$

From this, they can calculate all the intrinsic moments appearing in equation 3.4. Electron excitation of nuclear rotations is also discussed in (98–104). A good survey of recent work on rotations is contained in (24). (See, in particular, the paper by G. Ripka in this volume.) One of the most exciting prospects with the new high energy, high resolution, electron linacs is a systematic study of the structure of deformed nuclei.

3.2   Particle-Hole States

In this section we consider the electro-excitation of nuclear states whose character involves essentially a particle-hole excitation of a closed-shell ground state. We include in this category only states seen as relatively sharp features in the electron scattering excitation functions. These may be particle-hole states below particle emission threshold or states such as the familiar giant electric dipole resonance whose particle wave functions are unbound, but resonant, and so quasi-stationary. The general nature of the electron scattering response surfaces for such states has been discussed in Section 1.1 (Figures 1.3 and 1.4). A representative selection of pertinent references for the period between (1) and the present is contained in (105–145).

The inelastic form factors for the excitation of a discrete state with quantum numbers $J_f^\pi T_f$, M_{T_f}, from the ground state with quantum numbers $J_i^{\pi_i} T_i$, M_{T_i} (where $M_{T_f} = M_{T_i} = M_T$) are given by equations 1.29. The operators $\hat{M}_J$, $\hat{T}_J^{el}$ and $\hat{T}_J^{mag}$ have been discussed in Section 1.2. Conservation of parity requires that for $\pi_i \pi_f = +1$ only even-J electric and odd-J magnetic multipoles enter, and the reverse for $\pi_i \pi_f = -1$. The total form factor F^2 may then be written in terms of the longitudinal and transverse form factors F_L^2 and F_T^2 as indicated in equation 1.29. Each of the operators has a part that is isoscalar $\hat{T}_{J;00}$ and a part that is the third component of an isovector $\hat{T}_{J;10}$ (see Section 1.2), where $\hat{T}_J = \hat{T}_{J;00} + \hat{T}_{J;10}$ is any one of $\hat{M}_J$, $\hat{T}_J^{el}$ or $\hat{T}_J^{mag}$. Using the Wigner-Eckart theorem on isospin, we may write their matrix elements as

$$\langle J_f^{\pi_f} T_f, M_T \| \hat{T}_{J;T0}(q) \| J_i^{\pi_i} T_i, M_T \rangle$$

$$= (-)^{T_f - M_T} \begin{pmatrix} T_f & T & T_i \\ -M_T & 0 & M_T \end{pmatrix} \langle J_f^{\pi_f} T_f \vdots \hat{T}_{J;T}(q) \vdots J_i^{\pi_i} T_i \rangle, \qquad 3.6$$

that is, in terms of doubly-reduced matrix elements. Using the results of Section 1.3, equation 1.43, on one-body operators we may expand these doubly-reduced many-body matrix elements in terms of single-particle matrix elements (which are assumed to be reasonably well known) with expansion coefficients $\chi_{J;T}^{(i)}(ab)_{fi}$, where a and b stand for complete sets of single-particle quantum numbers. These one-body density matrix elements may in turn be expressed in terms of the quantities $\psi_{J;T}^{(i)}(ab)$ according to equation 1.44. Thus knowledge about the ψ's from some dynamical model of the ground and excited states involved leads directly to the form factors, as the single-particle matrix elements of the multipole operators are easily computed. Conversely, experimental knowledge about the form factors may lead to information on the ψ's and so constrain the possible nuclear models. We explore both points of view. In most situations studied thus far only a few coefficients χ are really important, and we may truncate the sum over a and b to a small subspace.

We are most interested in comparing theoretical results with data taken at large angles where the transverse contributions dominate over the longitudinal ones (see equation 1.29); we therefore concentrate on the transverse form factor $F_T^2(q)$. The analysis for $F_L^2(q)$ follows similar lines. For the most tractable model of the single-particle wave functions, the harmonic oscillator model, we may write for matrix elements of the transverse operators [see (1)]

$$\langle a \vdots T_{J;T}(q) \vdots b \rangle = [p_c(y) + p_m(y)] y^k e^{-y}, \qquad 3.7$$

where $y = (\frac{1}{2} bq)^2$ with b the oscillator parameter, $k = \frac{1}{2}(J-1)$ for electric multipoles, $k = \frac{1}{2} J$ for magnetic multipoles and $p_c(y)$ and $p_m(y)$ are simple polynomials in y coming from the convection current and magnetization current pieces of the operators (see equations 1.25 and 1.37). It is generally true that $p_m(y)$ is of higher degree in y than $p_e(y)$; therefore, if experiments are performed at large momentum transfer q, the magnetization current contributions are dominant. A further consequence of the large isovector single nucleon magnetic moment $[(\mu^v/\mu^s)^2 \approx 28]$ is that *isovector excitations are strongly enhanced over isoscalar ones at high q and large θ.* We shall see in discussing experimental results that this may lead to a much simplified excitation spectrum.

Further observations may be made using equation 3.7. Note first that at low q the electric multipoles go roughly as $y^{(J-1)/2}$, whereas the magnetic multipoles are proportional to $y^{J/2}$. Consequently, the dominant multipoles at low q are the electric dipoles, a fact well known from photoexcitation where $q = \omega$ (and is thus generally of order 20–30 MeV or smaller) and the electric dipole is dominant. For both electric and magnetic operators the ratio of the matrix elements for multipolarity $J+1$ to those for multipolarity J is roughly of order y. Taking for the oscillator parameter the estimate $b \approx 2$ fm, this means the ratio is of order $[q(\text{MeV}/200)]^2$. For low q, such as in photoexcitation, this means that generally E1 $\gg$ E2 $\gg$ E3 ... and M1 $\gg$ M2 $\gg$ M3 ... However, at medium to large values of momentum transfer in electron scattering, such as we are concerned with here ($q \gtrsim 200$ MeV), there is generally no inhibition of the higher multipoles. We shall see examples of direct excitation in electron scattering of states of large angular momentum (e.g. $0^+ \rightarrow 4^-$ (M4) in ^{12}C and ^{16}O, $0^+ \rightarrow 6^-$ (M6) in ^{28}Si). These high spin states are very hard to study by any other means. Note that although these results were discussed within the context of the harmonic oscillator model, they are more generally valid. Calculations with other single-particle bases such as those with Woods-Saxon wave functions yield very similar results (cf. Sections 2.1 and 2.2 for elastic scattering).

We turn now to specific nuclear models for the one-body density matrix coefficients ψ, following the detailed presentation in (25). For simplicity we consider only nuclei with ground states having $J_i^{\pi_i} T_i = 0^+ 0$. Then only one multipole enters in electroexciting a state with $J_f^{\pi_f} T_f$. Note that 0^- states cannot be excited in first Born approximation. The nuclear Hamiltonian $\hat{H}$ may be taken to have two parts, a spherical Hartree-Fock piece $\hat{H}_{HF}$ (a c-number plus a diagonal one-body operator) with HF single-particle energies $\varepsilon_\alpha = \varepsilon_a$ and a two-body residual interaction $\hat{V}$. We may construct particle-hole creation and annihilation operators $\hat{\xi}_{\alpha\beta}^\dagger \equiv a_\alpha^\dagger b_\beta^\dagger$ and $\hat{\xi}_{\alpha\beta} \equiv b_\beta a_\alpha$ where $\varepsilon_a > \varepsilon_F$ and $\varepsilon_b \leq \varepsilon_F$ with ε_F the Fermi energy. The matrix elements of these operators between the ground state $|i\rangle$ and some excited state $|f\rangle$ define particular elements of the one-body density matrix:

$$\psi_{\alpha\beta}^{(fi)} \equiv \langle f|\hat{\xi}_{\alpha\beta}^\dagger|i\rangle$$
$$\phi_{\alpha\beta}^{(fi)} \equiv \langle f|\hat{\xi}_{\alpha\beta}|i\rangle. \qquad 3.8$$

Equations of motion may be obtained by commuting the operators with $\hat{H}$ and again taking matrix elements:

$$\langle f|[\hat{H}, \hat{\xi}_{\alpha\beta}^\dagger]|i\rangle = (E_f - E_i)\psi_{\alpha\beta}^{(fi)} \equiv \omega_{fi}\psi_{\alpha\beta}^{(fi)}$$
$$\langle f|[\hat{H}, \hat{\xi}_{\alpha\beta}]|i\rangle = \omega_{fi}\,\phi_{\alpha\beta}^{(fi)}. \qquad 3.9$$

If we substitute for $\hat{H}$ the two pieces $\hat{H}_{HF} + \hat{V}$, the first yields simply $\varepsilon_a - \varepsilon_b \equiv \varepsilon_{ab}$ times $\psi_{\alpha\beta}^{(fi)}$ or $\phi_{\alpha\beta}^{(fi)}$ in the two equations respectively; however, the second term $\hat{V}$ is more complicated. An approximation commonly made is to retain only the terms in the commutators that are proportional to $\hat{\xi}^\dagger$ and $\hat{\xi}$ and drop all others (for example, proportional to $a^\dagger a$ or $b^\dagger b$). This constitutes the random-phase approximation (RPA). The basis may be reduced in angular momentum and isospin; we therefore consider reduced density matrix elements (forward- and backward-going amplitudes respectively)

$$\psi_{J;T}^{(fi)}(ab) \equiv \sum_{m_\alpha m_\beta} \sum_{\mu_\alpha \mu_\beta} \langle j_a m_\alpha j_b m_\beta | j_a j_b J M_J\rangle \langle \tfrac{1}{2}\mu_\alpha \tfrac{1}{2}\mu_\beta|\tfrac{1}{2}\tfrac{1}{2}T M_T\rangle \psi_{\alpha\beta}^{(fi)}$$

$$\psi_{J;T}^{(fi)}(ba) \equiv \sum_{m_\alpha m_\beta} \sum_{\mu_\alpha \mu_\beta} \langle j_a m_\alpha j_b m_\beta | j_a j_b J M_J\rangle \langle \tfrac{1}{2}\mu_\alpha \tfrac{1}{2}\mu_\beta|\tfrac{1}{2}\tfrac{1}{2}T M_T\rangle S_\alpha S_\beta\, \phi_{\alpha\beta}^{(fi)} \qquad 3.10$$

which are independent of M_J, M_T by the Wigner-Eckart theorem. Here we define $S_\alpha \equiv (-)^{j_a - m_\alpha}(-)^{1/2 - \mu_\alpha}$. In the random-phase approximation we then have coupled equations-of-motion

$$(\omega_{fi} - \varepsilon_{ab})\psi_{J;T}^{(fi)}(ab) - \sum_{a'b'}\left[v_{ab;a'b'}^{J;T}\,\psi_{J;T}^{(fi)}(a'b') + v_{ab;b'a'}^{J;T}\,\psi_{J;T}^{(fi)}(b'a')\right] = 0 \qquad 3.11$$

$$(\omega_{fi} + \varepsilon_{ab})\psi_{J;T}^{(fi)}(ba) + \sum_{a'b'}(-)^{j_a - j_b}(-)^{j_{a'} - j_{b'}}$$
$$\times \left[v_{ab;a'b'}^{J;T}\,\psi_{J;T}^{(fi)}(b'a') + v_{ab;b'a'}^{J;T}\,\psi_{J;T}^{(fi)}(a'b')\right] = 0,$$

where

$$v_{ab;a'b'}^{J;T} \equiv -\sum_{J'T'}(2J'+1)(2T'+1)\begin{Bmatrix}j_{b'} & j_a & J'\\ j_b & j_{a'} & J\end{Bmatrix}\begin{Bmatrix}\tfrac{1}{2} & \tfrac{1}{2} & T'\\ \tfrac{1}{2} & \tfrac{1}{2} & T\end{Bmatrix}$$
$$\times \left[\langle a'bJ'T'|v|ab'J'T'\rangle - (-)^{1/2 + 1/2 + T'}(-)^{j_a + j_{a'} + J'}\langle a'bJ'T'|v|b'aJ'T'\rangle\right]. \qquad 3.12$$

Given a model residual interaction $\hat{V}$ we may solve these simultaneous matrix equations to obtain the ψ's and, through equation 1.44, the χ coefficients.

A special subcase of the random-phase approximation is the Tamm-Dancoff approximation (TDA) in which the ground state is taken to be a closed shell. This means that there are no particle-hole correlations present in the ground state and that consequently all backward-going amplitudes are zero (see equations 3.8 and 3.10). The forward-going amplitudes are easily interpreted in this case, as we have

$$|f; JM_J, TM_T\rangle = \sum_{a \geq b} \psi_{J;T}^{f,i}(ab)|ab; JM_J, TM_T\rangle, \qquad 3.13$$

where $|ab; JM_J, TM_T\rangle = [a_a^\dagger \odot b_b^\dagger]_{JM_J, TM_T}|i\rangle$ are pure 1p-1h states. So in TDA the excited states $|f\rangle$ are simply linear combinations of pure 1p-1h states with the forward-going amplitudes $\psi_{J;T}^{f,i}(ab)$ as admixture coefficients.

As an application of these ideas we consider the electroexcitation of particle-hole states in ^{16}O. [This discussion is based on (119); other work on ^{16}O is contained in (105, 108, 109, 114, 123, 125, 128, 136, 137, 139)]. We compare with experimental data taken at large scattering angles, $\theta \approx 135° - 180°$ and generally medium to high electron energy, $\varepsilon \approx 100$–400 MeV. Consequently, the momentum transfer is large and, using the arguments given above, the $T = 1$ states are preferentially excited. A typical excitation spectrum, seen in Figure 3.4, shows several strongly excited features below or in the vicinity of particle threshold

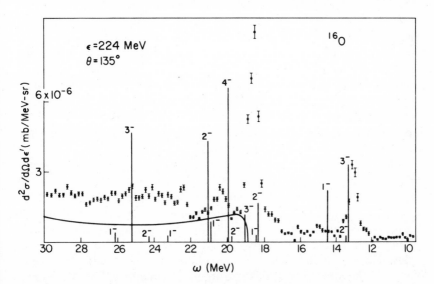

Figure 3.4 Excitation spectrum for ^{16}O from (119). The calculated $T = 1$ spectrum (arbitrary overall normalization) is also shown.

(the neutron separation energy is 18.7 MeV). We use the TDA to analyze the data, at least initially, and then later refer to extended calculations involving the RPA and other improvements. We take the ground state of ^{16}O to contain closed $1s$ and $1p$ shells and describe the excited states as linear combinations of $1p$-$1h$ states where the particle is in the $2s$-$1d$ shell and the hole in the $1p$ shell. The residual interaction was taken to be a Serber-Yukawa force with parameters determined by fitting low-energy nucleon-nucleon scattering. Configuration-mixed states that lie nearby in energy were grouped together into complexes and the total electron scattering form factors were compared with experiment. Following the idea of considering the single-particle-hole states to be the doorway states in electroexcitation, we make this comparison only with data averaged over energy intervals on the order of 1 MeV. Of course in addition to the strong, well-defined peaks observed, the experimental excitation spectrum contains fine structure; but this should be considered only within the framework of a model having more complicated admixtures of many particle-many hole states.

The oscillator parameter was taken to be 1.77 fm, the value obtained from elastic electron scattering (see Section 2.1). The unperturbed configurations and their corresponding (Hartree-Fock) energies in ^{17}O and single-hole energies in ^{15}O are the following: $1d_{5/2}(1p_{1/2})^{-1}$ at 11.52 MeV $(2^-, 3^-)$, $2s_{1/2}(1p_{1/2})^{-1}$ at 12.39 MeV $(0^-, 1^-)$, $1d_{3/2}(1p_{1/2})^{-1}$ at 16.60 MeV $(1^-, 2^-)$, $1d_{5/2}(1p_{3/2})^{-1}$ at 17.68 MeV $(1^-, 2^-, 3^-, 4^-)$, $2s_{1/2}(1p_{3/2})^{-1}$ at 18.55 MeV $(1^-, 2^-)$ and $1d_{3/2}(1p_{3/2})^{-1}$ at 22.76 MeV $(0^-, 1^-, 2^-, 3^-)$. These are all the $1p$-$1h$ states that can be formed by promoting a particle up one oscillator spacing from a closed $1p$-shell. Using the Serber-Yukawa residual interaction these pure $1p$-$1h$ states are mixed and shifted in energy and we find at least four clearly defined groupings of levels: complexes at 13 MeV, 17 MeV, and 19 MeV and the giant resonance region all having $T = 1$. These complexes are labeled by their experimental energies, which are roughly one MeV lower in energy than the values calculated with this simple model.

The 13 MeV complex contains 1^-, 2^-, and 3^- states with experimental energies 13.09 MeV, 12.97 MeV and 13.26 MeV respectively and calculated energies 14.38 MeV, 13.59 MeV and 13.57 MeV respectively. In addition, there is a 0^- state at 12.80 MeV that cannot be electroexcited in first Born approximation (see Section 4.3). The form factor for this complex is shown in Figure 3.5. Here the calculated form factors have been reduced in amplitude by a factor 1.7 to obtain agreement with experiment. Reduction factors of about this size are usually required when describing $T = 1$ states with the simple $1p$-$1h$ model, a point we return to below when we again consider the 13 MeV complex.

The 17 MeV complex contains 1^- and 2^- states at calculated energies 18.46 MeV and 18.45 MeV respectively. The dipole state is seen in photoexcitation at 17.3 MeV. In Figure 3.5 we show the form factor for this complex where a reduction in amplitude by 1.4 has been employed.

In Figure 3.5 the form factor for the 19 MeV complex is shown with no reduction in amplitude, although a reduction factor of about 1.4 appears to be necessary. The states in this complex and their calculated energies are $1^-(20.73$ MeV), $2^-(19.77$ MeV and 20.96 MeV), $3^-(19.17$ MeV) and $4^-(19.86$ MeV). Here the 2^- giant magnetic quadrupole resonance (20.96 MeV) is seen experimentally in electron scattering at 20.4 MeV. At low q the giant quadrupole is dominant; at intermediate values of q, however, the 4^- state is by far the most strongly excited. This *direct excitation* of a high-spin state is a unique feature of electron scattering.

In Figure 3.5 we also show the sum of the form factors for the $1^-(20.73$ MeV) and $2^-(20.96$ MeV) states reduced in amplitude by 1.4 and compared with data in the region $\omega = 20.0$–20.8 MeV where a subcomplex appears in the excitation spectrum (see Figure 3.4).

Finally, in Figure 3.5 we show the form factor in the giant resonance region (20.8–26.0 MeV) which is seen in photoexcitation as two major peaks at 22.3 and 24.3 MeV. The calculated energies are 23.26 MeV and 26.13 MeV (1^- states), 24.28 MeV (2^- state) and 25.30 MeV (3^- state). The calculated form factor goes from being all giant electric dipole resonance (23.26 MeV) at low q to being more spin-flip dipole resonance (26.13 MeV) at intermediate q and finally to being primarily 3^- at high q (see also the discussion of Figure 1.4 in Section 1.1). In addition to the form factors for these discrete levels in this energy range, the quasi-elastic form factor has been included [see Section 3.3 where the square-well shell model approach taken in (154) has been used].

The form factors for these discrete particle-hole states and for the quasi-elastic cross section are indicated on Figure 3.4. The line features are to be regarded as the integrated strengths under peaks with finite widths, where the widths may be several MeV wide in the region above particle threshold.

These calculations based on the simple TDA can be extended to the full RPA; the results, however, are not drastically changed and the rather sizeable reduction factors are still required. It is possible to view the TDA or RPA as a starting point in a description of the $T = 1$ states on ^{16}O. We may in fact modify the coefficients χ from their TDA or RPA values in such a way as to produce agreement with the experimental form factors. The usefulness of obtaining such a set of χ coefficients is made apparent in Section 4 on weak interactions.

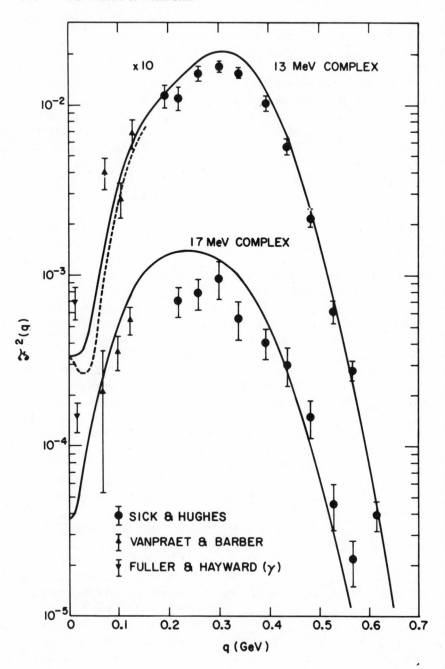

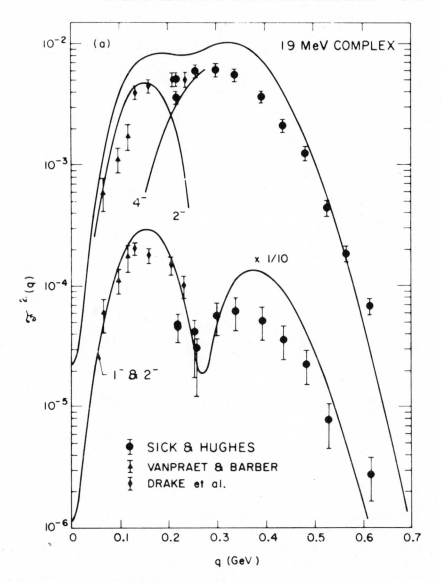

Figure 3.5 Inelastic form factors $\mathscr{F}^2(q) \equiv F^2(q)/(\frac{1}{2}+\tan^2\theta/2)$ at large θ for the 13 and 17 MeV complexes (*opposite*), the 19 MeV complex (*above*), and the giant resonance region (*following page*) of ^{16}O (119).

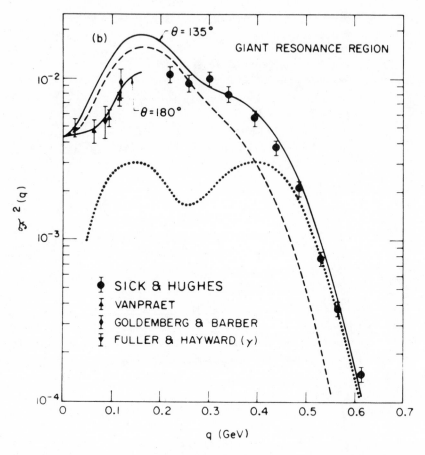

Figure 3.5   (*continued*)

We proceed to a brief application of these ideas to transitions involving the 0^+0 ground state and the $T = 1$, $1^-(13.09$ MeV$)$, $2^-(12.97$ MeV$)$ and $3^-(13.26$ MeV$)$ excited states in ^{16}O [see (139)]. These states are among the best studied in electron scattering, being reasonably isolated states below particle threshold in ^{16}O. As before, the basis of single-particle states involved is truncated to include only the $1p$ and $2s$-$1d$ shells. At this level of approximation we still have very complicated states present involving many particles in the $2s$-$1d$ shell and many holes relative to a closed $1p$ shell. In principle we should now proceed with an analysis of inelastic electron scattering, using the relatively small number of amplitudes $\chi_{J;T}^{(fi)}(ab)_n$ discussed above as adjustable parameters. The (e, e') data, however, are not at present sufficiently accurate to enable us to carry out this program fully and we must take a somewhat cruder point of view. We instead consider particle-hole states in the TDA and RPA as a *starting point,* and alter the amplitudes from these model values in two ways: (*a*) by including an overall reduction factor ξ for all states considered and (*b*) by allowing the individual amplitudes $\chi_{J;T}^{(i)}(ab)_n$ to vary up to 10% from this overall reduced amplitude. We consider three sets of amplitudes determined to yield a good fit to the electron scattering data: Set 1 uses the TDA results (discussed above). Set 2 uses the RPA instead of the TDA, but is otherwise the same as set 1. This allows us to examine the importance of backward-going amplitudes at least within the framework of the TDA and RPA. Finally, set 3 is the same as set 1 except that the harmonic oscillator wave functions in the transition matrix elements are replaced by Woods-Saxon wave functions where the well parameters are chosen to yield a reasonable fit to the single-particle (neutron) energies and elastic electron scattering (cf. Section 2.1). All of these sets have the model amplitudes multiplied by a/ξ where $\xi = 1.7$ for all states and where $a = 1.1$ for the dominant configurations in the 0^-, 1^- and 2^- states ($2s_{1/2}$ or $1d_{5/2}$ particles and $1p_{1/2}$ holes) and $a = 0.9$ for the dominant configuration in the 3^- state and for the small components of all four states. Using these sets of amplitudes we obtain the agreement with the (e, e') data and the γ points shown in Figure 3.6. We emphasize that, having altered the amplitudes with the reduction factor a/ξ, we are *no longer using* the TDA and RPA, but are in principle including many-particle-many-hole components of the wave functions not included in these approximations. Once we have determined a set of altered amplitudes from electron scattering, we are in a position to calculate the reduced matrix elements of any other one-body operator of the same multipolarity between the same nuclear states (or their isobaric analogues). We return to these ideas in Section 4.

Although ^{16}O is the only case we have discussed in detail, similar results have been obtained by the same group in ^{12}C and ^{28}Si [see (110, 126–128)]. The conclusions to be drawn from this work are the following: 1. The particle-hole structure in the region 10–30 MeV seems to become clearer at large q where well defined groupings of states show up strongly. In particular, high-spin states can dominate the spectrum at large q. 2. the q dependence of the peaks serves as a valuable tool in identifying the levels, and some of the complexes still show diffraction minima. 3. The particle-hole model is very successful in predicting the location of the states, generally to better than 1 MeV, and in predicting the inelastic form factors out to $q \approx 700$ MeV; however, the amplitudes may be too large by factors of the order of 1.4–1.7.

374

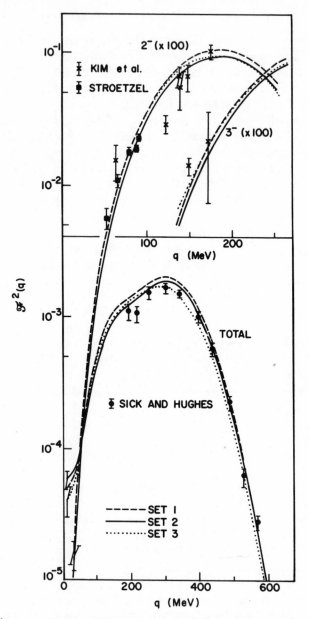

Figure 3.6 (above) Same as Figure 3.5, but showing details of the 13 MeV complex in ^{16}O (139).

Figure 3.7 (opposite) Cross sections for quasi-elastic electron scattering with $\varepsilon = 500$ MeV and $\theta = 60°$ (157). The solid lines are the results of Fermi gas calculations with parameters indicated on the figure.

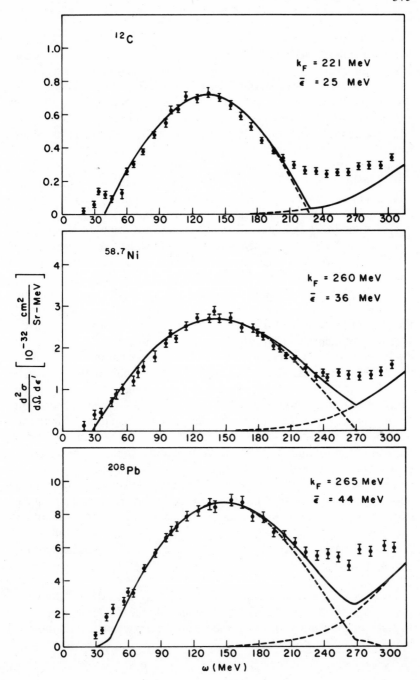

3.3   Quasi-Elastic Scattering

In this section on inelastic electron scattering we turn finally to the region where a nucleon is given sufficient energy and momentum in the scattering process to escape from the nucleus [see (1, 3)]. There is no sharp distinction between this region and the parts of the response surfaces discussed in the last section in that the states excited there may be unbound and may subsequently decay with nucleon emission. The distinction we make here is in the excitation spectrum. For example, the giant dipole resonance appears as a reasonably sharp peak and although the particle wave functions involved in a $1p$-$1h$ description of this state may be unbound they are resonant (or quasi-bound) and frequently may be replaced with truly bound wave functions. On the other hand quasi-elastic scattering involves direct emission with no intermediate resonant state and so has a qualitatively different behavior in the q-ω plane (see Figures 1.3 and 1.4). Basic material on quasi-elastic scattering in the period between (1) and the present is contained in (146–170).

The simplest description of a nucleus is a collection of free nucleons at rest (1). Electron scattering from this idealized nucleus is then simply the sum of the elastic scatterings from the individual nucleons. In this case if a momentum q is transferred to the nucleus, i.e. to one of the nucleons, then that nucleon will recoil with momentum q and energy $q^2/2M$. However, the nucleon is a real, not a virtual particle and must absorb all the energy ω transferred in the scattering as well. Thus in this simple model the response surface would contain a δ-function peak along the line $\omega = q^2/2M$. The nucleus, however, is not so simple. Firstly, the nucleons are in motion and have a spread of momenta characterized by the Fermi momentum k_F. Thus the δ-function feature is spread into a peak whose width is determined by k_F. Secondly, the nucleons are interacting and form a self-bound system. There is a mean separation energy $\bar{\varepsilon}$ for a given nucleus that must be overcome before a nucleon can be emitted. So the approximate relation between q and ω that determines the position of the peak becomes $\omega = q^2/2M + \bar{\varepsilon}$ [see (148, 154)]. Additional nuclear interactions may be incorporated to some extent into the simple model by using an effective mass M^* rather than the full nucleon mass M.

To make these ideas more quantitative we may use a degenerate Fermi gas for the nucleus [see (1, 25)] and consider only the Coulomb interaction. In that case the (dimensionless) response function is given (see equation 1.3) by

$$(4\pi/M_T)S_L(q,\omega) = \sum_f |\langle f|\hat{\rho}(q)|i\rangle|^2 \,\delta[\omega - (\mathrm{E}' - \mathrm{E})], \qquad 3.14$$

where $|i\rangle$ is the degenerate Fermi gas and $|f\rangle$ has a single particle (a proton) above the Fermi sea and a single hole below. We substitute the plane wave single-particle wave functions to obtain

$$(4\pi/M_T)S_L(q,\omega) = (3Z/4\pi k_F^3) \int_0^{k_F} d\mathbf{k}\, \theta(|\mathbf{k}-\mathbf{q}|-k_F)$$

$$\times\, \delta(\omega-[(\mathbf{k}-\mathbf{q})^2/2M^* -k^2/2M^*+\bar\varepsilon]). \qquad 3.15$$

The integral may be evaluated to give

$$(4\pi/M_T)S_L(q,\omega) = \begin{cases} (R/Q)[1-(v/Q-Q/2)]^2 & \text{region I} \\ (2R/Q)v & \text{region II} \end{cases} \qquad 3.16$$

where $v \equiv (\omega-\bar\varepsilon)M^*/k_F^2$, $Q \equiv q/k_F$, $R \equiv 3ZM^*/4k_F^2$ and where region I corresponds to the union of $Q \geq 2$, $Q(\tfrac{1}{2}Q+1) \geq v \geq Q(\tfrac{1}{2}Q-1)$ and $Q < 2$, $Q(1+\tfrac{1}{2}Q) \geq v \geq Q(1-\tfrac{1}{2}Q)$ and region II corresponds to $Q < 2$, $Q(1-\tfrac{1}{2}Q) \geq v \geq 0$. Region II represents the Pauli suppression of the cross section for low momentum transfer. At high q, that is $Q \geq 2$, the quasi-elastic peak is parabolic in shape with a width which is directly proportional to the Fermi momentum k_F. In Moniz et al (157) a refined version of this simple model was applied to a range of nuclei with results indicated in Figure 3.7, and Table 3.1. In their work $M^* = M$ and a minimization program was used in comparing the data with the Fermi gas calculation in order to obtain best-fit values for k_F and $\bar\varepsilon$. No overall normalization factor was employed, and consequently, once the

Table 3.1 Nuclear Fermi momentum k_F and average nucleon interaction energy $\bar\varepsilon$ determined by least-squares fit of theory to quasi-elastic peak (157).

Nucleus	k_F (MeV)[a]	$\bar\varepsilon$ (MeV)[b]
${}^{6}_{3}$Li	169	17
${}^{12}_{6}$C	221	25
${}^{24}_{12}$Mg	235	32
${}^{40}_{20}$Ca	251	28
${}^{58.7}_{28}$Ni	260	36
${}^{89}_{39}$Y	254	39
${}^{118.7}_{50}$Sn	260	42
${}^{181}_{73}$Ta	265	42
${}^{208}_{82}$Pb	265	44

[a] The fitting uncertainty in these numbers is approximately ± 5 MeV.
[b] The fitting uncertainty in these numbers is approximately ± 3 MeV. Simple estimates for $\bar\varepsilon$ give numbers in reasonable agreement with those in the table.

peak position was fixed, the single parameter k_F was varied to fit both the height and width of the quasi-elastic peak. The model provides a surprisingly good description of the data in both shape and magnitude. The nuclear Fermi momentum is roughly constant at 260–265 MeV for the target nuclei from nickel through lead, whereas it increases from lithium up through calcium. This behavior is a reflection of the saturation of nuclear forces in nuclei; that is, nuclear densities are roughly constant for the heavier nuclei and decrease as one goes to lighter nuclei. From elastic electron scattering, one can infer a nuclear matter density $\rho \approx 0.17$ fm^{-3}, giving an equivalent Fermi momentum $k_F = (3\pi^2\rho/2)^{1/3} = 270$ MeV. The small difference between this and the extracted Fermi momentum for lead, 265 MeV can be attributed to a lower "local Fermi momentum" in the nuclear surface. The main point is that this method provides a determination of k_F completely independent of that obtained from the ground-state density as measured in *elastic* electron scattering. This is a *dynamical* determination of the Fermi momentum and the agreement with the values obtained from the static ground-state densities confirms the essential validity of this simple picture of the nucleus.

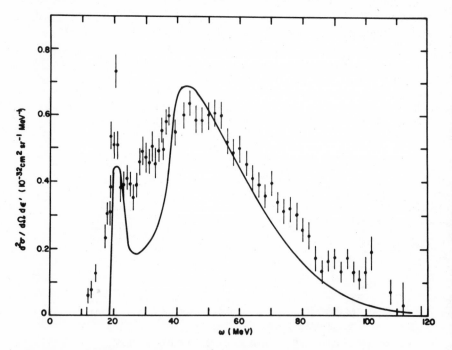

Figure 3.8   Quasi-elastic (e, e′) from ^{12}C at $\varepsilon = 148.5$ MeV and $\theta = 135°$ showing the square-well shell model results (154 and data referenced therein).

Note that the Fermi-gas calculation systematically underestimates the cross section at very large electron energy loss, a fact that is not understood and that bears further attention both experimentally and theoretically. At lower values of momentum transfer the simple Fermi gas model is less able to represent the data. The quasi-elastic peak moves to lower excitation energies and becomes entangled with the other inelastic features discussed in the last section. In particular, the giant dipole resonance and the quasi-elastic peak can overlap. An attempt to describe both the resonant and nonresonant $1p$-$1h$ excitations of the nucleus was made in

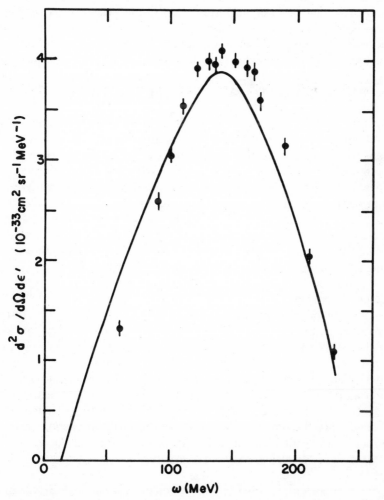

Figure 3.9   Same as Figure 3.8, but for ^{40}Ca with $q = 500$ MeV and $\theta = 120°$ (154).

(154). Here a finite square-well potential was used with bound-state solutions below particle threshold for the occupied (single-hole) states, and for discrete unoccupied (single-particle) states and with unbound solutions above threshold for the continuum (single-particle) states. Included in the latter set may be resonant wave functions—for example, the d-wave in light nuclei such as ^{12}C or ^{16}O. All significant matrix elements of the multipole operators $\hat{M}_J$, $\hat{T}_J^{el}$ and $\hat{T}_J^{mag}$ were computed (this sometimes entailed $J \approx 10$) and used to obtain the longitudinal and transverse response functions. Some of the results are shown in Figures 3.8 and 3.9, where in the first case the competition between the giant dipole resonance near $\omega = 20$ MeV and the quasi-elastic peak is evident. Results for ^{16}O were also presented in Figure 3.4. Note that at high momentum transfer given in Figure 3.9 the results of the finite-well calculation are very similar to the predictions of the Fermi gas model.

4   RELATIONS TO WEAK INTERACTIONS

We now turn to the extension of this analysis to include semileptonic weak interactions in nuclei: neutrino reactions (v_l, l^-), antineutrino reactions $(\bar{v}_l, l^+)$, charged-lepton capture (l^-, v_l), and $\beta^\pm$ decay. Here l^- is a lepton, either electron or muon, and v_l is the corresponding neutrino. These are all charge-changing reactions for the nuclear target. We also consider neutrino and antineutrino scattering (v_l, v_l), $(\bar{v}_l, \bar{v}_l)$ through weak neutral currents later in this section. The general structure of these processes is that indicated in Figure 1.1 (except for $\beta^\pm$ decay, where the initial lepton line is crossed into the final state). In contrast to the electromagnetic interactions, which are mediated through the photon propagator $\sim 1/q_\lambda^2$, the weak interactions proceed, as far as we know, through a contact interaction.

The topic of semileptonic weak interactions in nuclei is of interest for two reasons. First, most of the fundamental tests of our ideas on weak interactions (parity nonconservation, the conserved-vector-current (CVC) theory, the helicity of the neutrino, μ-e universality, the current-current interaction, etc) have involved nuclei, and if nuclei are to serve as laboratories in which to conduct experiments on weak interactions, it is essential that the nuclear physics of these processes be well understood. Second, once the fundamental nature of the weak interaction is understood, it can also be used as a known probe for testing our theoretical ideas on nuclear structure and for exciting new and unusual states in nuclei.

It is important to discuss semileptonic weak and electromagnetic processes together since the CVC theory states that half of the matrix elements in the weak processes, those coming from the vector current, are identical to those measured in electron scattering. In addition, electron scattering data at all momentum transfer q can be used, in many cases, to determine nuclear wave functions; this allows us to have some confidence in predicting new processes, such as neutrino reactions, and serves to eliminate nuclear physics uncertainties when examining the basic structure of the weak interaction itself, for example, in investigations of weak coupling constants, of second-class currents, or of neutral weak currents. The many nuclear targets and transitions available make nuclei the most convenient laboratory for studying the nature of the weak interactions.

The subject of semileptonic weak interactions with nuclei has been reviewed recently (171). We do not attempt to repeat all the material and references contained therein, but merely give a brief summary of this topic and discuss a few recent applications which emphasize the close connection with electron scattering.

4.1   Multipole Analysis and the Nuclear Current Operator

The principal additional feature in the weak processes is the presence of an axial-vector current $J_{\lambda 5}$, giving for the total current $\mathscr{J}_\lambda = J_\lambda + J_{\lambda 5}$. In analogy with the arguments presented in Section 1.2, the S-matrix element for the process of interest can be written

$$S = +2\pi i \delta(E' + \varepsilon' - E - \varepsilon)[G/(2)^{1/2}]\langle f|l_\mu|i\rangle \mathscr{J}_\mu(\mathbf{q}),   \qquad 4.1$$

where

$$\mathscr{J}_\mu(\mathbf{q}) = \int \exp(-i\mathbf{q}\cdot\mathbf{x})\langle f|\hat{\mathscr{J}}_\mu(\mathbf{x})|i\rangle \, d\mathbf{x}   \qquad 4.2$$

is the Fourier-transform of the transition matrix element of the nuclear current operator, and $\langle f|l_\mu|i\rangle$ is the appropriate transition matrix element of the lepton current $l_\mu = i\bar{\psi}_l\gamma_\mu(1+\gamma_5)\psi_{\nu_l}$ or $i\bar{\psi}_{\nu_l}\gamma_\mu(1+\gamma_5)\psi_l$. Here G is the weak coupling constant, with the value obtained from μ decay $M_p^2 G_\mu = 1.023 \pm 0.002 \times 10^{-5}$, where M_p is the proton mass.

A multipole analysis of the nuclear current, which allows us to make use of the Wigner-Eckart theorem and all the nuclear selection rules, can again be carried out in direct analogy with Section 1.2. The rates and cross sections can be computed in a similar manner.[11] As two representative results (171) that show the close connection with the previous analysis, we give the cross section for neutrino (antineutrino)

---

[11] Since ν_l and $\bar{\nu}_l$ are completely polarized, one does not *average* over initial neutrino (antineutrino) spins in computing the cross section for neutrino (anti-neutrino) reactions.

reactions when the final lepton is relativistic (ERL)

$$
\frac{d\sigma^{\text{ERL}}}{d\Omega_{\nu(\bar{\nu})}} = \left(G^2 \varepsilon'^2 \cos^2 \frac{\theta}{2} \Big/ 2\pi^2 \right)\left(1 + 2\varepsilon \sin^2 \frac{\theta}{2} \Big/ M_T \right)^{-1} [4\pi/(2J_i+1)]
$$

$$
\times \left\{ \left[\sum_{J=0}^{\infty} |\langle J_f \| \hat{\mathcal{M}}_J(q) - (q_0/q)\hat{\mathcal{L}}_J(q) \| J_i \rangle|^2 \right] + \left[(q_\lambda^2/2q^2) + \tan^2 \frac{\theta}{2} \right] \right.
$$

$$
\times \left[\sum_{J=1}^{\infty} \left(|\langle J_f \| \hat{\mathcal{T}}_J^{\text{mag}}(q) \| J_i \rangle|^2 + |\langle J_f \| \hat{\mathcal{T}}_J^{\text{el}}(q) \| J_i \rangle|^2 \right) \right] \qquad 4.3
$$

$$
\mp \tan \frac{\theta}{2} \left(q_\lambda^2/q^2 + \tan^2 \frac{\theta}{2} \right)^{1/2}
$$

$$
\left. \times \left[\sum_{J=1}^{\infty} 2\text{Re} \langle J_f \| \hat{\mathcal{T}}_J^{\text{mag}}(q) \| J_i \rangle \langle J_f \| \hat{\mathcal{T}}_J^{\text{el}}(q) \| J_i \rangle^* \right] \right\}
$$

and the rate for capture of a muon from the $1S$ Bohr orbit is

$$
\omega_\mu = (G^2 \varepsilon'^2/2\pi)(1+\varepsilon'/M_T)^{-1}[4\pi/(2J_i+1)]\left\{ \sum_{J=0}^{\infty} |\langle J_f \| \hat{\mathcal{M}}'_J(\varepsilon') \right.
$$

$$
\left. - \hat{\mathcal{L}}'_J(\varepsilon') \| J_i \rangle|^2 + \sum_{J=1}^{\infty} |\langle J_f \| \hat{\mathcal{T}}_J'^{\text{mag}}(\varepsilon') - \hat{\mathcal{T}}_J'^{\text{el}}(\varepsilon') \| J_i \rangle|^2 \right\}, \qquad 4.4
$$

where ε' is the final neutrino energy. In the first expression the multipole operators are precisely those of equations 1.20 and 1.25. They are computed from the current operator $\hat{\mathcal{J}}_\lambda(\mathbf{x})$. Since this current has both a vector and axial vector part, the multipole operators will similarly be a linear combination of two terms with opposite parity, $\hat{\mathcal{T}}_{JM} = \hat{T}_{JM} + \hat{T}_{JM}^5$. In equation 4.4 the prime denotes the fact that the multipole operators still contain the (spherically symmetric) muon wave function $\phi_{1S}(r)$. (The value of this wave function averaged over the nucleus can be removed from the multipoles to a very good approximation.)

As for isospin, the electromagnetic current is known to consist of an isoscalar and isovector part $J_\lambda^\gamma = J_\lambda^s + J_\lambda^{v3}$. CVC identifies the charge raising and lowering parts of the vector current entering into the weak interactions through the following relation: $J_\lambda^{(\pm)} = J_\lambda^{v1} \pm iJ_\lambda^{v2}$. The axial vector current is also an isovector operator and has the structure $J_{\lambda_5}^{(\pm)} = J_{\lambda_5}^{v1} \pm iJ_{\lambda_5}^{v2}$.

The nuclear current operators can again be constructed as in equations 1.33–1.37. For the single-nucleon matrix element, we have from Lorentz invariance and isospin

$$
\langle \mathbf{p}'\sigma'\rho' | \hat{\mathcal{J}}_\mu^{(-)}(0) | \mathbf{p}\sigma\rho \rangle = (i/\Omega)\bar{u}_{\sigma'}(\mathbf{p}')\eta_{\rho'}^\dagger [F_1\gamma_\mu + F_2\sigma_{\mu\nu}q_\nu + iF_S q_\mu
$$

$$
+ F_A\gamma_5\gamma_\mu - iF_P\gamma_5 q_\mu - F_T\gamma_5\sigma_{\mu\nu}q_\nu]\tau_- \, \eta_\rho u_\sigma(\mathbf{p}), \qquad 4.5
$$

where $p_\lambda = p'_\lambda + q_\lambda$. CVC implies that $F_{1,2}$ can be obtained from electron scattering from the nucleon (see Section 1.2). This implies that $F_1(0) = F_1^v(0) = 1$ and $2MF_2(0) = 2MF_2^v(0) = 3.706$. F_A can be determined, at least at $q_\lambda^2 = 0$, from the β decay of the neutron, $F_A(0) = -1.23 \pm 0.01$. We assume the universal shape given in Section 1.2. We also keep the $1 - \pi$ pole contribution to F_P, $F_P(q_\lambda^2) \cong 2MF_A(q_\lambda^2)/(q_\lambda^2 + m_\pi^2)$.

A reduction similar to equation 1.36 yields

$$\langle \mathbf{p}'\sigma'\rho' | \mathscr{J}_\mu^{(-)}(0) | \mathbf{p}\sigma\rho \rangle = (1/\Omega)\eta_{\rho'}^\dagger \chi_{\sigma'}^\dagger [\mathscr{M}_\mu - q_\mu(F_S + F_P\boldsymbol{\sigma} \cdot \mathbf{q}/2M)]\tau_- \chi_\sigma \eta_\rho$$

$$\mathscr{M} = F_A\boldsymbol{\sigma} - (F_1 + 2MF_2)i\boldsymbol{\sigma} \times \mathbf{q}/2M + F_1(\mathbf{p}' + \mathbf{p})/2M \qquad 4.6$$

$$\mathscr{M}_0 = F_1 + \boldsymbol{\sigma} \cdot [F_A(\mathbf{p}' + \mathbf{p})/2M - F_T\mathbf{q}]$$

up through order $(v/c)_{\text{nucleon}}$. We again assume that F_T is at least of $0(1/M)$. In fact, F_S and F_T vanish unless the weak interaction contains second-class currents, and we initially assume that such currents are not present. The nuclear current now follows exactly as in Section 1.2.

4.2   Unified Analysis Based on One-Body Densities

If hermiticity, time reversal invariance (to the present level of accuracy), and isotopic-spin invariance are assumed, then equations 1.41–1.44 *continue to hold for the full weak current*. This presents the interesting possibility of using electron scattering to determine the coefficients $\chi_{J;T}^{(i)}(ab)_n$ in equation 1.43 by exploiting the very different q dependence of the single-particle matrix elements and then computing all the weak nuclear matrix elements from these same coefficients. Thus we can try and use electron scattering to determine the one-body transition density matrix, including the spin dependence and spatial distribution, and use this density matrix to compute the semileptonic weak processes. We present four brief examples of this approach to a unified analysis of semileptonic weak and electromagnetic interactions with nuclei.

4.3   Examples

First, consider transitions between the $J = 0^+0$ ground state of ^{12}C and first $T = 1$ excited state [1^+ at 15.11 MeV] and its isobaric analogues, the ground states of ^{12}B and ^{12}N. The simplest description is obtained in the particle-hole model (TDA) where the ground-state is a closed $1p_{3/2}$ shell and the excited state is a $(1p_{3/2})^{-1}(1p_{1/2})_{1+1}$ excitation. In this case, there is a single matrix element $\langle 1p_{1/2}, \frac{1}{2} :: T_{1;1}(q) :: 1p_{3/2}, \frac{1}{2} \rangle$ in equation 1.43, with unit coefficient. The RPA, where backward-going graphs are included in the polarization propagator, only serves to reduce the current multipoles (equation 1.25) by a factor $1/\xi_{\text{RPA}}$. (The weak processes are insensitive to the M_1^5 charge multipole in this case.) More extensive calculations by O'Connell, Donnelly & Walecka (172) in the open-shell-RPA (OSRPA) indicate that the predominant effect is to further decrease these multipoles by a factor $1/\xi_{OSRPA}$. These authors have taken the TDA calculation using harmonic-oscillator single-particle wave functions and parametrized the one-body nuclear density through the oscillator parameter b and a reduction factor ξ as shown in Figure 4.1. Having completely specified the necessary nuclear physics, one is now in a position to compute any weak process. The computed partial weak rates are shown in Table 4.1. The correction to the shape of the allowed $\beta^{\pm}$ spectra is the classic test of weak magnetism in β decay. The calculated

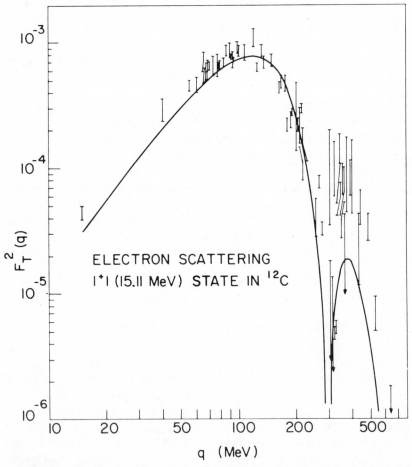

Figure 4.1 $F_T^2(q)$ for the 1^+1 state in ^{12}C (15.11 MeV) (126 and data referenced therein, 172). The curve is a best fit with $b = 1.77$ fm and $\xi = 2.25$ (172).

Table 4.1 Partial weak rates with ^{12}C. [References to the experimental work are contained in (172).]

Process	Experiment	Theory
β^- decay rate	$32.98 \pm 0.10 \ \mathrm{sec}^{-1}$	$33.8 \ \mathrm{sec}^{-1}$
β^+ decay rate	$59.55 \pm 0.22 \ \mathrm{sec}^{-1}$	$66.9 \ \mathrm{sec}^{-1}$
μ^- capture rate	$6.75^{+0.30}_{-0.75} \times 10^3 \ \mathrm{sec}^{-1}$	$6.64 \times 10^3 \ \mathrm{sec}^{-1}$

shape correction factor of O'Connell et al is compared with the data of Lee et al (173) in Figure 4.2. In Figure 4.3 the theoretical and experimental partial μ-capture rate are compared as a function of $\mu^V \equiv F_1 + 2MF_2$. The weak magnetism value of CVC is $\mu^V(0)_{\mathrm{CVC}} = 4.706$. Here we have a demonstration of weak magnetism in μ capture.

Next consider transitions between the 0^+0 ground state of ^{16}O and

Table 4.2 Partial weak rates with ^{16}O. [References to the experimental work are contained in (139). Also J. Deutsch, private communication.]

Experimental		Theory		
		Set 1	Set 2	Set 3
μ capture ($10^3 \ \mathrm{sec}^{-1}$)				
0^-	1.1 ± 0.2	0.86	0.86	0.70
	1.6 ± 0.2			
	$0.85^{+0.145}_{-0.060}$			
1^-	1.88 ± 0.10	1.42	1.28	1.16
	1.4 ± 0.2			
	$1.85^{+0.355}_{-0.170}$			
2^-	6.17 ± 0.71	7.54	6.65	7.44
	7.9 ± 0.8			
3^-	≤ 0.08	0.060	0.054	0.077
β decay ($10^{-2} \ \mathrm{sec}^{-1}$)				
2^-	2.53 ± 0.20	2.18	1.92	2.29
0^-	43 ± 10	42	46	$-$

the lowest $T = 1$ multiplet $[0^-(12.80$ MeV$)$, $1^-(13.09$ MeV$)$, $2^-(12.97$ MeV$)$, $3^-(13.26$ MeV$)]$ and its analogues, which include the ground states of ^{16}N and ^{16}F (Donnelly & Walecka 139). This was discussed

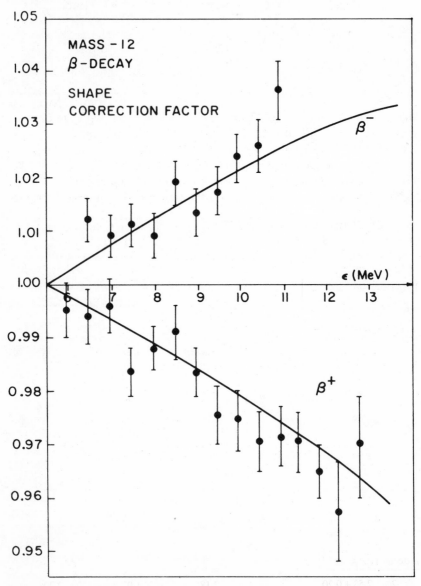

Figure 4.2 Shape correction factor (normalized to unity at $\varepsilon = 5.45$ MeV) for the ground-state β decays in mass 12 (172–173).

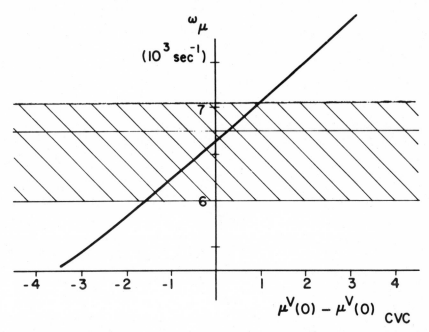

Figure 4.3 Muon capture rate to the ground state of ^{12}B as a function of $\mu^V(0) = F_1 + 2MF_2$, where $\mu^V(0)_{\text{CVC}} = 4.706$. The shaded region corresponds to the experimental capture rate quoted in (172).

in detail in Section 3.2 (see, in particular, Figure 3.6). The weak rates calculated with the amplitudes of Section 3.2 are shown in Table 4.2. Note that the parameters ξ and a for the 0^- state are not directly determined by (e, e'), but the values were assumed to behave similarly to those for the other three states. The present approach allows us to calculate the weak rates in an almost model-independent fashion to an accuracy of the order of 15–20%.

Figure 4.4 shows the total cross section for neutrino (antineutrino) excitation of these states. Figure 4.5 shows the differential neutrino and antineutrino cross sections for the 0^- state with and without a second-class tensor current where the coupling constant $MF_T = +5$ is the maximum value needed to produce agreement in the ^{12}B–^{12}N β-decay rates using harmonic oscillator wave functions. We can see that a comparison of neutrino and antineutrino cross sections at low-to-medium momentum transfer would yield a very sensitive and unambiguous test of the presence of second-class currents. If $\hat{\mathscr{J}}^{(\pm)}$ correspond to the raising and lowering components of an isovector operator, then isospin invariance implies $(d\sigma/d\Omega)_\nu = (d\sigma/d\Omega)_{\bar{\nu}}$ in this case. Second-class currents destroy

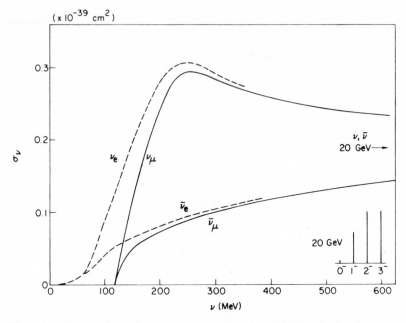

Figure 4.4 Neutrino reaction cross sections for the 13 MeV complex in mass-16 using Set 1 amplitudes from Table 4.2. The arrow indicates the (common) cross section at $\nu = 20$ GeV and the inset shows the relative contributions at this energy (139).

this equality (171). Alternatively, the observation of a large cross section at high momentum transfer is a direct measure of second-class currents, since, like the high multipoles in elastic magnetic scattering, they are almost totally responsible for the contribution there.

, A more definitive application of this unified approach (174) involves ^{6}Li (1^+0) and the first 0^+1 multiplet, which includes the ground states of ^{6}He and ^{6}Be. Here harmonic oscillator single-particle wave functions are used with the space for the two valence particles truncated to the p shell. The ground state can be written $|1^+0\rangle = A|(p_{3/2})^2;$ $1^+0\rangle + B|(p_{3/2}p_{1/2});$ $1^+0\rangle + C|(p_{1/2})^2;$ $1^+0\rangle$ and the excited state $|0^+1\rangle = D|(p_{3/2})^2;0^+1\rangle + E|(p_{1/2})^2;0^+1\rangle$ with $A^2 + B^2 + C^2 = D^2 + E^2 = 1$. A minimum χ^2 fit is then made to all the experimental electromagnetic properties of these two levels with the parameter set $\{A,\ldots,E,b\}$, where b is the oscillator parameter. The authors use the quadrupole moment Q, magnetic moment μ, the magnetic form factor

$$\langle 1^+0\|\hat{T}_{1;0}^{\text{mag}}\|1^+0\rangle/(3)^{1/2}(q/M)\,e^{-y}\,f_{\text{sn}}\,f_{\text{CM}} \equiv p(y) = \alpha_e + \beta_e\,y \qquad 4.7$$

for the ground state, and a similar magnetic form factor with coefficients

α_i, β_i for the transition (the accurate data for $q < 200$ MeV are used). Here $y = (bq/2)^2$, and f_{sn} and f_{CM} are the single-nucleon and center-of-mass form factors respectively. The linear dependence $p(y) = \alpha + \beta y$ of

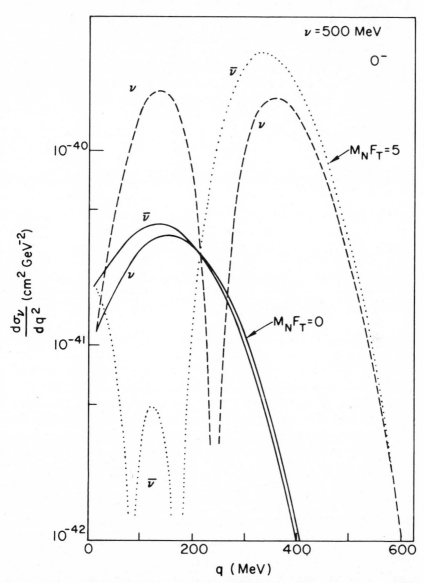

Figure 4.5 Neutrino reaction cross sections for the 0^-1 state of the 13 MeV complex in mass-16 using Set 1 amplitudes from Table 4.2 and showing results for two values of $M_N F_T$ (see text) (139).

this ratio is a consequence of the use of harmonic oscillator wave
functions and corresponding truncation scheme. The resulting best fit to
the elastic and inelastic magnetic form factors is shown in Figure 4.6.
The experimental data is from (50, 53, 175–180). The resulting set of
coefficients is shown in Table 4.3 (the uncertainties are determined by
the condition $\chi^2 \leqq 1$). Having completely determined the one-body
nuclear densities from the electromagnetic properties within this framework, one may now calculate the weak rates and these are also shown
in Table 4.3. The close agreement between the calculated and experimental
β-decay rates suggests that any weak exchange currents must behave the

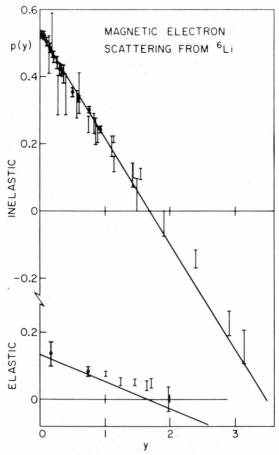

Figure 4.6 Elastic and inelastic magnetic electron scattering from ^{6}Li (50, 53,
175–180) in terms of $p(y)$ (equation 4.7). The straight lines represent a minimum
χ^2 fit to the accurate data at $q < 200$ MeV (*heavy error bars*) (174).

Table 4.3 Results for ^{6}Li. [References to the experimental work are contained in (174, 175–180).]

A	B	C	D	E	b(FM)
0.810	−0.581	0.084	0.80	0.60	2.03
±0.001	±0.001	±0.002	±0.03	±0.04	±0.02

Process	Experiment	Theory
β^- decay (sec^{-1})	0.864 ± 0.003	0.877 ± 0.023
μ^- capture (10^3 sec^{-1})	$1.6^{+0.33}_{-0.13}$	1.39 ± 0.04

same as the electromagnetic exchange currents, to about the 5% level, and provides a justification of the basic assumptions made in this unified analysis.

Let us now turn to the phenomena of neutral weak currents. Recent high-energy experiments give strong indications that such currents are present in the weak Hamiltonian. These results have aroused great interest since neutral weak currents play an important role in certain unified gauge theories of the elementary weak and electromagnetic interactions proposed by Weinberg, Salam and Ward, and others. Such currents would lead to a new class of experiments in which nuclear energy levels could be excited through the inelastic scattering of neutrinos [see (171) for references].

In Weinberg's theory (181), there is an additional neutral current interaction with neutrinos of the form

$$H_\nu = -[iG/(2)^{1/2}][\overline{\psi}_{\nu_e}\gamma_\lambda(1+\gamma_5)\psi_{\nu_e} + \overline{\psi}_{\nu_\mu}\gamma_\lambda(1+\gamma_5)\psi_{\nu_\mu}]\mathscr{J}_\lambda^{(0)} \qquad 4.8$$

giving rise to neutrino scattering processes. According to Weinberg, the isospin structure of the neutral hadronic current is $\mathscr{J}_\lambda^{(0)} = \mathscr{J}_\lambda^{v_3} - 2\sin^2\theta_w J_\lambda^{v} = J_\lambda^{v_3} + J_{\lambda 5}^{v_3} - 2\sin^2\theta_w(J_\lambda^s + J_\lambda^{v_3})$. This effective Hamiltonian density for the neutral weak interactions of leptons with nucleons may be taken over directly to nuclear processes, using the analysis we have discussed here. In the long-wavelength limit, the monopole operator corresponding to allowed Fermi transitions in Weinberg's theory is $\hat{M}_0^{(0)} \underset{qR \to 0}{\to} \hat{T}_3 - 2\hat{Z}\sin^2\theta_w$ where Z is the charge. This operator is diagonal in the nuclear Hilbert space and cannot cause transitions. Therefore, the strongest transitions in nuclei in this limit will be those with large allowed Gamow-Teller matrix elements. The predicted inelastic neutrino scattering cross sections in the Weinberg theory for the transitions in ^{12}C and

^{6}Li that we have examined are shown in Figure 4.7 (Donnelly et al 182). If neutral currents in fact exist, it should be possible to detect γ rays from nuclear excitation caused by neutrinos with either reactor neutrinos or those from an accelerator such as LAMPF. By making use of nuclear selection rules, one can isolate particular spin and isospin components of the neutral weak currents for examination.

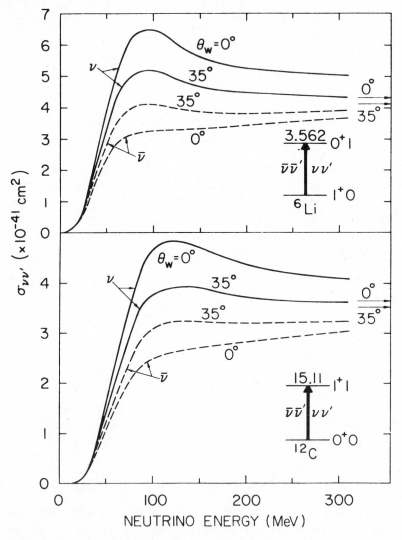

Figure 4.7 (ν, ν') cross sections for two values of the Weinberg angle θ_w. The arrows indicate asymptotic cross sections (182).

Finally, we very briefly discuss one example where the nucleus remains within the ground-state isotopic multiplet ("isoelastic"). The previous discussion (Section 1.3) implies that $J + \eta$ must be even in this case. Using the additional parity selection rules, we conclude that only the even $\hat{M}_{JM}$ and odd $\hat{T}_{JM}^{mag}$, $\hat{L}_{JM}^{5}$, $\hat{T}_{JM}^{el_5}$ multipoles contribute to the isoelastic processes. Thus, in principle, all the necessary coefficients in equation 1.43 can be obtained from electron scattering. We discuss a calculation for the

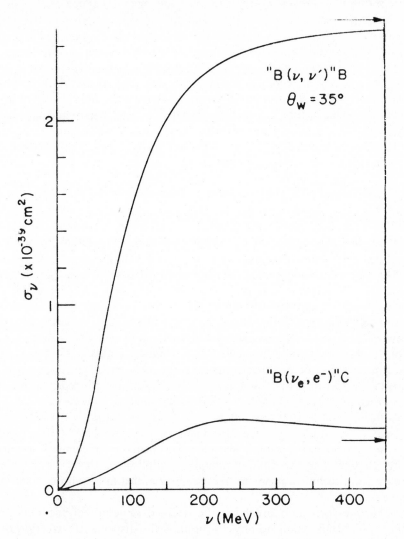

Figure 4.8   Calculated neutrino cross sections for isoelastic processes with ^{11}B–^{11}C

^{11}B–^{11}C isodoublet (Donnelly and Walecka, to be published). Harmonic-oscillator single-particle wave functions are used with the valence space truncated to the p shell. Again, the coefficients in equation 1.43 are chosen to fit the measured electromagnetic properties, μ and Q for *both* ground states, and the elastic magnetic form factor for ^{11}B (Figure 2.4) including both magnetic dipole and magnetic octupole contributions; the observed rate for the β^+ transition is also used. Since the electron scattering has been measured only on one member of the isodoublet there is not enough information available to completely separate the isovector and isoscalar contributions in the magnetic scattering. The authors assume that $\psi_{1;0}(1p^2_{1/2}) \cong -\psi_{1;1}(1p^2_{1/2})$ for the small components in the dipole matrix element and also $\psi_{3;1}(1p^2_{3/2}) \cong -\psi_{3;0}(1p^2_{3/2})$ as is true, for example, in a one-hole model. One is helped in determining the isovector amplitudes in both cases by the large lever arm provided by the ratio of isovector to isoscalar magnetic moment. The predicted cross sections for the (ν_e, e^-) process and *elastic neutrino scattering* (ν, ν) on ^{11}B are shown in Figure 4.8. Here there is an appreciable coherent contribution to the elastic process (183).

5   SPECIAL TOPICS

5.1   *Dispersion Corrections*

There are many types of corrections to the analysis of electron scattering we have presented. Perhaps the most important is the Coulomb correction where the wave function of the electron is distorted by the static Coulomb field of the nucleus. Since this correction to the Born approximation is of order $Z\alpha$, it becomes important for medium and high Z nuclei. The dominant effect is a shift in the diffraction patterns. The inclusion of this effect is a technical problem, involving the solution of the Dirac equation for the electron in an extended charge distribution, and is straightforward although it may be computationally quite complicated. The approaches to this problem are discussed in Überall (23) and are not repeated here. Another important correction in extracting nuclear data from electron scattering is due to radiation by the electron during the scattering process. These radiative corrections involve the quantum electrodynamics of the electron and depend on specific experimental arrangements. Again, though they may be computationally complicated, these corrections represent a technical problem that can be handled in a well-understood fashion (1).

The one correction involving the detailed structure of the nuclear target, and thus requiring physics input, is the dispersion correction or nuclear polarization effect. This correction is due to the virtual excitation and deexcitation of the nucleus during the scattering process. Since each such excitation and deexcitation costs one power of e, the dispersion correction to the amplitude is of order α and is therefore small. The basic theoretical ideas and work done up until 1966 are summarized in (1) and are not repeated here. We briefly comment on the recent work on this topic.

There are essentially three different approaches to the dispersion correction. One is to construct a complex optical potential for the electron that includes the effects of the absorption by all the inelastic channels on the elastic scattering. This is used in (184–186). A second approach is to use closure on the sum over intermediate nuclear states and relates the nuclear excitation and deexcitation amplitudes to a ground-state correlation function. This approach involves an estimate of the mean nuclear excitation energy and is used in (184, 185, 187). Friar & Rosen (187) emphasize that it is essential to use a completely consistent nuclear model when evaluating the required correlation functions. The third approach involves an explicit term-by-term sum over intermediate nuclear states. This is discussed in (188–197).

The most reliable method probably involves the explicit sum over intermediate nuclear states. One is aided here by the observation that the inelastic nuclear vertices for any nuclear state required in the calculation of the dispersion correction are the inelastic electron scattering amplitudes themselves. Thus the contribution of any single nuclear state to the dispersion correction can be evaluated in a model-independent way [Cole (198), see also (196)]. In general, it is important that any model used to evaluate the contribution of a region of nuclear excitation to the dispersion correction must reproduce the inelastic electron scattering response surface for that region. deForest (193) has carried out such a sum over states retaining just the Coulomb interaction and working within the harmonic oscillator model. He observes that high-lying nuclear states could make a sizable contribution and that it is important to retain the actual nuclear excitation energies. Lin (197) has carried out a calculation of the contribution to the dispersion correction of virtual excitation through the quasi-elastic peak. He uses the same Fermi gas model, which gives a good description of the observed quasi-elastic scattering (Section 3.3) and retains both the Coulomb and transverse interaction. His calculation should provide a good description of the contribution of high-lying nuclear excitations to the dispersion correction. His correction only depends on k_F and has very little angle dependence. The correction to the elastic cross section for ^{40}Ca is estimated to be $\sim -2\%$ at 500 MeV and $\sim -3\%$ at 1 GeV.

It has been observed [see e.g. Brown & Kujawski (185)] that the energy dependence of the effective nuclear charge density can be used to study dispersion correction since a true static charge density $\rho(x)$ must be capable of describing elastic scattering at all energies, while the additional effective charge density coming from the dispersion correction $\delta\rho(x, E)$ will depend on the electron energy. It has also been emphasized (199) that the dispersion corrections can be expected to play the largest relative role in the vicinity of diffraction minima where the basic amplitude is small.

The main theoretical conclusions concerning the dispersion corrections are summarized by Bethe & Molinari (184): "All three methods agree that the correction to the elastic scattering process is small. In particular it is very small near the minima of elastic scattering, so that these minima are neither substantially shifted nor filled in." Though small, these dispersion effects must be included when one wants to improve the accuracy with which charge and current distributions are extracted from electron scattering.

5.2 Exchange Currents

Exchange currents, that is, two-nucleon contributions to the electromagnetic current originating from the exchange of mesons between nucleons, have long been considered as important corrections to the simple one-body analysis of electromagnetic and weak interactions with nuclei [see e.g. (200, 201)]. Early phenomenological models of the exchange currents have been developed and extended [see e.g. (202–204)]. Recently the most common approach has been to deduce the exchange currents from meson exchange Feynman diagrams in the presence of an electromagnetic field. Although any attempt to include all meson exchanges in this way would be very complex, one believes that the anti-correlations between nucleons introduced by the repulsive core of the nuclear interaction may limit the important parts of the meson currents to the longer-range exchanges shown in Figure 5.1.

Perhaps the simplest place to look for exchange current effects is in the deuteron (^2H); however, electron scattering from the deuteron is most often used to explore the spatial distribution and structure of the *neutron* itself. Meson exchange effects have been considered in the analysis of e-d scattering [see e.g. (205, 206)]. Unfortunately, exchange current effects in ^2H represent smooth, relatively small corrections to smoothly varying form factors and therefore it is difficult to disentangle the exchange currents from other factors (e.g. the short-distance behavior of the wave functions, relativistic corrections, the structure of the neutron, etc) in any model-independent way [see, however (206, 207)]. Furthermore, in elastic scattering from ^2H ($T = 0$), there is no contribution from isovector currents. We may then be forced to consider nuclei more complex than the deuteron. One would like to exploit the many different *nuclear* transitions possible in order to use the nucleus as a laboratory for studying exchange currents. We eventually seek a situation where exchange currents do not represent small corrections, but are a large (possibly dominant) effect and may be more easily identified.

The effects of exchange currents are seen in magnetic moments [see the many papers in (18) and also (208, 209)]. For example, in the trinucleon system Harper et al (210), employing the diagrams in Figure 5.1 and using Fadeev wave functions, calculate a value of 2.571 n.m. for the isovector magnetic moment, 0.401 n.m. of this arises from exchange currents. The experimental isovector moment is 2.553 n.m. These calculations are sensitive to the amount of *D*-state present and they find the *S-D* contribution from Figure 5.1*c* is surprisingly large. Riska & Brown (211) find similar *S-D* dependence in triton β decay.

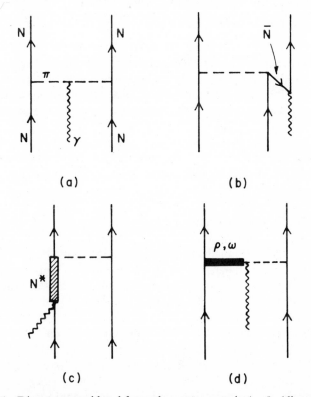

Figure 5.1   Diagrams considered for exchange currents in (e, e′). All permutations of the diagrams should be included.

Since electron scattering is capable of measuring the spatial distribution of the current densities within the nucleus it should provide a strong test of any description of exchange currents. At the present time, relatively few calculations of exchange current effects in electron scattering have been completed. Perhaps the most thorough of these is in (212, 213) in which all of the diagrams in Figure 5.1 are included in a calculation of the magnetic form factors of ^{3}He and ^{3}H. Kloet & Tjon (214) have examined exchange current contributions to the charge form factors of these same nuclei. They state that they are able to calculate the position of the first minimum and second maximum in ^{3}He, but are more than a factor of two below the experimental amplitude of the second maximum. These calculations did not include diagram c in Figure 5.1, and such a term should be examined before the full effect of exchange currents in the charge form factor of ^{3}He can be judged.

Relatively little effort has been made towards examining exchange current effects in electron scattering from nuclei with $A > 4$. One notable exception is the work of Chemtob & Lumbroso (215) who evaluated diagrams a and b of Figure 5.1 in examining the electroexcitation of $T = 1$ states in ^{12}C above 14 MeV using harmonic oscillator wave functions. They find the exchange corrections to be small at low momentum transfers and to dominate the one-body contribution at $q \gtrsim 600$ MeV. T. W. Donnelly, J. Dubach, and J. Koch (personal communication) have used the basic approach of (212) with some modifications to calculate similar effects for the $1^+0 \rightarrow 0^+1$ transition in ^{6}Li (see Figure 4.6). The same oscillator parameter and $1p$-shell amplitudes were used. The new theoretical curve in Figure 4.6 remains essentially a straight line over the indicated range. The intercept at $y = 0$ is increased by about 2%. The zero is pushed out from $y = 1.69$ to $y = 1.72$ and the amplitude at $y = 3$ is reduced from -4.08 to -4.22. By $q \approx 650$ MeV, however, the exchange current contributions dominate the one-body term in F_T. Clearly in this case, one cannot yet draw any model-independent conclusions about exchange currents. Donnelly et al have included a two-body correlation function to mock-up the expected hard core and have found a reduction in the exchange contribution of only about 20 or 30%. We note that to leading order in $1/M_{\text{nucleon}}$, their model contains no corrections to the charge operator [cf (214)].

It is clear that the contribution of exchange current effects is very much an active problem in electron scattering theory at this time. These effects seem to be very important in the mass-3 system and at high momentum transfer in heavier nuclei. Even at smaller momentum transfer the exchange effects represent a correction that must be included if an accurate analysis of experiment is to be achieved. One final possibility being explored by Donnelly et al is to search for transitions that are predominantly two-particle excitations. Such transitions have little one-body amplitude and would therefore be quite sensitive to the meson currents.

5.3   Relation to High-Energy Hadronic Scattering

The scattering of high-energy hadrons from nuclei takes places through the strong interactions. A remarkably successful theory of such processes due to Glauber (216) views the scattering as taking place from A stationary target particles through the elementary projectile-nucleon scattering amplitude f_{pN}. (We here assume f_{pN} is independent of spin and isotopic spin.) The elastic scattering amplitude for the many-body nuclear target is then obtained by averaging over the positions of the target particles using the ground-state wave function. With certain assumptions about the multiple-scattering processes, one can show the elastic scattering amplitude is identical to that generated from the one-body wave equation

$$(\nabla^2 + k^2)\psi = v\psi, \qquad\qquad 5.1$$

using an optical potential (216–219)

$$v_{\text{opt}} = -4\pi A \rho(\mathbf{x}) f_{pN}(0), \qquad\qquad 5.2$$

where $\rho(\mathbf{x})$ is the normalized $[\int \rho(\mathbf{x})\,d\mathbf{x} = 1]$ one-body matter density for the target and $f_{pN}(0)$ is the forward, elementary scattering amplitude. In the limit where the elementary amplitude is purely absorptive, the optical theorem $f_{pN}(0) \cong iImf_{pN}(0) = (ik/4\pi)\sigma_{pN}^{\text{Tot}}$ can be used to relate $f_{pN}(0)$ to the elementary p-N total cross section.

If the nucleus has isospin $T = 0$ then the one-body charge and matter densities must be identical, and $\rho(\mathbf{x})$ can be inferred from elastic electron scattering (Section 2.1). If $f_{pN}(0)$ is known from elementary-particle scattering, the hadronic optical potential (equation 5.2) is then *completely determined*. One still has to solve the wave equation 5.1 and thus sum the relevant strong-interaction multiple-scattering graphs, but the resulting hadronic elastic-scattering diffraction pattern reflects the elastic electron-scattering amplitude. Further corrections to the optical potential involve the two-body matter density $\rho^{(2)}(\mathbf{x}_1, \mathbf{x}_2)$ (216–219). The two-body charge density, a closely related quantity, can be obtained from inelastic electron scattering sum rules (1, 219).

Inelastic hadronic scattering from nuclei can be described within a similar framework. In the distorted wave Born approximation (DWBA), the incoming and outgoing hadronic wave functions are distorted by the optical potential in equation 5.2. The transition matrix element for collective states involves the one-body transition matter density $\rho_{fi}(\mathbf{x})$. For a $T = 0 \to T = 0$ transition, this is identical to the transition charge density (Section 3.1). Application of these ideas to high-energy inelastic nucleon scattering process $[(N, N')]$ has proven very successful [see (220) for recent review of this topic]. These ideas can be extended to more complex multiple-scattering processes than just those contained in the DWBA. Also, known nuclear transition densities determined from electron scattering (see 3.2) can be used to sort out the actual spin and isospin components of f_{pN} (221). Once these are known, the (N, N') process can be used to study new nuclear transitions. This should be one of the most exciting classes of results from LAMPF, for example. Clearly electron scattering *complements* high-energy hadronic scattering and plays an essential role in the theory of such processes.

ACKNOWLEDGMENTS

We are grateful to J. Dubach for the material in Section 5.2. We also wish to thank Mrs. V. Bonnici and the physics library staff for their assistance in the literature survey, and Vicki LaBrie for her careful typing.

Literature Cited

1. deForest, T. Jr., Walecka, J. D. 1966. *Adv. Phys.* 15:1–109 [also 15:491; 17:479]
2. Goldemberg, J., Pratt, R. H. 1966. *Rev. Mod. Phys.* 38:311–29
3. Jacob., G., Maris, T. A. J.. 1966. *Rev. Mod. Phys.* 38:121–42; 1973. 45:6–21
4. Tzara, C. 1968. *J. Phys. (Paris) Suppl.* No. 1:17–24
5. Überall, H. 1969. *Springer Tracts Mod. Phys.* 49:1–89. Berlin: Springer. 146 pp.
6. Peterson, R. J. 1971. *Ann. Phys.* 65:125–40
7. Drake, T. E. 1972. *Proc. Mont. Tremblant Summer School on Dyn. Struc. of Nucl. States 1971*, 420–48. Toronto: Univ. Toronto Press. 585 pp.
8. Walker, G. E. See Ref. 7, 509–41
9. Rowe, D. J., Donnelly, T. W. See Ref. 7, 101–53
10. Theissen, H. 1972. *Springer Tracts Mod. Phys.* 65:1–57. Berlin: Springer. 145 pp.
11. Bishop, G. R. 1964. *Congr. Int.*
    Phys. Nucl., Paris, 1964, 341–70. Paris: Cent. Nat. Rech. Sci. 1227 pp.
12. Walecka, J. D. 1967. *Proc. Int. Nucl. Phys. Conf., Gatlinburg, 1966*, 289–313. New York: Academic. 1121 pp.
13. Isabelle, D. B. 1967. *Proc. Int. Conf. High Energy Phys. Nucl. Struct., 2d, Rehoroth, 1967*, 126–42. Amsterdam: North-Holland. 489 pp.
14. Goldemberg, J. 1968. *Proc. Int. Conf. Nucl. Struct. Tokyo, 1967*. In *J. Phys. Soc. Jap. Suppl.* 24:379–92. 755 pp.
15. Walecka, J. D. 1970. *Proc. Int. Conf. High Energy Phys. Nucl. Struct., 3rd, New York, 1969*, 1–17. New York: Plenum. 859 pp.
16. Isabelle, D. B. See Ref. 15, 18–32
17. Walecka, J. D. 1972. *Proc. Int. Conf. High Energy Phys. Nucl. Struct., 4th, Dubna, 1971*, 543–64. Dubna: D1-6349. 670 pp.
18. Walecka, J. D. 1973. *Proc. Int. Conf. Nucl. Mom. Nucl. Struct., Osaka, 1972*. In *J. Phys. Soc. Jap. Suppl.* 34:302–16. 637 pp.
19. Donnelly, T. W. 1973. *Proc. Int. Conf. Photonucl. React. Appl.,*

*Asilomar,1973,*1303–12. Oak Ridge: US AEC CONF-730301. 1392 pp.
20. Bishop, G. 1973. *Proc. Int. Conf. Nucl. Phys., Munich, 1973,* 2:460–507. Amsterdam: North-Holland, 837 pp.
21. Bertozzi, W., Kowalski, S., eds. 1967. *Med. Energy Nucl. Phys. Elect. Accel., MIT Summer Study, 1967.* TID 24667. Cambridge, Mass.: MIT. 661 pp.
22. Eisenberg, J. M., Greiner, W. 1970. *Excitation Mechanisms of the Nucleus.* Amsterdam: North-Holland. 370 pp.
23. Überall, H. 1971. *Electron Scattering from Complex Nuclei.* New York: Academic. 2 vols. 425 pp., 867 pp.
24. Shoda, K., Ui, H., eds. 1972. *Nucl. Struct. Studies Using Electron. Scatter. and Photoreact. Proc. Int. Conf., Sendai, 1972.* In: *Res. Rep. Lab. Nucl. Sci. Tokoku Univ., Suppl.* Vol. 5. 544 pp.
25. Fetter, A. L., Walecka, J. D. 1971. *Quantum Theory of Many-Particle Systems.* New York: McGraw-Hill. 601 pp.
26. Drell, S. D., Walecka, J. D. 1964. *Ann. Phys.* 28:18–33
27. Hofstadter, R. 1963 *Nuclear and Nucleon Structure.* New York: Benjamin. 690 pp.
28. Blatt, J. M., Weisskopf, V. W. 1952. *Theoretical Nuclear Physics.* New York: Wiley. 864 pp.
29. Edmonds, A. R. 1957. *Angular Momentum in Quantum Mechanics.* Princeton, NJ: Princeton Univ. Press. 146 pp.
30. Bertozzi, W., Friar, J., Heisenberg, J., Negele, J. 1972. *Phys. Lett. B* 41:408–14
31. Barrett, R. C. 1974. *Rep. Prog. Phys.* 37:1–54
32. Donnelly, T. W., Walker, G. E. 1969. *Phys. Rev. Lett.* 22:1121–24
33. Lenz, F. 1969. *Z. Phys.* 222:491–503
34. Friedrich, J., Lenz, F. 1972. *Nucl. Phys. A* 183:523–44
35. Lenz, F., Rosenfelder, R. 1971. *Nucl. Phys. A* 176:513–25
36. Friar, J. L., Negele, J. W. 1973. *Nucl. Phys. A* 212:93–137
37. Sick, I. 1974. *Nucl. Phys. A* 218:509–41
38. Heisenberg, J. et al 1969. *Phys. Rev. Lett.* 23:1402–5
39. Sick, I. 1973. *Phys. Lett. B* 44:62–64
40. Hetherington, J. H., Borysowicz, J. 1974. *Nucl. Phys. A* 219:221–31
41. Pratt, R. H., Walecka, J. D., Griffy, T. A. 1962. *High Energy Phys. Lab.* [Stanford Univ.] *Rep.* 272; 1965. *Nucl. Phys.* 64:677–84
42. Griffy, T. A., Yu, D. U. L. 1965. *Phys. Rev. B* 139:880–85
43. Kabachnik, N. M., Grishanova, S. I. 1966. *Yad. Fiz.* 4:819–24
44. Kruger, J., Van Leuven, P. 1969. *Nucl. Phys. A* 139:418–24
45. Li, G. C., Sick, I., Walecka, J. D., Walker, G. E. 1970. *Phys. Lett. B* 32:317–20
46. Donnelly, T. W., Walecka, J. D. 1973. *Nucl. Phys. A* 201:81–106
47. Kerimov, B. K., Aruri, T. R., Safin, M. Ya. 1973. *Izv. Akad. Nauk SSSR Ser. Fiz.* 37:1768–74†
48. Engfer, R. 1973. *Proc. Int. Conf. Nucl. Phys., Munich, 1973,* 2:438–58. Amsterdam: North-Holland. 837 pp.
49. Brandenburg, R. A., Kim, Y. E., Tubis, A. 1974. *Phys. Rev. Lett.* 32:1325–27
50. Peterson, G. A. 1962. *Phys. Lett.* 2:162–63
51. Goldemberg, J., Torizuka, Y. 1963. *Phys. Rev.* 129:312–15
52. Vanpraet, G. J., Kossanyi-Demay, P. 1965. *Nuovo Cimento* 39:388–90
53. Rand, R. E., Frosch, R., Yearian, M. R. 1965. *Phys. Rev. Lett.* 14:234–7; 1966. *Phys. Rev.* 144:859–73; 148:1246
54. Chertok, B. T., Jones, E. C., Bendel, W. L., Fagg, L. W. 1969. *Phys. Rev. Lett.* 23:34–37
55. DeVries, H., Van Niftrik, G. J. C., Lapikás, L. 1970. *Phys. Lett. B* 33:403–6
56. Van Niftrik, G. J. C., Lapikás, L., DeVries, H., Box, G. 1971. *Nucl. Phys. A* 174:173–92
57. Lapikás, L., Dieperink, A. E. L., Box, G. 1973. *Nucl. Phys. A* 203:609–26
58. Collard, H. et al. 1965. *Phys. Rev. B* 138:57–65

59. McCarthy, J. S., Sick, I., Whitney, R. R., Yearian, M. R. 1970. *Phys. Rev. Lett.* 25:884–88
60. Bernheim, M. et al 1972. *Lett. Nuovo Cimento* 5:431–34
61. Murphy, J. D., Überall, H. 1973. *Phys. Lett. B* 44:347–50
62. Bohr, A., Mottelson, B. 1969. *Nucl. Struct. Vol. I.* New York: Benjamin. 471 pp.; *Vol. II.* to be published
63. Heisenberg, J. H., Sick, I. 1970. *Phys. Lett. B* 32:249–52
64. Gillet, V., Green, A. M., Sanderson, E. A. 1964. *Phys. Lett.* 11:44–46; 1966. *Nucl. Phys.* 88:321–43
65. Blomqvist, J. 1968. *Phys. Lett. B* 28:22–25
66. Phan Xuan Ho, Bellicard, J., Leconte, Ph., Sick, I. 1973. *Nucl. Phys. A* 210:189–210
67. Friedrich, J. 1972. *Nucl. Phys. A* 191:118–36
68. Litvinenko, A. S. et al 1972. *Ukr. Fiz. Zh., Russian* ed. 17:1210–12
69. Groh, J. L., Singhal, R. P., Caplan, H. S., Dolbilkin, B. S. 1971. *Can. J. Phys.* 49:2743–53
70. Heisenberg, J., McCarthy, J. S., Sick, I. 1971. *Nucl. Phys. A* 164:353–66
71. Itoh, K., Oyamada, M., Torizuka, Y. 1970. *Phys. Rev. C* 2:2181–99; 1973. 7:458–59
72. Afanas'yev, V. D. et al 1970. *Yad. Fiz.* 12:885–91*
73. Torizuka, Y. et al 1969. *Phys. Rev.* 185:1499–1507
74. Curtis, T. H., Eisenstein, R. A., Madsen, D. W., Bockelman, C. R. 1969. *Phys. Rev.* 184:1162–77
75. Barreau, P., Bellicard, J. B. 1967. *Phys. Rev. Lett.* 19:1444–46
76. Peterson, G. A., Ziegler, J. F. 1966. *Phys. Lett.* 21:543–45; 1968. *Phys. Rev.* 165:1337–51
77. Bernheim, M., Stovall, T., Vinciguerra, D. 1967. *Phys. Lett. B* 25:461–63
78. Miller, H. G., Dreizler, R. M. 1974. *Nucl. Phys. A* 225:137–40
79. Erikson, T. 1973. *Nucl. Phys. A* 211:105–24
80. Theis, W., Werner, E. 1973. *Phys. Lett. B* 44:481–83
81. Penner, S. 1973. *J. Soc. Jap. Suppl.* 34:387–88
82. Antoine, B., Gillet, V., Normand, J. M., Bellicard, J. B. 1973. *Phys. Lett. B* 44:55–58
83. Hosoyama, K., Torizuka, Y., Kawazoe, Y., Ui, H. 1973. *Phys. Rev. Lett.* 30:388–91
84. Lee, H. C. 1972. *Phys. Lett. B* 41:421–24
85. DeJager, J. L., Boeker, E. 1972. *Nucl. Phys. A* 186:393–416
86. Hammerstein, G. R., Larson, D., Wildenthal, B. H. 1972. *Phys. Lett. B* 39:176–78
87. Lightbody, J. W. Jr. 1972. *Phys. Lett. B* 38:475–79
88. Theis, W. 1972. *Z. Phys.* 250:99–119
89. Lukyanov, V. K., Pol, Ya. S. 1970. *Yad. Fiz.* 11:556–66*
90. Bouten, M., Van Leuven, P. 1967. *Ann. Phys.* 43:421–27
91. Strehl, P. 1970. *Z. Phys.* 234:416–42
92. Donnelly, T. W., Federman, P. 1974. *Phys. Rev. C* 10:12–18
93. Horsfjord, V. 1973. *Nucl. Phys. A* 209:493–519
94. Gerace, W. J., Sparrow, D. A. 1970. *Nucl. Phys. A* 154:576–86
95. Onley, D. S. 1970. *Nucl. Phys. A* 149:197–208
96. Boeker, E. 1967. *Phys. Lett. B* 24:616–18
97. Bertozzi, W. et al 1972. *Phys. Rev. Lett.* 28:1711–13
98. Hollowell, P. et al 1973. *Phys. Rev. C* 7:1396–1409
99. Nakada, A., Torizuka, Y. 1972. *J. Phys. Soc. Jap.* 32:1–13
100. Lees, E. W. et al 1973. *J. Phys. A* 6:L116–20
101. Singhal, R. P., Caplan, H. S., Moreira, J. R., Drake, T. E. 1973. *Can. J. Phys.* 51:2125–37
102. Slight, A. G., Drake, T. E., Bishop, G. R. 1973. *Nucl. Phys. A* 208:157–95
103. Horikawa, Y., Nakada, A., Torizuka, Y. 1973. *Prog. Theor. Phys.* 49:2005–20
104. Horikawa, Y. 1972. *Prog. Theor. Phys.* 47:867–79
105. Grünbaum, L. 1966. *Nucl. Phys.* 83:528–44
106. Friar, J. L. 1966. *Nucl. Phys.* 84:150–60
107. Überall, H. 1966. *Phys. Rev. Lett.*

17: 1292–94
108. Holder, F. D., Eisenberg, J. M. 1967. *Nucl. Phys. A* 106: 261–74
109. Kabachnik, N. M. 1968. *Yad. Fiz.* 7: 823–31*
110. Donnelly, T. W., Walecka, J. D., Sick, I., Hughes, E. B. 1968. *Phys. Rev. Lett.* 21: 1196–1200
111. Kelly, F. J., Überall, H. 1968. *Phys. Rev.* 175: 1235–42
112. Gul'karov, I. S. et al 1969. *Yad. Fiz.* 10: 694–98; 9: 478–86*
113. Vanpraet, G. J., Kossanyi, P. 1969. *Z. Phys.* 222: 455–61
114. Torizuka, Y. et al 1969. *Phys. Rev. Lett.* 22: 544–46
115. Tomaselli, M. et al 1969. *Phys. Lett. B* 29: 579–80
116. Albert, D. J., Wagner, R. F., Überall, H., Werntz, C. See Ref. 15, 89–91
117. Rowe, D. J., Wong, S. S. M. 1969. *Phys. Lett. B* 30: 147–49
118. Chertok, B. T. 1969. *Phys. Rev.* 187: 1340–47
119. Sick, I., Hughes, E. B., Donnelly, T. W., Walecka, J. D., Walker, G. E. 1969. *Phys. Rev. Lett.* 23: 1117–21
120. Torizuka, Y. et al 1969. *Phys. Rev.* 188: 1745–46
121. Gul'karov, I. S. 1970. *Yad. Fiz.* 12: 694–99
122. Goncharova, N. G., Yudin, N. P. 1970. *Yad. Fiz.* 12: 725–31
123. Kim, J. C., Singhal, R. P., Caplan, H. S. 1970. *Can. J. Phys.* 48: 83–93
124. Lightbody, J. W. Jr., Penner, S. 1970. *Phys. Rev. Lett.* 24: 274–76
125. Goldmann, A., Stroetzel, M. 1970. *Phys. Lett. B* 31: 287–88
126. Donnelly, T. W. 1970. *Phys. Rev. C* 1: 833–46
127. Donnelly, T. W., Walecka, J. D., Walker, G. E., Sick, I. 1970. *Phys. Lett. B* 32: 545–49
128. Donnelly, T. W., Walker, G. E. 1970. *Ann. Phys.* 60: 209–72
129. Torizuka, Y. et al 1970. *Phys. Rev. Lett.* 25: 874–77
130. Grishanova, S. I., Titarenko, N. N. 1971. *Yad. Fiz.* 13: 712–18*
131. Yang, C. S., Tomusiak, E. L., Gupta, R. K., Caplan, H. S. 1971. *Nucl. Phys. A* 162: 71–81
132. Yamaguchi, A., Terasawa, T.,

Nakahara, K., Torizuka, Y. 1971. *Phys. Rev. C* 3: 1750–69
133. Bely, Yu. I. et al 1971. *Nucl. Phys. A* 170: 141–52
134. Bergstrom, J. C. et al 1971. *Phys. Rev. C* 4: 1514–32
135. Fagg, L. W. et al 1972. *Phys. Rev. C* 5: 120–24
136. Donnelly, T. W., Lo Iudice, N. 1972. *Nucl. Phys. A* 179: 569–93
137. Goswami, A., Graves, R. D. 1972. *Phys. Lett. B* 39: 499–502
138. Antony-Spies, P. 1972. *Nucl. Phys. A* 188: 641–65
139. Donnelly, T. W., Walecka, J. D. 1972. *Phys. Lett. B* 41: 275–80
140. Fukuda, S., Torizuka, Y. 1972. *Phys. Rev. Lett.* 29: 1109–11
141. Buskirk, F. R. et al 1972. *Phys. Lett. B* 42: 194–96
142. Suzuki, T., Hinohara, C. 1973. *Nucl. Phys. A* 204: 289–306
143. Nagao, M., Torizuka, Y. 1973. *Phys. Rev. Lett.* 30: 1068–71
144. Chkalov, I. I. et al 1973. *Yad. Fiz.* 18: 945–49*
145. Fivozinsky, S. P., Penner, S., Lightbody, J. W. Jr., Blum, D. 1974. *Phys. Rev. C* 9: 1533–42
146. Henry, G. R. 1966. *Phys. Rev.* 151: 875–78; 1967. 157: 871–73
147. deForest, T. Jr. 1969. *Ann. Phys.* 132: 305–21
148. Moniz, E. J. 1969. *Phys. Rev.* 184: 1154–61
149. Stachan, C., Watt, A. 1969. *Proc. Phys. Soc. London A* 2: 547–58
150. Dementii, S. V. et al 1970. *Yad. Fiz.* 11: 19–28*
151. Sitenko, A. G., Pasichyi, A. A., Tartakovskii, V. K. 1970. *Yad. Fiz.* 12: 1208–17*
152. Freiberg, D. H. 1970. *Phys. Rev. C* 1: 1735–39
153. Antony-Spies, P. et al 1970. *Phys. Lett. B* 31: 632–34
154. Donnelly, T. W. 1970. *Nucl. Phys. A* 150: 393–416
155. Wong, C. F., Hutcheon, R. M., Shin, Y. M., Caplan, H. S. 1970. *Can. J. Phys.* 48: 1917–23
156. Vlasenko, V. G. et al 1971. *Yad. Fiz.* 13: 259–67*
157. Moniz, E. J. et al 1971. *Phys. Rev. Lett.* 26: 445–48

158. Pasichnyi, A. A., Tartakovskii, V. K. 1971. *Ukr. Fiz. Zh.,* Russian ed. 16:177–82
159. Stanfield, K. C., Canizares, C. R., Faissler, W. L., Pipkin, F. M. 1971. *Phys. Rev. C* 3:1448–65
160. Ciofi degli Atti, C. 1971. *Lett. Nuovo Cimento* 1:590–95
161. Balashov, V. V., Grishanova, S. I., Kabachnik, N. M., Kulikov, V. M. 1971. *Phys. Lett. B* 36:325–27
162. Jackson, D. F. 1971. *Nucl. Phys. A* 173:225–38
163. Valdrodov, G. M., Gorchakov, V. V. 1972. *Izv. Akad. Nauk SSSR Ser. Fiz.* 36:680–85†
164. Vlasenko, V. G. et al 1972. *Ukr. Fiz. Zh.,* Russian ed. 17:1440–45
165. Berthot, J., Isabelle, D. B. 1972. *Lett. Nuovo Cimento* 5:155–59
166. Bernabéu, J. 1972. *Nucl. Phys. B* 49:186–205
167. Klawansky, S., Kendall, H. W., Kerman, A. K., Isabelle, D. B. 1973. *Phys. Rev. C* 7:795–800
168. Fursa, A. D., Tartakovski, V. K., Koslovski, I. V. 1973. *Ukr. Fiz. Zh.,* Russian ed. 18:372–77
169. Kuplennikov, E. L. et al 1973. *Yad. Fiz.* 17:20–23*
170. Titov, Ya. I. et al 1974. *Yad. Fiz.* 19:479–82*
171. Walecka, J. D. In *Semi-Leptonic Weak Interactions in Nuclei,* ed. V. W. Hughes, C. S. Wu. New York: Academic. In press
172. O'Connell, J. S., Donnelly, T. W., Walecka, J. D. 1972. *Phys. Rev. C* 6:719–33
173. Lee, Y. K., Mo., L. W., Wu, C. S. 1963. *Phys. Rev. Lett.* 10:253–58
174. Donnelly, T. W., Walecka, J. D. 1973. *Phys. Lett. B* 44:330–34
175. Barber, W. C., Goldemberg, J., Peterson, G. A., Torizuka, Y. 1963. *Nucl. Phys.* 41:461–81
176. Hutcheon, R. M., Drake, T. E., Stobie, V. W., Beer, C. A., Caplan, H. S. 1968. *Nucl. Phys. A* 107:266–72
177. Neuhausen, R. 1969. *Z. Phys.* 220:456–65
178. Eigenbrod, F. 1969. *Z. Phys.* 228:337–52
179. Hutcheon, R. M., Caplan, H. S.

180. 1969. *Nucl. Phys. A* 127:417–28
180. Neuhausen, R., Hutcheon, R. M. 1971. *Nucl. Phys. A* 164:497–512
181. Weinberg, S. 1972. *Phys. Rev. D* 5:1412–17
182. Donnelly, T. W., Hitlin, D., Schwartz, M., Walecka, J. D., Wiesner, S. J. 1974. *Phys. Lett. B* 49:8–12
183. Freedman, D. Z. 1974. *Phys. Rev. D* 9:1389–92
184. Bethe, H. A., Molinari, A. 1971. *Ann. Phys.* 63:393–431
185. Brown, W. D., Kujawski, E. 1971. *Ann. Phys.* 64:573–98
186. Petkov, I. 1971. *Izv. Fiz. Inst. ANEB Bulg. Akad. Nauk.* 20:159–64 (in English)
187. Friar, J. L., Rosen, M. 1972. *Phys. Lett. B* 39:615–19
188. Rawitscher, G. H. 1966. *Phys. Rev.* 151:846–52
189. Bottino, A., Ciocchetti, G., Molinari, A. 1966. *Nucl. Phys.* 89:192–208
190. Toepffer, C. 1968. *Phys. Lett. B* 26:426–28
191. Toepffer, C., Greiner, W. 1969. *Phys. Rev.* 186:1044–58
192. Toepffer, C., Drechsel, D. 1970. *Phys. Rev. Lett.* 24:1131–34
193. deForest, T. Jr. 1970. *Phys. Lett. B* 32:12–14
194. Kujawski, E. 1970. *Phys. Lett. B* 32:75–77
195. D'Auria, R., Molinari, A. 1971. *Lett. Nuovo Cimento* 1:83–86
196. Bottino, A., Ciocchetti, G. 1971. *Phys. Lett. B* 34:187–90
197. Lin, W.-F. 1972. *Phys. Lett. B* 39:447–49; 1973. *Nucl. Phys. A* 199:14–22
198. Cole, R. K. 1967. *Phys. Lett. B* 25:178–80; 1969. *Phys. Rev.* 177:164–83
199. Rawitscher, G. H. 1970. *Phys. Lett. B* 33:445–48
200. Villars, F. 1947. *Helv. Phys. Acta* 20:476–90
201. Miyazawa, H. 1951. *Prog. Theor. Phys.* 6:801–14
202. Osborn, R. K., Foldy, L. L. 1950. *Phys. Rev.* 79:795–802
203. Bosco, B., Piazza, A. 1970. *Phys. Rev. C* 1:787–95

204. Lock, J. A., Foldy, L. L., Buskirk, F. R. 1974. *Nucl. Phys. A* 220: 103–13
205. Adler, R. J., Drell, S. D. 1964. *Phys. Rev. Lett.* 13:349–52
206. Hockert, J., Riska, D. O., Gari, M., Huffman, A. 1973. *Nucl. Phys. A* 217:14–28
207. Chemtob, M., Moniz, E. J., Rho, M. 1974. *Phys. Rev. C* 10:344–52
208. Chemtob, M. 1970. *Nucl. Phys. A* 123:449–70
209. Chemtob, M., Rho, M. 1971. *Nucl. Phys. A* 163:1–55
210. Harper, E. P., Kim, Y. E., Tubis, A. 1972. *Phys. Lett. B* 40:533–36
211. Riska, D. O., Brown, G. E. 1970. *Phys. Lett. B* 32:662–64
212. Hadjimichael, E., Barroso, A. 1973. *Phys. Lett. B* 47:103–6
213. Barroso, A., Hadjimichael, E. 1975. *Nucl. Phys. A* 238:422–36
214. Kloet, W. M., Tjon, J. A. 1974. *Phys. Lett. B* 49:419–22
215. Chemtob, M., Lumbroso, A. 1970. *Nucl. Phys. B* 17:401–29
216. Glauber, R. J. 1959. *Lectures in Theoretical Physics, Vol. 1*, 315–414. New York: Interscience. 414 pp.
217. Feshbach, H. 1958. *Ann. Rev. Nucl. Sci.* 8:49–104
218. Fetter, A. L., Watson, K. M. 1965. *Advances in Theoretical Physics, Vol. 1*, 115–94. New York: Academic. 323 pp.
219. Foldy, L. L., Walecka, J. D. 1969. *Ann. Phys.* 54:447–504
220. Saudinos, J., Wilkin, C. 1974. *Ann. Rev. Nucl. Sci.* 24:341–77
221. Viollier, R. D. 1974. Preprint ITP-483. Stanford Univ. To be published in *Ann. Phys.*

* English translation: *Sov. J. Nucl. Phys.*
† English translation: *Bull. Acad. Sci. USSR Phys. Ser.*

NUCLEAR SAFEGUARDS   ✕5565

Theodore B. Taylor

International Research and Technology Corporation, 1501 Wilson Boulevard,
Arlington, Virginia 22209

CONTENTS

INTRODUCTION

One of the most critical tasks facing the world is to establish and maintain worldwide controls of nuclear fuels and radioactive materials that will as much as practicable reduce the risks of purposeful destructive uses of these materials. Without much more stringent worldwide controls of nuclear materials than now exist or are definitely planned, many responsible people find it credible that nuclear violence—nuclear war, use of nuclear weapons by terrorists or other criminals, or release of dangerous amounts of radiactive materials by sabotage of nuclear operations—may soon reach levels that most people would consider intolerable.

This paper is a brief overview of these risks and the actions that are being or could be taken to reduce them. Since this is an extremely complex subject, some important parts of which remain classified, it is impossible to cover it here in technically satisfying detail. For more detailed analyses of the subject, the reader is referred to various recent publications that deal with many of the technical, economic, legal, social, and political issues (1–7). The nature of the worldwide nuclear industry and of both international and national actions to prevent purposeful abuse of nuclear materials and facilities is changing rapidly and can be expected to change both qualitatively and quantitatively in the years to come. Partly for this reason, the primary purpose of this paper is to present nuclear scientists and engineers with some of the basic facts and issues that are likely to be relevant to this subject for some time, rather than to review in detail the situation as it exists at the time of the writing (early 1975).

For purposes of this paper, the often confusing phrase *nuclear safeguards* is taken to mean actions designed to accomplish one or more of the following objectives:
1. to detect or prevent the *national* diversion of nuclear fuel materials from peaceful purposes to their incorporation into nuclear explosive weapons;
2. to prevent clandestine diversion or overt theft of nuclear materials by *non-national* organizations or individuals for purposes of (*a*) making nuclear explosives, (*b*) making radiological weapons, or (*c*) supplying an illegal market in nuclear materials;
3. to assure the recovery of dangerous nuclear materials diverted or stolen by non-national organizations or individuals before they are actually used for destructive purposes;
4. to prevent sabotage of nuclear facilities or transport systems in ways that would expose the public to dangerous levels of radiation.

Emphasis is on safeguards to prevent the illicit production of fission explosives. Safeguards to prevent diversion of fissionable materials will automatically prevent illicit production of thermonuclear explosives as long as practical pure fusion explosives remain undeveloped. The success of current intensive efforts to heat small pellets of thermonuclear fuels to explosive ignition temperatures by implosions driven by intense laser or electron beam pulses, however, is likely to raise important new issues regarding proliferation of nuclear weapons, at least on a national scale.

Gram quantities of plutonium suspended in the air as dusts or aerosols might conceivably be used to serve the purposes of terrorists or other criminals, since only a few micrograms of inhaled plutonium can cause eventual death. This should be kept in mind when considering alternative ways to safeguard plutonium against theft or diversion.

Safeguards specifically designed to prevent sabotage of nuclear facilities in ways that could release dangerous quantities of radioactive materials are not explicitly discussed in this paper. This is not because such risks are considered to be unimportant. Rather, it is because recent studies, which show that reactors, nuclear fuel reprocessing facilities, and high level radioactive waste repositories and transport systems are vulnerable to such sabotage, are classified.

Clearly, no conceivable safeguards systems can reduce the risks of nuclear violence to zero. It is partly for this reason that detailed information related to the risks and to actions being taken to reduce them should be available to the general public, despite the obvious risk that public discussion of the subject may stimulate attempted acts of nuclear violence. If the risks cannot be removed altogether, how far should safeguard actions be taken? Who should decide? How is the public to be convinced that all practicable precautions have been taken? It is inconceivable that such questions can be answered to the satisfaction of some consensus of the public, either in individual nations or in the world as a whole, without extensive public understanding and discussion of the issues involved.

RESOURCES REQUIRED TO MAKE FISSION EXPLOSIVES

Basic Requirements

The time and resources required to design and make fission explosives depend on the type of explosive wanted. It is much more difficult to make large numbers of reliable, efficient, and lightweight nuclear warheads for a national military system than to make crude, inefficient nuclear explosives with unpredictable yields in the range, say, equivalent to one hundred to several thousand tons of chemical explosive. In all cases, appropriate information, skilled people, nuclear materials, non-nuclear materials and components, and construction and assembly facilities are required. Each of these basic requirements is discussed briefly below.

Information

All of the basic information required for the design and construction of a wide variety of types of fission explosives is published in the open technical literature. Abstracts and titles of these publications are systematically organized by subject in such publications as *Nuclear Science Abstracts,* and these publications are now distributed in technical libraries throughout the world. Nevertheless, the following allusions to the kinds of information available are made without specific references, in the interests of not oversimplifying the task of retrieving such information.

The theory of chain-reacting systems, in forms ranging from simple diffusion theory and approximate rules of thumb to computer programs for solving the neutron transport equations, is widely published, primarily for use in nuclear reactor design, but also in forms that are applicable to the design of nuclear explosives.

Extensive data on the critical masses of a wide variety of combinations of fissionable materials and other materials, in many types of reflectors, are published. These critical mass data can be used both for relatively "quick and dirty" designs of fission explosives and for experimental tests of detailed neutron transport calculations.

In forms suitable both for approximate estimates and for detailed computer calculations in one or more dimensions, the theory of hydrodynamics, used in calculating both the assembly and disassembly of fission explosives is now available in the open literature. Some forms of the equations of state of high explosives, fissionable materials, and materials that might be used as reflectors are all published.

The phases through which a nuclear explosive proceeds, during both assembly and nuclear disassembly, are described in the open literature, so that a designer can determine what he should think about, and how he should think about it, when he analyzes a given design concept. Included in these phases are initial assembly of the core; compression, if there is any; achievement of criticality; increase in supercriticality; initiation of a chain reaction by a neutron source; buildup of both internal and kinetic energy; disassembly of the core; and reduction of the neutron multiplication of the core back to negative values. Specific design concepts for fission explosives are also described in varying degrees of detail in open publications.

The procedures and safety precautions required for conversion of uranium or plutonium compounds into metallic or other forms suitable for fission explosives are all published.

Fission explosive assembly systems can require the use of high explosives. Especially since the end of World War II, there have been extensive unclassified experimentation and theoretical study of the properties and uses of high explosives for a variety of purposes that relate directly to their use for fission explosives. Information on techniques for making high explosive lenses and for accelerating metal plates to velocities as high as several millimeters per microsecond have been widely reported. Methods for casting and fabrication of a wide variety of especially energetic high explosives are described in the literature. Techniques for measurements of shock wave and plate acceleration uniformity, and of detonation wave, shock, and free surface velocities have been thoroughly described.

All the information necessary for the design of fuzing and firing systems required for many different types of fission explosives can be either obtained from the open literature directly or derived by reasonably competent high explosives and electronics specialists.

In summary, the information required for the design and construction of fission explosives of various types, given the required amounts of fissionable materials, is available in the open literature. The performance of the explosives designed by a group will depend upon their access to information about certain concepts that are now classified, or on their levels of understanding and ingenuity, or both.

Skilled People

Three kinds of people would be useful to an organization that wants to develop and build nuclear explosives: people with direct experience in designing, building, or testing nuclear explosives; people with highly developed technical skills and basic knowledge of the specific technical fields required for such a project; and people with necessary basic skills, but without specific knowledge or experience in the specific fields required for the project (e.g. people with training and experience in related engineering fields).

There are thousands of people in the first category, concentrated in those nations that have built and tested nuclear weapons. Perhaps several hundred of these have recently lived or are now living for extended periods in other countries.

There are at least tens of thousands of people in the second category, distributed throughout the world in the industralized nations, and concentrated in those nations that have extensive programs for research and development related to the development of civilian nuclear technology. Especially relevant professional disciplines include theoretical and experimental reactor physics and nuclear engineering, theoretical and experimental plasma physics, chemistry and metallurgy of uranium and transuranium elements, hydrodynamics, and explosives technology.

There are millions of people in the last category—scientists, engineers, and skilled technicians with education and experience in the physical sciences and engineering. In 1971 more than 90,000 people were granted bachelor's degrees in the physical sciences, mathematics, or engineering in the United States alone.

At the beginning of 1943 there were no people who had experience in the detailed design or the construction of nuclear weapons. Two and a half years later the first implosion bomb was successfully tested. It should be expected, therefore, that a group of professionals and technicians from this last category would be sufficient for construction of efficient fission weapons. The entire group assembled at Los Alamos in 1943, though exceptionally capable, consisted of people in this category.

Nuclear Materials

Three types of fissionable materials that are used or produced in components of nonmilitary nuclear technology can be used for making fission explosives: highly enriched uranium, plutonium, and ^{233}U.

The critical masses of spheres of metallic uranium (density = 19 g cm^{-3}) surrounded by a 15-cm thick reflector of natural uranium, for different levels of ^{235}U enrichment, are shown in Table 1 (8). For comparison, the critical masses of 100% ^{235}U spheres inside 15-cm thick reflectors of aluminum, water, nickel, and beryllium are 28, 23, 20, and 11 kg, respectively (8).

The term *highly enriched uranium* is often taken to mean uranium enriched above 20% in ^{235}U, and has been used by the U.S. Atomic Energy Commission as a cutoff in enrichment, below which it was considered that further enrichment would be required to make a practical fission explosive. It is clear from Table 1 that such a cutoff is somewhat arbitrary.

It should not be concluded that the minimum amount of ^{235}U from which a fission explosive can be made is in the range of 10–20 kg, corresponding to the critical mass at normal density in a reasonably good neutron reflector. It is well known that "implosion" types of fission explosives achieve supercriticality by compression of the core substantially above normal density and that the critical mass varies inversely with the square of the compression if it is uniform in the core and reflector (8, 9). How little fissionable material can be used in a practical fission explosive depends on the knowledge, ingenuity, and skills of the explosive designers and fabricators. In thinking about such limits, it should be kept in mind that the complete fissioning of 1 kg of ^{235}U or plutonium would release the equivalent of 17 kilotons of high explosive, somewhat more than the yield of the Hiroshima weapon. This lack of clear defini-

Table 1   Critical mass of uranium vs ^{235}U enrichment

Enrichment (% of ^{235}U)	Critical Mass (kg of ^{235}U)	Critical Mass (kg of U)
100	15	15
80	17	21
60	22	37
40	30	75
20	~50	~250
10	~130	~1300

tion of what one might call a "minimum strategic quantity" of a particular fissionable material is especially troublesome when it comes to setting quantitative objectives for nuclear safeguards.

Highly enriched uranium, plutonium, or ^{233}U do not have to be in metallic form to be usable as core material in a fission explosive. As is the case of uranium metal of intermediate enrichment, however, the acceptable extent of dilution by any particular material, such as oxygen, carbon, or thorium is not well defined. It can be stated categorically, on the other hand, that oxides or carbides of these fissionable materials could be used directly as core materials in practical fission explosives (10). The critical masses of the oxides or carbides of these metals are in the range of 20–30% greater than those of the pure metals.

Plutonium with all isotopic compositions contemplated for use in nuclear power systems could be used as the core material for fission explosives. The spherical critical mass of α phase ($\rho = 19.5$ g cm^{-3}) ^{239}Pu in a thick natural uranium reflector is 4.4 kg. The critical mass of plutonium metal as a function of the contained isotopes that are not fissionable by thermal neutrons (^{240}Pu and ^{242}Pu) can be estimated from simple diffusion theory corrected to give results that agree with measured critical masses. The results of this estimate are shown in Table 2.

The volume fractions of nonfissile plutonium used in the nuclear industry vary over wide ranges—less than 10% for plutonium produced in fast breeder reactor blankets; 20–35% for plutonium produced in light water reactors (LWRs); and perhaps as high as 40–50% for plutonium in fast breeder reactor cores.

Besides increasing the critical mass, the presence of ^{240}Pu and ^{242}Pu can affect the performance of fission explosives as a result of the neutrons they emit by spontaneous fission. These spontaneous fission rates are of the order of 10^6 neutrons per kilogram per second. Neutron generation

Table 2 Critical mass of plutonium vs isotopic composition

Volume Fraction of ^{240}Pu + ^{242}Pu	Critical Mass of α-phase ^{239}Pu in a Thick U Reflector (kg)	Total Pu Critical Mass (kg)
0	4.4	4.4
10	4.5	5.0
20	4.5	5.6
30	4.6	6.7
40	4.7	7.8
50	4.8	9.6

times in typical fission explosives are of the order of 10^{-8} sec, and something like 40–50 generations of fissions are required to build up a fast chain reaction to an explosive level. It is therefore clear that, at least under some conditions, such a neutron source strength would cause an assembly to explode before it has reached maximum criticality. Such situations cannot be discussed in satisfying technical detail without getting into classified areas.[1] However, it can be stated categorically that such high neutron source strengths do not prevent the construction of practical fission explosives. Under some circumstances, yields of explosives of identical designs could vary substantially.

As a material for use in fission explosives, ^{233}U more nearly resembles plutonium than ^{235}U, with one important exception: the neutron release rates from ^{233}U in the form it would be in when extracted from reprocessed nuclear fuel are several orders of magnitude lower, per kilogram, than the neutron release rates from typical forms of plutonium used in reactors. The critical mass of ^{233}U metal or compounds is about 30% greater than that of the same chemical forms of α-phase ^{239}Pu, or about 5.8 kg of metal in a thick uranium reflector.

Non-Nuclear Materials and Components and Related Facilities

Given the required fissionable materials, knowledge, and skilled people, the additional materials, equipment, and facilities required to make fission explosives can vary over a huge range of degrees of accessibility and complexity, depending on the desired explosive characteristics, the degree of concern for the safety of the people involved, the available time to complete the process, the perceived need for secrecy of the operations, and a number of other factors.

Conceivably, one person who possessed about one normal density "fast" critical mass of fissionable material and a substantial amount of chemical high explosive could design and build a crude fission bomb. By a "crude fission bomb" is meant one that would be likely to explode with a yield equivalent to at least 100 tons of high explosive and that could be transported in an automobile. This could be done using materials and equipment that could be purchased at a hardware store and from commercial supliers of scientific equipment for student laboratories.

New national programs to develop and build fission weapons with characteristics specifically adapted to national needs could now result in any of a wide variety of highly sophisticated types of nuclear

---

[1] This is not to say, however, that relevant information that is considered to be classified cannot be *derived* from unclassified sources by competent people.

explosives, not necessarily along lines that have previously been followed by other nations. Much of the required work could be done before the nuclear core materials were made available. If it were very important for a nation to do so, it could acquire militarily useful fission weapons for which design, construction, and non-nuclear tests had been carried out in advance, and arm them within a few days of the time when it gained access to the required amounts of concentrated fissionable materials.

OVERVIEW OF INTERNATIONAL CIVILIAN NUCLEAR TECHNOLOGY

In 1974, 20 nations had operating nuclear power reactors. An additional 15 nations without nuclear power plants had research reactors that either contained highly enriched uranium or could annually produce plutonium sufficient for at least one fission explosive. By 1980 the number of nations in either or both categories is expected to exceed 45 (11).

Light water moderated and cooled reactors fueled with slightly enriched uranium (2–4% ^{235}U) accounted for about 80% of the world's nuclear electric capacity at the end of 1974 (12). Most of the remaining capacity was provided by natural uranium fueled reactors, either gas cooled, graphite moderated, or heavy water cooled and moderated. High temperature, gas cooled reactors (HTGR) initially fueled with highly enriched uranium have been built in the United States and Western Europe, but only one or two are expected to be operating by 1980. Liquid sodium cooled, plutonium fueled fast breeder reactors are under intensive development in several countries, several are newly in operation, and several more are expected to operate by 1980. Thus, until at least the mid-1980s, more than 95% of the world's nuclear power will be provided by reactors fueled with natural or slightly enriched uranium, some of which will be supplemented by recycled plutonium made by neutron capture in the ^{238}U. For this reason, this paper focuses attention on safeguards applied to fuel cycles for thermal fission reactors that use natural or slightly enriched uranium, and, even more sharply, on safeguards applied to plutonium produced in these reactors.

Table 3 presents estimates, by country, of total installed nuclear power, annual net rates of production of plutonium, and cumulative net quantities of plutonium produced, at the end of 1974 and 1980. These estimates use projections of nuclear generating capacity made by the International Atomic Energy Agency in 1974 as a point of departure (12). The conversion factors used to relate each 1000 MW(e) of installed electric generating capacity to net kilograms of plutonium produced per year in the reactors are listed, for the main types of reactors, in Table 4. These

Table 3 Installed nuclear power and plutonium production, by country—1974 and 1980

Country	Installed Nuclear Electric Power [MW(e)]		Rate of Plutonium Production (kg per year)		Cumulative Plutonium Produced (kg)	
	1974	1980	1974	1980	1974	1980
Argentina	320	920	59	340	59	1,100
Austria	—	690	—	160	—	720
Belgium	400	1,660	46	380	59	2,920
Brazil	—	600	—	140	—	620
Bulgaria	440	1,760	48	300	48	1,310
Canada	2,510	6,120	930	2,260	2,590	13,000
Czechoslovakia	110	1,770	41	370	82	1,090
Finland	—	1,540	—	350	—	1,130
France	2,870	15,170	1,100	3,750	4,880	18,000
Germany, F.R.	4,230	21,770	770	4,660	2,450	19,080
Germany, D.R.	440	1,760	100	410	100	1,920
Hungary	—	440	—	51	—	51
India	780	1,580	200	460	480	2,700
Ireland	—	600	—	69	—	69
Italy	610	3,390	180	820	1,740	4,530
Japan	5,020	18,400	950	4,420	2,420	21,200
Korea	—	1,160	—	240	—	820
Mexico	—	1,310	—	300	—	900
Netherlands	530	530	120	120	220	950
Pakistan	130	130	46	46	87	360
Spain	1,070	8,550	340	1,950	1,070	8,240
Sweden	2,610	8,270	350	1,800	510	7,200
Switzerland	1,010	5,690	230	1,090	580	3,800
Taiwan	—	3,110	—	720	—	1,640
Thailand	—	500	—	69	—	69
USSR	3,510	10,010	810	2,300	5,200	17,900
United Kingdom	5,800	10,750	1,970	3,000	20,200	36,600
USA	40,460	137,800	7,430	18,400	18,300	93,600
Yugoslavia	—	1,400	—	240	—	240
South Africa	—	2,000	—	230	—	230
Totals	73,000	270,000	16,000	49,000	61,000	260,000
Totals (less USA, USSR, UK, France)	20,000	96,000	4,400	22,000	12,500	96,000

factors include an assumed ratio of 0.7 between the average electric power output and the rated power output.

Even a brief scanning of Table 3 will give the reader an idea of the

Table 4   Conversion factors for estimating plutonium production rates in reactors

Reactor Type	Conversion Factor [kg of Pu per 1000 MW(e)/yr]
Light water moderated and cooled (LWR), slightly enriched uranium fuel	230
Heavy water moderated and cooled (HWR), natural uranium fueled	370
Gas cooled, graphite moderated (GCR), natural uranium fueled	430

magnitude and complexity of the task of effectively safeguarding plutonium in the immediate future. But the detailed nature of this task will depend to a large extent on specifically what is done in each country with the plutonium produced in its power reactors.

It is presumed that most of the plutonium produced in power reactors in the United Kingdom and France through 1974 has been used for making nuclear weapons, but that the plutonium made in new power plants will not.

Roughly 1% of the mass of a typical LWR fuel assembly is plutonium when it is removed from the reactor for substitution with fresh fuel. To be usable in fission explosives it must be separated from the fission products and uranium at a reprocessing plant, where massive shielding is required to protect workers until the separation is complete. The plutonium nitrate solution from this separation can be stored for future use or converted to plutonium oxide for subsequent incorporation into fresh fuel, to supplement ^{235}U, at a fuel fabrication plant.

Irradiated reactor fuel is largely self-protected against theft by criminals because of its intense gamma-ray activity, but it could be diverted to weapons use by a nation that has reprocessing facilities. After separation from the fission products, this self-protecting feature disappears until the plutonium has been placed back in an operating reactor. The disposition of irradiated fuel removed from a reactor thus has an important effect on the types of safeguards that are subsequently required.

At the end of 1974 no commercial fuel reprocessing plant had been operating in the United States since early 1972, and the first new plant was expected to start operations in 1976 or later. Routine recycling of plutonium in this country is not expected to start for several years after that, perhaps not until the early 1980s. On the other hand, reprocessing plants are now operating, or soon will be, in at least six other countries— the United Kingdom, France, the Federal Republic of Germany, India,

Italy, and Japan. Thus large quantities of plutonium are likely to be separated, and perhaps recycled, outside the United States sooner than in this country.

PRESENT SAFEGUARDS

The International Atomic Energy Agency has the responsibility for safeguards to detect diversion of nuclear materials from peaceful to unauthorized purposes by nations that are parties to the Treaty on Non-Proliferation of Nuclear Weapons or that have otherwise agreed to place their civilian nuclear materials under international safeguards. It is not responsible, however, for applying physical security safeguards to prevent overt theft or clandestine diversion by non-national groups, such as terrorists or other criminals. This responsibility is left to individual countries.

The IAEA's safeguards are based primarily on materials accounting procedures and, to an increasing extent, on inspection, tamper-proof surveillance, and seal techniques designed to detect unauthorized operations that could lead to diversion of nuclear materials. National safeguards, on the other hand, are primarily designed to detect and prevent diversion of nuclear materials by employees or overt theft by outsiders, and to deter or prevent acts of sabotage that could endanger the public.

Considerable concern has been expressed about the inadequacy of national safeguards to prevent theft of fissionable materials by determined criminal groups with resources and skills similar to those used for successful bank robberies or hijacking of valuable shipments (1–6). The following section briefly sketches actions that might be taken to improve such safeguards.

POSSIBLE FUTURE SAFEGUARDS

A guiding principle, called the "Principle of Containment" has been proposed for the design of national nuclear safeguards systems (5). According to this principle, the overall objective of a safeguards system would be to detect attempts at theft of nuclear materials from authorized channels as early as possible, and to place sufficient impediments in the way of people attempting a theft to assure that whatever forces are required can be brought to the scene in time to control any credible theft operation. Measures that could be used for applying this principle fall into five broad categories:

1. Containment of nuclear materials at fixed sites having physical barriers designed to prevent unauthorized penetration long enough to

insure that on-site or reserve guard forces arrive on the scene before the theft is completed, with the capacity to deal with any credible type of attempted theft.

2. Shipment of nuclear materials in massive containers and vehicles designed to resist penetration, transfer of the shipment to another vehicle, or commandeering of the vehicle itself long enough for accompanying or reserve guard forces to be able to deal effectively with any credible theft attempt.

3. Provision of automatic alarms that will immediately detect attempts to remove materials from authorized channels and sound alarms at several specified control points.

4. Provision of on-site and in-transit guard forces for the purpose of denying access of unauthorized people to places where nuclear materials exist.

5. Provision of on-call law enforcement forces that can be brought to the scene of an attempted theft before a theft can be completed, along with secure communications between the control points where the alarms are sounded and the off-site forces.

A few examples of specific actions that conform to this principle are presented below.

Some particularly vulnerable transportation links could be removed if nuclear fuel cycle facilities were located at the same sites. Shipments of concentrated plutonium oxide or plutonium nitrate could be avoided if fuel reprocessing facilities and fuel fabrication plants that use their output were next to each other. Facilities for intermediate conversion of plutonium or highly enriched uranium, which are often now located by themselves, could be integral parts of fuel reprocessing or fabrication facilities. As an extreme, but generally attractive possible option, *all* fuel cycle components that handle fissionable materials, including very high capacity power reactors, could be located at the same site. This would not only remove the need for transportation of fissionable materials over large distances but would also reduce, through economies of scale, the overall costs of safeguards for a complete fuel cycle.

It is likely that rather massive gamma-ray and, in some cases, neutron shielding will be required to insure insignificant radiation exposure to workers at all points in the fuel cycle for systems that use recycled plutonium or ^{233}U. This would mean that, for reasons not connected with safeguards, heavy containers and barriers will have to be used in the storage, transport, and fabrication of nuclear fuels from the time they are separated at a reprocessing plant to the time they are placed in reactors for refueling. These barriers and containers will make theft much more difficult.

Another possibility is the establishment of a national nuclear materials accounting system that contains electronically reported, accurate data from all nuclear facilities concerning the quantities, physical and chemical forms, and locations of all safeguarded nuclear materials. Present materials accounting requirements by the U.S. Nuclear Regulatory Commission call for periodic reporting of nuclear material balance accounts and input-output material balance accuracies for special nuclear materials that are between 0.5% and 1%, depending on the types of materials and facilities. These uncertainties could amount to several dozen kilograms or more of plutonium per year at a large fuel reprocessing or fuel fabrication plant, enough material for at least several nuclear explosives. One way to considerably improve these accuracies would be to monitor all points of access to a facility that contains special nuclear materials with instruments designed to detect the removal of very small quantities of special nuclear material through channels (such as personnel exits) that are *not* authorized channels for the flow of these materials. For example, instruments that can detect the presence of one gram or less of unshielded plutonium, at a distance of a few feet, and in about a second, are commercially available. Use of such devices, along with as accurate as possible measurements of the flows and inventories of nuclear materials in *authorized* channels, could provide a high assurance that even very small quantities of materials have not been surreptitiously removed from nuclear facilities. Any recorded significant changes in the amounts or forms of nuclear materials could be immediately reported by direct communication links to a central, national data storage and processing system.

Finally, there is the possibility of providing the nuclear industry with expert protective security services in such a way as to assure the greatest practicable security as well as an equitable means of paying added costs. If this service—including the required professional personnel, equipment, communications systems, and so on—were provided by the federal government, where and as needed, the industry would not have to deal with the problems of arming its own employees or depending on private protective services whose effectiveness may vary considerably from place to place. Furthermore, if the costs of such a service were paid for out of a federal fund raised by taxes on the utilities that were proportional to their annual production of nuclear power, and these taxes were included in the utilities' pricing structure, the costs of the required protective services could be equitably passed on to the consumers of nuclear electric power.

Preliminary estimates of the operating costs of safeguarding the components of an 80,000 MW(e) LWR plutonium recycle power system against plutonium theft or diversion and against sabotage, to a level of security considerably greater than that called for by present (early 1975) Nuclear

Regulatory Commission requirements, correspond to less than 1% of the costs of generating electric power with the system (7). These cost estimates include a 16% per year allocation for investment charges for capital equipment for safeguards, the capital costs of which are estimated to be about 0.5% of the capital costs of the power system. The power system consists of 20 4000 MW(e) power plants, a reprocessing plant, a fuel fabrication plant, a uranium enrichment plant, and a high level waste storage facility, all separately sited, along with rail transportation systems between all components of the fuel cycle. The Principle of Containment has been applied at all sites. All shipments of fuel and plutonium oxide are in approximately 100-ton casks specifically designed to resist penetration by sophisticated means long enough to allow law enforcement and, if necessary, military forces to arrive at the scene of a theft before plutonium can be removed from the casks. The total number of physical security personnel employed for safeguards purposes is about 800, of which 160 are on duty at any given time.

It thus appears that the technical problems in achieving highly effective nuclear safeguards can be solved at acceptable costs. Whether or not the related institutional and political problems can be solved, worldwide, in time to prevent nuclear violence from reaching unacceptable levels remains to be seen.

Literature Cited

1. Leachman, R. B., Althoff, P., eds. 1972. *Preventing Nuclear Theft: Guidelines for Industry and Government.* New York: Praeger. 377 pp.
2. Willrich, M., ed. 1973. *International Safeguards and Nuclear Industry.* Baltimore, Md: Johns Hopkins Univ. Press. 307 pp.
3. Comptroller General of the United States. 1973. *Improvements Needed in the Program for the Protection of Special Nuclear Materials.* Rep. to Congr. by Gen. Account. Office, Washington DC; 1974. *Protecting Special Nuclear Materials in Transit: Improvements Made and Existing Problems (B-184105)*
4. McPhee, J. 1974. *The Curve of Binding Energy.* New York: Farrar, Straus & Giroux. 232 pp.
5. Willrich, M., Taylor, T. B. 1974. *Nuclear Theft: Risks and Safeguards.* Cambridge, Mass: Bailinger. 252 pp.
6. Rosenbaum, D. M. et al 1974. *Special Safeguards Study.* Prepared for the AEC's Director of Licensing, spring 1974
7. U.S. Atomic Energy Commission. 1974. *Tech. Rep. WASH-1535. Proposed Final Environmental Statement, Liquid Metal Fast Breeder Reactor Program.* Washington DC: US AEC
8. Paxton, H. C. et al 1964. *Tech. Rep.* LAMS-3067 (Los Alamos Sci. Lab.) Washington DC: US AEC
9. Glasstone, S., ed. 1962. *The Effects of Nuclear Weapons.* Washington DC: US AEC.
10. Hall, D. See Ref. 1, 275–8
11. Taylor, T. B. 1974. *Tech Rep.* IRT-363-R. Arlington, Va: Int. Res. Technol. Corp.
12. International Atomic Energy Agency. 1974. *Power and Research Reactors in Member States.* Vienna: IAEA

LOW LEVEL COUNTING TECHNIQUES ×5566

H. Oeschger and M. Wahlen

Physikalisches Institut, Universität Bern, Bern, Switzerland

CONTENTS

1 INTRODUCTION

1.1 *Spectrum of Problems Requiring Low Level Counting Techniques*

The term *low level counting* (LLC) is widely used whenever "small" activities of radionuclides are to be measured by direct observation of the radioactive decay. The unique assay of a specific radioactive transition requires the identification of the emitted radiation and its energy (and possibly its half-life). It is impossible to specify a general boundary of activity below which the measuring procedure requires LLC techniques. We reserve the term LLC for problems and applications

423

where results of sufficient precision may not readily be obtained with conventional radiation detection systems in reasonable counting times. This is generally the case when source count rates are of the order of ≤ 1 count per minute (cpm). Special effort is needed to choose and adapt detector systems to attain high counting sensitivity and to keep instrument background as low as possible.

The most important application of LLC is in the field of natural radioactivity. The determination of radioisotope concentrations in terrestrial and extraterrestrial environments is, however, not only a matter of sensitive counting. Since the radioisotopes are highly diluted in their natural reservoirs, radioactivity measurements are generally possible only after considerable efforts in physical and chemical concentration and sometimes even isotopic enrichment.

Though similar in methodological aspects the expression LLC is

Table 1   Studies based on measurement of natural and man-made radioactivities

Field	Topics	Isotopes
Cosmic rays	Fluxes and spectra of galactic CR and solar energetic protons; erosion rates of lunar samples and ablation of meteorites; radial gradient in the CR intensity away from the sun; lunar gardening and surface effects; irradiation ages (in conjunction with stable isotopes)	Isotopes with different half-lives in moon and meteorites; depth profiles, e.g. ^{37}Ar, ^{7}Be, ^{56}Co, ^{54}Mn, ^{55}Fe, ^{22}Na, ^{60}Co, ^{39}Ar, ^{59}Ni, ^{81}Kr, ^{36}Cl, ^{26}Al, ^{10}Be, ^{53}Mn, (^{129}I, ^{244}Pu)
Cosmic dust	Global influx rate; irradiation history in solar system	^{53}Mn, ^{59}Ni, (and ^{26}Al) measurements on samples from ice and sediments
Atmospheric phenomena	Interaction of cosmic rays with atmosphere; history of cosmic rays; modulation of cosmic rays (solar cycle, earth's magnetic field); atmospheric mixing; behavior of water vapor in atmosphere; behavior of aerosols in atmosphere	CR-produced isotopes in atmosphere, hydrosphere, biosphere, and lithosphere: ^{32}P, ^{33}P, ^{37}Ar, ^{7}Be, ^{35}S, ^{22}Na, ^{3}H, ^{39}Ar, ^{32}Si, ^{14}C, ^{81}Kr, ^{36}Cl, ^{26}Al, ^{10}Be; ^{55}Fe, ^{85}Kr, ^{90}Sr, ^{137}Cs, and other man-made activity; Rn isotopes
Oceanography	Air-sea interchange; ocean mixing and circulation; behavior of tracers	^{7}Be, ^{3}H, ^{39}Ar, ^{32}Si, ^{14}C, ^{55}Fe, ^{90}Sr, ^{137}Cs; ^{222}Rn, ^{226}Ra, and other isotopes of U and Th series
Hydrology	Quantitative discussion of hydrological systems; dating of groundwater, mixing of lakes	^{3}H, ^{85}Kr, ^{39}Ar, ^{32}Si, ^{14}C, (^{234,238}U)
Glaciology	Dating of glaciers and polar ice; accumulation rates; behavior of impurities	^{3}H, ^{210}Pb, ^{39}Ar, ^{32}Si, ^{14}C, ^{36}Cl; fission products
^{14}C-dating	Archaeology, paleobotany, geomorphology, sedimentology	^{14}C
Geological dating	Earth's crust	^{87}Rb/^{87}Sr, ^{40}K/^{40}Ar, ^{187}Re/^{187}Os, U/Pb
Biology	Pathway of radioactive tracers in living organisms; nuclear medicine; radiological effects	Man-made radioisotopes
Environment	Study of behavior of tracers in vicinity of nuclear power plants and nuclear fuel reprocessing plants; global levels of radioactivity	^{3}H, ^{14}C, noble gas isotopes, fission products

generally not used to identify and measure particles in accelerator experiments; therefore these problems are not discussed here.

Following are some of the major problems and studies that require LLC techniques; we summarize these applications in Table 1.

Radioisotopes surviving from the time of nucleosynthesis and their daughter products   Here U and Th isotopes and ^{40}K, together with their daughter nuclei, are of special interest. Most of the isotopes of high proton number show α (and/or β decay), followed by γ emission. In connection with stable isotope analyses, these long-lived radioisotopes and their short-lived daughter nuclides allow age determinations on geological time scales. In addition, radiochemical studies of these isotopes yield information on the differentiation processes in the evolution of planetary bodies (1, 2).

Cosmogenic radionuclides   Specifically, we mean cosmogenic radio-nuclides that are produced by interaction of cosmic radiation with meteorites, the lunar surface, cosmic dust, and the earth's atmosphere. Two sources are known to produce such radionuclides: galactic and solar cosmic ray particles.

The dominant mechanism of radionuclide production from galactic cosmic rays is spallation reactions induced by high energy particles incident on meteorites, the moon, and the earth's atmosphere. The most abundant secondary nucleons, the neutrons, give rise to a variety of radionuclides by (n, x) or, after being moderated, by (n, γ) reactions. The effect of solar cosmic rays (generally much lower in energy) can be observed in samples from the top of the lunar surface and in cosmic dust by the detection of radioisotopes produced in low energy reactions, whereas this effect is essentially absent in the atmosphere (because of magnetic shielding) and in meteorites (because of ablation of surface layers). Most of these cosmic ray-produced isotopes can be identified by the detection of the emitted radiation (β^+, β^-, X rays, Auger electrons, γ rays, and annihilation quanta).

Studies on these cosmogenic nuclides yield information on the intensities and spectral shapes of the radiations. The scatter in half-lives (half-lives of several days to some 10^6 yr) of the detectable radionuclides allows the investigation of possible variations in the production sources, i.e. the history of the intensity of cosmic radiations (CR) over the last few million years. On the other hand, irradiation ages can be determined for the bombarded bodies, such as meteorites and lunar samples (3–9). Sun-earth phenomena (solar cycle variations of CR intensity and their possible relationship to climatic changes) and terrestrial phenomena (history of earth's magnetic field) can be studied by investigations of atmospheric radiogenic nuclides (10).

Man-made radioisotopes produced by nuclear weapon tests   The natural equilibrium activity of radionuclides in the earth's atmosphere has been drastically disturbed by atmospheric and underground tests of nuclear weapon devices since 1945 (first fission bomb test) and 1952 (first fusion bomb test). Large quantities of radionuclides have been released and consequently spread globally. Some of these isotopes are fission products; others originate from neutron reactions with constituents of the earth's atmosphere and crust. Even assuming a continuing complete test ban, the global ^3H and ^{14}C inventory would remain altered by the man-made input for at least the next few decades according to the residence times of these isotopes in terrestrial reservoirs; measurement of natural isotope production is therefore impossible. The artificial input of man-made radionuclides may be used, however, to study atmospheric circulation and mixing, and exchange with other reservoirs (11–16).

Radioactive isotopes released into the atmosphere by the nuclear industry Nuclear power plants and reprocessing facilities for nuclear fuel are known to release considerable quantities of radioisotopes. Reliable surveillance of the dispersion of these isotopes into the environment requires sensitive detection techniques (11, 12).

Applications in biological sciences   There is a growing need for sensitive techniques for measuring radioactive isotopes used in biological studies. In addition to being required for basic research on radiological health effects, such techniques are becoming more and more necessary in tracer experiments in which work with very low levels of activity is carried out.

Studies on rare processes in nuclear physics   LLC methods are often needed in studies in nuclear physics, such as determination of long half-lives [spontaneous fission, double β, or electron capture (EC) decay], decay schemes, and identification of solar neutrinos.

1.2   Summary of Recent Technological Developments

A number of improvements in LLC techniques have recently been achieved that enable more and more delicate applications to be undertaken. There have been several improvements in gas proportional counters. The realization of the so-called rise-time-discrimination method (RTD) was a big step forward in background reduction (37–47). By carefully analyzing the timing (pulse shape) of the acquired signals, one is able to separate background events from actual sample decays. The resulting background reduction obtained by this method ranges from a factor of 5 to ~ 20 (see Section 3). Operation of proportional counters at very high pressure increases efficiency. Furthermore, the "on-line" data acquisition by small computers allows the operation of "batteries" of counters under stable and controlled conditions over long periods of time and more readily allows the recognition and rejection of spurious counts by repetitive readout and consequent statistical tests.

In solid-state detectors, LLC techniques have profited, with respect to the background problem, from the excellent energy resolution of present devices (for X- and γ-ray applications). The sensitivity is being greatly improved as larger and larger Ge crystals become available.

The liquid scintillation technique has been considerably advanced by improvements in low noise photomultiplier tubes and the development of scintillating solutions that yield high counting efficiencies even at low scintillator concentrations.

In addition to such technical developments and refinements, the background count rates of LLC systems have been drastically reduced by the operation of these devices in deep underground laboratories. This is especially true when underground facilities are carefully built (using special low activity concretes) to avoid any radiations contributed by construction materials.

1.3   Outline of Organization

It is impossible to discuss in one article all the fields of LLC applications along with the associated literature. The selection of topics presented here is of course influenced by the authors' fields of specialization. We do not give a detailed review of all LLC applications, techniques, and methods in use, but we do present the basic principles and guidelines that must be considered in work in this field (Section 2). Representative topics of major concern are therefore discussed with selected examples. In Section 3 we summarize the most common detector systems and discuss the parameters important to their application, such as sensitivity and background count rates. LLC techniques are also compared to other analytical procedures.

In Section 4 we give a representative sample of the spectrum of applications of LLC in various scientific disciplines. Several typical LLC systems are reviewed, together with some of the results obtained. Finally, we present an overview of future possibilities for applications of low level counting techniques.

2   PRINCIPLES OF LOW LEVEL COUNTING

Most LLC applications require combined optimization of both the radioactivity measurement and the sampling and sample preparation.

Generally, the first requirement is a counting device of high quality performance, specifically tailored for the radiation to be detected. The accuracy and thus the sensitivity of the measurement is usually determined by the Poisson error, which depends on the achievable net source count rate and is influenced by the presence of extraneous counts (or background). The rate of background counts is determined by measuring a "dummy source" that is identical to the actual unknown source in all respects but the presence of the isotope in question. The activity of the unknown sample is then measured by comparing its net source count rate to that obtained from a standard sample of known activity, which is identical to the unknown sample with respect to chemistry and geometry.

The occurrence of natural isotopes is highly diversified, and they are mostly very much diluted in their natural environment. The sampling technique must ensure sufficient amounts of activity and guarantee an accurate and representative collection from the parent material. Any subsequent physical or chemical treatment should increase the specific activity of the sample and lead to a form suitable for counting in the chosen system. The yield must be controlled and kept at maximum, and any contamination must be avoided.

In the following we discuss some criteria for counting statistics, which allow the estimate of the lowest sample count rates that can be determined in the presence of a given background count rate. On the basis of background count rates of typical LLC systems we then define the requirements of sampling and sample processing.

2.1   Statistical Considerations

The general problem of counting statistics has been treated among others by Loevinger & Berman (17) and Weise (18). Special attention to the subject of detection limits has been given by Currie (19).

A commonly used simple concept for evaluating detector systems is the minimum combined counting time T that must be spent counting source and background to obtain a specified precision in the net source counting rate. The *figure of merit* (FOM) of a counting system can then be defined as $1/T$.

First we assume that the accuracy is determined only by the Poisson error of counting and exclude additional sources of error, for example, from radioactive contamination, unstable counting efficiency, isotopic fractionation, background variations with fluctuating muon flux, and spurious counts.

The following terms are defined: S = net counting rate of unknown source; B = counting rate of blank (background); T_{S+B} = counting time for unknown source, T_B = counting time for blank, $\varepsilon = \sigma_S/S$ relative standard deviation in S (resulting from statistical variations in counting). The standard deviation in S is equal to

$$\sigma_S = [(S+B)/T_{S+B} + B/T_B]^{1/2}. \qquad 1.$$

It can easily be shown that for any given counting precision the combined counting time $T = T_{S+B} + T_B$ becomes minimum if $T_{S+B}/T_B = [(S+B)/B]^{1/2}$. The FOM then becomes

$$1/T = \varepsilon^2 \cdot \frac{S^2}{[(S+B)^{1/2} + (B)^{1/2}]^2} = \varepsilon^2 \cdot \frac{S^2}{4B} \cdot \left(\frac{2}{(1+S/B)^{1/2} + 1}\right)^2 \qquad 2.$$

The following special cases are of interest: If we assume that the net counting rate is small compared to the background counting rate, we get for the FOM $1/T \approx \varepsilon^2 \cdot S^2/4B$, i.e. proportional to S^2/B. If we compare alternative counting systems in a range where $S \ll B$, the statistically most favored system is the one with the maximum value of S^2/B. On the other hand, when $S \gg B$ the background count rate does not affect the statistical error and we get $1/T \approx \varepsilon^2 S$.

The FOM generally used in low level counting is $S^2/4B$. The FOM of a system can be improved either by background reduction, or, more importantly, by an increase in the sample count rate, i.e. by an increase in the absolute counting efficiency and/or by use of a larger sample. In many LLC problems the amount of sample material (with activity A) is limited and the appropriate FOM becomes $e^2 f^2 A^2/4B$, or $e^2 f^2/4B$, where ef is the overall counting efficiency of the sample (e = detection efficiency cpm/dpm, f = yield of sampling and source preparation procedure). Using $S^2/4B$ as an estimate of the FOM is incorrect insofar as, compared with equation 2, it favors those systems with a relatively low ratio of S/B but a high net count rate S for the sample. If we compare two counting systems with equal FOM = $S^2/4B$ but different S/B values, e.g. 1 and 0.1, we see first that the actual FOM (equation 2) is 1.4 times higher for the system with the higher S/B ratio. Second, assuming equal overall efficiencies ef for the two systems, only one third of the sample material is needed for the system with the higher ratio S/B to obtain the same precision. Third, we can have more confidence in the measurements if $S \approx B$ than if $S = B/10$ because of the non-Poisson errors involved.

According to equation 2 one may calculate the minimum net sample count rate S, which must be achieved to obtain a measurement of prescribed precision ε, in the counting time T, in presence of the background count rate B:

$$S = (1/(\varepsilon T^{1/2}) + 2B^{1/2})/(\varepsilon T^{1/2}).$$

Figure 1 gives calculated minimum count rates $S(B)$ for $\varepsilon = 0.1$ and for T of 20–20,000 min.

It is difficult to treat the non-Poisson errors with a simple formalism. Spurious counts (e.g. electric breakdowns in insulators, noise in the power supply, pickup of electromagnetic noise) are generally not distributed stochastically with time. The following formalism is therefore used only to demonstrate in principle further error problems in LLC. We assume that $S \ll B$ and therefore that $T_{S+B} \approx T_B \approx T/2$. Again we analyze the relative standard deviation in S, but include a non-Poisson error component in the background that is randomly distributed with a standard deviation σ_{BN}. The total standard deviation σ_{stot} of the net counting rate then becomes

$$\sigma_{\text{stot}} = \left(\frac{(S+2B)}{T/2} + \sigma_{BN}^2 \right)^{1/2} ,$$

as long as $T \ll 2(S+2B)/\sigma_{BN}^2$ the Poisson error dominates and the above considerations hold. If, on the other hand, the total counting time is increased, the non-Poisson error starts to contribute significantly to the total errors. We see that increasing the counting time much above $2(S+2B)/\sigma_{BN}^2$ is of no use.

Part of σ_{BN} may be proportional to the background count rate B (e.g. due to CR flux variations); another part may be proportional to S (e.g. due to isotopic fractionation effects); still another component (e.g. spurious counts) may be independent of the sample to be analyzed. This component is called σ_{BNI}. The non-Poisson error of the background may then be written as

$$\sigma_{BN}^2 = (\alpha B)^2 + (\beta S)^2 + \sigma_{BNI}^2$$

and the relative total standard deviation of S becomes

$$\frac{\sigma_{\text{stot}}}{S} = \left[\frac{2(S+2B)}{T \cdot S^2} + \left(\alpha \frac{B}{S} \right)^2 + \beta^2 + \frac{\sigma_{BNI}^2}{S^2} \right]^{1/2} .$$

The non-Poisson error components become more and more important with increasing T. With regards to the error component that is assumed to be proportional to B, the system with the higher S/B ratio is preferable, for equal α values. As for the error component proportional to S, both systems are equal, for equal β values. If both systems have equal values of σ_{BNI}, the system with the higher S is preferable. In designing LLC systems it is imperative to eliminate spurious counts by shielding against pickup of any kind of electromagnetic noise and by properly solving all isolation problems.

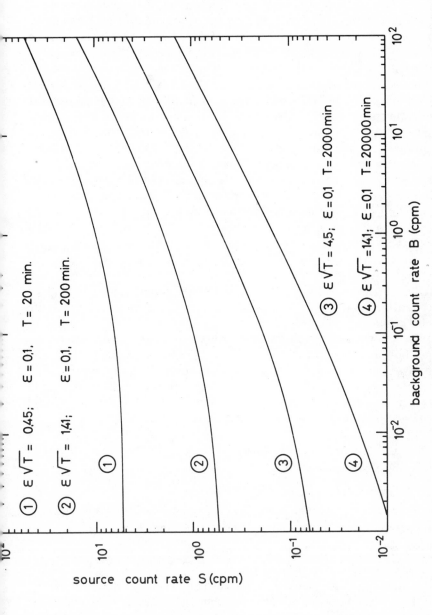

Figure 1 Minimum count rates $S(B)$ for relative standard deviation ε, in total counting time T in presence of background count rates B.

Another approach to evaluating counting systems is through detection limits. Currie (19) considered in detail the various possible definitions. He gives three limits—critical level, detection limit, and determination limit; we briefly discuss the first and third. The *critical level* $S_c = k_c \sigma_0$ is the lowest net count rate that one can statistically distinguish from zero with a given confidence level. Here σ_0 is the standard deviation of the net signal for a sample of zero net activity; $k_c = 1.64$ for a 95% confidence level (one-sided confidence interval). For very long counting times the Poisson errors vanish, and $\sigma_0 = \sigma_{BNI}$, i.e. the minimum counting rate one can statistically distinguish from zero is determined by non-Poisson background count variability. For quantitative work, one must know a result to a reasonable accuracy. The *determination limit* is defined as the lowest activity that can be measured with a prescribed relative error, i.e. $S_Q = k_Q \sigma_Q$, where S_Q is the corresponding net count rate, σ_Q its standard deviation, and $1/k_Q$ the desired relative standard deviation.

The conclusions from these error considerations can be summarized as follows: From a purely Poisson-statistical point of view, the FOM's $S^2/4B$ or $S^2/B[(1+S/B)^{1/2}+1]^2$ can be derived. If the amount of sample material activity is limited, $(ef)^2/4B$ is the appropriate FOM, and since (ef) is in many cases close to 1 (internal proportional counting) or cannot be considerably increased, it is essential to aim for a minimum background.

In extreme LLC problems (background count rates of a few counts per hour and less) the counting limits are determined by non-Poisson background contributions like spurious counts. An important criterion for the quality of an LLC system is therefore how well it is designed to avoid or eliminate counts that are not caused by the interaction of charged particles with the sensitive material of the counter.

In the following section on sampling we emphasize that the net count rates S of the samples to be analyzed should, if possible, be at least of the same order of magnitude as the background count rate B.

2.2 Sampling and Counting Strategy

In evaluating the feasibility of an LLC experiment, the sampling strategy, sample preparation procedures, and detection methods are chosen according to the following criteria:

1. decay scheme of the isotope [emitted radiation(s), energy, half-life];
2. nature of sample material (occurrence and environment, availability);
3. range of achievable counting efficiencies and background count rates;
4. amount of sample to be collected.

Nondestructive external γ-ray counting, requiring minimum sample preparation, is performed on samples with relatively high specific activity. Typical photopeak efficiencies for noncoincident γ-ray counting range from $\sim 50\%$ to $\sim 10\%$ (NaI, 0.3–1.3 MeV) and from about 15% to $\sim 1.5\%$ (GeLi). The background count rates are of the order of ~ 1 cpm cm^{-3} of detector material for the energy interval of 0.1–2 MeV. Energy resolutions are typically 8% (NaI) and $\sim 1.5\%$ (GeLi).

Internal counting should be considered whenever a sample can be prepared so that it can be used as the counting gas of a proportional counter, or as all or part of the sensitive liquid of a liquid scintillation counting system. Internal gas counting is advantageous especially when the samples are of low specific activity and emit radiations of short range, i.e. X rays, α particles, and low energy β particles (e.g. ^{14}C, ^{3}H, radioactive noble gas isotopes). This technique offers high counting efficiencies generally of 80% or more. As the background of internal counting systems increases to some degree with increasing amount of sensitive detector material, the background count rates must be expressed in cpm per gram of detector substance and are typically of the order of 1 cpm g^{-1} for internal gas counting, but in some cases can be as low as 0.1 cpm g^{-1}. For internal liquid scintillation counting the efficiencies obtained are similar and the backgrounds are of the order of 1–5 cpm g^{-1}.

When internal counting is not applicable the samples must be counted externally, i.e. outside the detector not forming a part of its sensitive material. For example, samples can be put on the inside wall of Geiger counters or proportional chambers; or, if they are placed outside, the radiation can be allowed to penetrate into the detector through thin windows. External counting is applicable when the range of radiation to be analyzed exceeds some 10 mg cm^{-2}. Sample and detector geometry (2π, 4π arrangement) and self-absorption (of the radiation by the sample itself) are important parameters affecting counting efficiency. Sample preparation techniques should provide sources of the highest possible specific activity. We can analyze α, β and X rays by external counting with efficiencies between 50% and 100%. Typical overall backgrounds are between 1 and 0.1 cpm (or 10^{-1} and 10^{-3} cpm cm^{-2}). Energy resolution is about 15–20%.

It should be noted that the generalized background values and efficiencies given above are somewhat artificial. The discussion is valid, however, for establishing the order of magnitude of the absolute and specific activities that are desirable for counting a given isotope in a given sample. From Figure 1 we see that for background count rates of 0.1–5 cpm, a relative accuracy between 0.01 and 0.03 can be obtained in a measurement carried out for 20,000 min, if the source count rate S is 1 cpm. From this we conclude that for overall efficiencies ef close to

100%, the absolute activity of a sample should be of the order of 1 dpm and the specific activity of the order of 1 dpm g^{-1}.

The specific activity of radionuclides found in terrestrial and extra-terrestrial matter (air, water, ice, rocks, organic material, etc; meteorites, cosmic dust, lunar samples) is generally very small and only in a few cases can the samples be measured as originally collected without any further concentration.

In lunar and meteoritic material the concentrations are between 1 and 10^{-3} dpm g^{-1}. In the air (troposphere) the concentrations are between several times 10^{-2} dpm (g air)$^{-1}$ (fission-produced ^{85}Kr, ^{3}H in ^{3}HH and in ^{3}HHO, ^{14}C in CO_2) and 10^{-7} dpm (g air)$^{-1}$ (^{81}Kr, ^{22}Na). For hydrological and oceanographic studies, isotopes with specific activities between 10^{-2} (^{3}H) and 10^{-8} (^{32}Si and ^{39}Ar) dpm g H_2O^{-1} are of interest; and in glaciology one has to deal with concentrations of the order of 10^{-1} dpm (g ice)$^{-1}$ (^{3}H and fission products), 10^{-7} dpm (g ice)$^{-1}$ (^{39}Ar, ^{14}C, ^{32}Si), and even 10^{-9} dpm (g ice)$^{-1}$ (^{53}Mn in cosmic dust). More details may be found in (3, 4, 20).

The necessary increase in specific activity must be achieved by bulk concentration. Currently used extraction techniques for traces from solid material are, among others:

1. vacuum melting of extraterrestrial material to extract noble gas isotopes and ^{3}H;
2. vacuum melting of tons of ice in bore holes to extract gases (noble gases, CO_2 for ^{14}C dating);
3. dissolution and chemical separation of elements from sediments, rocks, and extraterrestrial material;
4. combustion of organic material to produce CO_2 and concentrate solid impurities in the ash (e.g. fission products).

Gases and solid and dissolved matter must be isolated from water and melted ice for hydrological and glaciological studies. Commonly used procedures are:

1. extraction of gases under reduced pressure or by purging with a carrier gas;
2. extraction of particulates by filtration (e.g. aerosols, cosmic dust in ice);
3. ion exchange "filtering" of elements that are present in ionic form;
4. use of natural biological processes for sample concentration (e.g. ^{32}Si extracted from SiO_2 of sponges or ^{131}I and ^{90}Sr concentrated in milk).

Procedures for bulk concentration of atmospheric constituents are:

1. filtration of aerosols;

2. adsorption on (cooled) charcoal or molecular sieves;
3. separation based on differing boiling and solidification points.

In addition to bulk separation, chemical purification procedures are necessary, especially when interfering activities may still be present in the sample. Separation according to elements is often mandatory. Procedures are volatilization, precipitation and coprecipitation, solvent extraction, chromatography and selective electrodeposition, etc.

Isotope separation may be necessary in ultimate LLC applications. The most common isotope enrichment methods in use for LLC are thermal diffusion and electrolysis. Thermal diffusion has been used in ^{14}C dating to extend the conventional dating range to 75,000 years in order to cover the last interglacial period (20a). Enrichment by a factor of 16 allows for an additional 23,000 years of dating. Special attention, however, must be given to the contamination problem. An admixture of only 0.1% of recent carbon completely falsifies the results in this dating range. Extremely low specific activities appear in ^{37}Ar measurements on atmospheric Ar samples. The accuracy of these measurements can be increased by thermal enrichment. ^{3}H in water and ice samples originating from the time before the fusion-bomb tests can only be measured (with satisfactory accuracy) if the ^{3}H is enriched prior to counting. The most widely used technique is electrolysis, which gives enrichment factors in the range $10-10^{3}$. Another separation technique of growing importance is the use of electro-magnetic separator machines. A throughput of $\sim$mg h^{-1}, a yield of $\sim 10\%$, and a mass resolution of $\sim 10^{3}$ are reported (20b).

In all these sampling and concentrating procedures, which are necessary for a quantitative measurement of the required accuracy, extreme care must be taken to avoid any contamination by naturally and artificially produced radionuclides. The latter are frequently found in our present environment. Therefore all reagents and carriers must be checked for radioactive contaminants prior to use.

3 EXPERIMENTAL METHODS

There is a growing literature on experimental solutions to LLC problems. Excellent reviews of this subject and many experimental details may be found (e.g. 21–23).

The basic problem is to find an experimental array that allows both the quantitative measurement of low activities of radionuclides in a given sample and the unique identification of the radioactive transition according to radiation and energy. The decay mode and physical properties of the sample determine the choice of the detector system. The guiding principles are the achievement of high sensitivity, low background, and good energy resolution. No general recipe can be given, because many determining factors are interrelated and the specific experimental solution depends on the given problem. Nevertheless, in the following sections we summarize some basic aspects to be considered when assaying low level radioactivities.

3.1 Detection Methods

3.1.1 GAS DISCHARGE COUNTERS
An ionizing particle moving through the gas of a Geiger-Müller counter (GMC) or a proportional counter (PC) gives rise to a number of electrons and an equal number of positive ions along its track. The electrons are drawn to the positively charged anode wire. Close to the anode wire the electric field strength increases rapidly in magnitude and the electrons acquire enough energy between collisions with atoms or molecules of the counting gas to produce multiplication. In the case of the GMC a gas discharge is started along the anode wire, which results in a pulse that is independent of the number of track electrons (24, 25). Already one electron can produce a signal that is well above the amplifier noise. In PC each track electron is multiplied close to the anode wire by the same gas amplification factor and a signal proportional to the number of track electrons is produced. The energy required to produce an ion pair in gases is nearly independent of the energy of the charged particle and ranges between 20 and 30 eV (Table 2). The fact that in PC the signal is proportional to the energy deposited by the particle to be detected is essential for radiation identification and background reduction.

PCs are used for the assay of X-ray and low energy γ-ray emitters from external sources. For a 1 atm argon-methane mixture, frequently used

Table 2 Relevant data for media used for γ- and X-ray detectors

Material	Atomic Number	Density (g cm^{-3})	Average Energy Required to Create Charge Carrier (eV)	Energy Resolution for Photons (%) in Energy Range of MeV or keV
NaI	11, 53	3.7	300	8 (MeV)
CsI	55, 53	4.5	400	10 (MeV)
Ge	32	5.5	2.9	0.15 (MeV)
Si	14	2.3	3.7	(keV)
Ar	18	1.8×10^{-3}(STP)	26	15–20 (keV)
Xe	54	5.9×10^{-3}(STP)	22	15–20 (keV)

in proportional counting, the useful energy range is below 20 keV. The energy range may be increased by using heavier gases like Kr and Xe (with higher cross sections for photoionization) at higher pressures.

For the measurement of solid external β sources, GMC and PC compete with plastic scintillation counters and other solid state detectors, which can easily be built according to special requirements regarding sample size and geometry. Such counters may be constructed from lucite with windows of gold- or aluminum-coated mylar (~ 1 mg cm^{-2}). They are operated with a flow gas at atmospheric pressure. When operated in a γ shield and in anticoincidence with a muon detector they show background count rates of a few counts hr^{-1} cm^{-2}. Sandwich arrays together with absorbers are able to resolve β radiations of different energies originating from the same source. A further application may be found in coincidence systems (β-γ, β-γ-γ, etc) together with NaI crystals, plastic scintillators, or GeLi detectors.

The most important LLC application of PC is the measurement of β- and X-ray (EC) emitters in samples of low specific activity. If such samples (e.g. natural ^{3}H and ^{14}C) are counted externally, only a small portion of the emitted short range radiation can reach the detector, and the counting efficiency and FOM $e^2/4B$ are small. If such samples, however, are transferred into a gas that can serve (with admixtures of other gases) as the counting gas of a PC or GMC (internal counting), high efficiencies between 80% and 95% are obtained. GMC and PC for such gas sources are usually metal, plastic, or quartz cylinders with volumes ranging from 0.1–10 liters according to the sample sizes. They are operated in a shield and are surrounded by an anticoincidence system to reduce the γ and muon background. For very low energy β emitters like ^{3}H ($E_{max} = 0.0186$ MeV), Auger electrons, and photoelectrons from electron-capture X rays, the particle tracks are small compared with the dimensions of the counter. Therefore the energy deposited in the counter by these radiations is in most cases equal to their total energy. For EC events, lines with a resolution of 15–20% are observed in the pulse height spectra, whereas the low energy β particles occupy well-defined energy bands. Since the background radiation spreads over a wide energy band it is possible to raise the ratio of source counts to background counts by examining the pulses in the source band only.

Internal proportional counting was used for example for the detection of ^{37}Ar and ^{81}Kr in atmospheric Ar and Kr (26). Both decay by EC. Pulse height analysis enabled the identification of the energy of X rays and electrons emitted in these EC transitions. For isotopes with higher β energies [0.157 MeV (^{14}C), 0.57 MeV (^{39}Ar), 0.67 MeV (^{85}Kr)] counted in PC at pressures of one to several atmospheres, the β-particle tracks are not small compared to the counter dimensions, and pulse height spectra are obtained that cover most of the energy range of the background spectra. Restriction to a portion of the pulse height spectrum may improve S/B but not necessarily the FOM $S^2/4B$. Nevertheless examination of the pulse height spectra is for these cases highly recommended. Comparison of the net pulse height spectrum with that of a standard source confirms that the observed pulses are due to the isotope of interest. The appearance of an increased number of small pulses would indicate the presence of spurious counts. Additional pulses in the energy band above 1 MeV are mostly due to α particles probably emitted by ^{222}Rn and daughters, which often cause contamination problems in internal gas source counting.

The simple working principle of PC inspired a number of authors to attempt an improvement of the S/B ratio for special counting problems. In these constructions straight wires or wire loops are used as anodes; the cathodes are not only cylinders, but also parallel plates, concentric cylinders, single wires, and wire grids (26a,b).

3.1.2 γ-RAY COUNTING   If a radionuclide does not decay directly into the ground state of the daughter nuclide, but the transition occurs with sufficient yield via one or several excited states, the quantitative measurement of the abundance is preferentially performed by γ-ray spectroscopy. This is because the penetrating nature of γ rays facilitates source preparation, and the high resolution of γ-ray detectors allows conclusive isotope identification and easy discrimination among nuclides.

The most important media for γ- and X-ray spectrometers are listed in Table 2, together with densities and atomic numbers. A γ ray of a given energy will be recorded in the "full energy peak" or "photopeak" if it is totally absorbed within the detector, independent of whether the interaction occurs in a single photoelectric absorption or in a sequence of Compton and photoelectric interactions. If the γ ray is not totally absorbed, its record will be found in the Compton continuum at correspondingly lower energies. Absolute and relative probabilities for the various interactions depend on the material and dimensions of the detector and the energy of the incident photon. Geometry strongly affects the intrinsic efficiency (number of photons detected/number of photons incident on the detector), whereas the photofraction or peak/total ratio is more dependent on the material than on geometry.

The energy resolution is primarily determined by the statistical fluctuations in the number of charge carriers created upon interaction of γ rays with the detector medium, and thus depends on the average energy required for the production of a photoelectron or ion pair, or electron-hole pair (see Table 2). Additional effects come from charge carrier collection and/or from the multistage conversion mechanisms into a final electrical signal.

Prior to about 1965 the assay of γ rays was performed mainly with scintillation detectors of the NaI or CsI type. These detectors offer a very high efficiency and can be built to extreme sizes, whereas the energy resolution $\Delta E/E$ is moderate (8–10%). When high resolution is not essential (no interfering isotopes) but ultimate detection sensitivity is required, these devices are preferable for many LLC applications, especially when the radiation can be detected using the coincidence technique or when samples are voluminous or bulky. One of the largest and most sensitive γ-ray spectrometers (NaI) is described by Eldridge et al (27 and references therein) for the nondestructive counting of radio-nuclides in lunar samples.

In recent years the performance of semiconductor detectors has continuously improved in both efficiency (i.e. volume) and resolution. These semiconductor detectors have replaced scintillation detectors in many applications. LLC techniques profit from the excellent resolution of present devices (≤ 2 keV FWHM at 1.3 MeV). γ rays from different or interfering isotopes can easily be separated, so that sample purification efforts can be reduced. In addition the background count rates are greatly reduced as a result of the high resolution (energy discrimination). Today GeLi detectors 50–100 cm^3 in volume are easily obtained (although their price is high), and they display considerable counting efficiencies. A rough comparison between a $3'' \times 3''$ NaI and a 65 cc GeLi detector (14% efficiency relative to a $3'' \times 3''$ NaI-detector, point source at 25 cm distance) for $E_\gamma = 1.3$ MeV according to the FOM $= S/(R \cdot B)^{1/2}$ ($S =$ net source count rate/unit energy, $R =$ resolution, $B =$ background count rate/unit energy) shows that the semiconductor detector is favored by a factor of more than 5 for noncoincident γ-ray counting.

In selecting a GeLi detector for a given LLC application, the ratio of the cross-sectional area facing the sample to the volume should be maximized. This is because sensitivity depends more on the area, while background count rates are roughly proportional to the sensitive volume.

FOMs for different GeLi systems are discussed in (28–30). The figure of merit FOM (PF) for finding a peak at all in a continuum is $S/RB^{1/2} \approx e/RB^{1/2}$ ($e =$ counting efficiency). It is strongly influenced by the resolution R, since the height of a normal distribution is inversely proportional to its width. The figure of merit FOM (Q) for quantitative determination of an activity is $S/(RB)^{1/2} \approx e/(RB)^{1/2}$. Both expressions are of course energy dependent. In Table 3 FOMs are summarized in terms of practical detector properties according to (30).

Table 3 FOM of GeLi spectrometers as function of detector properties

	Low Energy	High Energy
FOM (Q)	$\dfrac{(\text{area})^{1/2}}{(\text{resolution})^{1/2}(\text{length})^{1/2}}$	$\dfrac{(\text{volume})^{0.79}}{(\text{resolution})^{1/2}}$
FOM (PF)	$\dfrac{(\text{area})^{1/2}}{\text{resolution }(\text{length})^{1/2}}$	$\dfrac{(\text{volume})^{0.79}}{\text{resolution}}$

Generally, for low energy applications planar GeLi detectors up to 40 cm^3 are superior to closed-end coaxial detectors, whereas in high energy applications the performance of planar and coaxial detectors is about equal below 40 cm^3. Above 40 cm^3 coaxial detectors are preferable because of the better resolution.

If samples are of small size (high specific activity), improved sensitivity (for equal background) can be obtained with well-type detectors, which are available for both scintillation and semiconductor devices. As an example, an absolute counting efficiency of 5.5×10^{-2} cpm dpm^{-1} at 835 keV is reported for a GeLi well detector of 45 cm^3 (31).

The background of γ-ray detectors consists mainly of two sources: environmental radioactivity and cosmic ray-induced radiation. Natural radioactivity may be present in the surroundings of the detector or in the detector assembly itself (NaI—K^{40} and products of the natural decay series in wall material of the assembly and photomultiplier tube; GeLi— radioactivity from cryostat, mainly contained in the molecular sieve). Cosmic ray-induced effects, apart from direct muon interaction, result from bremsstrahlung, energetic electrons, annihilation radiation, muonic X rays, and evaporation neutrons.

A shield of Pb and distilled Hg effectively reduces the background due to external radioactivity and partly reduces the cosmic ray-induced component. The most prominent reduction (energy dependent) is obtained when operating the detectors in conjunction with an anticoincidence array made from plastic scintillator material or NaI. This measure in addition strongly reduces the Compton-scattered source count events (anti-Compton shield).

Whenever a radionuclide decays by the emission of associated radiations, it should be detected by the coincidence technique. The coincidence method for γ-γ or β-γ transitions (in conjunction with gaseous β detectors) yields the best results, because the background count rates may be reduced by orders of magnitude, whereas the counting efficiency is only moderately affected. In coincidence applications NaI detectors, because of their high intrinsic efficiency, are superior to semiconductor devices.

Planar semiconductor detectors are the best choice for assaying X-ray emitters of energies above ~ 10 keV. They offer high intrinsic efficiencies and unique resolution. For lower energies the assay should be performed with proportional counters. This is especially true for nuclides of low atomic number that decay by electron-capture transitions directly into the ground state (low fluorescence yield).

3.1.3 LIQUID SCINTILLATION COUNTING The liquid scintillation technique has proven to be a very powerful tool in assaying α, β, and X-ray emitters of moderate to low activity levels, such as ^{3}H, ^{14}C, ^{45}Ca, ^{55}Fe, ^{90}Sr-^{90}Y, and plutonium isotopes (32). Automatic sample-changing liquid scintillation spectrometers are widely used in biological applications to measure labeled compounds. A common feature of such systems is their high counting efficiency and the minimum time and skill required for sample preparation. When specially adapted detector systems are used, this technique can also be extended to detect low level activities in favorable cases.

Modern liquid scintillation systems have the following components: The sample container, containing the scintillation mixture, is placed between two low noise photomultiplier tubes, which are operated in coincidence to reduce photomultiplier noise. This array is packed into a lead shield to suppress γ radiation from the surroundings; the equipment can be operated inside a temperature-controlled freezer. Electronic circuitry allows processing of the coincidence signal according to the decay energy of the isotope in question by one or more single-channel analyzers. External standardization systems are provided.

Although commercially available liquid scintillation systems generally exhibit rather high background count rates, their specific advantages of high sensitivity (close to 100% for β particles) and the large amount of sample that can be dissolved in a few milliliters of modern scintillation fluids, make these devices suited for low activity applications (internal counting). The main sources of background counts are:

1. γ radiation from components and surroundings;
2. noise in photomultipliers;
3. radiation from muon interaction (γ rays and Cerenkov radiation).

The first component can be reduced by using radioactively pure construction material, a lead shield of 5–10 cm, and quartz-walled photomultiplier tubes; noise pulses are effectively eliminated by the coincidence technique with short resolving times; γ-ray and Cerenkov radiation following muon interaction can be partially removed when an anticoincidence counter is operated on top of the array. Typical background count rates of commercially available systems range between 10 and 20 cpm for 25 ml of scintillation mixture in the low energy range (^{3}H and ^{14}C). A summary of background composition may be found in (33).

Major difficulties in operation (stability, reproducibility, variations in background) arise from the effects of phosphorescence, chemiluminescence, and quenching. The effect of quenching is mostly accounted for by measuring an external γ standard, but it is safer to run identical blank and standard samples.

A figure for the sensitivity of liquid scintillation systems has been proposed by Moghissi (32):

$$Y \text{ (pCi g}^{-1}) = B^{1/2}/2.22 \cdot e \cdot M.$$

This value, the minimum limit of detection at σ confidence level and a 1-min counting time, correlates the background count rate B (cpm), efficiency e (cpm/dpm), and the amount of sample M (g) contained in the scintillation mixture.

The following ultimate values of Y (in pCi g^{-1}) can be reached according to (32): ^{3}H, 1; ^{14}C, 0.12; ^{55}Fe, 0.33; ^{85}Kr, 2.0; ^{90}Sr-^{90}Y, 2.0. As an illustration we quote actual LLC values from (34): 5.3 pCi g^{-1} for ^{3}H and 0.32 pCi g^{-1} for ^{14}C have been reached. As can be seen, the latter figure is comparable to that obtained by internal gas counting of about 0.5 pCi g^{-1}.

3.2 The Background of Low Level Counting Systems

3.2.1 BACKGROUND COMPONENTS   The background components of the different types of LLC systems are the same, though their relative importance may differ. One can distinguish two main sources of background counts. The first contribution results from ionizing radiation interacting with the detector material. The second contribution is more varied and consists of pulses of differing origin, e.g. electronic noise, electromagnetic disturbances, spontaneous emission of electrons from photocathodes of photomultipliers, and breakdowns in high voltage insulators.

The effects due to the interaction of ionizing radiations with the detector material are as follows:

1. Cosmic ray muons pass through the detector. A rough estimate for their contribution at sea level is 1 cpm cm^{-2} of detector cross section.

This penetrating background component is only slightly reduced by the shields.

2. An essential background component is the γ rays that interact with the detector material and its walls. They may originate from the soft component of CR; from natural radioactivity of the detector and shielding material and laboratory walls; from bremsstrahlung and energetic electrons produced by muons and other charged particles; from deexcitation of nuclei after CR collisions; from muonic atoms and n capture.

3. A third contribution is from the nucleonic component (protons and neutrons) of cosmic radiation. Protons traversing the detector are registered, as are also recoil nuclei resulting from neutron collision.

4. An additional source of background is α and β particles and photons from radioactive impurities of the detector substance (scintillator, counting gas, semiconductor) or of adjacent detector components and sample contamination.

3.2.2 GENERAL BACKGROUND REDUCTION PRINCIPLES

Shielding Shielding of LLC equipment starts with the choice of the location of the laboratory and with its construction. If the laboratory is situated in the cellar of a multistory building, the effects of the soft and nucleonic components of CR are greatly reduced. With 5 m (or 1000 g cm^{-2}) of concrete above the laboratory, the nucleonic component is reduced by more than a factor of 400, since the attenuation length is ~ 165 g cm^{-2}. On the other hand the 1000 g cm^{-2} of concrete reduce the muon flux only by a factor of 1.4. Significant reduction of the muon (and muon-induced) background component is obtained only in underground laboratories. A main source of the γ component of the background is ^{40}K and the isotopes of the U and Th series in the walls of the laboratories. By selection of special low radioactive material, the γ background in the laboratory can be reduced by a factor of 20 and more (35). In addition the emanation rate of ^{222}Rn from this concrete should be considerably lower than in laboratories built with "normal" construction materials. ^{222}Rn and its daughters are known as a nonconstant source of background. Another advantage of the special concrete is its low production rate of neutrons that result mainly from spontaneous fission of ^{238}U.

LLC systems are generally surrounded by radioactively pure, specially selected material to substantially reduce the γ flux. Materials used are, among others, steel, lead, bismuth, and mercury. A shield consisting of 15 cm of lead or 25 cm of steel provides for a reduction of the γ flux (in the critical energy range) by at least a factor of 10^3. Additional shielding generally results in no additional background reduction, since the background from γ rays of radioactive impurities in the shielding material and the background induced by cosmic radiation interaction in the shield become predominant.

In the search for radioactively pure shielding material the following considerations are important: An excellent shielding material is lead that was mined decades or centuries ago because the radiogenic ^{210}Pb ($T_{1/2} = 20$ yr) has decayed. New lead shows γ contributions from bremsstrahlung from the decay of this isotope. Concerning impurities, different grades of lead may be found, according to the content of natural radionuclides in ores. An inner shield of 2–3 cm of distilled mercury in containers of plastic or stainless steel is sometimes used, because of its high purity. In buildings of light construction, steel is preferable to lead because of the lower multiplicity for neutron production from cosmic ray interactions.

Artificially produced radioactivity may be present in a number of materials. An important example is ^{60}Co. It is used to monitor the wearing of the wall-lining of blast furnaces. Radioactivity used in luminescent dyes may enter materials during recycling.

Evaporation neutrons produced in shields of high Z by proton and muon interactions are commonly removed by borated paraffin wax. About 15 cm are needed so that evaporation neutrons are thermalized and subsequently absorbed.

Background reduction by anticoincidence After successful shielding against the γ component the residual background is largely due to muons traversing the detector. This is especially true for large proportional counters. Pulse height analysis generally does not allow signals induced by minimum ionizing muons ($dE/dx = 2$ keV mg^{-1} cm^2) to be distinguished from actual source signals. This background component is very effectively eliminated if the main detector is surrounded with guard detectors, operated in anticoincidence with the main counter.

The most common guard system is that using GM counters. The disadvantages of a GMC anticoincidence system are the long rise and dead times of their pulses. Guard counters operated in the proportional regime have the advantage of much shorter rise times and practically no losses resulting from dead time. If operated in conjunction with proportional counters, both types show identical sensitivity for spurious counts, for example, which are thus rejected.

A disadvantage of anticoincidence systems working with gas guard counters is their low γ sensitivity. In this respect organic scintillators that can be obtained in large size and virtually any shape are preferred. γ

radiation emitted from radioactive impurities in the anticoincidence detectors is partly rejected too, because the accompanying α or β particles trigger blocking pulses.

The γ background component can be further eliminated by introducing additional shielding material between guard and actual counter. If the detector is surrounded by very pure shielding material a considerable part of the remaining background (after muon elimination) is due to muon-induced γ rays from that part of the shield that is close to the main detector. Signals from the γ rays of the inner shield are accompanied by a muon-induced signal in the anticoincidence array and are therefore canceled.

Further background reduction is possible by discrimination methods, using the following criteria: total deposited energy (pulse height analysis); ionization density (rise-time analysis); range of radiation to be analyzed; simultaneous measurement of more than one particle (coincidence technique). Another improvement comes from underground laboratories. These points are discussed below.

Discrimination methods based on the range of radiation   The range of β radiations of ^{3}H ($E_{max} = 18.6$ keV) and ^{14}C ($E_{max} = 158$ keV) is short compared to that of γ-induced electrons of the background. This difference is used to reduce background in gas source counters. For this purpose the anticoincidence system is incorporated into the counter cylinder as follows: A cylindrical mylar foil, metalized on both sides, is placed concentrically in the counter tube. The inner surface of the mylar cylinder, together with the central anode wire, is the actual counter tube, while the inner wall of the main tube and the outer wall of the mylar cylinder, together with multiple anode wires, serve as the anticoincidence detector. For ^{14}C counting the inner cylinder consists of a 7 mg cm^{-2} mylar foil, which is thick enough to prevent β particles resulting from ^{14}C decay in the inner counter from reaching the anticoincidence volume. Part of the electrons induced by γ rays in the gas or the wall of the inner counter penetrate the foil and are eliminated because they give rise to an anticoincidence pulse. In addition, γ-induced electrons from the wall of the outer cylinder have to cross the anticoincidence before reaching the inner counter and are cancelled too. Oeschger (36) reported a background reduction by a factor of 5 when compared with a counter of similar construction, but with β-thick mylar foil. For ^{3}H counting the inner cylinder can even be replaced by a grid of parallel wires that are at the same voltage as the outer cylinder. The background reduction then is even higher (factor of 10) and because of their very low range only a small percentage of the β particles emitted by ^{3}H nuclei in

Table 4   Background of an Oeschger-type ^3H counter (1.5-liter sensitive volume, 1200 Torr CH_4)

Mode of Operation	Background (cpm)
No shield, no anticoincidence	400
In shield (Fe + Pb)	170
With anticoincidence	0.45
^3H window (energy discrimination)	0.25
In deep Laboratory (60 mwe)	0.05

the inner counter reach the anticoincidence volume and are lost. As shown in Table 4, extremely low backgrounds can be obtained in these systems.

Semiquantitative discussion of background problems and analysis of residual background of gas source counters   The following formulas are of value for the discussion of background problems:

1. The flux F of particles with range R through a surface area A of a semi-infinite medium with source strength S for radioactive particles is $F = (A \cdot R \cdot S)/4$, where [$F$ = flux (particles min^{-1}), R = range (g cm^{-2}), S = source strength (particles $min^{-1} g^{-1}$), A = surface area (cm^2)].

2. The mean path $\bar{s}$ of particles isotropically traversing a convex body is $\bar{s} = 4V/O$, where V = volume of convex body, O = surface of convex body).

These formulas are now used for the calculation of the rate of γ interactions in detectors. We assume the range of the γ rays in the shield to be short compared to the linear dimension of the shield.

The flux $F_{\gamma,D}$ of γ rays from the shield through the surface of the detector is: $F_{\gamma,D} = (O_D/4) \cdot R_{\gamma,s} \cdot S_{\gamma,s}$ (O_D = surface of detector, $R_{\gamma,s}$ = range of γ rays in shield, $S_{\gamma,s}$ = source strength of γ rays in shield). The number $I_{\gamma,D}$ of first interactions of γ rays in the detector then becomes $I_{\gamma,D} = (O_D/4) \cdot R_{\gamma,s} \cdot S_{\gamma,s}\overline{[1 - \exp(-s\rho/R_{\gamma,D})]}$, where s = path of γ rays in detector, ρ = density of detector, $R_{\gamma,D}$ = range of γ rays in detector. In scintillation crystals of huge dimensions $\bar{s} \cdot \rho$ is large compared to $R_{\gamma,D}$ and the number of first γ interactions per unit time is $(O_D \cdot R_{\gamma,s} \cdot S_{\gamma,s})/4$. In gas counters, $\bar{s} \cdot \rho$ is generally small compared with $R_{\gamma,D}$ and we obtain $I_{\gamma,D} \approx (O_D/4) \cdot R_{\gamma,s} \cdot S_{\gamma,s} \cdot \bar{s}\rho/R_{\gamma,D} = S_{\gamma,s} \cdot V_D \cdot \rho \cdot R_{\gamma,s}/R_{\gamma,D}$.

The source strength $S_{\gamma,e,D}$ of secondary electrons in the detector is $S_{\gamma,e,D} = I_{\gamma,D}/V_D \cdot \rho = S_{\gamma,s} \cdot R_{\gamma,s}/R_{\gamma,D}$. In detectors with $\bar{s}\rho \ll R_{\gamma,D}$ the source strength of γ-induced electrons is equal to the γ-source strength in the shield times the ratio of the ranges of the γ rays in the shield and in the detector.

Whereas in solid and liquid detectors the direct interaction of the γ rays with the sensitive material dominates, in gas source counters the γ interaction with the detector walls in general is more important for the background counting rate than interaction with the gas.

The background count rate $B_{\gamma,e}$ by γ-induced electrons based on the above considerations is

$$B_{\gamma,e} \approx S_{\gamma,s}\left(V_D \cdot \rho \cdot \frac{R_{\gamma,s}}{R_{\gamma,D}} + \frac{O_D \cdot R_{e,w}}{4} \cdot \frac{R_{\gamma,s}}{R_{\gamma,w}}\right), \qquad\qquad 3.$$

where $R_{e,w}$ = range of γ-induced electrons in counter wall and $R_{\gamma,w}$ = range of γ rays in wall. This equation is a good approximation if $\bar{s} \cdot \rho \ll R_{\gamma,D}$ and $R_{e,w} \ll d_w \ll R_{\gamma,w}$, with d_w = thickness of detector wall. If we assume $R_{\gamma,s} = R_{\gamma,D} = R_{\gamma,w}$, the ratio of the γ-induced volume effect to the wall effect becomes $4 \cdot V_D \cdot \rho/(O_D \cdot R_{e,w})$. For a typical cylindrical counter for ^{14}C dating with radius $r = 3$ cm and length $L = 30$ cm, filled with 3 atm of CO_2 ($\rho = 5.9$ mg cm^{-3}) the ratio of volume to wall effect is 0.32, if for $R_{e,w}$ a value of 100 mg cm^{-2} is assumed (36). This example shows that a great portion of the γ background in gas source counters is due to the γ interaction with the counter wall. Equation 3 is used for the discussion of the composition of the remaining background of gas source counters. If we apply the above considerations to the effect of n-induced recoil nuclei we see that the wall effect can be neglected because of the short range of the recoil nuclei.

A decomposition of the residual background of gas source counters is possible by measuring its pressure dependence. The three most important background components (not taking into account spurious counts) are:

1. Muons not detected by the anticoincidence system. This component of background is essentially independent of pressure.

2. γ-induced electrons from detector wall and volume. This background component can be described by equation 3. Assuming $R_{\gamma,s} \approx R_{\gamma,D} \approx R_{\gamma,w}$ and expressing the pressure dependence of ρ by $\rho_0 p/p_0$ we obtain $B_{\gamma,e}(p) \approx S_{\gamma,s}[(V_D\rho_0 \cdot p/p_0 + (O_D \cdot R_{e,w})/4]$. The wall effect is independent of pressure and can be estimated by extrapolation of the measured values of $B_{\gamma,e}(p)$ for $p \to 0$. If $R_{e,w}$ is known, the ratio of volume effect to wall effect can easily be calculated.

3. The background contribution by recoil nuclei increases linearly with pressure. Because of the short range of the recoil nuclei the wall effect can be neglected.

4. The background contribution by β and α particles from the counter wall is independent of pressure.

From an experimentally determined pressure-dependence of the background of a gas source counter system we can now analyze its composition in the following way. First we determine the background for $p = 0$ by extrapolation and calculate the pressure-dependent part under the assumption that the background for $p = 0$ is due solely to γ interaction with the wall. If the actually determined background shows a steeper slope we conclude that there is a considerable contribution by recoil nuclei, if radioactive contamination of the gas is excluded. Such relatively strong volume effects are observed in laboratories with little construction material on top (and therefore little attenuation of the nucleonic component of CR) and without neutron shielding. If on the other hand the measured slope is less steep than that calculated for γ interaction, there is an indication of high contributions by muons or by α and β particles from the counter wall. If the measured and calculated slopes agree, the residual background is probably due mainly to γ rays.

The rise-time-discrimination method   The method of rise-time-discrimination (RTD) has successfully been applied to LLC proportional systems in the last few years (9, 37–47). Thus in gaseous detectors for α particles, X rays, and low to moderate energy β particles, considerable background reduction can be achieved without substantial loss in counting efficiency, with a consequent improvement in sensitivity and precision. The method makes use of the fact that in a proportional counter the rise time of a pulse observed at the central anode depends on the extension of the initial ionization track produced by a charged particle interacting with the counter gas. Low energy X rays or low energy electrons following electron capture decay or β decay cause very short ionization tracks, and the consequent signal at the anode is of short rise time. On the other hand, "background events" such as charged cosmic ray particles or energetic Compton electrons produce extended ionization tracks and thus signals with considerably longer rise times, although the same energy may have been deposited. If, in identifying a certain isotope, one measures the pulse rise time along with the pulse height, one is able to effectively distinguish between source events and background events.

The experimental realization of this principle is preferentially done in a two-parameter mode: pulse height versus rise time. The signals from the proportional counter are amplified by a fast charge-sensitive pre-amplifier and then routed in parallel into a "fast" branch and "slow" branch. In the fast branch the pulse is fed into a timing filter amplifier where it is differentiated with a short time constant of typically 10–100 ns (depending on counter geometry, pressure, and filling gas), then stretched and digitized by an ADC. The digital information (rise time) is displayed on the y axis of a two-parameter plot. In the slow branch the signal is amplified with a time constant of typically 200–3000 ns and subsequently digitized by a second ADC. This information (pulse height) is plotted on the x axis. One then finds the signals originating, for example from an X-ray source in an area determined by the corresponding energy and rise time, whereas background events of the same energy are displaced towards longer rise time, that is, they give smaller signals from the fast differentiating branch. The discrimination against pulses produced by γ rays may be measured with an external γ source, and rejection ratios (indicating background reduction) up to nearly 100% may be reached. As the absolute counting efficiency is generally inversely proportional to the rejection ratio, the FOM of a system may be optimized for a given application.

More information on background reductions obtained by this technique may be found in (9, 41, 44–46). Bradley & Willes (44) report a background reduction by a factor of 4.5 (with an absolute counting efficiency reduced by 10%) for a tritium counter ($CH_4 + H_2$) of 2.2 liters of sensitive volume, which is equivalent to doubling the FOM $ef/B^{1/2}$. Wahlen et al (9) obtained a background reduction by a factor of 5 for an argon-filled counter for ^{55}Fe (7) (solid source on the wall) without loss in efficiency, thus also doubling the FOM. Frommer (46) applied the RTD method to an Oeschger-type tritium counter (CH_4). The same background count rate as in the conventional mode was achieved, when operated with RTD inside an external guard counter, with the internal anticoincidence array removed (i.e. doubled counting efficiency).

Finally, RTD is not effective against one background component, namely recoil nuclei caused by fast neutron interaction. On the other hand, RTD can be used for γ-ray background suppression in recoil detectors for fast neutrons (47).

Underground laboratory   An appreciable reduction in background count rates is obtained for all LLC detectors when they are operated underground. As the muon intensity decreases with depth, the emission of secondary particles and γ rays from the interaction of muons in material surrounding the detector also decreases (48). After a few meters of water equivalent (mwe) the neutral and hadron component of CR is completely absent. Unfortunately, the data on muon intensity attenuation are not in agreement. One can say approximately, however, that at 5 mwe underground the muon intensity is reduced by a factor of 1.4, at 60 mwe by 12, and at 5000 mwe by 10^6 (49, 50).

Neutrons and γ rays (48) generated by stopped and subsequently captured μ^- are drastically reduced according to the decrease of the muon-stop rate with depth (~ 1 μ-stop g day^{-1} at sea level): the μ-stop rate drops by a factor of ~ 3, ~ 40, and $\sim 10^9$ at 5, 60, and 5000 mwe underground.

Special attention must be devoted to radioactively clean materials in the construction and operation of an underground laboratory. First, measures must be taken against the natural radioactivity (U, Th, and their daughters; ^{40}K; Rn emanation) of the surrounding soil or rock material. This can successfully be done by coating the whole cavern with a thick layer of radioactively pure concrete. Our newly built underground laboratory (60 mwe below surface) is equipped with a layer of 40 cm of special concrete, low in natural activities. This concrete, made up of serpentine and Danish cement, exhibits a total γ activity of 20 times less than common concrete (35). As a result a bare GeLi detector (65 cm^3) shows a γ-count rate (100–2000 keV) reduced by a factor of 13 when it is operated in the center of our shielded cavern, as compared to an adjacent identical cavern constructed from common concrete. The signal from ^{40}K is 40 times lower. The background for the same detector within a shield of 5 cm Hg and 5 cm Pb operated in the underground laboratory at 60 mwe is 5 times lower (at 800 keV) than in a laboratory 5 mwe below surface. Any material brought into such a laboratory should be regarded as a possible source of radioactive contaminants, which may affect background count rates. The background of an LLC system operated in a clean deep laboratory is dominated by the activity of counter material and its close surroundings, because the cosmic ray-induced contributions are heavily depressed. Therefore the need for specific selection of clean material is obvious.

In the following we compare some preliminary results obtained for different counters in our underground laboratory (60 mwe versus 5 mwe underground): A 12 cc counter (4 atm of C_2H_4) with external anti-coincidence used for work on lunar tritium showed a reduction of the background count rate to 0.02 cpm (factor of 3) for the tritium window, whereas the integral gross count rate dropped by a factor of 10 (proportional to muon intensity). For an Oeschger-type tritium counter with internal anticoincidence (1.5-liter sensitive volume, 1000 Torr CH_4), we measured in the tritium window an 0.05 cpm background in the underground laboratory versus 0.25 cpm at 5 mwe. Stuiver (51) obtained a background count rate of 0.85 cpm for a ^{14}C counter of 2.3 liters (3.3 atm CO_2) and 1.40 cpm for a similar counter of 4 liters, both operated with external anticoincidence in a laboratory of 30 mwe underground.

Summary   The background of LLC systems is composed of contributions from radioactive impurities and from CR. The γ shield removes about $\frac{2}{3}$, the anticoincidence system about $\frac{1}{3}$, of the background of an unshielded gas source counter, with a residual background of 1%. The additional strong background reduction by about a factor of 3–4 in an underground laboratory indicates that in present low level gas source counting the CR-induced background dominates. The advantages of underground counting can be fully exploited only if additional consideration is given to the selection of extremely pure material for shield and detector construction.

A general principle in LLC is to identify radiations as specifically as possible. Generally it can be said that the more specifically the given radiation is identified, the lower is the corresponding background.

3.3   *Sensitivity Comparison of Isotope Detection Methods*

Measuring the decay rate is not necessarily the most sensitive way to determine the number of radioactive nuclei present in the sample. Especially for long-lived isotopes other methods of isotope identification are competitive, since during the radioactivity measuring period only a small percentage of the nuclei present in the sample decay, and will be observed. For a comparison of competitive methods of isotopic measurements we calculate the minimum number of nuclei N that can be detected by sensitive counting systems with an accuracy of 10% in a total counting time of 20,000 min. Knowing the background count rate B of the system, the corresponding minimum sample count rate S can be obtained from equation 2, Section 2, or Figure 1, and from this, according to half-life and efficiency, the number of nuclei N. For samples with low specific activity the ratio of this number to that of all nuclei of the element N_{tot} present in the sample is also important. These values are given in Table 5 and compared with those valid for other methods.

The number of nuclei required, for example, in internal proportional counting varies according to the half-life from 400 for ^{37}Ar to 2×10^{10} for ^{81}Kr. The most sensitive mass-spectrometric methods allow the determination of 10^7 nuclei and become competitive for half-lives above 100–1000 years for samples with relatively high N/N_{tot} ratios. However, these high sensitivity mass spectrometers are not designed for high resolution and therefore are not directly applicable because of interference by neighboring masses of higher abundance. Measurements of ^{81}Kr by mass spectrometry would allow a considerable reduction of the sample amount if, by isotopic enrichment, the ratios of ^{81}Kr/^{80}Kr and ^{81}Kr/^{82}Kr could be increased by a factor of 10^5 with a high yield for ^{81}Kr.

Table 5 Minimum number of nuclei N which can be detected by LLC with an accuracy of 10% in a counting time of 20×10^3 min; comparison with the minimum number required by other techniques

Isotope	$T (2)^{1/2}$ (years)	B (cpm)	S (cpm)	Sample Amount	Counting Efficiency e (percent)	N	N/N_{tot}	Counting Method	Occurrence	Ref.
^{37}Ar	0.0961	0.1	0.05	3 liter Ar	80	2.9×10^3	3.6×10^{-20}	proportional counting	atmosphere	26
		3×10^{-5}	6×10^{-3}			420		proportional counting	neutrino detection	43
^{3}H	12.26	0.1	0.05	3.2 liter CH_4	50	9.3×10^5	2.7×10^{-18}	proportional counting	hydrosphere	—
		0.05	0.035		80	2.6×10^5		proportional counting	moon, meteorites	—
^{3}He	∞	—	—	—	—	10^7–10^8	10^{-3}–10^{-4}	mass spectrometry	moon, meteorites	52
^{39}Ar	269	0.27	0.08	2 liter Ar	80	2.0×10^7	3.8×10^{-16}	proportional counting	atmosphere hydrosphere	53
^{14}C	5730	1.4	0.17	13.2 liter CO_2	95	7.8×10^8	2.2×10^{-15}	proportional counting	biosphere atmosphere	51
^{81}Kr	2.1×10^5	0.45	0.1 (0.017)	2 liter Kr	80	2.0×10^{10}	3.7×10^{-13}	proportional counting	atmosphere	26
						10^6	10^{-3}–10^{-4}	mass spectrometry	moon, meteorites	54
^{53}Mn	3.7×10^6	—	$\sim 10^{-3}$	3×10^{-4} g Mn	—	3×10^9	$\sim 10^{-9}$	neutron activation	moon, meteorites	31
^{7}Li	∞	—	—	$< 10^{-15}$ g	>90	$\sim 10^8$	—	Ion sputt. or thermion source mass spectrometry	meteorites	55
									meteorites	56

The measurement of natural ^{14}C is being attempted by high resolution negative-ion mass spectrometry (57). $C^{15}N^-$ (pure ^{15}N) is produced, so that ^{14}C shows up with mass 29 where there is little interference. Theoretically, measurements of $^{14}C/C$ ratios of $\leq 10^{-14}$ (natural $^{14}C/C$ ratio $\cong 10^{-12}$) should be possible. In special cases neutron activation allows very sensitive measurements of long-lived isotopes. The minimum ^{53}Mn activities determined by this method on lunar samples (10^{-3} dpm) are considerably below the lower limits for counting (31). A sensitive method for ^3H determination in large water samples could be the mass-spectrometric analysis of its decay product ^3He, if all ^3He originally present is removed from a sample.

4 REVIEW OF APPLICATIONS AND TECHNIQUES

Conservative and Nonconservative Tracers in the Atmosphere-Hydrosphere System

Our knowledge of the dynamic behavior of air and water masses and their contained particulate matter is to a great extent based on environmental isotope studies (58).

Isotopes that behave identically to the bulk mass of air and water are called conservative tracers. Conservative tracers for air are the noble gas radioisotopes ^{37}Ar, ^{39}Ar, ^{81}Kr, and ^{85}Kr. ^3HHO molecules behave like the bulk of the water molecules except for isotopic fractionation. ^3H is therefore a conservative tracer for water vapor, liquid water, and ice. Other conservative tracer isotopes for water are ^{39}Ar and ^{85}Kr in dissolved Ar and Kr. $^{14}CO_2$ behaves like the bulk of the natural CO_2 and is therefore an ideal tracer for studying the CO_2 exchange between air, biosphere, and oceans. Isotopes of solids, such as ^7Be, ^{35}S, ^{22}Na, ^{32}P, ^{33}P, ^{90}Sr, and ^{137}Cs are nonconservative tracers and allow study of the behavior of ions and particulate matter in air and water.

The above-mentioned conservative tracers, noble gas radioisotopes, and ^3H and ^{14}C, are highly diluted by stable isotopes of their elements and present samples of low specific activity for LLC. The nonconservative tracer isotopes belong to elements of low abundance in the atmosphere-hydrosphere system and so the collection of samples with relatively high specific activity is possible. Natural ^3H, ^{14}C, $^{37,39}Ar$, and $^{81,85}Kr$ samples are measured by internal counting.

^3H ^3H, produced by CR and in tests of nuclear weapons, is found as ^3HHO in the hydrosphere and is therefore ideally suited for water-cycle studies. The specific ^3H activities are expressed in tritium units (1 TU corresponds to $[^3H]/[H] = 10^{-18}$). The average specific activity of CR-produced ^3H in precipitation is of the order of 10 TU. This level is now masked by bomb-produced ^3H. In 1963 an increase by a factor of 10^2–10^3 was observed in northern hemispheric precipitation and at present the average specific activities in precipitation are still around 100 TU.

Important results of ^3H studies are values for the residence times of water vapor in the stratosphere (years) and in the troposphere (months to weeks). The measurements of bomb-produced ^3H in the upper ocean layer helps to develop models for oceanic mixing and circulation. In hydrology, ^3H measurements allow in certain cases age determinations of ground water, important information for water resource problems (15). In Table 6 sample and background count rates are given for typical ^3H counting systems. In proportional counting higher efficiencies and lower backgrounds are obtained, but the sample preparation is more complicated than in scintillation counting. Direct counting of ^3H samples with an accuracy of 10% is possible for activities of 20–50 TU in PC and of 100–300 TU in liquid scintillation counters. For lower specific activities the samples must be enriched by electrolysis or/and thermal diffusion.

^{14}C Radiocarbon atoms are produced in the atmosphere by neutrons in the reaction ^{14}N (n, p) ^{14}C and oxidized to $^{14}CO_2$. CO_2 (and $^{14}CO_2$) exchanges with the dissolved CO_2, bicarbonate, and carbonate in the oceans, and with the biosphere by photosynthesis, respiration, and decomposition. ^{14}C analyses of atmospheric and oceanic CO_2, dead plants and animals, carbonate deposits, etc have contributed to the solution of problems in earth and planetary, biological, and even historical sciences. Only a few of these applications are mentioned here (10):

1. Radiocarbon dating allows us to establish the time scale for the history of man, animals, plants, and the earth's surface during and after the last glaciation.
2. Based on measurement of natural and bomb-produced ^{14}C in the atmosphere, the biosphere, and the oceans, dynamic parameters for the CO_2 exchange between these reservoirs can be determined.
3. $^{14}C/C$ variations observed in tree rings that cover the last 8000 years are probably a record of the modulation of CR by the solar cycle and by variations of the dipole moment of the earth's magnetic field.

The specific prebomb ^{14}C activity of carbon in modern wood is 13.5 dpm $(gC)^{-1}$. ^{14}C production by neutrons from atmospheric H-bomb tests resulted in an increase of the atmospheric ^{14}C level in the northern hemisphere by a factor of 2 in the years 1963 and 1964. Natural ^{14}C

Isotope	Decay	Reference Sample	S (cpm)	B (cpm)	e (%)	Sample Preparation and Detection Method	Ref.
^3H	β^- 0.0186 MeV	100 TU (activity in recent precipitations)	2	1	90	internal proportional counting as CH_4, C_2H_2, C_2H_6; H_2+CH_4	13 59 60
			2	5	20	liquid scintillation counting of mixture of water and scintillator	59
^{14}C	β^- 0.158 MeV	13.5 dpm $(gC)^{-1}$ (activity of "modern" carbon)	20	1	90	internal proportional counting as CH_4, CO_2, or C_2H_2	13
			20–60	2–10	70	liquid scintillation counting of synthesized benzene and scintillator	61
^{37}Ar	EC 2.8 keV (K energy)	2.5×10^{-3} dpm (liter Ar)$^{-1}$ (natural level)	~0.01	0.09	65	internal proportional counting of X rays and auger electrons; counting gas $Ar+CH_4$; prepared by fractional distillation and gas chromatography	26
^{39}Ar	β^- 0.57 MeV	0.1 dpm (liter Ar)$^{-1}$ (natural level)	0.14	0.27	~70	internal proportional counting, counting gas $(Ar+CH_4)$ prepared by fractional distillation and gas chromatography	53
^{81}Kr	EC 12.6 keV (K energy)	0.1 dpm (liter Kr)$^{-1}$ (natural level)	0.13	0.45	65	internal proportional counting, counting gas $(Kr+CH_4)$, commercial Kr from before 1952 (prebomb)	26
^{85}Kr	β^- 0.67 MeV	35 dpm (liter Kr)$^{-1}$ (present level)	0.5	0.02	80	internal proportional counting, counting gas $(Kr+CH_4)$ prepared by fractional distillation and gas chromatography or chemically with hot Ca	62
			0.4	0.1	60		80

was discovered in samples that were enriched by thermal diffusion. The introduction of the anticoincidence principle by Libby and co-workers made direct measurements possible and was the first important step in the development of LLC. The early stage of radiocarbon dating is described by Libby (63, 64).

Natural ^{14}C is measured in PC and liquid scintillation counters. In Table 6 typical sample and background count rates for the two methods are given. In PC the samples are measured as CH_4, CO_2, or C_2H_2; for liquid scintillation counting usually C_6H_6 is synthesized. PC are mostly superior regarding FOM and sample preparation procedures. Especially for samples with little carbon (< 1 g), PC are preferable. At present most of the radiocarbon dating laboratories use PC.

^{37}Ar This isotope is produced by CR and in underground tests of nuclear weapons, whence it may escape into the atmosphere. The specific activity of CR-produced ^{37}Ar is in tropospheric samples ~ 0.0025 dpm (liter Ar)$^{-1}$ (65). After underground H-bomb tests (with ventilation), specific activities as high as 0.1 to 2 dpm (liter Ar)$^{-1}$ were measured far away from the test sites. From measurements of vertical and latitudinal profiles of natural and artificial ^{37}Ar, values for vertical and latitudinal eddy diffusion coefficients have been determined (65). For details on the measuring technique see Table 6.

^{39}Ar ^{39}Ar in atmospheric Ar was measured for the first time by Loosli & Oeschger (53). The specific activity is (0.112 ± 0.012) dpm liter^{-1} Ar and the contribution from nuclear tests is probably less than 10%. This isotope is used for radioactive dating of polar ice (occluded Ar) and ground water (dissolved Ar) (66). Further applications will be in oceanography to establish the time scale of oceanic mixing and circulation processes. The dating range of ^{39}Ar is 50–1000 yr and is between those of ^{3}H ($T_{1/2} = 12.3$ yr) and ^{14}C ($T_{1/2} = 5730$ yr).

^{81}Kr Until now the specific activity of ^{81}Kr in atmospheric Kr has been determined in two laboratories. The specific activities reported are (0.10 ± 0.01) dpm (liter Kr)$^{-1}$ (26) and (0.046 ± 0.010) dpm (liter Kr)$^{-1}$ (67). Applications in hydrology and glaciology (dating of old water and ice) seem prohibitive at present since they would require sampling of the Kr contained in 10^3 m^3 of water or ice. The sample amount may be reduced by one or two orders of magnitude when a measuring technique based on isotopic enrichment and mass spectrometry is developed.

^{85}Kr The CR-produced level of this isotope is completely masked by ^{85}Kr escaping from nuclear power and nuclear fuel reprocessing plants.

The specific activity has risen during the last 20 years almost linearly with time. At present it has the highest specific activity (35 dpm cm^{-3} Kr) of all atmospheric gases. Its increase is monitored at a series of stations all over the world. The data are interpreted in terms of atmospheric mixing. First results have been obtained on samples (of dissolved Kr) from ocean profiles (62) that give information on the mixing of the upper ocean layers. An important application will be ground water dating in the range < 20 yr.

Radiogenic nuclides in extraterrestrial matter   Studies on radiogenic nuclides found in meteorites and lunar samples have yielded information on the nature of galactic and solar CR.

Investigations on radionuclides produced by galactic CR interactions with meteorites have led to the conclusion that the galactic CR intensity was essentially constant within a factor of 2, over the last 5 million years (3, 6). Recently Ar isotopes with different half-lives (^{37}Ar, $T_{1/2} = 35$ days and ^{39}Ar, $T_{1/2} = 269$ days) were studied in a meteorite whose elliptic orbital trajectory is known from photographic observations of the fireball produced when the meteorite entered the earth's atmosphere (68, 69). From this study the radial gradient of the galactic CR intensity away from the sun could be inferred. The ^{37}Ar activity reflects the CR intensity incident on the meteorite during the very last part of its trajectory before the encounter with the earth [i.e. at 1 AU (astronomical unit)], whereas the ^{39}Ar activity represents the CR intensity at the location where the meteorite spent most of its time (i.e. at a distance much further away from the sun). From this study an increase in the CR intensity of $(60 \pm 25) \%/$AU away from the sun was obtained.

Studies on depth profiles of the activities of radionuclides (such as ^{56}Co, ^{54}Mn, ^{55}Fe, ^{22}Na, ^{26}Al, and ^{53}Mn) in lunar rocks and soils have resulted in detailed information about solar CR (7–9). Solar CR, predominantly protons in the energy range between 10 and 100 MeV, are sporadically emitted during solar flare events and the flux intensity strongly varies according to the 11-year solar cycle. Because of the short range of the solar CR, the activity of the nuclides they produce in the outer surface of a sample exhibits a strong depth dependence.

When compared to theoretical calculations based on measured cross-section data, the analysis of the depth profiles for isotopes of different half-lives allowed determination of the intensity and spectral shape of the flux of solar CR for different periods in the past. As a striking fact it was found that both intensity and spectral shape were essentially the same during the past few years as they were in the past few million years. In addition, proton fluxes of the individual flares occurring immediately prior to sampling were obtained from the measurement of the 77-day isotope ^{56}Co. The experimental realization of this study is a fine example of LLC applications. Extensive chemistry was needed to isolate the isotopes by chemical separation of the various elements followed by specific radiochemical purification. The counting of ^{56}Co, ^{22}Na, and ^{26}Al was performed by coincidence techniques. The counting of ^{56}Co was performed by detecting the X rays from the electron capture decay in coincidence with the following γ rays. The isolated Co was electroplated onto the walls of a small pressurized proportional counter (sample area 10 cm^2) operated inside the well of a NaI detector. An external flat guard counter was operated on top of the array. Typical background count rates were in the range of 4×10^{-3} cpm whereas the efficiency was between 1 and 2%.

^{22}Na and ^{26}Al were counted externally by β-γ coincidence techniques. The samples were mounted on both sides of a flat GMC operated inside the well of a NaI detector. Here the background of the coincident events was around 0.005 cpm, with a counting efficiency of 8%. The same figures hold for the (n, γ)-induced isotope ^{60}Co, which allowed the determination of the present lunar neutron production rate (70).

A single-wire pressurized proportional counter (volume, 0.5 liter), operated inside a GM guard counter, was used for the assay of the X rays from ^{55}Fe. Samples of 1 g Fe were electroplated onto a copper sheet (central active area, 150 cm^2) which served as counter wall. A counting efficiency of 7%, and background rates of 9×10^{-2} cpm, and 1.6×10^{-2} cpm with RTD, were achieved.

Pure β emitters like the galactic CR-produced ^{10}Be and ^{36}Cl were detected by a coincidence-type β spectrometer according to (71). This array consisted of a small flat PC (identification of charged particle), operated in coincidence with a plastic scintillation block (energy assay) and in anticoincidence with a muon guard tray. Representative values for counting efficiencies were around 40%, and background values for the energies of the β particles from these isotopes were about 0.05 cpm.

The "spinner" detector   This detector was developed especially for the detection of rare nuclear processes (72). The detector functions by producing negative pressures in a liquid through the action of centrifugal force. The metastable state thus created can be destroyed, as in a bubble chamber, by localized energy deposition of particles. In contrast to a bubble chamber, however, this detector is continuously sensitive, and has 4π geometry when the radiation source is dissolved in the liquid. It operates as a threshold detector with high discriminatory power for radiations with different energy deposition densities; it is especially suited to investigations of rare spontaneous or induced fission. A new upper limit for the decay constant for spontaneous fission of ^{235}U has been obtained by Grütter et al (73), using the "spinner" technique; this limit $(<3.9 \times 10^{-18} \text{ yr}^{-1})$ is a factor of 10 below the value previously reported. In another study the spinner technique was used in the search for superheavy elements (74).

"Track" technique   A procedure similar to the photographic emulsion technique is frequently used in determining cosmogenic effects in extra-terrestrial bodies (75). The so-called track technique is based on a natural detector method, which utilizes changes in solids caused by CR inter-actions. Although radioactivity is not involved here, it may nevertheless be called a low level technique because each interaction is recorded. A high energy charged particle $(Z \geqq 20)$ penetrating into a sample of, say, meteoritic or lunar material will damage its crystalline structure. This damage becomes visible in form of a track that can be seen when the crystal is etched with a suitable agent. By counting the tracks and measuring their properties (diameter, length, etc) one is able to get information about the flux and the energy spectrum, as well as the elemental composition of the relevant cosmic rays incident on the sample over the period of its exposure (exposure age). For example, very interesting information has been obtained about the evolution of planetary bodies, as in several meteorites individual grains have been found that were irradiated prior to their agglomeration. Fission products from U fission are also recorded in solid material and can be analyzed by the track technique. Measuring the abundance of fission tracks in terrestrial and extraterrestrial samples together with the U content (induced fission tracks observed after subsequent artificial n irradiation of the samples) allows the determination of solidification ages.

Neutron activation of radioisotopes   The answers to many problems in geophysics and cosmophysics can be obtained from investigation of long-lived radioisotopes. In favorable cases the abundance of such isotopes may be determined by mass spectrometry (e.g. ^{81}Kr from extraterrestrial samples, $T_{1/2} = 2 \times 10^5$ yr), but often one must rely on direct observation of the radioactive decay. Often in this case only upper limits are obtained, or, if high sensitivity techniques do enable quantitative analyses at all, the results are of poor precision and therefore do not allow a unique answer to a given problem. In the case of the 3.7 million year nuclide ^{53}Mn and the 17 million year ^{129}I these difficulties were overcome by neutron activation, that is, by transforming these isotopes into radionuclides of much shorter half-lives. Accurate results were obtained for the CR-produced isotope ^{53}Mn from small amounts of lunar samples with the neutron activation technique [^{53}Mn (n, γ) ^{54}Mn, $T_{1/2} = 312$ d (31)]. A long-duration irradiation with well-thermalized neutrons (total dose $\sim$ 2×10^{19} thermal n cm^{-2}) yielded an amplification factor of about 6000 dpm ^{54}Mn/dpm ^{53}Mn. The results from these studies were compared to the data on the flux of solar cosmic rays (intensity and spectral shape) inferred by shorter-lived isotopes (^{56}Co, ^{54}Mn, ^{22}Na, ^{55}Fe, ^{26}Al; $T_{1/2}$ from several months to 7.5×10^5 yr) and conclusions could be drawn about the solar energetic particle flux in the past. In addition, for the first time a complete depth profile of a galactic CR-produced isotope was obtained from the surface of the lunar regolith down to a depth of 400 g cm^{-2}. Recently ^{129}I has been positively identified in terrestrial tellurium ores (76). This was possible with neutron activation of ^{129}I (n, γ) ^{130}I. The γ radiation of the 12.3 h ^{130}I was then measured by a sum-coincidence technique. The production of terrestrial ^{129}I is attributed to muon interaction with tellurium, and the data suggest a constant CR muon intensity during the past 10^9 yr.

Neutrino astronomy   A solar neutrino experiment has been carried out since 1967 by Davis and co-workers (43, 77, 78), which has pushed the LLC technique to its limits. The goal is to observe neutrinos from the sun by the neutrino capture reaction ^{37}Cl (v, e$^-$) ^{37}Ar. The result was surprising in that the flux of solar neutrinos anticipated from standard solar model calculations was not observed. Instead, after numerous runs under continually improving conditions, it was possible to set an upper limit on solar neutrino capture in ^{37}Cl of 1 SNU [1 SNU $= 10^{-36}$ captures sec^{-1} (^{37}Cl atom)$^{-1}$], which is 9 times less than expected.

In this experiment 6.1×10^5 liters of perchloroethylene were repeatedly exposed for several months as a target in a mine at 4400 mwe underground. The ^{37}Ar produced, together with a small amount of dissolved ^{36}Ar carrier, was then purged with He gas. The recovered Ar was transferred, after purification, into a small quartz proportional counter for observation of the ^{37}Ar decay. This counter had an active volume of ~ 0.6 cm^3 and was operated inside the well of a NaI crystal surrounded by a cylindrical proportional-counter guard array. The assembly was shielded by 30 cm of steel. Pulse height analysis and rise-time analysis were performed to identify the signals from the ^{37}Ar decay; in addition the half-life was monitored. The background of this system amounts to ~ 1.5 counts per 35 days, for the photopeak region at 2.9 keV (FWHM). The results leading to the above-mentioned upper limit reflect a production rate of (0.13 ± 0.20) atoms d^{-1} in the tank, which is equivalent to the detection of about 10 atoms of ^{37}Ar. To verify this result, the overall performance of the experiment has been carefully checked in many ways and the contributions to the production by other reactions have been examined. The result has caused reconsideration of the basic processes and nuclear mechanisms occurring in the sun. However, this puzzle has not yet been satisfactorily resolved.

5 CONCLUSIONS

Radioisotope analyses play a more and more important role in many fields of earth and planetary sciences. Many new interdisciplinary problems have become evident through environmental isotope studies. These studies provide fundamental information for environmental problems (behavior of tracers in the atmosphere-ocean system; partitioning of CO_2 from fossil fuel combustion in the atmosphere-biosphere-ocean system).

The spectrum of radioisotopes applicable in earth and planetary science problems has been extended by new developments in sampling and LLC techniques:

1. The development of GeLi detectors of relatively large volume has significantly improved the sensitivity.
2. The background for internal proportional counting has been considerably improved by the rise-time discrimination method.
3. By operating counting systems in underground laboratories, significant background reductions have been obtained.

These improvements can be fully exploited only if the counting time for individual samples is considerably increased. Long sample measurement times have been made more convenient by modern electronics, which allows computer-controlled measurements by a series of similar counters in parallel.

For long-lived isotopes, determinations by other analytical methods such as mass spectrometry, activation, and infrared spectroscopy (79) are or may become competitive and may help to fill out the spectrum of radioactive isotope methods available.

In addition to its applications in the above-mentioned fields, LLC is an important tool for nuclear medicine and the control of environmental radioactivity levels above the natural baselines.

ACKNOWLEDGMENTS

We would like to acknowledge most valuable discussions with Dr. R. C. Finkel, Dr. U. Siegenthaler, and Dr. L. A. Currie.

462   OESCHGER & WAHLEN

Literature Cited

1. Wainerdi, R. E., Uken, E. A. eds. 1971. *Modern Methods of Geochemical Analysis.* New York: Plenum
2. Hamilton, E. I., Farquhar, R. M., eds. 1968. *Radiometric Dating.* London: Interscience
3. Honda, M., Arnold, J. R. 1967. *Handb. Phys.* 46(2):613. Berlin: Springer Verlag
4. Lal, D., Peters, B. 1967. *Handb. Phys.* 46(2):551. Berlin: Springer Verlag
5. Lal, D. 1972. *Space Sci. Rev.* 14:3
6. Geiss, J., Oeschger, H., Schwarz, U. 1962. *Space Sci. Rev.* 1:197
7. *SHRELLDALFF.* 1970. *Proc. Apollo 11 Lunar Sci. Conf., Geochim. Cosmochim. Acta* 2: Suppl. 1, 1503
8. Finkel, R. C. et al 1971. *Proc. 2nd Lunar Sci. Conf., Geochim. Cosmochim. Acta* 2: Suppl. 2, 1773
9. Wahlen, M. et al 1972. *Proc. 3rd Lunar Sci. Conf., Geochim. Cosmochim. Acta* 2: Suppl. 3, 1719
10. Olsson, I. U., ed. 1970. *Nobel Symp. 12. Radiocarbon Variations and Absolute Chronology.* Stockholm: Almqvist & Wiksell
11. *Proc. Noble Gas Symp., Las Vegas, Sept. 1973.* In press
12. Slade, D., ed. 1968. *Meteorology and Atomic Energy.* Washington DC: AEC, Div. Techn. Info.
13. International Atomic Energy Agency. 1967. *Radioactive Dating and Methods of LLC,* Proceeding series. Vienna: IAEA
14. International Atomic Energy Agency. 1970. *Isotope Hydrology,* Proceeding series. Vienna: IAEA
15. International Atomic Energy Agency. 1974. *Isotope techniques in Groundwater Hydrology,* Proceeding series. Vienna: IAEA
16. *Proc. Int. Symp. Trace Gases (CACGP), Mainz, 1973.* 1974. *Tellus* 26(1, 2)
17. Loevinger, R., Berman, M. 1951. *Nucleonics* 9(1):26
18. Weise, L. 1971. *Statistische Auswertung von Kernstrahlungsmessungen.* München: Oldenburg Verlag
19. Currie, L. A. 1968. *Anal. Chem.* 40: 586

20. Oeschger, H. et al 1972. *Proc. 8th Int. Conf. Radiocarbon Dating* 1: D70
20a. Grootes, P. M. et al 1975. *Z. Naturforsch. A* 30:1
20b. Currie, L. A., De Voe, J. R. 1971. *Nuclear Techniques in Environmental Pollution,* p. 183, Proceeding series. Vienna: IAEA
21. Watt, D. E., Ramsden, D. 1964. *High Sensitivity Counting Techniques.* New York: Pergamon
22. Int. Commission on Radiation Units and Measurements. 1972. *Measurement of low level radioactivity,* ICRU Rep. 22. Washington DC: ICRU
23. *Nucl. Instr. Methods* 1973. 112(1, 2)
24. Curran, S. C. 1958. *Handb. Phys.* 45:174. Berlin: Springer Verlag
25. Siegbahn, K. 1965. *Alpha-, Beta- and Gamma-Ray Spectrometry.* Amsterdam: North-Holland
26. Loosli, H. H., Oeschger, H. 1968. *Earth Planet. Sci. Lett.* 7:67
26a. Lal, D. et al 1967. *Radioactive Dating and Methods of LLC.* p. 615, Proceeding series. Vienna: IAEA
26b. Wahlen, M., Oeschger, H. 1968. *Nucl. Instr. Methods* 66:193
27. Eldridge, J. S. et al 1973. *Nucl. Instr. Methods* 112:319
28. Cooper, J. A. 1970. *Nucl. Instr. Methods* 82:273
29. Cooper, J. A. 1971. *Nucl. Instr. Methods* 94:289
30. Armantrout, G. A., Bradley, A. E., Phelps, P. L. 1972. *IEEE Trans. Nucl. Sci.* 19(1):107
31. Imamura, M., Finkel, R. C., Wahlen, M. 1973. *Earth Planet. Sci. Lett.* 20: 107
32. Moghissi, A. A. 1970. *The Current Status of Liquid Scintillation Counting,* ed. E. D. Bransome Jr., p. 86. New York & London: Grime & Stratton
33. Boyce, I. S., Cameron, J. F. 1962. *Tritium in the Physical and Biological Sciences,* 1:231. Vienna: IAEA
34. Schwarz, P. (IAEA, Vienna), Rajner, V. (Geotechn. Inst., Vienna). Personal communications
35. Pulfer, P. 1974. Lizentiatsarbeit. Univ. Bern, Bern, Switzerland
36. Oeschger, H. 1963. *Radioactive*

Dating, p. 13, Proceeding series. Vienna: IAEA
37. Mathieson, E., Sanford, P. 1964. Proc. Int. Symp. Nucl. Electron., Paris, 1963, p. 65. Paris: ENEA
38. Campell, J. L. 1968. Nucl. Instr. Methods 65:333
39. Gorenstein, P., Mickiewicz, S. 1968. Rev. Sci. Instr. 89:816
40. Ricker, G. R., Gomes, J. J. 1969. Rev. Sci. Instr. 40(2):227
41. Lewyn, L. L. 1970. Nucl. Instr. Methods 82:138
42. Harris, T. J., Mathieson, E. 1971. Nucl. Instr. Methods 96:397
43. Davis, R. et al 1972. Conf. "Neutrino 72", Europhys. Conf. Balatonfured, Hungary, June 1972, ed. A. Frenkel, G. Marx, 1:77
44. Bradley, A. E., Willes, E. H. 1973. IEEE Trans. Nucl. Sci. 20(1):80
45. Culhane, J. L., Fabian, A. C. 1972. IEEE Trans. Nucl. Sci. 19(1):569
46. Frommer, H. 1973. Lizentiatsarbeit. Univ. Bern, Bern, Switzerland
47. Cuttler, J. M. et al 1968. Nuclear Instr. Methods 75:309
48. Gorshkov, G. V., Zyabkin, V. A. 1973. Sov. At. Energy 34(3):269
49. Bergamasco, L. et al 1970. Lincei Rend. Sci. Fis. Mat. nat. 48:423
50. Grupen, C. et al 1972. Nuovo Cimento B 10(1):144
51. Stuiver, M. 1974. Personal communication
52. Eberhardt, P. 1974. Personal communication
53. Loosli, H. H., Oeschger, H. 1968. Earth Planet. Sci. Lett. 5:191
54. Eugster, O., Eberhardt, P., Geiss, J. 1967. Earth Planet. Sci. Lett. 2:387
55. Gradsztajn, E. et al 1968. Earth Planet. Sci. Lett. 3:387
56. Eugster, O. 1974. Personal communication
57. Schnitzer, R., Aberth, W., Anbar, M. Proc. 22nd Conf. Mass Spectrometry and Allied Topics, Philadelphia, May 1974. In press
58. Lal, D., Suess, H. E. 1968. Ann. Rev. Nucl. Sci. 18:407
59. Proc. 6th Int. Conf. Radiocarbon and Tritium Dating, CONF-650652, Washington DC, 1965. Washington DC: Pullman/US AEC
60. Moghissi, A., Carter, W., eds. 1973.

Proc. Tritium Symp., Las Vegas, 1971
61. Rafter, T. A., ed. 1972. Proc. 8th Int. Conf. Radiocarbon Dating, Wellington, New Zealand, 1972. Wellington, NZ: Roy. Soc. New Zealand
62. Münnich, K. O., Schröder, J. Personal communication
63. Libby, W. F. 1955. Radiocarbon Dating, Chap. 3. Chicago: Univ. Chicago Press
64. Libby, W. F. 1967. Radioactive Dating and Methods of Low-Level-Counting, Proceeding series. Vienna: IAEA
65. Loosli, H. H., Oeschger, H., Studer, R., Wahlen, M., Wiest, W. Proc. Noble Gases Symp., Las Vegas, 1973. In press
66. Oeschger, H., Gugelmann, A., Loosli, H., Schotterer, U., Siegenthaler, U., Wiest, W. 1974. Isotope Techniques in Groundwater Hydrology, 2:179, Proceeding series. Vienna: IAEA
67. Pomansky, A. A. Personal communication
68. Fireman, E. L., Spannagel, G. 1971. J. Geophys. Res. 76(17):4127
69. Forman, M. A., Stoenner, R. W., Davis, R. 1971. J. Geophys. Res. 76(17):4109
70. Wahlen, M. et al 1973. Earth Planet. Sci. Lett. 19:316
71. Tanaka, E. et al 1967. Int. J. Appl. Radiat. Isot. 18:161
72. Hahn, B., Spadavecchia, A. 1967. Nuovo Cimento B 54:101
73. Grütter, A. et al 1974. Physics and Chemistry of Fission, 1973, 1:305, Proceeding series. Vienna: IAEA
74. Behringer, K. et al 1974. Phys. Rev. C 9(1):48
75. Lal, D. 1972. Space Sci. Rev. 14:3
76. Tagaki, J., Hampel, W., Kirsten, T. 1974. Earth Planet. Sci. Lett. 24:141
77. Davis, R., Evans, J. C. 1973. Proc. 13th Int. Cosmic Ray Conf., Denver, August 1973, 3:2001
78. Davis, R., Harmer, D. S., Hoffman, K. C. 1968. Phys. Rev. Lett. 20(21):1205
79. Lehmann, B., Wahlen, M. 1975. Helv. Phys. Acta 48:33
80. Loosli, H. H. 1974. Personal communication

BLOCKING MEASUREMENTS OF NUCLEAR DECAY TIMES

×5567

W. M. Gibson
Bell Laboratories, Murray Hill, New Jersey 07974

CONTENTS

INTRODUCTION

Determination of transition matrix elements for decay of excited nuclear states provides an important test of the validity or scope of nuclear models. This can be done by measurement of the energy width, Γ, or through use of the uncertainty relationship $\Gamma = \hbar/\tau$, by measurement of the decay time, τ, of the state. Indeed decay time measurements have played an important role in nuclear structure studies [see (1) for reviews of previous nuclear lifetime work]. The two principal techniques traditionally employed for lifetime measurements are: (*a*) delayed coincidence between population and decay of the state (2–4) where the state is populated either by radiation or by pulsed beam excitation, and (*b*) Doppler shift of γ-ray de-excitation radiation as the excited recoil nuclei are stopped (5, 6) or slowed down (1, 7). In recent years there has appeared a new technique that extends the accessible time range from the limit of $\sim 10^{-12}$ sec obtainable by coincidence and $\sim 10^{-14}$ sec by Doppler shift attenuation methods to a new limit of $\sim 10^{-18}$ sec. The purpose of this report is to indicate the basis and present status of this new technique, which, for reasons that will become evident, has been called the Blocking Technique for Nuclear Lifetime Measurement, or, for brevity, the Blocking Lifetime Technique (BLT).

Relationship to Other Nuclear Lifetime Measurement Techniques

Before examining this technique in detail it is perhaps useful to briefly compare the various lifetime (or level width) measurement techniques. Figure 1 shows the time (or level width) range covered by the various direct or indirect methods that have been used or proposed. For lifetimes

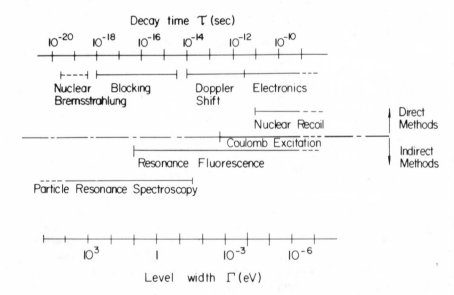

Figure 1 The time (or level width) window accessible through blocking lifetime (BLT) measurements and other direct or indirect measurement techniques.

longer than $\sim 10^{-14}$–10^{-15} sec, decay rates are dominated by electromagnetic transitions. For shorter lifetimes (or for level widths $\gtrsim 1$ eV), matrix elements for nucleon emission begin to dominate. Because of this transition, the BLT differs from previous direct measurement methods in a qualitatively vital way. It provides the first decay time measurement of transition matrix elements for decay by nucleon emission rather than matrix elements depending entirely on electromagnetic transitions. In a sense, therefore, BLT affords the first opportunity for direct measurement of a "nuclear" lifetime (as contrasted to an electromagnetic transition time).

The importance of lifetime measurements in this time range has been recognized for a long time, not only to obtain transition rates but also to provide unambiguous discrimination between direct interaction and compound nuclear mechanisms in nuclear reactions (8). Indeed there have been a number of proposals for nuclear lifetime measurements in the 10^{-16}–10^{-20} sec range (9–11), but aside from a preliminary and subsequently unconfirmed report (12) of feasibility tests of the nuclear bremsstrahlung technique (9) (shown as a dotted line on Figure 1) these suggestions have not been demonstrated experimentally.

It is noteworthy that the level width range accessible to BLT studies occurs for most nuclei at excitation energies where the level density is high enough that single levels cannot in general be resolved because of detector resolution, target thickness, and beam energy instability effects. On the other hand the level density is usually too low to allow statistical fluctuation analysis [Ericsson fluctuations (13)] to obtain the mean level width. Also, as we shall see, techniques are being developed to obtain more than "mean" values from BLT measurements. The BLT is therefore more than just a complementary and alternative method of determining transition matrix elements: it is a virtually unique tool in an important lifetime region. There do exist overlaps with width measurements of isolated resonances in light nuclei and with fluctuation analysis at the short lifetime extreme of the method. No examples of the latter comparison are yet available but the former already provides important tests of the BLT, as we discuss later.

The significance of this new tool extends beyond specific tests of present nuclear reaction models because it has begun to stimulate reconsideration of nuclear reactions and decay processes in terms of their time evolution. The first detailed discussions of this type have already appeared (14, 15) and provide a useful framework for consideraton of the present limited experimental results and for planning of future experiments.

Relationship to Particle Channeling Experiments

The blocking effect used to determine short nuclear lifetimes is a particular example of a more general phenomenon involving the steering of charged

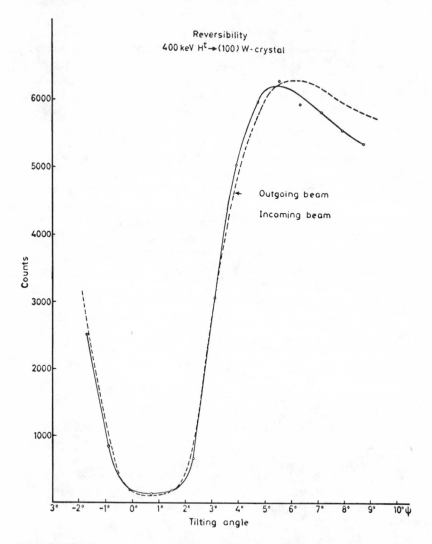

Figure 2 Test of reversibility. The solid curve corresponds to channeling (incident beam direction varied relative to ⟨100⟩ axis, emergence in random direction) and the dashed curve corresponds to blocking (emergent yield relative to ⟨100⟩ axis with incidence in random direction) (from reference 18).

particles by rows or planes of atoms in a crystal. This phenomenon has come to be referred to as *particle channeling* [see (16) for recent general reviews]. The steering of particle trajectories takes place because of correlated scattering of particles moving at small angles to rows or planes by successive small deflections of the particle as it passes atoms in the atomic row or plane. If the particle is positively charged, it is steered gently along through the open spaces in the lattice.

Because of this steering (or channeling), positive particles incident on a crystal parallel to atomic rows or planes have a very much reduced probability for striking a lattice atom (for negative particles such as electrons, or negative pions, the effect is reversed because they experience an attractive rather than repulsive potential from the rows or planes). Soon after the discovery of particle channeling it was recognized that this effect of concentrating the particle flux distribution into the open spatial regions of the crystal could be used to determine the location of impurity atoms in the crystal. Impurities located at normal lattice sites (substitutional impurities) experience the same reduced probability for close interactions (characterized by large angle scattering, X-ray production, nuclear reaction) as atoms of the crystal, whereas those situated in the open spaces between rows or planes (interstitial impurities) experience an enhanced interaction probability. It is precisely the same effect that can be used for nuclear lifetime determination. This, however, was not apparent until it was shown that a detailed reciprocity or reversibility exists (17) when comparing particles moving into the crystal from the outside and particles moving out of the crystal from the inside. This means that a positively charged particle emitted from a lattice position in a crystal will have a probability for emergence from the crystal in a direction parallel to atomic rows or planes that is reduced in the same way that a particle incident on the crystal parallel to atomic rows or planes has a reduced probability for striking a lattice atom. The quantitative equivalence of these two effects has been demonstrated experimentally (18) as shown on Figure 2. In this figure, the protons originating from lattice positions were actually incident beam protons scattered through a large angle. Over the small range of emergent angles examined, the initial Rutherford scattering can be considered to be uniform. The yield reduction takes place because those protons that move along the row or plane direction after the scattering are deflected or blocked by the other atoms in the row or plane; hence the term "blocking effect."[1] The importance of recognizing the relationship

[1] The term "shadowing effect" has also been used, especially in the countries of Eastern Europe; but although this is perhaps more descriptive and certainly easier to translate, the term "blocking" is the generally accepted form in English.

(equivalence) between channeling and blocking will become apparent as we consider analysis of measured emergent particle angular distributions (blocking dips) because it means that the extensive analytical techniques developed for channeling can be used quantitatively and that the same problems and limitations also apply. The converse, of course, is also true.

Basis of the Technique

We now turn to a description of how the blocking effect can give information on nuclear lifetimes. The principle is very simple and was recognized (19, 20) almost immediately after discovery of the blocking effect (21). If an excited compound nucleus is produced by a nuclear reaction between a lattice atom and an incident particle, the excited nucleus will recoil along the incident particle direction. If the nucleus decays by emission of charged particles before it has a chance to move, a normal angular distribution of particles emerging from the crystal will result. This is the normal or "prompt" blocking dip shown schematically on Figure 3. If, however, some of the nuclei live long enough to recoil

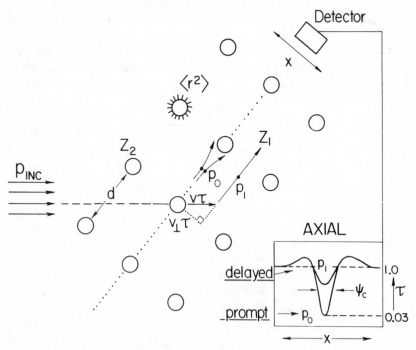

Figure 3 Schematic representation of BLT measurements of charged particles emitted from lattice positions (curve P_0) and from recoiling nuclei displaced from the lattice (curve P_1).

into the open space between the rows or planes before decaying, the blocking dip will become narrower and more shallow as shown schematically by the curve labeled P_1 in Figure 3. Recoil velocities as high as $\sim 10^9$ cm sec^{-1} can be achieved if the incident particle beam is composed of high energy heavy ions. Mean recoil distances perpendicular to the row or plane as short as ~ 0.1 Å can produce measurable changes in the blocking distribution, leading to the lower measurement limit of $\sim 10^{-18}$ sec noted previously. The upper measurement limit is reached when the mean recoil distance becomes larger than the distance to the next atomic row or plane. The angle between the incident beam and the emission direction can be chosen so that the component of recoil velocity normal to the row or plane is controlled. The effective upper lifetime limit even for light and relatively low energy incident particles is $\sim 10^{-14}$ sec. The difference between the "prompt" and the "delayed" blocking distributions is related to the lifetime of the compound nucleus. In a subsequent section we consider the techniques and problems associated with quantitative analysis of this difference. There have been previous reviews of blocking lifetime measurements (22–24). Although we overlap those reports somewhat for completeness, we have placed most emphasis on recent results; the earlier reviews should be consulted for details of earlier work. For a review of the development and history of the technique, see especially (22).

Experimental Arrangement and Typical Examples

The emergent particle angular distribution is most conveniently determined by use of a position-sensitive detector placed at some distance from the target so that a particular spot on the detector corresponds to a well-defined emergence angle. In the experiments done to date, the detectors used have either been plastic or glass track detectors (25) or semiconductor position-sensitive detectors (26, 27). Track detectors have the advantage that they can be developed to show fission fragments (28–31) or helium ions (32, 33) selectively in the presence of a large background of lighter mass beam particles scattered from the crystal. They also allow the blocking distribution to be determined over a large range of angles (in two dimensions) simultaneously. At best, however, they give only very rough information on particle energies and do not allow determination of the blocking pattern on-line during the bombardment.

In cases where determination of the energy of the emergent particles is important, position-sensitive semiconductor detectors have been used. These have usually been one-dimensional in their position determination (26) although two-dimensional detectors have also been used (27). In a series of heavy ion induced fission studies (34) two-dimensional position-

Figure 4 Photograph of cathode ray tube display of blocking pattern of 40 MeV ^{16}O ions scattered from a ~ 3000 Å thick tungsten crystal and emergent along a $\langle 111 \rangle$ axis. The upper part of the photograph shows the (x, z) output of a thin (15 μm) two-dimensional position-sensitive detector. The lower part of the photograph shows a linear (with full scale 128 counts) representation of a vertical slice through the 2-D pattern taken at the position of the double dot on the bottom scale of the 2-D pattern (from reference 34).

sensitive semiconductor detectors have been used[2] that were also thin enough (15 μm) to allow the heavy ions (carbon, oxygen, fluorine) to pass through with low enough energy loss that their signal did not interfere with fission fragments that were stopped in the detector. Furthermore the heavy ions were then used to produce an anticoincidence signal from a detector behind the thin detector to give even more discrimination against scattered beam particles. Figure 4 shows a blocking pattern of 40 MeV oxygen ions scattered into the thin semiconductor position-sensitive detector from a thin (~ 3000 Å) tungsten crystal oriented with a $\langle 111 \rangle$ axis aimed toward the detector (34). It is expected that detectors such as this will play increasingly important roles in blocking lifetime measurements.

[2] E. Laegsgaard, to be published.

A typical experimental arrangement for a blocking lifetime measurement that utilizes a thick crystal target is shown in Figure 5. This was used by Sharma et al to determine the lifetime of ^{32}S compound nuclei produced by a ^{31}P(p, α) ^{28}Si reaction (32).

In practice it is not sufficient to measure an emergent particle blocking pattern and compare it with a theoretical distribution in order to obtain a lifetime. Scattering in the crystal or at the crystal surface can also modify the theoretical "prompt" distribution. Consequently a single measurement can generally allow only a lower lifetime limit to be set (28). In addition to crystal imperfections, thermal vibrations and detector angular resolution can lead to changes in the blocking pattern. Also the low reaction cross section often means that so much beam current is needed for a measurement that radiation damage and surface impurity film buildup can sometimes occur during the measurement. These effects can also change the blocking pattern. It is necessary therefore to obtain a reference blocking pattern that does not contain the lifetime effect. This should be obtained under the same conditions and preferably at the same time as the measurement. One way to do this is shown in

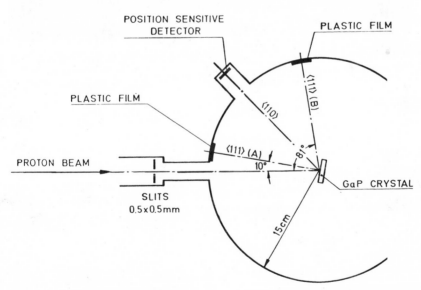

Figure 5 Experimental arrangement used for lifetime measurements of ^{32}S compound nuclei formed in ^{31}P(p, $\alpha)^{28}$S reactions. The position-sensitive detector was used to monitor crystal alignment and radiation damage by observing the blocking pattern produced by elastically scattered protons. This is an example of a general class of lifetime measurements which utilize backscattering geometry and thick targets (from reference 32).

Figure 5, where the crystal orientation was chosen so that the blocking pattern for emergence along a $\langle 111 \rangle$ axis at 170° to the incident beam direction, A, is measured simultaneously with a blocking pattern for emergence along a $\langle 111 \rangle$ axis at 99° to the beam, B. The compound nucleus recoil is nearly parallel to axis A so it will have very small velocity component away from this particular atomic row. It should therefore not contain an appreciable lifetime effect and can be used as a control for axis B. An example of the blocking patterns obtained with this apparatus for α particles emitted in the $^{31}P(p, \alpha)$ ^{28}Si reaction for 0.642 MeV protons incident on a GaP crystal is shown on Figure 6.

The other technique used to provide a "prompt" control measurement is used in thin target measurements with an experimental arrangement like that shown in Figure 7. This arrangement uses a single position-sensitive, energy-sensitive detector that differentiates the "prompt" elastically scattered beam particles, which can then be used as a control for the "slow" inelastically scattered particles as shown on Figure 8. Each of the lower energy peaks corresponds to inelastic proton emission to specific excited states of the daughter nucleus. Blocking distributions

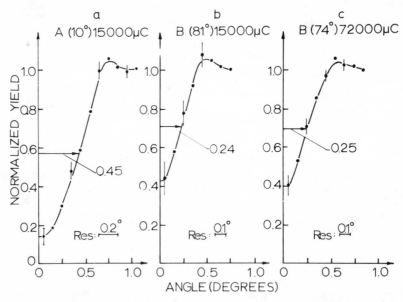

Figure 6 Blocking patterns for helium ions from $^{31}P(p, \alpha)^{28}S$ resonance reaction at 0.642 MeV near the surface (~ 1000 Å) of a single crystal of GaP. The dips were obtained by counting helium ion tracks in the plastic film in $\frac{1}{4}$ mm rings around a $\langle 111 \rangle$ direction. These results were obtained using the apparatus shown in Figure 5 (from reference 32).

for the elastic peak and a number of the inelastic groups are shown on Figure 9.

In each case the nuclear lifetime information must be derived from a measured blocking distribution and its associated control distribution. We now consider the present status and problems associated with such analysis.

ANALYSIS OF BLOCKING LIFETIME DATA

Particle channeling effects have been treated successfully by two techniques: an analytical technique based on a classical single-string continuum model (17) and by a binary scattering Monte Carlo technique (35). Because of the quantitative relationship between channeling and blocking, it is natural that these have initially formed the basis for analysis of blocking lifetime measurements.

Single-String Continuum Calculations

A theory of particle motion through a crystal lattice that assumes that the atomic rows can be represented by a continuum string of positive

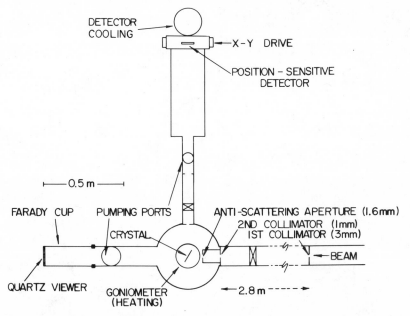

Figure 7   Schematic representation of experimental arrangement used to measure blocking distributions of protons scattered at 90° in thin germanium crystals (from references 50–52).

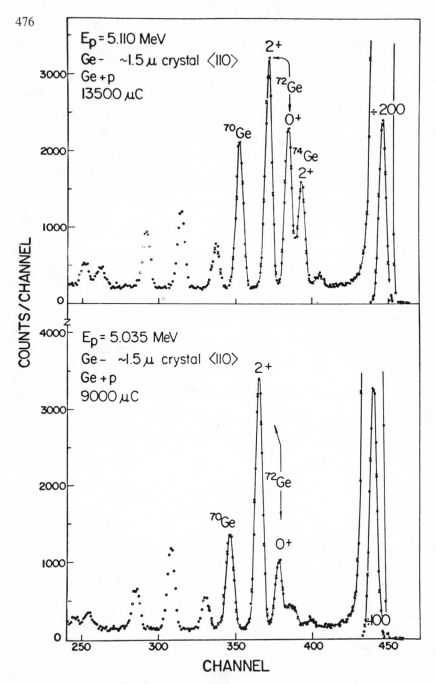

Figure 8 Pulse height spectra of scattered protons measured by the two-dimensional detector in the experimental setup of Figure 7. Spectra are shown for

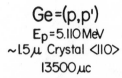

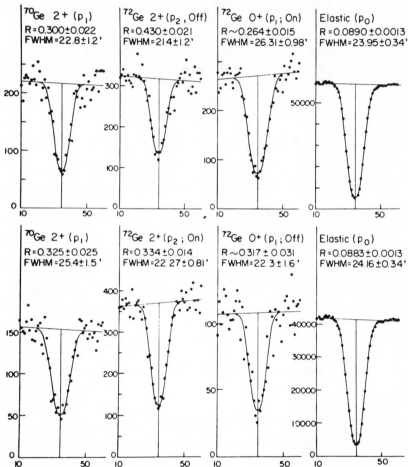

Figure 9 Blocking distributions for emission of protons along a ⟨110⟩ axis in a 1.5 μm thick Ge crystal for the elastic and selected inelastic proton groups from the spectra of Figure 8. The two groups of blocking dips correspond to the two incident proton energies of Figure 8. The solid curves are computer fitted. The *R* value indicated in each case is the same as the minimum yield χ discussed in the text (from reference 50).

---

two different incident proton energies chosen to correspond to analogue resonance energies in ^{71}As and ^{73}As compound nuclei formed by protons on Ge^{70} and Ge^{72} (from reference 50).

charge, and atomic planes by a continuum sheet of charge, has been enormously successful in describing many aspects of particle channeling (17). A further simplifying assumption is that particle trajectory distributions are in statistical equilibrium (that is, that they have uniform probability of being found in any spatial region of the crystal in which they are energetically allowed); in its simplest form the analysis assumes cylindrical symmetry of the potential around each atomic row. The usefulness of this approach lies in the very simple analytical expressions for important features of channeling or blocking distributions that result. For example, one of the distinctive and sensitive features of a blocking distribution is the yield of particles emergent from the crystal parallel to the atomic row or plane. This is called the minimum yield, χ, and is expressed as a fraction of the yield at large angles where the yield does not vary with the angle. Assuming a continuum potential, the minimum yield for an axis is given by

$$\chi = \int_s f(x,y)g(x,y)\,dx\,dy, \qquad\qquad 1.$$

where the x and y coordinates are perpendicular to the atomic row with their origin on the row and the area occupied by each row $S = 1/Nd = \pi r_0^2$, where N is the atomic density, d is the atomic spacing along the row, and r_0 is the effective radius of the cylinder occupied by each row. The normalized distributions of flux density and emitter atoms are represented by $f(x,y)$ and $g(x,y)$, respectively. Separating the flux density into contributions from channeled and random (nonchanneled) particle trajectories and assuming statistical equilibrium, we find that the flux density as a function of the distance r from the string is

$$f(r) = \pi Nd(C_1\rho^2 + C_2 r^2), \qquad\qquad 2.$$

which gives a value for the minimum yield of

$$\chi = \pi Nd(C_1\rho^2 + C_2\langle r^2\rangle), \qquad\qquad 3.$$

where $\langle r^2\rangle$ is the average of r^2 over the distribution g. If, in addition to thermal vibration, there is a recoil displacement perpendicular to the row (along the x axis) with an exponential decay of mean lifetime τ (or mean recoil distance $v_\perp\tau$ where $v_\perp$ is the velocity of the recoil normal to the row) then

$$g(x,y) = \frac{1}{\pi\rho^2(v_\perp\tau)}\int_{-\infty}^{x}\exp\left(-\frac{x'^2+y^2}{\rho^2}-\frac{x-x'}{v_\perp\tau}\right)dx', \qquad 4.$$

which gives a minimum yield

$$\chi = Nd\pi\{(C_1+C_2)\rho^2+2C_2(v_\perp\tau)^2\}. \hspace{3cm} 5.$$

Evaluations of the constants from experiments (36, 37) and computer simulations (38, 39) give

$$C = C_1+C_2 \simeq 3.0$$

and

$$D = C_2 \simeq 1.3\pm0.2.$$

Physically, the first term in equation 5 can be interpreted as being due to the dependence of the fraction of random (nonchanneled) particles on the thermal vibrational amplitude, whereas the second term, involving the constant C_2, arises from sampling by the distribution of emitter atoms of the region of higher flux that comes from the channeled trajectories. It has further been suggested (17) and confirmed experimentally (40) that other contributions to the minimum yield, such as multiple scattering from lattice imperfections, surface films, etc are additive, so the minimum yield is given by

$$\chi = \chi_1+2\pi NdD(v_\perp\tau)^2, \hspace{3cm} 6.$$

where χ_1 contains all effects of thermal vibration, multiple scattering, etc. It is important to use control measurements to avoid uncertainties due to the hard to control, or to evaluate χ_1 contributions. For a well-designed control measurement the second term of equation 6 will not contribute, but all other effects will be the same. The difference in minimum yield between the measurement and the control should therefore depend only on the mean recoil distance $(v_\perp\tau)$ of the compound nucleus

$$\Delta\chi = \chi-\chi_1 = 2\pi DNd(v_\perp\tau)^2. \hspace{3cm} 7.$$

For planar blocking a similar derivation gives

$$\Delta\chi' = C'\frac{v'_\perp\tau}{d_p} \hspace{3cm} 8.$$

where the primes indicate planar geometry, d_p is the spacing between planes, and $C' = 1.17+0.36$ is evaluated experimentally from α particles emitted from the ^{27}Al(p, α) ^{24}Mg reaction in aluminum crystal (37). The simplicity of equations 7 and 8 makes them very attractive for BLT determinations. It is apparent, however, from the form of these equations that they cannot be valid for large values of the mean recoil distance since the minimum yield could become arbitrarily large. This is a consequence of the assumption of a single-string potential. As we shall see, deviations become appreciable for recoil distances greater than $\sim$0.3 Å. With the recognition that at some distance, r_c, from the atomic row the particles are no longer steered by the row potential, a modified equation that has the proper asymptotic behavior has been derived (29)

$$\Delta\chi = \chi - \chi_1 = \frac{2D(v_\perp\tau)^2}{r_0^2}\left\{1 - \left(1 + \frac{r_c}{v_\perp\tau}\right)\exp\left(-\frac{r_c}{v_\perp\tau}\right)^2\exp\left(-\frac{r_c}{v_\perp\tau}\right)\right\}$$
$$+ \exp\left(-\frac{r_c}{v_\perp\tau}\right), \qquad 9.$$

where the first term includes those particles emitted inside r_c and the second, those emitted at distances larger than r_c. The appropriate value of r_c is expected to depend on the amount of multiple scattering, which in turn is dependent on crystal quality and other experimental parameters. Values of ~ 0.4 Å have been used (24, 29) based on qualitative arguments but recent comparisons to results of computer simulations (41) indicate

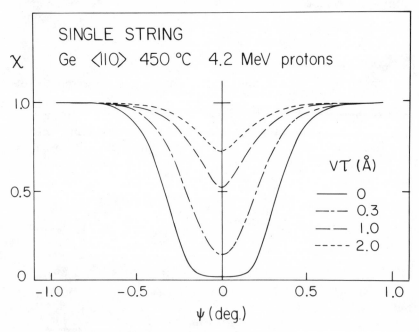

Figure 10 Blocking distributions for emission of 4.2 MeV protons along a $\langle 110 \rangle$ direction in germanium calculated for different values of the mean recoil distance, $v_\perp\tau$, according to the single-string continuum model of reference 17. The calculation assumes cylindrical symmetry of the potential about the atomic row and also assumes statistical equilibrium of particle trajectories. Surface transmissions effects are included. This corresponds to the single-string calculation indicated in Table 1.

that $r_c \approx r_0$ may be more appropriate. Other parameters of blocking distributions can also be calculated on the basis of the single-string continuum model. For example, the change in area of $\{111\}$ planar dips in germanium crystals was calculated for this model by Komaki & Fujimoto (42). Indeed, following the procedure suggested by Andersen (43) the entire blocking angular distribution can be calculated using the same model. A set of such calculated distributions for 4.0 MeV protons emitted from a germanium crystal along a $\langle 110 \rangle$ axis for different values of the mean recoil distance is shown in Figure 10.

Although the single-string continuum theory affords the most convenient, and until now the most used, framework for analysis of blocking lifetime data, it suffers from a serious and perhaps even fatal weakness, at least for axial blocking. This involves the assumption of cylindrical symmetry relative to the atomic row from which the recoil atom originates. For phenomena that take place within about 0.5 Å of the row, this is a reasonable assumption. At larger distances, however, the potential contours deviate appreciably from cylindrical symmetry, as shown on Figure 11 for a $\langle 110 \rangle$ axis in germanium. It is apparent from the impor-

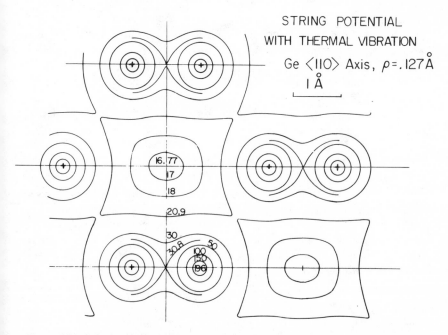

Figure 11   Calculated potential contours in the plane normal to a $\langle 110 \rangle$ direction in germanium. A Molieré approximation to the Thomas-Fermi screened atomic potential was assumed and averaged along the $\langle 110 \rangle$ atomic rows.

tance of the second term of equation 9 that the long tail on the exponential decay function plays an important role, especially in determining the emission yield parallel to the atomic row. Consequently the assumption of cylindrical symmetry becomes important even for mean recoil distances as short as 0.2–0.3 Å.

Monte Carlo Computer Simulations

The first indication that the assumption of cylindrical symmetry could be introducing significant quantitative errors in analysis of BLT results came from a large-scale computer simulation of the scattering of 4.0 MeV protons in germanium (38). This calculation considers charged particle motion through a crystal lattice as independent binary scattering of the particle from each successive lattice atom encountered. By such a Monte Carlo technique the time history of an individual particle trajectory is determined and recorded on magnetic tape. This is repeated over and over for additional particles until the statistical distribution of particle positions and directions at various points in the lattice and also outside the lattice is determined. For a long time such calculations have represented an alternative to analytical calculations based on the single-string continuum model and have been carried out in considerable detail in investigations of particle-channeling phenomena (35, 44, 45). Such calculations for particle channeling have shown good correlation with many features of single-string calculations (46) and have afforded useful insight into multiple scattering effects in the crystal, as such effects are included in a natural way in the computer simulation if the lattice atoms are allowed to vary in their position randomly over a range and distribution determined by the Debye temperature. It was primarily to investigate such thermal effects and to assess the values of the parameters D and r_c in equation 9 that the computer simulation of proton scattering in germanium was undertaken. Application of this approach to analysis of blocking determinations of nuclear lifetimes was first used by Massa (47) and Sona (48) and has been further developed by Clark et al (49). Unfortunately, the huge amount of computer time required often restricts such simulations to limited emission depths, numbers of particles, and emission conditions. In addition, if the blocking geometry is used where the particles originate from inside the lattice, each different initial condition, determined by different decay times, decay functions, recoil directions, etc, requires recalculation of an entirely new set of particle trajectories. It was recognized, however, (38) that an important reduction in computation effort could be realized by turning the calculation around, calculating the particle trajectories for a channeling geometry, and using the "rule of reversibility" (17) to apply it to the blocking case. In this way, the same set of particle trajectories could be used over and over for different lattice vibrations, compound nuclear decay functions, and recoil directions, as well as for investigations of particle flux distributions, depth effects, and other parameters.

A large-scale Monte Carlo computer simulation was then undertaken for proton scattering in a thin germanium crystal using the same conditions and geometry as have been used in an extensive experimental investigation of this same system (50–52). The crystal orientation used in the experiments was such that the recoil was nearly parallel to a $\{111\}$ plane. The minimum yield, χ, calculated for this recoil direction by the computer simulation, had the property that it approached a limiting value of ~ 0.4 instead of 1.0 at large recoil distances. Investigation of the origin of this surprising result indicated a strong dependence of the minimum yield on recoil direction (38). This is shown in Figure 12. Also shown in this figure are values calculated from equation 7 and equation 9, which have been normalized at $v_\perp \tau = 0$.

The recoil dependence of the minimum yield is a direct result of the

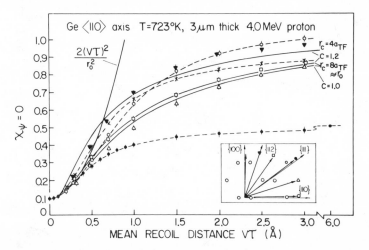

Figure 12 Calculation of the minimum yield, χ, for emission of 4.0 MeV protons along a $\langle 110 \rangle$ direction in germanium as a function of the mean compound nuclear recoil distance, $v_\perp \tau$. The points were obtained from a Monte-Carlo computer simulation calculation for different recoil directions in the transverse plane as shown in the inset. The points corresponding to recoil nearly parallel (1.5° away) from three major low index planes are joined by the dotted lines. Other directions are also shown. The recoil directions used for the open squares and open triangles were chosen to be random (as far as possible from planes) directions. The solid curves were calculated from the single-string continuum model using equation 7 or equation 9 in the text. These curves were normalized to the MC calculations at $v_\perp \tau = 0$ (from reference 39).

potential contour distribution shown in Figure 11 and is closely related to flux peaking effects exhibited by channeled particles (45). In fact the same set of calculated particle trajectories used to obtain the minimum yield values can be used to obtain the flux distribution of channeled particles. This is shown in Figure 13 for protons incident parallel to a $\langle 110 \rangle$ axis that have penetrated 3 μm into the crystal. The regions of high particle flux correspond to points from which particles emitted in the row direction can move easily through the crystal. Recoils originating at atomic lattice sites that move nearly parallel to $\{111\}$ planes avoid regions of high flux density, which accounts for the low value of χ for that recoil direction as compared with recoil along either $\{100\}$ or $\{110\}$ planes. It should be noted from Figure 12 that Monte Carlo calculations of the minimum yield for two different recoil directions chosen to avoid low index planes (so-called random directions) are in good agreement

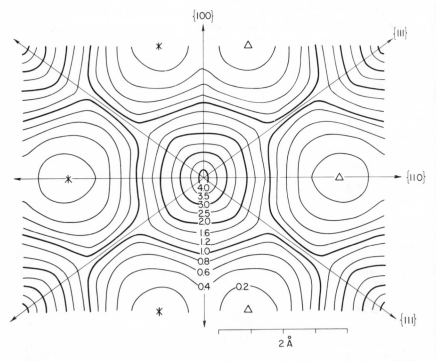

Figure 13 Flux density contours for 4.0 MeV protons incident parallel to a $\langle 110 \rangle$ axis in germanium evaluated for the depth range of 0–3.0 μm into the crystal by a large-scale Monte Carlo calculation. The triangles and stars correspond to mean lattice positions (in two different horizontal planes) for atoms in the $\langle 110 \rangle$ atomic rows.

with each other and with the results of single-string continuum model calculations using the "Gibson-Nielsen formula" (equation 9) for $D = 1.0$–1.2 in agreement with the discussion given above [as opposed to values of ~ 2.5 used previously (29)] and for $r_c \sim r_0$ [instead of $r_c \sim 3$–4 a_{TF} used previously (29)]. The Thomas-Fermi screening distance (a_{TF}) is defined by $a_{TF} = 0.8853\ a_0(Z_1^{1/2} + Z_2^{1/2})^{-2/3}$ (a_0 is the Bohr radius, and Z_1 and Z_2 are the proton numbers of the emitted particle and lattice atom respectively). It is not yet known if this apparent agreement is general and can be applied to other materials or particles, but it gives some encouragement that perhaps, under carefully controlled experimental conditions, the convenient single-string analysis can be used.

Computer simulations can be used not only to determine single parameters of the blocking distribution such as the minimum yield or dip width or area, but at the expense of increased computation time can also give the shape of the blocking distribution. Sona (48) has explored the shift in the blocking dip as a function of the mean recoil distance, target thickness, and target temperature for 12-MeV protons in silver crystals. Massa (47) has computed blocking dips for 4.3-MeV protons emitted along $\langle 110 \rangle$ axes in germanium for $(v_\perp \tau) = 0$ and $(v_\perp \tau) = 0.2$ Å. To reduce the computation effort required, it was necessary in the latter case to fold the distribution around the axial direction. In addition, the emission depth was limited to 2000 Å and a cutoff in the interaction potential of 0.5 Å was applied. Unfortunately, such restrictions have a strong effect on the results and may be the reason that the minimum yield calculations for protons in germanium of Fuschini et al (53) deviate from the relevant ($\{110\}$ recoil direction) curve of Figure 12. The large-scale Monte Carlo calculation of 4-MeV protons in germanium which takes advantage of a channeling geometry calculation of particle trajectories (38) has been used to calculate the emission patterns along a $\langle 110 \rangle$ axis corresponding to an azimuthal slice through the axis along a line tilted 16.5 degrees from a $\{100\}$ plane for different recoil directions and mean recoil distances. The results are shown in Figure 14. This azimuthal direction corresponds to the orientation of a one-dimensional position-sensitive detector used in a study of experimental proton scattering in germanium (50–52) using the apparatus shown in Figure 7.

The drastic changes in shape of the blocking distribution for different recoil directions shown in Figure 14 indicate the necessity to include the multistring potential of Figure 11 in analyzing any case in which the mean recoil distance exceeds about 0.3 Å. Even if it is possible to devise a general and simple analytical technique that could be used in particular situations such as for random recoil directions, the recoil direction dependence remains an intriguing additional tool for extending BLT

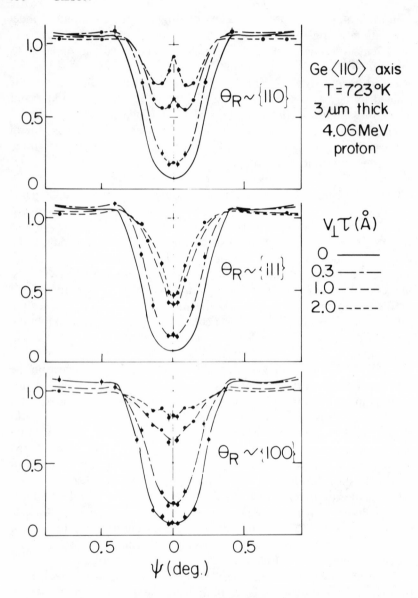

Figure 14 Blocking distributions for 4.06 MeV protons emerging from a 3.0 μm thick germanium crystal along a ⟨110⟩ direction for different recoil directions in the transverse plane, θ_R, and different mean recoil directions, $v_\perp \tau$. These were obtained from a large-scale computer simulation and are taken from reference 38. This calculation corresponds to the Monte Carlo calculation indicated in Table 1.

measurements beyond the determination of "mean" lifetimes to determination of the shape of the decay function of the compound nucleus.

Although large-scale Monte Carlo computer simulations can be used for sensitive and quantitative interpretation of experimental results, it is impractical to consider such simulations for every possible case of interest. It is therefore important to consider techniques intermediate between the efficient and simple single-string continuum calculation and the expensive and tedious Monte Carlo computer simulation.

Multistring Continuum Calculations

It is important to include the potential distribution in the lattice produced by other than just the single atomic row from which the recoil nucleus originates. In addition to the shape of the potential used, there are other differences between the single-string continuum calculations and the Monte Carlo computer simulations. Probably the two most important additional differences are the absence of multiple scattering effects and the assumption of statistical equilibrium in the single-string calculation. An attempt has been undertaken (54–57) to investigate the importance of each of these effects and hopefully to arrive at a compact, convenient, and general analytical technique that still contains all of the important elements for reliable and sensitive quantitative analysis. The various possible approaches to achieving this result can be defined in terms of the approximations used in each. They include the following in order of increasing convenience.

MULTISTRING NONSTATISTICAL EQUILIBRIUM   The multistring non-statistical equilibrium (MNE) calculation also involves calculation of individual particle trajectories (54). A reduction in computation effort of about a factor of ten relative to the full Monte Carlo calculation can be realized by solving the integrals for the equation of motion of the particle through the lattice at intervals larger than the lattice spacing interval for scattering used in the Monte Carlo calculation, when the particles are far away from the rows, which is most of the time. Calculated blocking distributions for 4-MeV protons in germanium are shown in Figure 15. Comparison with the curves of Figure 14 shows that the MNE calculations give blocking dips that are generally broader and deeper than the corresponding MC curves and exhibit more prominent shoulders. These differences are attributable to the fact that multiple scattering effects are present in the MC but not the MNE calculations. Figure 16 shows a direct comparison of the two types of calculation, which shows the differences to be most pronounced for small values of the recoil distance, $v_\perp \tau$, in agreement with this conclusion. Also

488

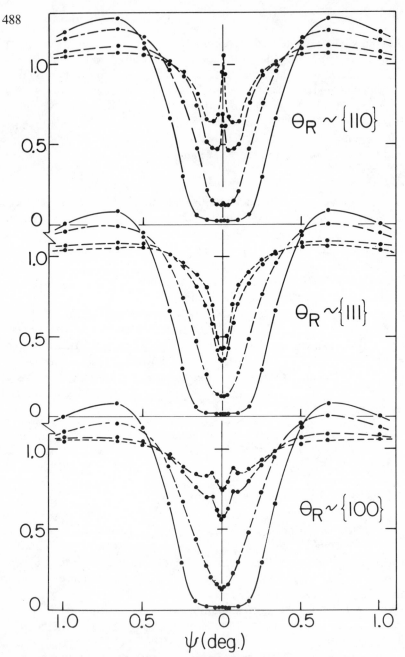

Figure 15 Blocking distributions for 4.0 MeV protons emerging from a 3.0 μm thick germanium crystal along a ⟨110⟩ direction for different recoil directions in the transverse plane, θ_R, and different mean recoil distances of 0, 0.3, 1.0, and 2.0 Å. Calculated from a multistring nonstatistical equilibrium model (from reference 54).

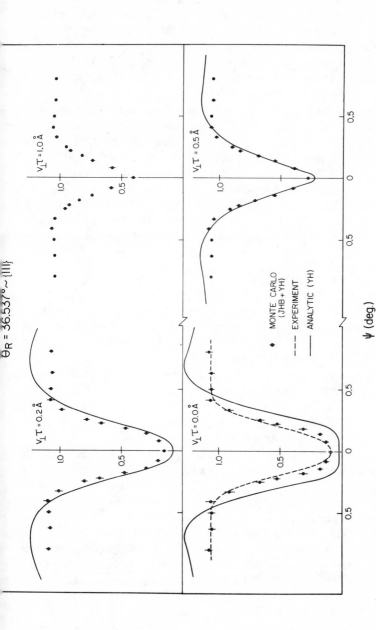

Figure 16 Comparison of blocking distributions for 4.0 MeV protons emerging from a 3.0 μm thick germanium crystal along a $\langle 110 \rangle$ direction for recoil 1.5° from a $\{111\}$ planar direction and for different mean recoil distances, $v_{\perp}\tau$. The points are from a large-scale computer simulation calculation, and the solid lines are from a multistring nonstatistical equilibrium continuum calculation. The dotted curve is an experimental measurement of a blocking distribution of emergent 4.0 MeV protons obtained from elastic proton scattering (from reference 57).

shown for $v_\perp \tau = 0$ is the experimentally measured blocking distribution for elastic scattering of 4.2-MeV protons in germanium, which illustrates the good agreement with the MC results. The increased computational efficiency of the MNE model has made possible a systematic and detailed investigation of many different aspects of particle motion in the crystal lattice (54, 55). One result of this systematic study is the observation that the particle flux distribution in the channel is not constant with depth but continues to exhibit oscillations even after the particle beam has traversed a 3 μm thickness. This indicates that at least in the absence of multiple scattering, statistical equilibrium is not established. Multiple scattering is expected to increase the rate of development of statistical equilibrium in the lattice, and the depth dependence of the flux distribution obtained from the MC calculations (38) is consistent with this expectation even though the number of events and thickness segments that could be studied was limited by the computational effort required. Even for this case, however, it was apparent that fluctuations persisted through much of the crystal thickness, indicating an absence of statistical equilibrium in particle trajectory distributions.

MULTISTRING STATISTICAL EQUILIBRIUM   Multistring statistical equilibrium (MSE) calculations represent the simplest analytical approach that retains the geometry of the potential contours in the crystal. Assumption of statistical equilibrium in the particle trajectory distribution allows the calculation of blocking distributions without calculation of individual particle trajectories, giving a very large increase in simplicity and a considerable decrease in the computation effort required. The calculation basically involves direct determination from potential distributions of the type shown in Figure 12 of a flux distribution for each particle direction of interest. The recoil direction and decay distribution of the excited nucleus is chosen and folded into the flux distribution to give the yield. Although evaluation of the arrays used in the calculation requires a computer, the computation time and effort is reduced perhaps another factor of ten. The main concern with these calculations is connected with the assumption of statistical equilibrium since the MC and MNE calculations indicated that statistical equilibrium was not achieved through much of the 3-μm target thickness. In the experiment, particles originate at all depths. This gives effective depth averaging, so the important question becomes: Does averaging over the fluctuations in flux density correspond to using the average flux density? To investigate this possibility, calculations have been carried out for the same case as described above, 4.0-MeV protons along a ⟨110⟩ axis in germanium (56, 57). The calculated blocking distributions are shown in Figure 17.

In this calculation the absence of shoulders on the distributions arises from use of the simplest form of the continuum theory (17) [as opposed to the so-called half-way plane calculation (43) used in connection with Figure 10]. It is clear, however, that in the important region of the minimum, the qualitative and, for most cases, quantitative features shown by the MNE calculations are well reproduced. Assumption of statistical equilibrium therefore appears to be a reasonable approximation, at least for this case. It may be important to average over at least one full period of the depth oscillations, which means that for much thinner samples (~ 1000 Å) the assumption may not be valid. The MSE calcula-

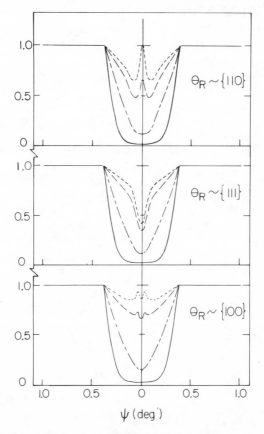

Figure 17 Blocking distributions for 4.0 MeV protons emerging from a crystal along a $\langle 110 \rangle$ direction for different recoil directions in the transverse plane, θ_R, and different mean recoil distances of 0, 0.3, 1.0, and 2.0 Å. Calculated from a multistring statistical equilibrium model (from reference 56).

tion has one additional advantage. Because of its simplicity and the form of the analytical expressions employed, it is more amenable to inclusion of multiple scattering effects than are the MNE calculations. Such inclusion will make use of the understanding and formalism of the analysis devised (58) to explain the depth dependence of particle dechanneling. An investigation of this possibility is now underway.

A summary of the similarities and differences between the four types of analyses discussed above is shown in Table 1. A similarly detailed study of planar blocking has not been carried out, although it is expected that many of the complications encountered with axial blocking will be absent in the planar case. For some cases planar geometry may be preferred (59). On the other hand, the weaker potential of planes makes them less sensitive to small changes in the decay distribution, and there is no possibility of using the recoil direction dependence to get information about the shape of the decay function.

Important Loose Ends

DEFLECTION OF RECOIL   In all of the analyses described above, the recoil nucleus has been assumed to move in a straight line. When the recoil

Table 1 Comparison of the assumptions and properties contained in the four analytical techniques discussed in the text

Technique	Single String (SS)	Multistring Statistical Equilibrium (MSE)	Multistring Nonstatistical Equilibrium (MNE)	Monte-Carlo Computer Simulation (MC)
Potential	$U_{SS}^T(r_\perp)$	$U_{MS}^T(\mathbf{r}_\perp)$	$U_{MS}(\mathbf{r}_\perp)$	$V(\mathbf{r}, t)$
Recoil Direction Dependence	No	Yes	Yes	Yes
Assumption of Statistical Equilibrium	Yes	Yes	No	No
Trajectory Calculation	No	No	Yes	Yes
Depth Dependent	No	No	Yes	Yes
Conservation of Transverse Energy	Yes	Yes	Yes	No
Multiple Scattering Included	No	No	No	Yes

distance becomes as large as the interatomic spacing in the lattice, however, the charged recoil can be deflected by neighboring atoms. The results of such an effect as calculated by the large-scale computer simulation (MC) described previously (38) are shown in Figure 18. This deflection has been found to be important in analysis of experimental measurements of inelastic proton scattering in germanium (50–52).

COMPLEX DECAY FUNCTION   Another and more serious limitation of all the analyses discussed up to this point is the assumption of a simple, single-component, exponential decay function. Such an assumption is justified in the case where a single isolated nuclear state is excited (32, 33) but this is limited to special cases in light nuclei. In most cases a large number of states can be excited in the compound nucleus, each with its own characteristic decay time (level width). The decay function expected from inelastic proton scattering of 5.4-MeV protons from ^{70}Ge has been calculated by Malaguti et al (60) from a statistical nuclear model of the

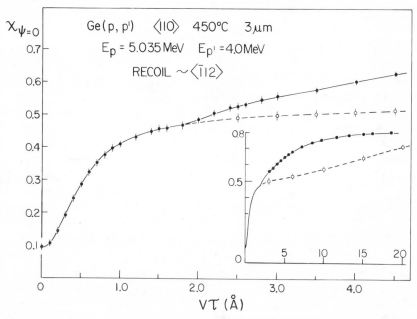

Figure 18   Calculation of the minimum yield, χ, for emission of 4.0 MeV protons along a $\langle 110 \rangle$ direction in germanium as a function of the mean compound nuclear recoil distance, $v_\perp \tau$. The initial recoil direction was $\sim 1.5°$ from a $\langle 112 \rangle$ axial direction and the open circles indicate the expected value if deflection from the adjacent atomic rows is neglected. This is a Monte Carlo calculation (from reference 57).

494   GIBSON

reaction. The calculated probability function decays more slowly than would be expected from a simple exponential (22) and has a prominent tail extending as much as a factor of two beyond the effective cutoff of the exponential form. Even more drastic deviations from exponential decay may be expected in other cases. An important challenge to future BLT studies will be to determine details of the form of the decay function. As we mention briefly in the next section, some initial attempts to accomplish this have already been made.

BLOCKING LIFETIME EXPERIMENTS

Resonance Reactions

The simplest experimental case to analyze, but not necessarily the simplest to carry out, involves cases in which a single state in the compound nucleus is excited which in turn decays through a single decay channel. In this case, the decay function is a simple exponential. If, at the same time, information on the lifetime of the state is independently available (as from resonance analysis), then a straightforward testing and calibration of the lifetime measurement technique becomes possible. Results from two such measurements have been reported. These are for resonant ^{27}Al(p, α) ^{24}Mg (33) and ^{31}P(p, α) ^{28}Si (32) reactions. Plastic track detectors were used in each case to observe the blocking patterns of the α-particle reaction products and to discriminate against the large background of scattered protons. We have already shown in Figures 5 and 6 the experimental arrangement and results of one such experiment. These results give a lifetime of $8^{\pm 4} \times 10^{-15}$ sec for the 9.486-MeV excited state of the ^{32}S compound nucleus, which corresponds to a level width $\Gamma = 8 \pm 4$ eV. This can be compared with the $\Gamma = 8 \pm 2$ eV calculated from the measured strengths of the ^{31}P(p, α), ^{31}P(p, γ), and ^{28}Si(α, γ) resonance reactions feeding the same 9.486-MeV level in ^{32}S. Resonance reactions have been used to demonstrate the effectiveness of changing the angle of recoil relative to an atomic row to extend the lifetime range investigated and to test the analytical technique over a range of effective recoil distances (32). Resonance reactions have also been used to investigate the physical origin of the constants in equation 5 (39) and equation 8 (37). It is clear that further systematic studies utilizing isolated resonances will be an important means of extending our understanding and the usefulness of blocking lifetime measurements. For example, the α-particle resonances in the ^{31}P(p, α) ^{28}Si reaction are sufficiently sharp and well separated that they can be made to take place at well-defined and shallow depth regions in the crystal by suitable choice of the incident proton energy. This depth selection can be used to investigate general depth (dechanneling) corrections and could also give information pertaining to the important statistical equilibrium questions discussed in the preceding section.

Fission

Nuclear fission studies were among the first to utilize the blocking technique (28–30) and continue to be of considerable current interest. The early studies used plastic or glass track detectors and have been reviewed previously (22). Recent measurements have also used glass track detectors but have also increasingly made use of two-dimensional position-sensitive detectors such as that used to give the blocking pattern shown in Figure 4. Recent interest has been centered on neutron induced fission of ^{238}U and on heavy ion induced fission of elements of $Z > 75$. We consider these in turn.

NEUTRON INDUCED FISSION   The early studies, particularly of the proton induced fission of uranium (29), showed that in fission of uranium by light particles, conditions are suitable for BLT measurements only when the compound nuclei have low excitation energies, near the fission barrier. This means that in charged particle induced fission, the decay of the initial compound nucleus could not be measured because of its high excitation energy (due to the energy necessary to get the bombarding particle over the coulomb barrier of the target nucleus) and that the effects observed were from decay of second or higher chance fission after one or more neutrons were emitted. The resulting uncertainty in excitation energy and fractional contribution of the compound nuclei involved precludes very sensitive or detailed analysis of the resulting blocking distributions. The most natural method for producing well-defined excitation energies for the fissioning nucleus and for avoiding uncertainty about the source of observed fission fragments is to use monoenergetic neutrons of energy up to a few MeV. Two groups of experimenters have carried out these very difficult but important experiments (61, 62). Neutrons of energy 1.7, 1.8, 2.3, and 3.6 MeV produced by the ^{3}H(p, n)^{3}He reaction were incident on a UO$_2$ crystal oriented with a $\langle 111 \rangle$ axis along the neutron direction (the "control") and another at $\sim 71°$ to the neutron direction. The observed differences in the blocking distributions for these two geometries have been analyzed assuming that the decay function could be approximated by a simple exponential giving mean lifetime values shown on Figure 19 as a function of the excitation energy of the ^{239}U compound nucleus. The theoretical (solid) curve was calculated from a Fermi gas model. The mean lifetimes were calculated by equation 7 with the value of the constant $D = 2.5$. This is a factor of two higher than has been

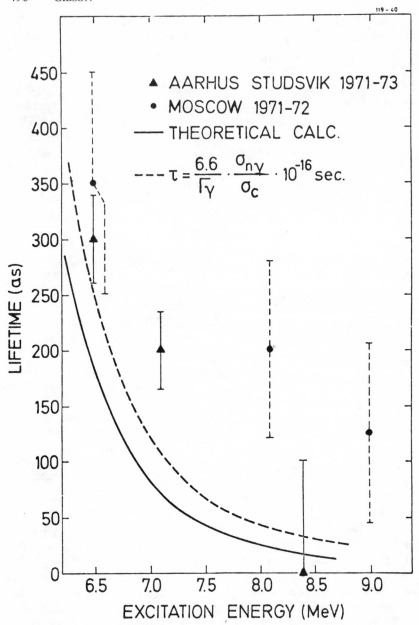

Figure 19 Mean lifetimes obtained from measured blocking distributions as a function of compound nuclear excitation energy for neutron induced fission of ^{238}U. The theoretical calculation is from a statistical nuclear Fermi gas model (from reference 61).

found from a variety of experiments and computer simulations, as discussed previously. Although the use of a thick crystal with possible increase of multiple scattering may increase the value to be used, it should be noted that the low absolute values of $\chi = 0.10$–0.15 obtained for the control measurement in the Danish-Swedish experiments (16), and the good agreement with the results obtained in the USSR that have higher χ values, imply that perhaps a lower value for D of 1.0–1.2 is still applicable. In addition, there is possible systematic error because of the single-string assumption used in the analysis and, probably more seriously, because of the assumption of a single exponential decay function. The uncertainties indicated for the experimental points in Figure 19 are statistical and do not contain possible systematic errors of this type. In view of these possible sources of error, it is not expected that the absolute difference between the experimental points and the theoretical curve is significant.

A similar qualitatively good agreement was obtained between these same experimental results and a "constant temperature" nuclear model in which the level density is constant with excitation energy over the range of excitation energies considered (24). The principle difference between the two nuclear models considered, *Fermi gas* and *constant temperature*, appears at excitation energies above ~ 8 MeV. At this excitation energy and above, the constant temperature model predicts that the lifetime is independent of excitation energy with a value of ~ 25asec $(2.5 \times 10^{-17}$ sec), whereas that predicted by the Fermi gas model continues to decrease with increasing excitation. It is also at the highest excitation energies that apparent deviations between the two sets of measurements begin to appear. Difficult as these measurements are, further and statistically more precise determinations at excitation energies close to the second chance fission threshold would be very desirable to help resolve these questions.

HEAVY ION INDUCED FISSION   Some of the most surprising and intriguing BLT results that have been reported are the observation of measurable lifetime effects for boron, carbon, oxygen, neon, and phosphorus ion induced fission of tantalum and tungsten isotopes (31, 40, 63). From a systematic and detailed study of the excitation energy and compound nuclear mass dependence of the lifetime (24), the results of which are shown in Table 2, it was concluded that for highly excited compound nuclei with $79 < Z < 82$ a Fermi gas model was appropriate, and for $Z > 85$ a constant temperature model was needed. In both cases, however, the expected level density yielded lifetimes at least a factor of ten shorter than those observed. In order to obtain agreement it was necessary to use level density parameters $a_f \sim a_n \sim A/4$–$A/5$ which give level densities

Table 2 Experimental measurements of minimum yield ($\Delta\chi$) changes for heavy ion fission of tungsten and tantalum (from ref. 24)

| | | | | $\tau_{\text{exp}} \times 10^{-19}\,\text{sec}^{\text{a}}$ | | $\tau_{\text{theor}} \times 10^{-19}\,\text{sec}$ | |
| | E_{ion} | E^* | | r_c | r_c | $a_n = a_c$ | $a_n = a_c$ |
Reaction	(MeV)	(MeV)	$\Delta\chi$	$= 0.4$ Å	$= 0.2$ Å	$= A/8$	$= A/4$
$^{186}\text{W}(^{11}\text{B},f)$	87	80	0.096 ± 0.014	69	38	0.79	16
$^{186}\text{W}(^{12}\text{C},f)$	80	64	0.087 ± 0.015	67	37	1.1	22
$^{186}\text{W}(^{12}\text{C},f)$	108	89	0.007 ± 0.015	$\leqq 19$	$\leqq 11$	0.29	4.2
$^{181}\text{Ta}(^{16}\text{O},f)$	96	63	0.104 ± 0.047	60	32	3.7	130
$^{181}\text{Ta}(^{16}\text{O},f)$	137	101	0.004 ± 0.035	$\leqq 33$	$\leqq 18$	0.45	8.1
$^{186}\text{W}(^{16}\text{O},f)$	97	67	0.083 ± 0.015	52	29	1.2	26
$^{186}\text{W}(^{16}\text{O},f)$	137	103	0.015 ± 0.008	22	12	0.22	3.1
$^{181}\text{Ta}(^{22}\text{Ne},f)$	116	69	0.044 ± 0.030	33	19	3.3	110
$^{181}\text{Ta}(^{22}\text{Ne},f)$	174	120	0.081 ± 0.030	37	21	0.35	4.4
$^{186}\text{W}(^{22}\text{Ne},f)$	116	68	0.080 ± 0.015	42	23	1.5	36
$^{186}\text{W}(^{22}\text{Ne},f)$	174	119	0.046 ± 0.015	27	15	0.14	2.0
$^{181}\text{Ta}(^{31}\text{P},f)$	155	60	0.023 ± 0.034	$\leqq 27$	$\leqq 15$	9.3	430
$^{181}\text{Ta}(^{31}\text{P},f)$	195	94	0.047 ± 0.030	22	12	0.81	17
$^{186}\text{W}(^{31}\text{P},f)$	155	59	0.091 ± 0.008	33	18	1.8	43
$^{186}\text{W}(^{31}\text{P},f)$	195	92	0.033 ± 0.009	20	11	0.27	3.6

^{a} Calculated using equation 9 of text.

much lower than expected. Clearly, in view of the high excitation energies involved (60–130 MeV), observation of any time delay is puzzling; as it leads to appreciable modification of long-accepted ideas, it is important to get as much information about this system as possible.

To investigate these interesting effects, a new series of experiments was recently undertaken (34) using a different detection system. Instead of track detectors, two-dimensional position-sensitive detectors were used to measure the fission fragment blocking patterns. Because ions from the particle beam that are scattered from the target have an energy overlapping the fission fragment energy, it was necessary to make the detector thin enough that the scattered beam particles would pass through without depositing all of their energy, but thick enough to stop the fission fragments. A thickness of ~ 15 μm was found to be suitable. For this thickness, the energy loss of scattered beam particles falls well below the broad fission energy spectrum. In addition, a second detector thick enough to stop the scattered beam particles was placed behind the thin detector and operated in anticoincidence in order to further discriminate against scattered ions. If a thick target is used, a continuous spectrum of scattered ion energies results, as ions can scatter from all depths down to the end of the ion range. The lower energy ions will also stop in the thin detector, producing an intense low energy pulse distribution that can interfere with the fission measurement. It was therefore desirable to use thin crystal targets.

Measurements have been carried out for 90 MeV and 96 MeV ^{16}O and 108 MeV ^{19}F induced fission of natural tungsten and 84 MeV, 90 MeV ^{12}C, 90 MeV, and 96 MeV ^{16}O fission of gold. Tungsten crystals 2000–3000 Å thick grown epitaxially on sapphire substrates and gold crystals 2000–4000 Å thick that were grown on sodium chloride and mica substrates were used. In all cases the substrate materials were of low enough Z that scattering of ions of the particle beam was sufficiently suppressed in both energy and yield that the crystals did not need to be removed from the substrate. The crystals were translated periodically in the course of the measurement to avoid radiation damage or distortion effects. Measured blocking curves for 96 MeV ^{16}O fission of tungsten are shown for the detectors at 170° and 130° to the particle beam in Figure 20. Also shown on each curve is a "prompt" blocking distribution determined by scaling (17) the measured distributions for 40 MeV ^{16}O elastically scattered from the tungsten crystal into the detector. The recoil was nearly parallel (within ~2°) to a $\langle 100 \rangle$ plane in each case. The existence of a very sharp central peak in the distribution was verified by a number of independent measurements, and measurements were also made using a recoil parallel to $\{111\}$ plane that showed a sharp central dip instead of the central peak.

A number of important conclusions can be drawn from these distributions. First, a majority of the fission events are faster than ~10^{-18} sec because the overall width of the distribution is in good agreement with the "prompt" curve, and after correction for the energy differences at the two detection angles are the same for the two measurements. Second, there is a very long-lived component present. Even the detector at 170° (with a recoil direction just 10° from the $\langle 111 \rangle$ axis) shows a "flux peak" of the type shown in Figures 14, 15, and 17 for long-range recoils into the center of an axial channel. Third, there appears to be no indication of appreciable contribution from fission of intermediate lifetime. The solid curves in Figure 20 were calculated from an MSE analysis and give agreement with the measured values for 80% fast ($<10^{-18}$ sec) and $20\pm5\%$ slow fission of lifetime ~3×10^{-16} sec. Similar structure has been detected for fission induced in gold crystals, although the amount of any slowly decaying component is significantly less, making resolution of the distribution into lifetime components less definite.

The origin of the long-lived fission component is not clear, although

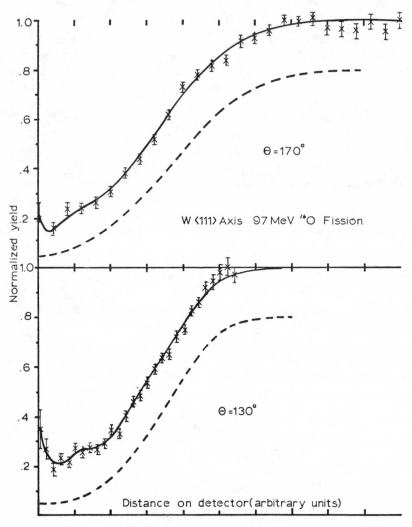

Figure 20 Blocking distributions for fission fragments emergent along a ⟨111⟩ axis of tungsten. The fission fragments were produced by 97 MeV ^{16}O bombardment of a ~3000 Å thick tungsten crystal mounted on a sapphire substrate. Two different emergent angles, θ, relative to the incident beam direction are shown. The blocking patterns were measured with two-dimensional detectors of the type used for the patterns shown in Figure 4 and were obtained by circular scans around the ⟨111⟩ direction. The dotted curves correspond to measured blocking distributions for 40 MeV ^{16}O ions elastically scattered in the same crystals and have been scaled in width by $(Z_{\text{fission}}/Z_{\text{oxygen}})^{1/2}/(E_{\text{oxygen}}/E_{\text{fission}})^{1/2} \sim 1.9$ and in yield by 0.8. The solid curves were calculated assuming that 80% of the fissions are prompt and follow the scaled oxygen curves and that 20% are slow with a mean lifetime of 3×10^{-16} sec (from reference 34).

it may be connected with fission of compound nuclei of angular momentum high enough that it cannot be effectively removed by multiple neutron emission, therefore leading to fission at a late stage in their de-excitation, and low enough that they do not fission with high probability at high excitation energy (64). It is clear that further measurements with good statistical accuracy and high angular resolution will be fruitful. Besides implying new information about decay of highly excited compound nuclei, even in their present preliminary form these results demonstrate the importance of the recoil direction and the possibility of getting information about the form of the decay function from the detailed shape of blocking distributions.

Inelastic Scattering

As indicated in the introduction, the first requirement for BLT measurements is a charged decay product. In certain special cases, such as those discussed in the previous two sections, this requirement can be met by α-particle or fission fragment reaction products. Inelastic scattering of charged particles, however, offers the possibility of studying the lifetime of excited states of a wide variety of compound nuclei. This is limited only by the target isotopes that can be obtained in suitable thin crystal form and by the available type and energy of ion beams. Thin targets are necessary to allow the inelastic particle groups to be separated from the strong elastic group, as shown in Figure 8, and to limit the range of energies excited in the compound nucleus.

Among the first BLT studies were the pioneering work of Maruyama et al (65) in which inelastic scattering of protons in germanium was studied using an experimental arrangement similar to that shown in Figure 7. Further studies of the same system have been carried out by a Harwell-Bologna group (49), who were the first to use two-dimensional position-sensitive detectors in such studies, and by a group at Rutgers University (50–52). It is clear from the previous discussion that inelastic proton scattering in germanium is the most completely analyzed system up to this time. It is also probably the most completely measured system. As much of the information on this system is readily available in the literature and has been previously surveyed, we do not repeat a detailed review of this case. It is perhaps of interest, however, to review the current status of the nuclear information available from this study.

The statistical model calculations of Malaguti et al (60) and Yoshida (14) indicate that the decay function is not expected to be a simple exponential, which makes a "mean lifetime" determination ambiguous. One way to overcome this ambiguity in comparing theoretical and experimental results is to calculate from the statistical nuclear model the expected change in minimum yield, summing over all of the spin and parity $(J\pi)$ values for excited states in the compound nucleus appropriately. This can then be compared directly with the experimental result.

The general expression used for this calculation is

$$(\Delta\chi)_{\text{calc}} = \sum_{J\pi} r(\Gamma_{J\pi}) \frac{\sigma_{J\pi}(90° \text{ lab})}{\Sigma\sigma_{J\pi}(90° \text{ lab})},$$

where $\Delta\chi$ is the difference between the minimum yield (χ value) for a given inelastic group and that measured for the elastic group (as shown by the blocking curves of Figure 9). A Hauser-Feshbach analysis was used to obtain cross sections, level densities given by Gilbert and Cameron were used, and transmission coefficients were determined using optical model parameters obtained from separate cross-section measurements. It is necessary in this analysis to use a relationship $r(\Gamma_{J\pi})$ between the minimum yield change and the level width (or mean displacement) for each level type $(J\pi)$ in the compound nucleus. This relationship was obtained from the MC calculated curve shown on Figure 12. Measured and calculated minimum yield changes are shown in Table 3.

The values calculated for the ^{72}Ge inelastic groups at the *off* resonance energies are in remarkably good agreement with the measured $\Delta\chi$ values. This is admittedly a particularly good case for application of the statistical model but still provides a useful test, as there are no adjustable parameters in the calculation. It was initially expected that the values obtained

Table 3   Observed and calculated changes in minimum yield (ΔR corresponds to $\Delta\chi$ in text) for protons inelastically scattered in germanium relative to the minimum yield for elastically scattered protons (from Ref. 52)

E_p(MeV)	Reactions		$(\Delta R)_{\text{exp}}$		$(\Delta R)_{\text{calc}}$	
5.035	^{72}Ge 0$^+$ P$_1$	off	0.250	0.033	0.225	
	^{72}Ge 2$^+$ P$_2$	on	0.245	0.014	0.372	Final state
	^{70}Ge 0$^+$ P$_1$	off	0.236	0.025	0.172	spin
						Analogue
5.110	^{72}Ge 0$^+$ P$_1$	on	0.186	0.015	0.227	Resonance
	^{72}Ge 2$^+$ P$_2$	off	0.341	0.021	0.376	
	^{70}Ge 0$^+$ P$_1$	off	0.211	0.022	0.162	Level density
5.210	^{72}Ge 2$^+$ P$_2$	off	0.336	0.012	0.372	Intermediate
	^{72}Ge 2$^+$ P$_3$	off	0.385	0.021	0.353	structure
	^{70}Ge 0$^+$ P$_1$	off/on	0.168	0.013	0.158	
	^{70}Ge 2$^+$ P$_2$	off	0.058	0.024	0.072	
	^{70}Ge 2$^+$ P$_3$	off	0.277	0.020	0.237	

for the ^{70}Ge inelastic groups would differ from the ^{72}Ge results primarily because of the different level density resulting from the lower Q value in the reaction. The poor agreement of the experimental values with the calculation for ^{70}Ge indicates that this is probably not the case. For this system, neglect of the compound elastic contribution to the total elastic cross section may be an important omission. Also the optical model parameters used in this case are less well determined.

Tests of particular aspects of the compound nuclear decay can be made by comparing various pairs of measurements as indicated in Table 3. For example, the effect of an analog resonance in the compound nucleus is clearly observed. In addition, measurements for decay into two different 2^+ states in the ^{72}Ge daughter show a small but significant effect that is in the opposite direction from that predicted from the energy dependence of the emergent proton transmission coefficients. This may arise from an intermediate structure in the compound nucleus not adequately characterized by simple spin and parity designations. It would be very informative to extend the analysis to consideration of the possibility of determining the shape of the decay function for each inelastic group. This will require improved statistical accuracy in the measurements and should be very much aided by measurements for different, carefully chosen recoil directions.

Use of inelastic proton scattering for investigation of the effect of an open neutron channel on the compound nuclear decay time has recently been reported (66). This extends the previous measurement of Maruyama et al (65) who showed that the lifetime became shorter at excitation energies above the neutron emission threshold.

The major difficulty in inelastic scattering measurements of nuclear lifetimes is involved with preparation of sufficiently thin (1–10 μm), crystallographically high quality, radiation-resistant, self-supporting crystals. This undoubtedly explains much of the concentration on germanium for which thin crystals are available. Recent results have been reported for proton scattering from ^{92}Mo (67) in which the level density parameter for the compound nucleus ^{93}Tc was determined

$$a(^{93}\text{Tc}) = (10.06 \pm 0.12) \text{ MeV}^{-1}.$$

It is hoped that as crystal growth and preparation techniques advance, more inelastic scattering results will appear.

Compound Elastic Scattering

Until now, the preponderance of coulomb or shape-elastic scattering has precluded direct measurements of compound elastic scattering. In principle the extension by BLT into the time regime where nucleon emission dominates the decay makes direct determination of compound elastic scattering possible. The first preliminary results of such a determination have recently been reported (68).

In those studies two two-dimensional position-sensitive detectors were placed at 35° and 145° to the incident proton beam direction, each aligned with the same $\langle 110 \rangle$ crystal axis in a thin (1.5 μm) germanium crystal. Because of the strong forward peaking of Rutherford (or shape-elastic) scattering, the elastic protons observed in the forward (35°) detector are almost entirely Rutherford, so the distribution obtained can be used as the "control." This distribution was compared to that obtained in the backward (145°) detector which was estimated from off-resonance angular distribution measurements to have a significant compound elastic contribution. For example at 5.0 MeV the ^{72}Ge elastic yield is estimated to be about 30% of the total elastic yield. This is very much diluted, however, by the presence of other unresolved germanium isotopes with lower (p, n) thresholds. The experiment was performed both *on* and *off* the $5/2^+$ isobaric analog resonance in ^{73}As. On resonance the compound elastic component is *enhanced* while its mean lifetime is shortened. The net effect, although not obvious, is calculable. Figure 21 shows the results of one of a set of three independent measurements, all of which gave the same result. Also shown are blocking distributions obtained at $E_p = 8.0$ MeV where the compound-elastic component is greatly diminished because of the opening of neutron channels in all isotopes. This serves as a useful check on the lower energy measurements.

Analysis of these results indicates an elastic enhancement of ~50%, roughly that predicted by current theoretical models (69) corrected for channel-channel correlations. Although crude, these results represent the first successful direct observation of compound-elastic scattering. The extent to which these experiments can be extended depends in large part on the success of attempts to prepare monoisotopic ^{72}Ge or other suitable crystals.

CONCLUSIONS AND PROGNOSIS

The continued rapid development of BLT studies has justified the initial expectation that this would become an important new tool in nuclear physics. It is unfortunate that the number of groups actively pursuing such studies continues to be small. It is hoped that the successes obtained as well as increased clarification of some of the experimental and analysis techniques will entice others to undertake such measurements. Preparation of suitable samples continues to be the principal bottleneck, but this should not be a deterrent to interested experimentalists. After all, many, if not most, of the thin film and detector advances during the past twenty years have been made by innovative nuclear physicists anxious to get on with the job.

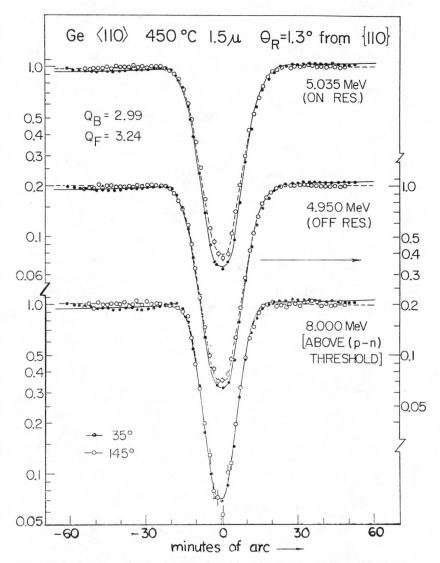

Figure 21 Blocking distributions measured for protons elastically scattered from a 1.5 μm thick germanium crystal and emergent along a ⟨110⟩ axis in a forward (35°) direction (solid points) and in a backward (145°) direction (open points). Results for three different incident proton energies are shown (from reference 68).

Extension of the technique to measurement of lifetimes of single particles such as $\pi°$, $\eta°$, $\Sigma°$ has been suggested (70). Useful steps in this direction are represented by recent observation of channeling of 1-GeV π^+ (71) and 4-GeV π^- beams (72) in germanium single crystals.

Literature Cited

1. Schwarzschild, A. Z., Warburton, E. K. 1968. *Ann. Rev. Nucl. Sci.* 18: 256; Fossen, D. B., Warburton, E. K. 1974. *Nuclear Spectroscopy and Reactions, Part C.* New York: Academic. 307 pp.
2. Bell, R. E. 1965. *Alpha, Beta and Gamma Ray Spectroscopy,* ed. K. Siegbahn. Amsterdam: North-Holland. 905 pp.
3. Ogata, A., Tao, S. J., Green, J. H. 1968. *Nucl. Inst. Methods* 60:141
4. Meiling, W., Stary, F. 1968. *Nanosecond Pulse Techniques.* New York: Gordon & Breach
5. Allen, K. W. 1966. *Lithium-Drifted Germanium Detector.* Vienna: IAEA. 142 pp.
6. Warburton, E. K. 1967. *Nuclear Research with Low Energy Accelerators,* ed. J. B. Marion, D. M. Van Patter. New York: Academic. 43 pp.
7. Broude, C. 1969. *Proc. Int. Conf. Properties Nucl. States,* ed. M. Harvey, R. Y. Cusson, J. S. Geiger, J. M. Pearson. 221 pp.
8. Clementel, E., Villi, C., eds. 1963. *Proc. Int. Symp. Direct Interactions and Nucl. React. Mech.,* Chap. IV. New York: Gordon & Breach. 382 pp.
9. Eisberg, R. M., Yennie, D. R., Wilkinson, D. H. 1960. *Nucl. Phys.* 18:338
10. Fox, R. 1962. *Phys. Rev.* 125:311
11. Gugelot, P. C. 1963. See Ref. 8, p. 382
12. Hansen, L. F. 1963. See Ref. 8, p. 367
13. Ericsson, T. 1960. *Phys. Rev. Lett.* 5:430; Ericsson, T., Mayer-Kuchuk, T. 1966. *Ann. Rev. Nucl. Sci.* 16:183
14. Yoshida, S. 1974. *Ann. Rev. Nucl. Sci.* 24:1
15. Yazaki, K., Yoshida, S. 1974. *Nucl. Phys. A* 232:249
16. Morgan, D. V., ed. 1974. *Channeling.* New York: Wiley; Gemmell, D. S. 1974. *Rev. Mod. Phys.* 46:129;

Gibson, W. M. 1971. *Proc. NATO Adv. Study Inst. Istanbul,* Brookhaven Nat. Lab. Rep. BNL-50336-P 332; Datz, S., Appleton, B. R., Moak, C., eds. 1975. *Atomic Collisions in Solids,* Vols. I, II. New York: Plenum; Andersen, S., Björkqvist, K., Domeij, B., Johansson, N. G. E., eds. 1972. *Atomic Collisions in Solids.* New York: Gordon & Breach; Palmer, D. W., Thompson, M. W., Townsend, P. D., eds. 1970. *Atomic Collision Phenomena in Solids.* Amsterdam: North-Holland
17. Lindhard, J. 1965. *Kgl. Danske Videnskab. Selskab, Mat.-Fys. Medd.,* Vol. 34 (14)
18. Andersen, J. U., Uggerhoej, E. 1968. *Can. J. Phys.* 46:517
19. Tulinov, A. F. 1965. *Dokl. Akad. Nauk SSSR* 162:546; 1965. *Sov. Phys. Dokl.* 10:463
20. Gemmell, D. S., Holland, R. E. 1965. *Phys. Rev. Lett.* 14:945
21. Domeij, B., Björkqvist, K. 1965. *Phys. Lett.* 14:129
22. Gibson, W. M., Maruyama, M. 1974. *Channeling,* ed. D. V. Morgan. New York: Wiley. 349 pp.
23. Nielsen, K. O. 1973. *Proc. Nucl. Phys. Solid State Phys. Symp. 15A,* Chandigarh, India
24. Karamyan, S. A., Melikov, Yu. V., Tulinov, A. F. 1973. *Fiz. El. Chast. Atom. Yad.* 4:456; 1973. *Sov. J. Particles Nucl.* 4:196
25. Fleischer, R. L., Price, P. B., Walker, R. M. 1965. *Ann. Rev. Nucl. Sci.* 15:1
26. Laegsgaard, E., Martin, F. W., Gibson, W. M. 1968. *Nucl. Inst. Methods* 60:24; 1968. *IEEE Trans. Nucl. Sci.* NS-15:239
27. Hofker, W. K., Oosthoek, D. P., Hoeberechts, A. M. E., Van Dantzig, R., Mulder, K., Oberski, J. E. J., Koerts, L. A. Ch. 1966. *IEEE Trans.*

Nucl. Sci. NS-13:208

28. Brown, F., Marsden, D. A., Werner, R. D. 1968. *Phys. Rev. Lett.* 20:1449
29. Gibson, W. M., Nielsen, K. O. 1969. *Int. Symp. Phys. Chem. Fission,* SM 122/129. IAEA: Vienna. 861 pp.; 1970. *Phys. Rev. Lett.* 24:114
30. Melikov, Yu. V., Otstavnov, Yu. D., Tulinov, A. F. 1969. *Zh. Eksp. Teor. Fiz.* 56:1803; 1969. *Sov. Phys. JETP* 29:968; 1970. *Yad. Fiz.* 12:50; 1971. *Sov. J. Nucl. Phys.* 12:27
31. Karamyan, S. A., Melikov, Yu. V., Normuratov, F., Otgonsuren, O., Solov'eva, G. M. 1971. *Yad. Fiz.* 13:944; 1971. *Sov. J. Nucl. Phys.* 13:543
32. Sharma, R. P., Andersen, J. U., Nielsen, K. O. 1973. *Nucl. Phys. A* 204:371
33. Komaki, K., Fujimoto, F., Naka-yama, H., Ishii, M., Hisatake, K. 1972. *Phys. Lett. B* 38:218; 1973. *Nucl. Phys. A* 208:545
34. Andersen, J. U., Forster, J. M., Gibson, W. M., Laegsgaard, E., Mitchell, I., Nielsen, K. O. *Phys. Rev. Lett.* In press
35. Barrett, J. H. 1971. *Phys. Rev. B* 3:1527
36. Altman, M. R., Feldman, L. C., Gibson, W. M. 1970. *Phys. Rev. Lett.* 24:464
37. Fujimoto, F., Komaki, K., Naka-yama, H., Ishii, M., Hisatake, K. 1973. *Radiat. Eff.* 20:141
38. Hashimoto, Y., Barrett, J. H., Gibson, W. M. 1973. *Phys. Rev. Lett.* 30:995
39. Barrett, J. H., Fujimoto, F., Komaki, K., Hashimoto, Y. To be published
40. Karamyan, S. A., Oganesyan, Yu. Ts., Normuratov, F. 1971. *Yad. Fiz.* 4:499; 1972. *Sov. J. Nucl. Phys.* 14:279
41. Barrett, J. H., Hashimoto, Y., Komaki, K., Gibson, W. M. To be published
42. Komaki, K., Fujimoto, F. 1969. *Phys. Lett. A* 29:544
43. Andersen, J. U. 1967. *Dan. Vid. Selsk. Mat. Fys. Medd.,* Vol. 36 (7):
44. Morgan, D. V., Van Vliet, D. 1968. *Can. J. Phys.* 46:503
45. Morgan, D. V. 1973. *Channeling,* ed.

D. V. Morgan. New York: Wiley. 79 pp.
46. Andersen, J. U., Feldman, L. C. 1970. *Phys. Rev. B* 1:2063
47. Massa, I. 1970. *Lett. al. Nuovo Cimento* 3:186
48. Sona, P. 1970. *Nuovo Cimento* 66:663
49. Clark, G. J., Poate, J. M., Fuschini, E., Maroni, C., Massa, I. G., Uguzzoni, A., Verondini, E. 1971. *Nucl. Phys. A* 173:73
50. Gibson, W. M., Hashimoto, Y., Keddy, R. J., Maruyama, M., Temmer, G. M. 1972. *Phys. Rev. Lett.* 29:74
51. Gibson, W. M., Maruyama, M., Mingay, D. W., Sellschop, J. P. F., Temmer, G. M., Van Bree, R. 1971. *Bull. Am. Phys. Soc.* 16:557 (Abstr.)
52. Gibson, W. M., Hashimoto, Y., Maruyama, M., Sellschop, J. P. F., Temmer, G. M. To be published
53. Fuschini, E., Malaguti, F., Maroni, C., Massa, I., Uguzzoni, A., Veron-dini, E. 1972. *Il. Nuovo Cim. A* 10:177
54. Hashimoto, Y. 1972. *Bull. Am. Phys. Soc.* 17:560 (Abstr.)
55. Hashimoto, Y., Barrett, J. H. 1973. *Bull. Am. Phys. Soc.* 18:119 (Abstr.)
56. Komaki, K. 1975. *Bull. Am. Phys. Soc.* 20:692 (Abstr.)
57. Barrett, J. H., Hashimoto, Y., Komaki, K., Gibson, W. M. To be published
58. Schiott, H. E., Bonderup, E., Andersen, J. U., Esbensen, H. 1975. *Atomic Collisions in Solids,* ed. S. Datz, B. R. Appleton, C. Moak. New York: Plenum. 843 pp.
59. Komaki, K., Fujimoto, F. 1970. *Phys. Status Solidi A* 2:875
60. Malaguti, F., Uguzzoni, A., Veron-dini, E. 1971. *Lett. al. Nuovo Cim.* 2:629
61. Andersen, J. U., Nielsen, K. O., Skak-Nielsen, J., Hellborg, R., Prasad, K. G. *Nucl. Phys.* In press
62. Melikov, Yu. V., Otstavnov, Yu. D., Tulinov, A. F., Chechenin, N. G. 1972. *Nucl. Phys. A* 180:241
63. Karamyan, S. A. et al 1970. *Yad. Fiz.* 13:944
64. Vandenbosch, R., Huizenga, J. R.

1973. *Nuclear Physics.* New York: Academic. 234 pp.

65. Maruyama, M., Tsukada, K., Ozawa, K., Fujimoto, F., Komaki, K., Mannami, M., Sakuri, T. 1969. *Phys. Lett. B* 29:414; 1970. *Nucl. Phys. A* 145:581

66. Leuca, I., Kanter, E. P., Temmer, G. M., Komaki, K., Gibson, W. M. 1975. *Bull. Am. Phys. Soc.* 20:692

67. Fuschini, E., Massa, I., Uguzzoni, A., Verondini, E., Petty, R. J., Clark, G. J. 1973. *Il. Nuovo Cim. A* 18:416

68. Temmer, G. M., Kanter, E. P., Hashimoto, Y., Leuca, I., Gibson,

W. M. 1973. *Proc. Int. Conf. Nucl. Phys.* Amsterdam: North-Holland. 1:512; 1973. *Nature* 245:240

69. Hofmann, H. M., Richter, J., Tepel, J. W., Weidenmüller, H. A. To be published

70. Temmer, G. M. Private communication

71. Fich, O., Golovchenko, J. A., Nielsen, K. O., Uggerhoej, E., Charpak, G., Sauli, F. 1975. *Phys. Lett. B* 57:90

72. Allen, D., Gibson, W. M., Gocley, D., Kanofsky, A., Lazo, G., Wegner, H. E. *Phys. Rev. Lett.* In press

COMPUTER INTERFACING FOR HIGH-ENERGY PHYSICS EXPERIMENTS[1]

✣5568

Frederick A. Kirsten
Lawrence Berkeley Laboratory, University of California,
Berkeley, California 94720

CONTENTS

[1] This work performed under the auspices of the United States Energy Research and Development Administration.

INTRODUCTION

Scope of the Article

Almost without exception, the data from high-energy physics experiments are analyzed by programs running in a digital computer. In a typical experiment, the data are generated by a large number of sources, as illustrated in Figure 1. These include not only the particle detectors— scintillation counters, spark chambers, multiwire proportional chambers, etc—but also devices such as scalers and beam-line monitoring instruments. A path for transmitting the data from the sources to the analysis computer must therefore be provided. There is a growing tendency to use a digital computer to control certain features of the equipment used in the experiment. This requires a path for flow of data from the computer to the equipment. In addition, information must be provided to the experimenter. This requires a path for flow of data from the computer to what is often called the "man-machine interface."

The total interfacing problem is that of providing all necessary data paths within the area enclosed by the dashed line in Figure 1. The paths may not be as simple as implied by the way the figure is drawn. Typically, the data flow through additional components, with appropriate interfaces, which are organized to perform some necessary part of the task. For example, most large-scale computers used for analysis are not well adapted for direct, on-line connection to data acquisition devices. Intermediaries are therefore needed.

The intent of this article is to outline the factors that influence the choice or design of the data-handling components, and their organization into systems useful for a class of high-energy physics experiments. As implied in the opening paragraphs, the class is the one sometimes described (1, 2) by that rather loose term, "on-line." It includes those experiments in which data are acquired electrically, and analyzed or recorded by electrical means. We therefore omit any discussion of equipment for recording or analyzing data on film from bubble chambers, optical spark chambers, etc. The interfacing of film-reading equipment to computers involves problems that are, to a large degree, different from those in on-line experiments (3).

Computers are also employed to control many accelerators used to supply the beams for high-energy physics experiments (4). Again, we do not consider this part of the physics program in this article.

Some Comments on System Design Techniques

The design of a system—a data acquisition system for a physics experiment, for example—involves, first of all, the choice or design of a number of individual components that adequately perform the specific tasks

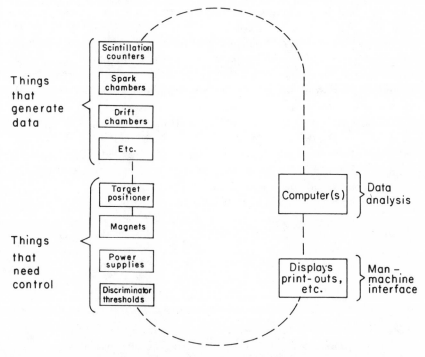

Figure 1 For a high-energy physics experiment, the generalized interfacing task is to provide all necessary data communication within the area enclosed by the dashed line.

required of the system. The components by themselves, however, do not make a system. The system is created by providing for all necessary communication among them. If each component is considered as a functional "black box," the communication is established by means of stimuli imposed and responses evoked via the terminals of the box. The functions are, of course, carried out by mechanisms internal to the box.

One of the very practical tasks of the designer is to effect an optimum trade-off between cost and performance of the total system. An interesting characteristic of the present state of electronics evolution is that cost and performance parameters are very often "quantized" in packages. These packages have various forms. Ten or fifteen years ago, the quantization was on a relatively fine scale; the components of electronics hardware were transistors, resistors, capacitors, etc. Nowadays, one deals mainly with packages containing complete circuits, where the internal transistors, resistors, capacitors, or other basic elements are virtually inaccessible. These packages are in the form of integrated circuits (of continually increasing functional complexity), hybrid or "potted" circuits having prewired and fixed combinations of integrated circuits or other components, CAMAC[2] modules, computers, computer peripherals, and so on. In addition, computer programs are packaged. Manufacturers have available complete real-time operating systems, file-handling systems, applications programs, etc. New packages that work faster, do more complex operations, or are easier to communicate with, continually appear. Complex as they are, their functions are described in terms of stimuli and response on a relatively small number of input and output ports. Again, the designer finds himself on the outside of the black box in trying to understand completely what is provided. It is generally impractical for him to go inside the box and modify its design to optimize performance for a particular application.

Curiously, cost is not always directly related to performance. Volume of production has become important. The larger the volume, the greater the base for amortizing development costs, and often the lower the unit production cost.

The point of this discussion is mainly to demonstrate that the contemporary system designer should have a knowledge of the existence of the various prepackaged designs, including their cost and performance data. This may, in fact, be more important than the ability to design for electrically optimum performance.

Another point is that the volume of usage generated by the total physics

---

[2] CAMAC is the name for a standard design for modular instrumentation, described in a later section.

program is usually quite small in relation to that generated by all users of electronic instrumentation. For example, it has been estimated that physics uses only a few percent of all the small computers manufactured. The percentage is undoubtedly lower in the case of integrated circuits. Therefore, most of the packages available to us will be tailored to satisfy a much broader set of needs than our own.

Definitions of Terms

For the benefit of those readers who are new to data handling techniques, some commonly used terms are defined as follows:

In an electrical circuit, numerical values may be represented by a voltage or current in *analog* or *digital* form. The amplitude of the voltage or current that carries analog information is continuously variable in a range between two predetermined limits. In physics instrumentation, voltage signals are normally used for transporting analog information between modules. For systems having bandwidths of 10 MHz or less (signal rise-times > 30 ns), the range is usually 5 or 10 V, but the limits may be different from one instrument to another. Typical ranges include: 0 to 10 V; 0 to 5 V; −5 to 5 V (5, 6). For systems of higher bandwidth (signal rise-times down to 1 ns), smaller ranges such as 0 to 1 V or −1 to 1 V are preferred (6).

The voltage (or current) amplitudes that represent digital information are expected to assume only certain values. We will consider only *binary* digital signals, which have two such values. The two voltage levels are often called *low* level and *high* level, referring to their relative algebraic value. The logic states represented by these levels may be called "1" or

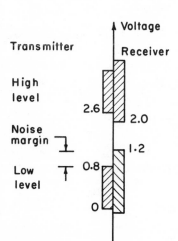

Figure 2   Example of definitions of high and low levels for binary logic circuits. On the left are ranges of voltages a transmitter is permitted to generate for the two levels. On the right are ranges a receiver will accept as representing the levels. Numbers are typical of TTL integrated circuits.

"0"; true or false; assertion or negation. The term *positive logic* is used to indicate that high level has been assigned to the logic 1 state. *Negative logic* indicates that low level represents 1. Many integrated circuit logic elements accept negative logic signals and generate positive logic signals, or vice versa.

In reality, high and low levels are themselves ranges of voltages, as shown in Figure 2, which is an example of acceptable ranges for circuits generating and receiving logic signals. A requirement that receivers be more tolerant than generators causes a *noise margin* to be created. If a circuit generates a "worst-case" low level signal, added noise of up to noise margin in amplitude should not influence the level discerned by the receiver.

Analog signals are less tolerant of noise. For example, consider a signal of 5 V range that will eventually be digitized into a 12-bit number, i.e. into one of 4096 resolvable levels. Each level has a range of 5 V ÷ 4096 ≈ 1.2 mV. Thus, added noise of ≥ 1.2 mV may cause an error of one level in the digitized amplitude.

In digital form, numbers are represented by groups of bits arranged according to a code. The most common is the *binary code,* in which bits are sequentially assigned coefficients of ascending powers of 2. All modern computers do arithmetic operations using binary code. However, since humans generally prefer a decimal representation of numbers and alphabetic representation of semantic information, other codes are used for these purposes. In the binary-coded decimal (BCD) code, the four bits of a group are given values 1-2-4-8 to represent the numerals 0–9. In the ASCII[3] code (7), groups of seven bits represent symbols such as alphabetic or numeric characters. This code is extensively used for teletypewriters, keyboard-display units, etc.

Groups of bits are given various names—character, byte, word—according to their uses. The terms tend to be used rather loosely. *Character* usually refers to a code for a symbol. A *byte* is a set of bits assembled into a convenient size for transmission or manipulation. Eight bits is the most commonly used size for a byte. Within a computer, arithmetic or logical operations are carried out on *words* of data. The information from a physical event usually occupies a number of words. In moving this information, we may consider those words to constitute a *block* of data, moved by means of a *block transfer.*

A typical experiment has many individual data sources. To assemble this data into a block which is transmitted to a single data sink, a device (or concept) called a *multiplexer* is used.

Words (or bytes, or characters) of data may be transmitted in *bit-serial*

---

[3] American Standard Code for Information Interchange.

or *bit-parallel*. In the former the bits of a word are transmitted in sequence over a single electrical circuit. Because the sources and sinks are themselves usually of bit-parallel persuasion, parallel-to-serial and serial-to-parallel converters may be found at the ends of the bit-serial transmission link. In bit-parallel transmission, the bits of an *n*-bit word are simultaneously sent over *n* electric circuits. Obviously, higher data transmission rates can be achieved with bit-parallel, but bit-serial uses simpler cabling.

The terms *hardware* and *software* are often used in computer-oriented systems: Software refers to the relatively easily changed programs that are executed by the computer's hardware; hardware denotes the relatively fixed assemblies of logic circuits or other electronics.

Computers exploit the *stored-program* concept. A fixed set of *instructions* is defined by the hardware design. These can be executed in an innumerable variety of sequences. A particular sequence is defined by the particular program that has been stored in the machine. The set of instructions includes some for achieving communication with devices external to the computer. The communication is carried over a logical and physical entity called the *input/output (I/O) structure*.

Computers are available in a wide range of sizes. We are concerned not so much with physical size as with its capabilities in four areas: (*a*) ability to perform operations on data; (*b*) facilities for man-machine communication; (*c*) facilities for communication with other electronic devices; and (*d*) capacity for data storage. We are especially concerned here with what are often called *minicomputers* (8). The boundaries between small- (mini), medium-, and large-scale computers are, of course, nebulous. For our purpose, a minicomputer at the bottom of the range has 4K (K = 1024) words of memory, a small operator's console, a send-receive teletypewriter and a basic I/O structure having approximately 1-μs cycle time. At the top of the range will be minicomputers of expanded memory (e.g. 64K of fast-access memory, one or more disk memories, and one or more magnetic tape units), additional man-machine communication links (fast printers, CRT displays, plotters, etc), and I/O structures capable of communication with several external devices simultaneously. The corresponding price range is approximately $5000 to $100,000.

SOME FUNDAMENTALS OF DATA TRANSMISSION

The need to move data from one location to another is omnipresent in the interfacing of physics experiments. It is therefore important to consider some characteristics of the mechanism by which data is transported.

Synchronism

The terms bit-serial and bit-parallel were introduced earlier. Actually, some degree of serialization is always present. (Bit-parallel, for example, implies word-serial.) Any serial transmission necessitates synchronization between sender and receiver; for successful communication they must agree on the instants at which the serial elements are, respectively, sent and received. Depending on the level of serialization in the transmission facility, synchronization is required at the bit, word, and message (block) levels. In other words, even after bit synchronization is achieved in a bit-serial system, it is still necessary to identify the reception of the first bit of a word.

In *synchronous* transmission, data are transmitted at a constant speed, and often accompanied by a separately transmitted clock signal. In *asynchronous* transmission, variable length gaps between data elements are permitted. An example is the bit-serial start-stop system used with teletypewriters. Both sender and receiver have clocks of (reasonably close to) the same frequency. By means of a special synchronization signal, the phase of the clock in the receiver is correlated to the transmitter clock; they stay in phase while the bit-serial character is sent and received. In this system, the eight bits that carry a character code are preceded and followed by a start bit and one or more stop bits. The state of the signal circuit during the stop bits and between character transmissions is logic 1. The start bit is logic 0. Recognition of the transition from 1 to 0 at the instant the start bit arrives is used to adjust the phase of the receive clock so that it properly times the arrival of the rest of the bits in the character. An obvious hazard is that a 1 to 0 transition may occur elsewhere in the character. Properly designed receiver circuits can cope with this: If character synchronization is lost, it is eventually recovered.

Bit-parallel systems are usually synchronized by separately sent signals. Since one already has n circuits to carry the n bits in parallel, the cost of a few additional circuits for synchronization purposes is not significant.

Noise

Noise is undesirable in many situations. Data transmission is no exception. We briefly consider two basic sources of noise: that created by or accompanying the signal itself, and that contributed by external sources.

The fundamental effect of noise is to cause errors in signal measurement. As explained earlier, all data transmission circuits have some tolerance to noise (noise margin, see Figure 2). If the noise exceeds this tolerance, errors can result. In a given situation, the effect of noise can often be reduced by lowering the rate of transmission. This is expressed mathematically by the Shannon-Hartley law (9),

$$C = W \log_2 (1 + S/N),$$

in which C is the signaling capacity of a data channel in bits per second; W is the frequency bandwidth of the channel in Hertz; and S/N is the ratio of signal power to (white) noise power at the receiver. This relation is derived under the assumption that noise is gaussian. In many practical situations this is not a good assumption (10).

TRANSMISSION LINE NOISE Physicists are familiar with the need for transmission line techniques for transporting nanosecond speed signals (11). They are accustomed to employing good quality, constant impedance coaxial cables for this purpose, and realize the need to properly terminate the cables to avoid reflections. Transmission line techniques are also needed in many data transmission situations, although some relaxation in the quality of the circuits is often permissible.

Signals traveling in transmission lines (coaxial cables, parallel wire, or twisted-pair cables) are subject to distortion by reflections at impedance discontinuities. One potential discontinuity is the termination—the impedance presented to the line by the transmitter or receiver circuit. Impedance discontinuities also exist at tees, branches, or changes in dimension. How "good" must our transmission line technique be for reliable data transmission?

A complete treatment of this problem is beyond the scope of this article, but some general observations can be stated. 1. If the ratio of signal rise-time to the electrical length at the transmission line (propagation time from one end of the line to the other) is large ($\gg 1$), the distortion resulting from poor termination is often negligible; if the ratio is small (< 1), the effects of reflections on the received waveform may be clearly evident. 2. The amount of distortion that can be tolerated depends on the noise margin of the system. If the distortion due to reflected signals is less than the noise margin, then no damage is done. Distortions of 10–25% can usually be tolerated in binary digital transmission. Analog systems are much less tolerant. 3. Distortion due to multiple reflections from a signal transition, often called *ringing*, diminishes with time. Thus, synchronizing the receiver to accept data only after the reflections have died away can be effective.

EXTERNAL NOISE SOURCES The locations in which high-energy physics experiments are performed often have high electrical power levels with accompanying high levels of electrical switching noise. Certain types of detectors such as spark chambers and streamer chambers generate noise.

Whatever its source, noise can be introduced into a transmission circuit in at least three ways. These are: (*a*) capacitive coupling of the electric field from a noise source into exposed (unshielded) parts of the transmission circuit; (*b*) inductive coupling when the transmission line passes through a magnetic field generated by the noise source; (*c*) conductive coupling, when the references (grounds) of the transmitter and receiver are at different potentials.

The prediction or measurement of noise fields to be found in a given location is difficult. Fortunately, (12) there are fairly simple and inexpensive ways of providing a good degree of noise immunity.

For example, most computer I/O structures use *single-ended* transmission systems. The circuit for each signal has a "signal" conductor; the circuit is completed by a ground return, which may be shared with other circuits. When a signal is impressed, the voltage of the signal conductor changes (with respect to some external reference); that of the ground return does not. When such a circuit responds to a magnetically or capacitively coupled noise field, unequal voltages are induced in the signal conductor and the ground return conductor. Thus the receiver sees a component of noise voltage that is indistinguishable from the signal. However, in a *balanced* system, each signal circuit has two conductors that are not shared with other circuits. The impedance of each conductor with respect to ground is carefully made the same value. The receiver is deliberately made sensitive to the difference in voltages of the two conductors. It is deliberately made insensitive to common-mode signals. These are the signals that cause both conductors to change voltage (with respect to the receiver ground) an equal amount. This insensitivity (common-mode rejection) is never perfect. Integrated circuit receivers that tolerate common-mode voltages of 5 V are available (e.g. Texas Instruments SN 75107). If these are augmented with balanced voltage dividers their common-mode rejection can be increased. Light-coupled devices can give common-mode rejection of up to 2500 V at low frequencies (e.g. Hewlett-Packard 5082-4360). Transformer coupling is also used (13).

Error Detection and Correction

There is apparently no such thing as a completely noise-immune data transmission circuit. The best one can do is to reduce the probability of error to a negligible value. The occurrence of an error can be made less serious if its presence can be detected. Again, foolproof systems are practically impossible, but the probability of *undetected* error can be made much smaller than the raw probability of error in transmission. Techniques have been developed both for detecting and for correcting errors in digital data transmission (10). The latter is a more complex procedure and is rarely applied to our problems.

Parity is a simple form of error detection. To each group of bits, such as a word, an extra parity bit is added. At the transmitting end, the parity bit is set to be 0 or 1 depending on a predetermined rule. In *odd* parity the sum of all bits, including the parity bit, is caused to be odd. The receiver checks the parity of the received word to see if it is still odd. *Even* parity is less often used. A single parity bit, illustrated in Figure 3*a,* can be used to detect an odd number of errors in the transmission of the bits of the word. Errors in even numbers of bits, however, cannot be detected.

The sensitivity to errors can be increased by such methods as the geometrical parity scheme in Figure 3*b.* Here, parity is calculated in two coordinates for a block of words. This technique is used in IBM-compatible magnetic tape units and in the CAMAC Serial Highway (14, 15).

Computer I/O Structures

Reliable communication between the computer and the outside world, via the I/O structure, is essential. Since economy of implementation is also an important factor, it is not surprising to find that the transmission systems used with minicomputer I/O structures are tailored to perform well within restricted areas of application. They are usually organized for bit-parallel transmission. Transmission of all the necessary signals

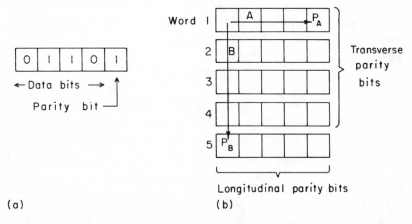

(a)   (b)

Figure 3   Examples of parity for error detection. (*a*) Odd parity for word of four data bits. (*b*) Geometrical parity: Sum of data bits along arrow *A* used to generate parity bit P_A, sum along arrow *B* to generate P_B.

(which are enumerated later) requires approximately 50 circuits. Typically, these are provided by flat, multiconductor cables of 50–100 wires. The structures must be able to accommodate the order of 30 external devices, each with a connection, or tap, onto the structure. This results in a physically complex, multiple-circuit transmission line structure, which is difficult to terminate properly. To avoid problems arising from signal reflections, the structures are generally limited to lengths of 10–20 m, and therefore operate as electrically short lines. (Their length can often be extended by addition of isolating repeaters.)

I/O structures usually use single-ended drivers and receivers. Because they are short, however, the noise coupling is correspondingly low. If the cable runs are restricted to the confines of a single metallic rack, the shielding afforded by the rack is often sufficient to prevent significant pickup from the ambient inductive or capacitive noise fields. Voltages induced in the metallic structure of the rack by ground currents are usually low enough not to cause problems from conductive coupling. The noise hazards increase if the cables extend between racks.

Error detection or correction techniques are almost never applied in minicomputer I/O structures.

Long Distance Systems

The need for data transmission over longer distances often exists in high-energy physics instrumentation. Extralaboratory communication is often done over commercial phone lines. Devices called *modems* (16) (*mod*ulator-*dem*odulators) are used as converters between the laboratory data transmission lines and the phone lines.

Communication within the laboratory usually involves distances less than 10 km, over dedicated cables. Although many successful transmission systems of this type have been built, little performance data have been published. Such data as are available always refer to measured error rates in a particular environment, since measurement of noise fields is difficult. We cite three examples:

1. A 12-bit parallel link at Lawrence Berkeley Laboratory gave a raw error rate of 10^{-7} (17). It used commercially installed, dedicated telephone cables of 1.2 km in length. Signaling rate was 80,000 words per second. The lines were driven in balanced mode with transformer coupling.

2. Bit-serial transmission over 3 km of twisted-pair conductors was found to give a raw bit error rate of 10^{-9} at CERN.[4] Balanced integrated circuit drivers of the TTL (Transistor-Transistor Logic) family were used.

[4] Personal communication from T. Bruins, CERN.

Differential receivers with balanced 20 to 1 attenuators gave a common-mode rejection of 100 V. Signaling rate was 10^6 bits per second.

3. A system for bit-serial transmission on coaxial cable is used at Fermilab. Over a 1.2 km link, at 10^6 bits per second, a raw word error rate of 10^{-9} was measured for words of 70 bit length (13).

For additional security, these longer links almost always use an error detection scheme.

FUNCTIONS OF DATA ACQUISITION SYSTEMS

Figure 4 shows, in simplified block form, the major functions performed in processing the data in typical high-energy physics experiments. The individual functions are introduced in this section, and discussed in more detail later.

Characteristics of an Event

It is characteristic of high-energy physics experiments that each event generates many words of data. Usually, the information from all the detectors in the total experimental complex is collected each time an event occurs. The group of signals from the transducers must be digitized, if they are not already in digital form, and stored or recorded to make room before the next event can be measured; the event detection

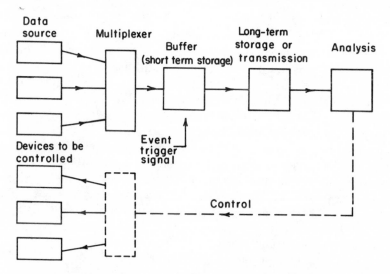

Figure 4   Major functions and data flow paths in a data acquisition system.

equipment is "dead" until this has been accomplished. For maximum utilization of the beam, it is desirable to minimize the dead time. To do this, the storage of the event should be accomplished in a small fraction of the short-time average spacing between events.

The occurrence of an event is indicated by a signal called *Event Trigger* in Figure 4. This signal is typically generated by a series of fast logic elements—discriminators, coincidence circuits, etc—that detect a certain relatively simple pattern of signals in an array of detectors. The existence of this pattern is considered sufficient evidence that an "event" has occurred to warrant initiation of the storage of data from the event. At some subsequent step, on the basis of more detailed examination of the data, further separation may be made into "good events" and "bad events." Filtering of data to discard bad events is sometimes done early in the system to reduce the data rate that must be accommodated by later components in the data path. Or it may be postponed and done completely in a single step at the analysis computer.

Signals From Detectors

Many types of detectors have been developed for use in high-energy physics experiments. All detectors have some associated electronics for the purpose of preprocessing the signals into a form acceptable to the data acquisition and processing system. Each detector has some unique characteristic; often the preprocessed signals reflect this. Some detectors naturally generate signals in digital form; others generate analog signals, which must be converted to digital at some point in the system.

A hodoscope is an example of a detector whose signals are naturally of the digital form. It consists of an array of detector elements that gives locational information. A coordinate of the trajectory of a particle is measured by noting which element of the hodoscope responded at the instant the particle passed through. The answer to the question, "Did a particle pass through this element?" is either "Yes" or "No." Such an answer is naturally represented by a binary 1 or 0.

Hodoscopes consisting of arrays of scintillation counters have often been used. A multiwire proportional chamber used with what may be called "amplifier-per-wire" sets of electronics is another example (18); a flip-flop associated with each wire of the chamber records whether that wire detected the passage of a particle during the event.

The data from the hodoscope may simply consist of the pattern of 1's and 0's in the set of flip-flops. Alternatively, if the number of 1's (bits) is small, a series of binary numbers, each representing the location of a flip-flop in the 1 state, may be generated.

Examples of sources of analog signals are dE/dx detectors and time-of-flight detectors. Both generate values that fall within some continuous range and can thus be represented by an analog signal. A value must be converted into a digital number before the analysis computer can use it. Two important performance parameters of analog-to-digital converters (ADCs) are resolution and speed. Resolution can be expressed in terms of the size of digital number developed by the ADC. A 12-bit ADC that develops a binary-coded number, for example, can recognize 2^{12}, or 4096, resolvable levels of analog voltage at its input. ADCs of up to 16-bit resolution are available. Beyond 12 bits, the cost goes up steeply, and the care required for noise protection of the analog signal also increases sharply.

There are a number of ways of implementing ADCs (11). The two types commonly found in physics instrumentation are sequential approximation and linear rundown (19). Both are available in a range of speeds and resolutions. The fastest sequential approximation ADCs require the order of n microseconds to do a conversion, where n is the number of bits in the output digital number. At present, the faster rundown ADCs require $m/100$ μs, where $m = 2^n$. For $n > \sim 8$, sequential approximation ADCs therefore perform the conversion faster, but rundown types have better differential linearity (i.e. better uniformity of resolution).

Multiplexers

Following an event, the data must be collected from the detectors and stored. Most storage devices have a single input port through which they accept data. However, there are many individual detectors and associated electronics as well as other sources of data. Multiplexing is the function of accessing the data from the many sources and steering it onto a single pathway. Conceptually, a multiplexer acts as a multiposition switch, connecting each source in turn to the path leading to the storage device.

ANALOG VS DIGITAL MULTIPLEXERS   Both digital and analog multiplexers exist. In most applications, the better type is obvious, but there are others in which a choice is possible. One such case is that in which a number of analog signals must be acquired, digitized, and stored for each event. Figure 5 shows two possibilities. In Figure 5a, the analog signals are multiplexed into a single ADC. One switch of the multiplexer is closed at a time. The ADC generates one digital word denoting the amplitude of the corresponding analog signal. This word must be stored before the ADC can do another conversion. The "data stored" signal indicates when this has been done. In response to it, the "switch control" logic opens the first switch and closes a second. Then, after a short wait for the signal on the analog bus to settle, it triggers the conversion process in the ADC. This sequence repeats until all channels have been digitized and the results stored.

In contrast, Figure 5b illustrates the use of a digital multiplexer. A separate ADC is used for each channel. The total data acquisition time in Figure 5a includes an aggregate time of $j \times$ (ADC conversion time + analog bus settling time). In Figure 5b, the comparable time is only $1/j$ as large, but j times as many ADCs are required.

DIGITAL MULTIPLEXERS   To simplify the presentation, Figure 5b denotes the digital paths by heavy lines. Each path consists of n wires, plus some extra wires for signal returns (grounds), for transmitting data in a bit-parallel (word-serial) mode. Each of the n wires carries one bit of data; the group of wires carries the word of n bits. Thus, to extend the switch analogy, one must specify that the multiposition switch has n poles, or contacts, at each position.

The interfacing problem would be simpler if n were a universal constant. The most convenient value would be equal to the number of bits in the computer word. Unfortunately, there is no one standard word size for computers. In fact the word size of the small computer used at the experiment site may be different from that of the large computer used to analyze the data. Small computers usually have word sizes of 12, 16, 18, or 24 bits. At the present, 16 bits is the most common. It may turn out that the multiplexer itself has a different word size than the computer. CAMAC, for example, has a 24-bit data pathway. It is often the case that the characteristics of the data source dictate the size of the words used to convey its data: An 8-element hodoscope will generate only 8 bits of data; a scalar of 10^7 count capacity needs at least 24 bits to transmit its contents.

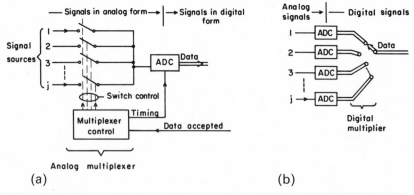

(a)   (b)

Figure 5   In many applications, either an analog or digital multiplexer may be used.

The positioning of the switch is, of course, controlled electronically—the speeds required are much too fast for any mechanical device. The electronic control needs to be able to cope with certain characteristics of the data. For example, the number of data words accessed at each position may be different. Some detectors—a small hodoscope, for example—may generate one word for each and every event. A larger hodoscope may generate two (or more) words, but always the same number per event. Many multiwire proportional chambers generate a number of words that depends on the number of particles detected in the particular event. In reading out that event from the chamber electronics, the multiplexer control may not know the total number of words until the last one passes through. The multiplexer control must therefore have a means of controlling the length of time (dwell) for which the switch remains at any position. It must wait until all words from that source have passed through before going on to the next position.

The end result of the multiplexer action is a block of data words that stream into the storage device. The origin and significance of each word must, of course, be preserved. This is information that the analysis computer must know. Conceivably, each word could carry, in addition to the datum, a code that denotes its origin. However, the word size is small enough that few, if any, bits are available for this purpose. Thus, the significance must be preserved by specifying a format for the list.

FORMAT   A fixed format can be specified if the number and order of data words is identical for every event. An example of a fixed format is given in Figure 6a. The first word for each event comes from a scaler. (The

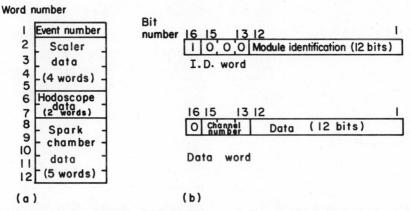

(a)   (b)

Figure 6   Format examples. (a) A fixed format of 12 words. Each source generates the same number of words for every event. (b) Example of word formats for source generating a variable number of words per event.

scaler is incremented each time an event is processed.) Following the first word are fixed numbers of words from three sources.

A fixed format cannot be specified if one or more sources generate a variable, unpredictable number of words. The block will then also be of unpredictable length. Solutions to problems of this type are limited only by the ingenuity of the systems designer or the characteristics of his equipment. Some examples follow.

If there is only one source generating a variable number of words, the multiplexer can be programmed to access that source last. The first part of the block is therefore of constant length and format, and the analysis program can know that the rest of the words in the block are all from the variable source.

This will not work, obviously, if there are two or more variable sources. One solution is to have the multiplexer count the words accessed from each variable source. It then places these word counts in a location known to the program, which gives it the information necessary to unravel the origin of the words in the block.

Another possibility is to insert a special, unique code (i.e. a unique number) as a marker to precede the group of words from each variable source. (One must then insure that this unique code never appears inadvertently within the data.) This technique is used in the digitizers (20) for the External Muon Identifier at the 15-ft bubble chamber of the Fermi National Accelerator. The Identifier uses an array of identical digitizers that collect data from a set of multiwire proportional chambers. Following an event, each digitizer contains from 0–16 words of data in each of its seven channels. In this case, a 16-bit computer is used, but a datum has only 12 bits of significance. This leaves 4 bits for identification purposes, which is insufficient for coding both the channel number and a digitizer serial number. Therefore, the digitizers were designed so that the first word always contains a 1 in the bit 16 position, and the digitizer serial number in the rest of the word, as shown in Figure 6b. The ensuing words carry a datum in the first 12 bits, channel number in the next three bits, and a 0 in the bit 16 position. The analysis program uses bit 16 to differentiate between the two types of words.

Buffers

The term *buffer* is used to mean "a storage device used to compensate for a difference in rate of data flow, or time of occurrence of events, when transmitting data from one device to another" (21). This attribute is essential in experiment interfacing, as indicated by the presence of the buffer block in Figure 2. At one end of the interface chain are arrays of instruments that generate data in a highly irregular way. Events occur at random points in time. The accelerator beams have a time structure owing to the radiofrequency accelerating field or because of the duty cycle of the accelerating process. At the other end of the chain is the analysis computer that accepts data at a rate that is relatively steady, but that may actually be quite different from the average rate at which data are generated. Thus, it is certainly true that buffering must occur between the two ends of the system; it may, in fact, need to be distributed at several points in the system.

The following questions may be used to help determine the suitability of a buffer device for a particular application. What is the rate at which it can accept data? Is this a synchronous (fixed) rate, or can it accept data asynchronously at various rates? What is the rate at which it can output data (usually the same as the input rate)? What is the data storage capacity? Can the device simultaneously accept and emit data, or must it alternately accept and output blocks of data? How easily is it interfaced to the existing data structure? Is there a time delay (latency) in accessing the device?

Buffers that are sometimes called *derandomizers* are often used close to the data sources to absorb data rate fluctuations due to short-term variations in beam intensity. The fastest buffers for this purpose are electronically active memories. Flip-flops, each of which stores one bit of information, can be arranged in various arrays of configurations. For example, inexpensive integrated-circuit memories capable of storing a single word in a few tens of nanoseconds, or a series of words at perhaps 50 ns per word, are available. For these speeds to be exploited, the memories must be located close to the data source: transmission of bit-parallel data becomes more expensive above about 10×10^6 words per second. The practical size of such fast memories is limited to the order of hundreds of words, partly by cost and partly by the organizational difficulties that increase with size.

Larger, but slower, buffers are available in a variety of costs and capabilities. They can be used to absorb fluctuations in data rate over larger increments of time, such as those fluctuations due to the duty cycle of the beam spill. Some examples of entities usable for large scale buffers follow.

1. *Core memory.* Typical minicomputers have a core memory with a capacity of 4K to 64K words. The portion of memory that is not necessary for program storage may be used for buffering. Core memories are convenient in that data words may be transferred asynchronously in blocks of any length from one word up to the complete memory capacity, and at any rate up to the maximum. Typical maximum rates are $1\text{--}10 \times 10^5$ words per second for programmed transfers, and 10^6

words per second for DMA transfers. (Programmed transfers and DMA transfers are described in a later section.)

2. *Disc storage.* On a typical disc memory, the storage format is organized into tracks each of which has a number of blocks, or records, of perhaps 256 words each. When a transfer to or from a disk is initiated, there may be a delay of several tens of milliseconds until the proper track is accessed, and until the desired record is in position under the recording and reading heads (latency time). When the transfer of words begins, it must continue synchronously for an entire record at a rate dictated by the speed of rotation of the disk and the density of storage on the magnetic medium. Typical transfer rates are 10^5 words per second. A disc may hold 10^6 words.

3. *Magnetic tape (IBM compatible).* Various magnetic tape formats are used with minicomputers, but the format known as IBM-compatible (22) is most often used for high volume buffering or storage. Data are written onto magnetic tape in records. Unlike disks, the record lengths can be variable. A latency time is also imposed with magnetic tape, because of the time required for the tape motion to start and stop. Once recording is begun, it must continue at a synchronous rate dictated by the speed at which the tape is moved. These rates range from 3000 to 50,000 words per second. Typical capacity is 10^7 words per reel of tape.

EVOLUTION OF DATA ACQUISITION SYSTEMS

In the early years of physics experiments performed with electronic sensors, the "data processing" often consisted of a few discriminators, coincidence circuits, and other elementary logic elements, with results collected on a few scalers and manually recorded. The advent of electronic computing facilities opened the door for more complex analyses and provided the capability for coping with larger volumes of input data. This raised the problem of transporting the data to the computer. Many man-hours were spent transcribing data from notebooks, scaler print-outs or oscilloscope photographs onto punched cards, since nearly every computer could read such cards. (Punched cards are, of course, still very important as an information medium for computer use; in high-energy physics, however, they are used mainly for program purposes, and very seldom for data.) Inevitably, a desire to transmit data with less manual intervention developed. In response to this desire, the card punches were initially connected directly ("on-line") to the experiment. Later, the card punches were replaced by faster devices such as paper tape punches. Still later, the even faster magnetic tape units were employed. In 1962, a direct wire tie from an experiment to an analysis computer was accomplished in several laboratories (23, 24).

The data multiplexers used with the early setups were hard-wired ensembles of logic elements. They were often custom designed to match the needs of a particular experiment. The capabilities of these devices to manipulate data were severely limited by economic considerations. Changes to cope with new situations or requirements usually required redesign and were expensive.

Relatively inexpensive, small-scale, programmable digital computers made their appearance as commercially available products about 1963. Physicists immediately recognized the many advantages of having a dedicated small computer connected directly (on-line) to their experiment (25). The impact of these computers therefore began to be felt in high-energy physics about 1964, although their use in low-energy physics had begun somewhat earlier (26–28).

The tremendous value of the small (mini-) computer in high-energy physics experiments is vividly illustrated by the many tasks in which it has been involved. These include the following: control of data acquisition; buffering or derandomizing data flow rates; data formatting; long-term storage of data on magnetic tape; transmittal of data to other computers; monitoring of beam raw-event rates and the state of experimental equipment; analysis or partial analysis of data; display of results of monitoring or analysis; and control of data acquisition equipment. As a stored-program device, the computer is trained or adapted for these tasks simply by being provided with the appropriate program. Now the question is not whether to use a minicomputer, but rather, which one?

Continuing Role of Hard-Wired Devices

One of the major contrasts between the "hard-wired ensembles of logic elements" referred to above and computers is the stored-program concept. The capability of storing a program consisting of thousands of instructions, and the provision of allowing the program to be easily changed, make the computer a tremendously powerful and flexible device, easily adaptable to many situations. At first glance, it might appear that computers would completely remove the need for hard-wired devices. Actually, this is not true. Hard-wired devices are still used, although usually to complement a computer rather than compete with it. Speed is the key parameter. Any operation a computer can do, a hard-wired device can do faster. The following three points illustrate this: 1. Before executing an instruction, a computer must fetch the instruction from its memory. This takes time. The "program" of a hard-wired device is stored in its wiring pattern—it proceeds without hesitation from one operation to the next. 2. The instruction set of a computer is fixed, whereas operations that are optimized to a particular need can be designed into the hard-wired device. 3. For various manufacturing reasons, most present-day minicomputers use TTL integrated circuits. A careful designer can successfully implement small systems using ECL (Emitter-Coupled Logic) integrated circuits,[5] which can be 10 times faster than standard TTL.

The point here is not to dispute the fact that computers are superior in the majority of cases. It is rather to point out that there are cases where it is necessary to do simple computer-like operations—data manipulation, control sequences, etc—at higher speeds than are attainable with computers. This is particularly true for preprocessors—devices intended to diminish the rate of data flow into a computer by means of filtering or condensing operations.

COMMUNICATION WITH THE ANALYSIS COMPUTER

Figure 7 shows three common data acquisition and analysis systems. There are some obvious similarities: each has a set of data sources connected via a multiplexer to a minicomputer. It is of interest to examine the differences regarding where the analysis is done.

In Figures 7b and 7c the data eventually arrive at a large computer, which presumably does the bulk of the analysis. In Figure 7a, the data flow terminates in the small computer, implying that all analysis is done there. The questions therefore arise: First, under what conditions is the small computer in Figure 7a inadequate to handle the complete analysis? Second, when it is inadequate, how should the data analysis tasks be partitioned between the large and small machines in Figures 7b and 7c?

Complete answers to these questions are outside the scope of this article on interfacing. However, they cannot be completely ignored, because the answers often have an important influence on the choice and arrangement of the hardware.

One significant point is associated with capability. Minicomputers are typically capable of performing logical operations at speeds comparable to those of the large machines. Thus, tasks such as data formatting or compressing and data sorting or filtering might be assigned to either. The large machines, however, are generally much faster at arithmetic

[5] Example of ECL circuits include the MECL III and MECL 10,000 series made by Motorola, Inc., Phoenix, Arizona, USA.

operations, or arithmetic operations requiring high (e.g. greater than 16-bit) precision. The second point is related to need. How urgently are the results of the analysis needed at the experiment? It is obviously necessary that the experimenter learn quickly of any equipment failures or other conditions that invalidate the data. Particularly in experiments

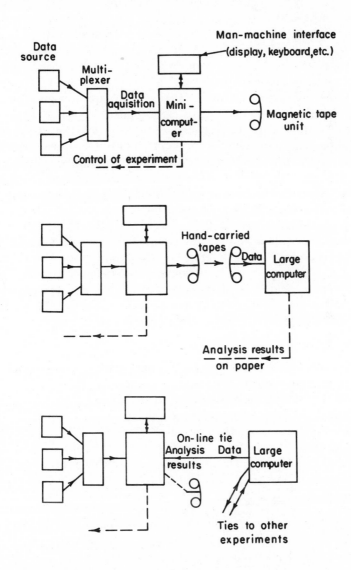

Figure 7   Three typical data acquisition and processing systems.

involving complex or large arrays of equipment, he may need to rely on the services of a computer to learn of these conditions. Often, even partial analysis of a fraction of the data is sufficient for this purpose. In other cases, it has been found sufficient to simply monitor certain aspects of the data that reflect the state of the equipment. Again, although this question cannot be fully covered in this article, there are hardware implications. For example, if the results of a complete analysis of all the data are needed urgently at the experiment, an on-line, bidirectional tie between experiment and large computer (Figure 7c) may be essential. If complete analysis is needed, but not so urgently, the scheme of Figure 7b may be adequate. In this case, the proverbial "boy on a bicycle" replaces the on-line tie; he carries magnetic tapes in one direction and computer output in the other. Some interesting arguments have developed concerning the value of fast turn-around. It has been argued that changes in the equipment of a high-energy physics experiment in response to results of analysis involve much more than simply adjusting a knob: such changes are made only after hours of thought, and may require hours of effort to implement. Experimenters who maintain this viewpoint are not willing to pay very much for the speed improvement of on-line ties.

The boy and his bicycle both deserve medals for valiant service. Although they still serve in many situations, there is a growing trend toward the on-line tie—i.e. a direct wire connection over which data are passed between computers under complete program control. On-line ties have been implemented at several laboratories, including Lawrence Berkeley Laboratory (17), Los Alamos Meson Physics Facility (29), Fermi National Accelerator Laboratory (30), and CERN (31).

On-Line Ties

Figure 7c depicts the situation in which the small computer has a direct-wire tie for transmission of data to and from the large computer. The stubs marked *To other experiments* in Figure 7c imply that more than one experiment and its small computer may be connected to the large machine. Perhaps the most obvious characteristic of this scheme is that it automates the process of transporting data from experiment to analysis computer. Two additional features are worth noticing. First, provision for transmission in both directions on the link means that the results of the analysis in the large computer can be transmitted back to the smaller. It, in turn, can organize displays or print-outs to make this information available to the experimenter. This permits him to see the results of the data-taking in the shortest possible time.

Second, efficient sharing of both hardware and software resources becomes possible (32, 33). It is possible to share the bulk storage facilities of the large computer, eliminating the need for magnetic tape or other bulk storage devices at the small computer. This would seem to be especially valuable when several small computers are tied to a single large computer. This concept can be expanded to include all types of hardware: line printers, disks, core memory, etc. There is an attendant disadvantage— that of placing a high value on the reliability of the larger machine. A failure in the larger machine can bring to a standstill all associated experiments. For this reason, users tend to insist on having bulk storage devices at the small machines, either continually, or as available back-up.

Software to be shared can include cross-assemblers and cross-compilers. These are programs that run in the larger machines, but whose purposes are to prepare programs to run in the smaller. It is, of course, common to have compilers and assemblers that run directly in small computers. However, it may be more convenient to do these operations in the larger. The larger computer probably has better features—card readers, line printers—for the man-machine interface. It may also have advantages in speed or program file storage.

The ability to share hardware or software is contingent on the existence of an appropriate *operating system*. Such an operating system includes hardware and software features that permit two or more computers to coordinate their activities. For the computers to be able to transmit "files" or blocks of information (be they data or programs), to retrieve or use these files, to synchronize their operations, and generally to understand (or at least not to confuse) each other requires that a significant amount of planning and preparation be done before the tie is usable. The protocol for certain operations must be carefully defined. For example, there generally must be a means of identifying such parameters as the source, destination (e.g. storage or analysis), size (number of words), and nature (data, source program, object program, etc) of files that are exchanged. The transmission of files is often preceded by an exchange such as: "Are you in running condition, and can you accept a data file of 1000 words?" "Yes, I am running and can accept the file. Please send it." "Here it is. [File is transmitted.] Did you get it?" "Yes, I received 1000 words with no errors." In addition, provision must be made for abnormal conditions at each step. For example, the called computer may not be in running condition, in which case the calling computer should not simply keep trying, but should output a message to the operator. Or there may have been an error in transmission of the file, for which procedures for recovery must be provided.

Certain limitations on the system may exist, depending on the dedication of the large computer. The large computers in the computation centers of laboratories and universities are usually batch oriented. This means that their operating systems are oriented toward maximizing through-put of a large number of individual jobs, which are individually submitted and individually run. When submitted, a job enters an input queue, where it waits until previous jobs have run to completion. When the job is finally started it is executed to completion and removed as quickly as possible.

In contrast, an on-line tie requires fast service and service for extended periods of time. When the buffer of a small computer is filled with experimental data, that buffer must be emptied quickly, or else the experiment is stalled. If the on-line tie is the means by which the buffer is emptied, then it may be intolerable to be forced to wait in the input queue of the large machine. Also, the buffer must be emptied periodically as long as the experiment is running. In this situation the large machine's operating system must be able to accommodate a program (to service the small computer's needs) that remains resident in the system for long periods of time, and that can be called into execution on short notice.

Thus, the on-line tie can provide very convenient and powerful features, but only to the extent permitted by the operating system or operating environment. The services of a professional programmer are also required. The cost can be minimized if the operating system is supplied by the computer manufacturer. However, this is usually possible only if all computers—large and small—are made by that manufacturer.

A system having this characteristic has been installed in the Bio-Medical facility at the Lawrence Berkeley Laboratory's Bevalac accelerator.[6] The central computer is a Digital Equipment Corporation (DEC) PDP-11/45, which is among the largest of the small computer class. It contains sufficient hardware—core memory, disk storage, and DEC tape—to support a sophisticated operating system with FORTRAN compiler, assembler, and file handling. The PDP-11/45 can support experiments directly but also provides for on-line ties to smaller PDP-11 computers that are themselves associated with specific experiments.

An important point is that the operating system is so sophisticated and well developed that the smaller PDP-11 machines can, in fact, be very small. The essentials are a minimum of 8K words of memory and an input-output device for man-machine interaction. The latter may be a teletypewriter or a keyboard-CRT terminal. Because of the support of the PDP-11/45, it appears to the user to be much larger. Whenever user demand exceeds the capacity of the smaller machine, it automatically calls upon the PDP-11/45 for support. In this sense, the small computer is "transparent." The user has the feeling that he is operating directly on the PDP-11/45.

[6] Personal communication from J. Gallup.

TYPICAL MINICOMPUTER I/O STRUCTURES

The Input/Output (I/O) structure is the means of data communication between the small computer and the outside world. Physically, the I/O structure is a set of wires by which the computer is electrically connected to other devices. Obviously, the significance of the signals on these wires must be defined. In addition, the computer instruction set must contain commands which, when executed, cause specific actions on the I/O structure. It is the existence of these instructions that enables the computer program to effect communications, coordination, and synchronization of processes between the computer and external devices.

It is essential, in designing an interface to a small computer, to understand both the hardware and software aspects. In analogy to a previous example, one can communicate bits if the wires are correctly attached and if the signal voltage levels are correct. But the usefulness of the communication depends on an agreement between software and hardware as to the significance of the bits.

The problem of interfacing experiments to small computers would be simplified if all computers used the same I/O structure. Then an interface for any particular instrument could be used on any computer. Un-

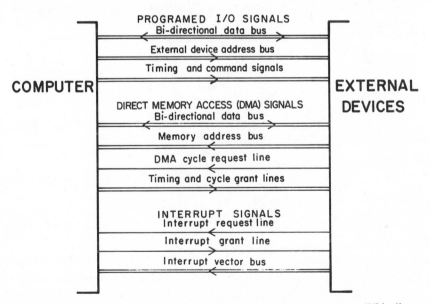

Figure 8 Signal pathways of a typical minicomputer I/O structure. Wide lines indicate multibit-parallel paths. Arrows indicate direction of signal flow.

fortunately, there are no standards for I/O structures. Those on the computers of one manufacturer are inevitably incompatible with those of another manufacturer. There are, however, similarities and typical features (34, 35). Most I/O structures contain at least the following: a programmed I/O facility; a direct memory access (DMA) facility; and an interrupt facility. Figure 8 shows schematically an example of the signal paths needed to implement these facilities. Various computers implement these paths in different ways. For example, a single bidirectional data bus may be used for both programmed I/O and DMA transfers.

Programmed I/O

The term *programmed I/O* refers to a process of transferring information via the I/O structure. In this process each individual transfer is initiated by the execution of one or more instructions in the computer program. Again, the problem of programming computers for such processes would be simplified if all computers had the same I/O instructions. This is not the case, although there are certain things which must be specified to execute an I/O operation. These are: the operation to be performed; the location (address) of the external device involved in the operation; and, in the case of data transfers, the location within the computer that supplies or accepts the data. During the execution of the instructions, the hardware of the computer automatically generates synchronization and control signals as required by the design of the I/O structure. Typical I/O operations that can be executed by computer instruction include the following:

1. input a word of data—move a word of data from the specified external device into the computer;
2. output a word of data—move a word in the opposite direction;
3. execute a dataless operation in an external device—e.g. generate a trigger pulse, enable or disable a process, etc.

The address of the external device involved in the operation is a number specified by the computer instruction. This number is carried by the address bus in Figure 8 to all external devices simultaneously. The device that has previously been assigned this number recognizes it and prepares to execute the operation. This implies, of course, that the programmer must know the addresses of all the external devices. The location in the computer that is its terminus for programmed I/O data transfers depends on the computer type. In many machines, it is a fixed location, such as the accumulator register. In other machines (the PDP-11, for example) it may be specified by the instruction as being in the central processor, in any location in memory, or even in another external device.

The control and synchronization signals on the I/O structure are, again, different for the various computer types. A common necessity is a means of alerting all external devices that an operation has been started, so that they know when to look at the I/O address bus. In the case of data output operations, the addressed external devices must be told when to accept the data from the data bus.

The rate at which programmed I/O operations can be executed largely depends on the number and execution times of the instructions necessary to set up the transfer. In computers such as the PDP-11, where execution of a single instruction (MOVe) can cause a word to be moved from any memory location to any external device, the set-up time is minimal. In other machines, separate instructions might be necessary to (a) move the word from memory to accumulator, and (b) move the word from accumulator to the external devices. The total time required varies from about 1–20 μs per word.

Direct-Memory Access Transfer

Direct-Memory Access (DMA) is a means of transferring blocks of data words directly between an external device and computer memory. The transfer of the individual words is *autonomous*—i.e. controlled entirely by hardware, which is usually located in the external device. The computer program need only issue a small number of initial instructions to the hardware. These usually include: the address of the external device; the address in computer memory at which the first word of the block is stored; and the number of words in the block. Once these instructions are issued (by programmed I/O), the program is free to perform other tasks while the block transfer is taking place.

When a block of data is being transferred into the computer, the sequence of steps that the hardware executes for each data word is as follows. 1. When the data word to be transferred is ready, the hardware requests a DMA cycle. 2. At the end of execution of the current program instruction, the computer program is automatically halted and a grant signal is issued. This is the signal to begin the transfer. 3. The hardware then specifies the memory address into which the data word is to be moved, and simultaneously (in this particular example) places the data word on the data bus. 4. The computer memory accepts the data word and stores it in the specified memory location. 5. The grant signal is removed and execution of program instructions resumes, precisely at the point where it was previously halted in step 2. 6. The hardware updates (increments by one) the memory address register and the word count register. The entire sequence is repeated each time a data word becomes available (e.g. from the MWPC electronics) until the number of words specified by the word count has been transferred. After the last transfer, the program is so notified. It can then process the data block.

Besides reducing the program's involvement, the use of DMA usually results in higher rates of data transfer as compared to programmed I/O. (Here is an instance in which a "hard-wired" DMA controller can execute the transfers faster than a stored program machine.) The transfer rate is limited by factors such as the rate at which the external device can prepare or absorb data; the rate at which the I/O structure can transport data; and the speed of the computer memories. Data transfer rates of 10^6 words per second are commonly achieved during a DMA block transfer.

The Interrupt Facility

There are many cases in which the processes within the computer need to be synchronized with external processes. The *interrupt* is a means by which such synchronization can be achieved automatically. An external device may generate an interrupt to cause the computer program to branch to a part of the program that provides the service desired by the interrupting device.

Two situations where synchronization is needed have already been mentioned. 1. By means of appropriate hardware, the Event Trigger can cause an interrupt. The interrupt in turn initiates execution of a segment of the program that organizes a DMA block transfer of the event's data into the computer memory. Of course, when the organization is done, the program returns to its former task. 2. When the block transfer is done, the DMA controller initiates a second interrupt to notify the program of this fact.

Of significance is the interrupt latency time, the elapsed time from the instant at which the interrupt is established until the service routine is entered. It is typically of the order of tens of microseconds.

INTERFACING TO SMALL COMPUTER I/O STRUCTURES

Since the small computer is central to nearly all data acquisition systems for physics experiments, it follows that a major part of the interfacing problem is that of electrically and physically connecting various devices to the computer. During the last decade, various systems have been developed to organize the electrical and mechanical tasks of implementing the connection of these devices to the I/O structure.

As described earlier, the I/O structure is essentially a set of wires together with definitions of the signals that traverse these wires. There is also some degree of mechanical specification in terms of cables and connectors. In principle, then, interfacing the computer to a peripheral device can be as simple as plugging them together. In fact, all computer manufacturers supply a certain range of peripheral equipment that is plug-compatible with their computers. In nearly all cases, however, this is limited to what might be called "standard" computer peripherals— magnetic tape units, line printers, keyboard-display units, etc. Beyond this there still exists the need to interface the detectors and other equipment of the experiment. Since physics represents a small percentage of the total small computer market, the computer industry has not attacked this problem with vigor. Most systems for this purpose have been developed at the laboratory level.

Let us consider some of the desirable characteristics of an interfacing system. First, experiments tend to have large numbers of diverse detectors and instruments to be interfaced. Thus the system should be capable of accommodating a large number of input ports. Interfaces for various types of detectors and instruments should be easily (commercially) available. The rules for designing new interfaces should be easily understood. Second, selecting the data acquisition equipment for an experiment is often a small experiment in itself. As the experiment progresses, changes or additions are often necessary. Thus, the process of adding or subtracting input ports should not be difficult. Third, the system should be usable on various models of computers. It is advantageous to be able to trade interfaces from one experiment to another—perhaps via an equipment pool—even though the experiments use different types of computers. Fourth, events often generate blocks of data, which need to be moved quickly, but there is also a need to access individual data sources, such as scalers. Thus, both DMA and programmed I/O modes should be supported. Fifth, there is a growing tendency toward computer control of data acquisition or beam-line equipment. Thus, the system should be bidirectional.

Extending The I/O Structure

Of the various ways of designing an interface system, perhaps the most obvious is to interface each device directly to the I/O structure of the computer. Because the I/O structures of different types of computers are incompatible, this immediately restricts the use of each device built in this way to the computer type for which it was designed. A group at Stanford Linear Accelerator Center has designed a set of equipment that falls within this category.[7] Their modules are compatible with the Unibus, which is the name given to the PDP-11 I/O structure. The design of the

[7] M. Schwartz, personal communication.

Unibus is such that equipment connected to it may function either as a master or a slave. As a master, it can request the use of the Unibus to send data directly to another (slave) device without going through the computer. The SLAC group has utilized this feature in the design of a series of modules that perform certain preprocessing tasks and can be programmed to send their output directly to another module. Other, general purpose modules have also been designed.

Data Highways

Several data multiplexing systems (it is convenient to call them data highways) have been developed that are independent of any particular computer I/O structure. In turn, they can be interfaced to the I/O structure of any particular computer. Adapting the system to various computers requires only that the respective computer interfaces be changed. Principal design considerations in highway systems include the following: (a) mechanical compatibility (modularity); (b) power distribution; (c) data and control signal compatibility. Most attempts at highway design have been concerned with all three; some deal with only the last.

HIGHWAYS IN NIM   One of the important standardization developments in nuclear instrumentation has been NIM. Specifications for NIM were written by the Nuclear Instrument Module (NIM) Committee of the USAEC in 1964 (5). Because it was designed before computers became so prominent, it provides mainly mechanical and electrical power compatibility. The NIM specification defines a module that is mechanically compatible with a bin that accommodates up to 12 modules. Mating connectors in the bin and module provide access to electrical power. Signal levels for transport of nanosecond speed signals between modules are defined. The NIM specification has been universally adopted by physics laboratories and manufacturers of physics equipment. It has also spread to various users outside the physics community. A graphic indication of its value is the fact that approximately 100,000 NIM bins have been manufactured since its inception.

Several highway systems were designed in which the third design consideration, signal and logical compatibility, was added to a NIM-based system by means of auxiliary connectors and cables. These include the NIDBUS system at LBL (36) and a system at the Princeton-Penn Accelerator (37). Several instrument manufacturing companies have built, or are still building, their particular design of highway for interconnecting systems of NIM-based modules. A system was designed at Bell Laboratories in which the multipin NIM power connector was replaced by a connector that accommodated both power and the highway signals (38). A system was designed by Le Croy Research Systems in which the mechanics were changed to include a card-edge connector for access to the highway. Of the above, only the Bell Laboratories' was bidirectional. However, it was designed for, and principally used in, low-energy physics. The other highways were designed basically for high-energy physics work and were optimized for transfers of data blocks.

STANDARDIZATION EFFORTS—CAMAC   All the highway systems mentioned above suffer from a common characteristic: each system is incompatible with the others. Each had (or has) a limited area of usage. In principle, any of them can interface any of a large number of instrument or detector types, but each has a limited repertoire of interface designs that have been implemented. Standardization is clearly needed. Perhaps the value of standardization can be demonstrated in the following way. Imagine M types of instruments or detectors, which might be used on any of N types of computers. A complete set requires $M \times N$ separate interfaces. Suppose instead that one defines a standard highway. It takes M interfaces to connect the instruments to the highway, and N interfaces to connect the highway to the computers, a total of $M + N$. If the standard is sufficiently well accepted that everyone—laboratory designers and commercial interests alike—designs to the standard, then the availability of compatible components is enhanced.

One of the groups that recognized the need for standardization is the ESONE (European Standards on Nuclear Electronics) Committee. They adapted a design that originated at Harwell, England. In 1969, they published a specification for this highway system that is now called CAMAC. In 1970, CAMAC was endorsed by the NIM Committee, who now participate with the ESONE Committee in maintaining CAMAC as an international, intercontinental standard.

There are inherent disadvantages to standardization. First, it takes time to establish a standard, so there is the possibility of its becoming obsolete before its benefits have been fully exploited. Second, its characteristics must be tailored to fit a reasonably wide spectrum of needs, so the characteristics may be less than optimum for a particular application. The standard is one of the "packages" referred to in the introduction.

CAMAC appears to have been fortunate in both these areas. For example, two basic principles on which it was based, TTL integrated circuit technology and bit-parallel transfers, are still very much in vogue. And, although the set of characteristics is by no means perfect, it has been applied in a variety of disciplines (39), including physics data acquisition, accelerator control (40), astronomy, laboratory instrumentation (41), and industrial control. CAMAC components are available from a number of manufacturers. A list of them appears periodically in the CAMAC Bulletin (42).

THE CAMAC DATA HIGHWAY

Because CAMAC has been extensively described in various publications, the discussion here is cursory.

The ESONE Committee in Europe and the NIM Committee in North America have jointly published several documents that specify various parts of the CAMAC System. The first document in this series (43) defines the heart of the CAMAC System—the modules and the crate that houses the modules. It specifies the features of modules and crates necessary to insure mechanical compatibility; the logical organization of the communication within the crate; the signal standards (voltage and current levels) for effecting this communication; and the power distribution. Two other documents specify highways for interconnecting crates and computers. The Branch Highway (44) is a bit-parallel highway for interconnecting up to seven crates. The Serial Highway (15) is a bit-serial (or byte-serial) highway for interconnecting up to 62 crates. A software language (IML) for CAMAC applications has been published (45). Other documents include Supplementary Information on preferred practices and power supplies (46), and analog signal standards (6).

The documents referred to above are primarily concerned with the specification of technical details that allow design of compatible and interchangeable components. Perhaps more important from the users' standpoint are publications that describe the features and use of CAMAC in less formal language. The CAMAC Tutorial issue of the IEEE Transactions on Nuclear Science (47) contains a series of papers that introduce various features and uses of CAMAC. Papers describing the Serial Highway and IML appeared later (48, 49). Individuals at various laboratories have written guides to the use of CAMAC (50, 51). A journal, the *CAMAC Bulletin,* publishes papers on CAMAC application (42).

Brief Characteristics

The fundamental entities in the CAMAC scheme are crates and modules. The physical form of these can be seen in Figures 9 and 10. A single-width module is 17 mm (0.67 in.) wide and 221.5 mm (8.72 in.) high. Larger modules are multiples of 17 mm in width. Each has a front panel that can be used for connections to external equipment, indicators, and controls. A small rear panel is also available. Behind the front panel is space for a printed circuit board approximately 290×200 mm ($11.4 \times$ 7.9 in.). Of the order of 10–15% of this area is needed for logic and wiring associated with the intracrate communication. The rest is available for "useful" purposes—for implementing the functional purpose of the module. When printed wiring techniques are used, this is sufficient for 40–60 dual in-line integrated-circuit packages. When wire-wrap techniques are used, up to 100 packages can be squeezed in.

The rear of the printed circuit is extended to provide an 86-way card-edge connector plug. When the module is inserted in the crate, the connector plug mates with a connector socket attached to the crate. All

Figure 9   A CAMAC module.

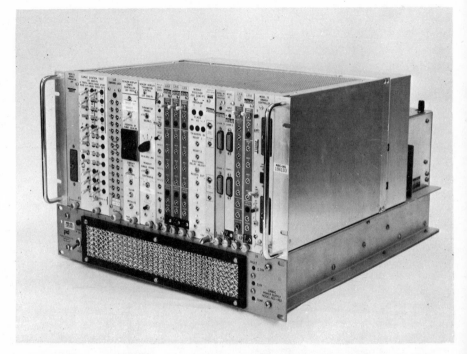

Figure 10   A CAMAC crate filled with modules. The crate controller is the right-most module.

connector sockets are prewired to a printed wire harness (the Dataway) that constitutes the highway within the crate. It also distributes power to the modules.

The standard crate contains 25 stations, each of which can accommodate a single-width module. Two stations are reserved for a Crate Controller. The rest are available for any other modules.

DATA AND CONTROL OPERATIONS   In a CAMAC crate, all modules are slaves and accept commands from the crate controller. The only spontaneous action on the part of a module is to request attention from the controller.

Addressed commands have the canonical form of NAF, where each of the symbols, N, A, and F, stands for a numerical code. The symbols N and A together specify an address. The value of N denotes a particular station number in the crate and the value of A specifies a particular feature (subaddress) of the module in that station. (Note that this is basically a form of addressing by physical location. It may be compared with a scheme in which each module is assigned a unique address to which it responds no matter where it is plugged in.) The subaddress A may be given values between 0 and 15. Since the crate contains 23 module stations, it follows that a crate has a capacity of $23 \times 16 = 368$ addressable locations—registers, in CAMAC parlance.

The value of F specifies the operation to be carried out by the addressed entity. Operations are of three classes: Read (data read from module), Write (data written into module), and Control (dataless operations). There are 32 possible F codes. Fourteen of these are either reserved or usable for miscellaneous purposes. The rest have been assigned specific meanings.

There are also two unaddressed commands: Clear and Initialize. Unaddressed means that these two commands are simultaneously sent to all stations of the crate. Although the acceptance and use of these is a function of the individual module design, Initialize is intended to be used to put the crate into some predetermined set of initial conditions—for instance, when power is first applied. Clear is a means of simultaneously clearing each of a group of registers—after a block of data has been read, for example.

A signal called Inhibit is also bussed to all module stations and may be used to control (gate) certain activities of a group of modules.

Addressed and unaddressed commands are executed synchronously in Dataway cycles, whose duration is determined by the Crate Controller. Two strobes, S1 and S2, are generated during each cycle to synchronize any irreversible actions, such as register clearing. The minimum cycle duration is 1 μs.

The 1-μs cycle period permits data to be moved between the crate controller and modules at up to 10^6 words per second. Each word may contain up to 24 significant bits. (Because of the preponderance of 16-bit word sizes in minicomputers, many CAMAC modules are designed with registers of 16-bit length to match the computer word size.) In most computer-controlled CAMAC systems, the data rates actually achieved are slowed by the time required for ancillary operations, such as computer I/O cycles and I/O instructions. For example, if 1-μs I/O cycle and 1-μs CAMAC cycle are required to move a word of data between a module and the computer memory, the total time can be 2-μs. Some interfaces have been designed in which I/O and CAMAC cycles occur simultaneously to increase the transfer rate.

LOOK-AT-ME   Each module in a CAMAC crate has the means for sending a service request signal directly to the crate controller. This signal is called Look-at-Me (LAM). It is the means by which the module can spontaneously call for attention, and is therefore analogous to the interrupt on the computer I/O structure. A LAM is often translated into a computer interrupt somewhere in the interface chain.

LAMs can be used for many purposes, but three uses predominate in physics interfacing. 1. They often are used to call attention to some isolated occurrence, such as an overflow in a scaler. In response to this, the computer program may make a mental note to take the overflow into account when next reading the scaler. 2. They are also used to synchronize processes. A CAMAC module might generate a LAM upon receiving an Event Trigger signal. In response, the computer organizes a block transfer of the event's data. 3. After a block transfer is begun, a LAM may be used to indicate the readiness of a module to transfer the next word of the block.

What Can CAMAC Do for an Experiment?

In essence, CAMAC performs the role of the multiplexer shown in Figure 4. It is the means by which the data from a number of transducers or detectors can be organized onto a single pathway leading to the computer. However, it can also accommodate communication used to control the experiment or beam-line (the dashed lines in Figure 4), and has also been used to connect computer peripherals and to implement data links between computers.

It may be instructive to consider a few examples of CAMAC module designs and how they are used. These examples do not necessarily correspond exactly with any existing designs, but are nevertheless typical of what can be done. It is hoped that the reader can extrapolate from these examples to gain an appreciation of how CAMAC may be used in the high-energy physics environment.

Some modules provide a direct interface between the detector and the digital highway (CAMAC Dataway). Examples of this include the following.

1. A Discriminator-Coincidence-Register module can be used to interface twelve elements of a scintillator hodoscope array. It contains twelve channels of logic, each with a pulse amplitude discriminator, a coincidence gate, and a register. The information in the register is accessible via the Dataway.
2. Multiwire Proportional Chamber (MWPC) Interface. The time-of-arrival of signals from a MWPC that uses delay-line readout (or signals from a magnetostrictively read out spark chamber) are digitized and stored in a scratch-pad memory. The memory is read out via CAMAC.
3. Analog-to-Digital Converter (ADC) module. This module contains

12 eight-bit ADCs. Each can digitize the electrical charge contained in a time-selected pulse from a scintillation detector or proportional chamber.

For various reasons, the processing electronics that directly receives the detector signals and digitizes them may need to be housed outside the CAMAC crate. It can still be interfaced to CAMAC with appropriate modules.

5. Dual, parallel, input register. This module can accept a word of digital data from each of two sources. Each is connected to it by a 24-wire cable. The connectors for these cables are on the front panel. Many different types of "boxes" that generate digital data can be connected to the highway by modules of this type.

6. First-in, First-out (FIFO) Buffer. This is similar to the register except that it can accept, in word-serial mode, a large number of digital words in its FIFO memory. At a later time, the words can be sequentially read out via CAMAC. Such a module can be used to store quickly the block of data from an event, to free the detector to be ready for the next event (i.e. reduce dead time). Or it may stack the data from several events so that they can be transferred to the computer in a single block transfer.

One of the "workhorses" in the stable is the scaler module. A number of configurations have been designed that have various maximum counting rates, numbers of scalers per module, and numbers of bits per scaler.

7. An example of the scaler module is one having eight scalers, each of 100 MHz capability, and each 24 bits long.

Modules are available that can be used for monitoring or controlling beam-line equipment.

8. Digital-to-Analog Converter (DAC) modules can be digitally commanded to generate analog voltages of specified values. These may be used to control power supply voltages or to set magnetic currents.

9. Digital Voltmeter (DVM) modules are used to measure the values of voltages or currents. Through use of a DVM with a CAMAC analog multiplexer module, a number (e.g. 16) of voltages or currents may be measured by the one DVM.

10. An interrupt register module can generate a LAM (computer interrupt) each time it receives an Event Trigger.

11. Switch register modules can be used to input information that changes infrequently, such as run number, date, run conditions, etc.

Computer Peripherals and On-Line Ties

The term "computer peripherals" is used here to describe devices which, although interfaced to the I/O structure of the computer, are not directly involved with the experimental data acquisition. They are concerned with data display, data storage, and the transmission of data to other computers.

All computer manufacturers offer some peripherals that are plug-compatible with their computers. The question therefore arises as to whether such types of peripherals should be interfaced directly on the computer I/O structure or via CAMAC. The consensus seems to be to buy peripherals from the manufacturer unless there are overriding considerations. For example, the peripherals may need to be moved from one computer type to another with a different I/O structure. Or the desired peripheral may not be available from the manufacturer. Peripherals such as magnetic tape units (52) and disks (53) have been successfully interfaced via CAMAC. In addition, interfaces to CRT display systems, teletypewriters, paper-tape punches, and magnetic tape cassette recorders are commercially available.

Several installations are using CAMAC as a medium to effect the transmission of data from one system to another (54). A data transmission link is established between a CAMAC module of one system and a module of the other system. Modules that effect this transfer on a word-by-word basis have been designed at the Clinton Anderson Meson Physics Facility (55), where they may be used to provide communication between data acquisition systems and a central minicomputer system with a rich supply of peripherals. The resources of the central computer can thereby be shared by the satellite systems. Transmission of blocks of data to the central computing facility can also be organized by the central minicomputer (29).

Modules that effect data transfer on a block-by-block basis have been designed at the Fermi National Accelerator Laboratory (30). In this case, each module contains a buffer that can be loaded with a block of words by the host CAMAC system. At the Fermi Laboratory, these will be used to implement transmission links between the experiments' data acquisition systems and the CDC6600 computer used for data analysis.

TOPOLOGY OF CAMAC SYSTEMS

It has been shown that CAMAC crates perform a bidirectional multiplexing. We must still describe how they can be interfaced to one or more points on the computer I/O structure. Three arrangements for interfacing

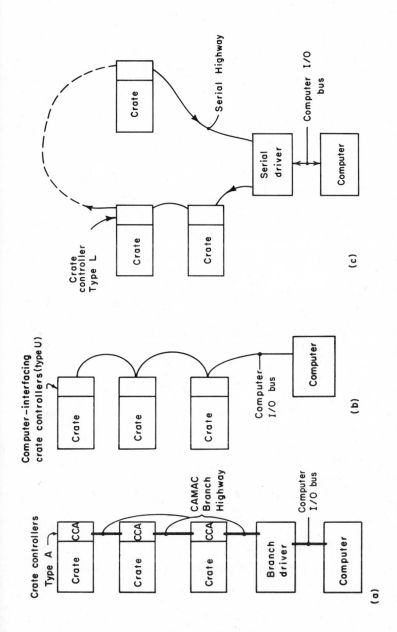

Figure 11 Three methods of interfacing groups of CAMAC crates to a computer I/O structure: (*a*) using Type U crate controllers; (*b*) via the branch highway; (*c*) via the serial highway.

groups of CAMAC crates to computers are shown in Figure 11, and described below (56).

The Type U Interface

Figure 11*b* shows a scheme in which each CAMAC crate is individually interfaced to the computer I/O structure. In this case, each crate controller is a port on the I/O structure. Such crate controllers (57, 58) are sometimes called Type U. Small systems can be built up economically in this way. One needs only a powered CAMAC crate, an appropriate Type U interface, and some modules to begin operation. As the number of modules exceeds the capacity of the crate, a second crate and Type U interface is added, and so on. The number of crates that can be accommodated is limited by the characteristics of the I/O structure, but this is generally not a significant limitation.

The fact that the DEC PDP-11 computer is widely used for high-energy experiments perhaps justifies using it as an example to illustrate the Type U crate controller. The PDP-11 I/O structure, called the Unibus, has three properties that are of interest here. 1. It is asynchronous—i.e. the time duration of an I/O cycle expands automatically to fit the needs of the addressed device. 2. External devices attached to the Unibus are assigned addresses that are an extension of the range of computer memory address. (This means that the software treats data transfers to external devices exactly like transfers to memory.) 3. It provides for vectored interrupts. This means that an interrupting device specifies a vector, or memory location, where access to the program that services the interrupt is located.

These three properties can be used to advantage in Type U controllers (58, 59). The asynchronism permits the CAMAC Dataway cycle to be "imbedded" in the I/O cycle. Thus an I/O cycle that moves a word of data from a module to computer memory contains three parts: (*a*) the command to execute the Dataway cycle moves from computer to the Type U; (*b*) the Dataway cycle is executed and moves the data from module to the Type U; and (*c*) the data is transferred from Type U to the computer memory.

The CAMAC Branch Highway

The branch highway (44) is a means of organizing a CAMAC system of up to seven crates, using bit-parallel communication. Each crate is connected to the branch highway by means of a rigidly specified crate controller, the Crate Controller Type A (CCA). As shown in Figure 11*a*, a single interface, the Branch Driver, suffices to interface the entire system to a computer.

The branch highway is very nearly transparent to the computer. That is, it is basically a vehicle for transporting CAMAC commands to the crates, and for bidirectionally transferring data. Thus, to cause an action' specified by NAF in a certain crate, the program specifies $CNAF$, where C is the address ($1 \leqq C \leqq 7$) of the crate.

The data transfer rate of a branch highway is slightly lower than that for the crate itself because of the extra time taken for the communication along the branch. Branch highway cycles of 1.5 μs or less can be achieved. The length of a branch is limited to about 25–50 m because of the signal standards employed. Branch extenders that extend this limit to a kilometer or more are commercially available.

The CAMAC Serial Highway System

The CAMAC Serial Highway System (15, 48) is another means of organizing multicrate systems. In contrast with the branch highway, it uses bit-serial (or byte-serial) communication, has a lower maximum operating speed, a much greater maximum length, and a capacity of 62 crates.

As shown in Figure 11c, crates are arranged on a loop on which signals travel unidirectionally. Data transmission may be bit-serial at rates up to 5×10^6 bits per second, or byte-serial at up to 5×10^6 bytes per second. A crate controller, the Type L, is specified for interfacing the crate to the serial highway.

Communication is carried in the form of messages. The computer interface, known as Serial Driver, directs command messages to crates. These messages carry $CNAF$ and, in the case of write operations, 24 bits of data. In response to a command message, the addressed crate executes a Dataway cycle, and returns a Reply message. It carries the results of the Dataway cycle, including data read from the addressed module during Read operations. The crate may spontaneously transmit a Demand message to notify the computer of the existence of a LAM in the crate. All messages carry a geometric parity error detection code, and certain error recovery procedures are provided for. The serial highway system has been defined so that messages to other than CAMAC crates can be carried without interference to the CAMAC communications.

The intercrate cabling is considerably simpler than that for the (parallel) branch highways, but the speed of operation is slower. In byte-serial, a minimum of 3.4 μs is required for a Command-Reply transaction (one Dataway cycle); 25 μs in bit-serial.

FUTURE TRENDS

As the state-of-the-art of electronics continues to advance, new devices and components useful to experimental physics will continue to become available. This raises at least two questions: How fast will present equipment become obsolete, and how will the new developments contribute to the capability to perform experiments?

Components that will enable future systems to have increased amounts of memory and decision-making power (intelligence) distributed throughout the system are becoming available. Integrated-circuit memory devices with a higher number of bits per package, at decreasing cost per bit of storage, are appearing. Integrated circuits with complete arithmetic and logic processors (microprocessors) are already available, and faster designs are being developed. Even now it is possible to put the equivalent of a modest minicomputer in the volume of a CAMAC module.

It seems clear that it is becoming feasible to incorporate larger buffer memories closer to the detectors. This can allow data taking at higher rates. It should also become possible to build more intelligence into data preprocessors. More complex electronics close to the detectors may extract more information from the detector signals. Perhaps track reconstruction or kinematic fits of the data will permit the discarding of "bad" events before they clog up the data transmission channels with useless data. More intelligence can be built into ancillary equipment, such as scaler displays, magnet current and beam monitors, photomultiplier high-voltage controls, etc. This will make complex arrays of detectors more easily manageable.

Since CAMAC is relatively young, and many laboratories are still building their inventories of CAMAC equipment, it is appropriate to speculate about its life span. The data-handling capabilities of the Dataway are based on TTL integrated-circuit technology and bit-parallel data transmission with 1-μs cycle times. These fundamental bases seem to be well chosen in the light of known and expected developments. Thus its continued use for interfacing to presently used types of detectors and computers seems assured. A critical question, however, is how usable CAMAC is for new systems that incorporate the distributed memory and intelligence facilities made possible by the new components. This is a situation that has not yet been adequately studied, although the beginnings are very promising. Microprogrammable branch drivers (60) have existed for several years. Crate controllers that incorporate microprocessors have been developed at several laboratories (61, 62) and by commercial interests. The serial highway system provides for multiple crate controllers—i.e. one for controlling local processes and one for communication with computers. Modules for providing computer-to-computer communication have been developed (30, 55).

One thing that is certain is that high-energy physics experimentation will continue to demand and effectively utilize the latest electronic advances.

Literature Cited

1. Levrat, B. C. 1971. *Proc. 1970 CERN Comput. Data Process. Sch. CERN Rep.* 71-6:315–34
2. Russell, R. D. 1972. *Proc. 1972 CERN Comput. Data Process. Sch. CERN Rep.* 72-21:275–340
3. Gelernter, H. 1965. *Adv. Comput.* 6:229–96
4. Gore, R. A., Machen, D. R. 1973. *Proc. IEEE* 61:1589–96
5. US AEC NIM Committee. 1974. *Standard Nuclear Instrument Modules.* TID-20893 (Rev. 4). Washington DC: GPO
6. US AEC NIM Committee. 1974. *CAMAC—Specification of Amplitude Signals within a 50 Ohm System.* Y3AT7:22,TID-26614. Washington DC: GPO (EUR-5100e)*
7. Martin, L. 1969. *Telecommunications and the Computer,* 108–12. Englewood Cliffs, NJ: Prentice-Hall
8. Koudela, J. 1973. *Proc. IEEE* 60:1526–34
9. See Ref. 7, 204–15
10. Burton, H. O., Sullivan, D. D. 1972. *Proc. IEEE* 60:1293–1301
11. Tutorial Issue on High-Speed Pulse Instrumentation Techniques. 1973. *IEEE Trans. Nucl. Sci.* NS-20(5):1–66
12. Morrison, R. 1967. *Grounding and Shielding Techniques in Instrumentation.* New York: Wiley. 144 pp.
13. Smith, S. R., Goodwin, R. W., Storm, M. R. 1973. *IEEE Trans. Nucl. Sci.* NS-20(3):536–40
14. Sebestyen, L. G. 1973. *Digital Magnetic Tape Recording for Computer Applications.* London: Chapman & Hall. 157 pp.
15. US AEC NIM Committee. 1973. *CAMAC Serial System Organization A Description.* Y3AT7:22 TID-26488. Washington DC: GPO (ESONE/SH/01)*
16. Davey, J. R. 1972. *Proc. IEEE* 60:1284–92
17. Andreae, S. W., Lafore, R. W. 1968. *IEEE Trans. Nucl. Sci.* NS-15(1):103–8
18. Olson, S. R. et al 1974. *IEEE Trans. Nucl. Sci.* NS-21(1):851–56
19. Herbst, L. J., ed. 1970. *Electronics for Nuclear Particle Analysis,* p. 258 ff. London: Oxford Univ. Press
20. Binnall, E., Kirsten, F. A., Lee, K., Nunnally, C. 1973. *IEEE Trans. Nucl. Sci.* NS-20(1):367–74
21. See Ref. 7, p. 439
22. See Ref. 14, pp. 106–11
23. Foley, K. J. et al 1964. *Nucl. Instr. Methods* 30:45–60
24. Blieden, H. et al 1964. *Proc. Informal Meet. Film-less Spark Chamber Tech. Assoc. Comput. Use. CERN Rep.* 64-30:49–56
25. See Ref. 24, pp. 3–9
26. Spinrad, R. J. 1964. *IEEE Trans. Nucl. Sci.* NS-11(3):324–29
27. Atomic Energy Commission. 1969. *Proc. Skytop Conf. Comput. Syst. in Exp. Nucl. Phys. Conf.* 690301, US AEC Div. Tech. Info., Oak Ridge, Tenn.
28. Fulbright, H. W. et al 1969. *On-Line Data-Acquisition Systems in Nuclear Physics.* Washington DC: Nat. Acad. Sci.
29. Anderson, H. L. et al 1970. *Rep. LAMPF Data-Acquisition Study Group, Los Alamos Sci. Lab. Rep.* LA-4504MS
30. Brenner, A., Droege, T. F., Martin, R. G. 1975. *IEEE Trans. Nucl. Sci.* NS-22(1):494–98
31. Special Issue on Computers. 1972. *CERN Courier* Vol. 12, No. 3:58–91
32. Dimmler, D. G. 1974. *IEEE Trans. Nucl. Sci.* NS-21(1):838–50
33. Stubblefield, F. W., Dimmler, D. G. 1975. *IEEE Trans. Nucl. Sci.* NS-22(1):473–79
34. Soucek, B. 1972. *Minicomputers in Data Processing and Simulation,* 151–291. New York: Wiley-Interscience
35. Korn, G. A. 1973. *Minicomputers for Engineers and Scientists,* 131–76. New York: McGraw Hill
36. Kirsten, F. A. 1970. *IEEE Trans. Nucl. Sci.* NS-17(1):452–57
37. Droege, T., McFadden, J., Wash, S., VonColln, R. 1970. *IEEE Trans.*

* European editions of these documents available from Office for Official Publications of the European Communities, Luxembourg, PO Box 1003.

Nucl. Sci. NS-17(1):445–51
38. Gere, E. A., Lie, H. P., Miller, G. L. 1970. *IEEE Trans. Nucl. Sci.* NS-17 (1):436–44
39. *Proc. First Int. Symp. CAMAC in Real-Time Comput. Appl.** Dec. 1973, Luxembourg
40. Crowley-Milling, M. C. et al. See Ref. 39, 121–28
41. Brill, A. B. et al 1974. *IEEE Trans. Nucl. Sci.* NS-21(1):892–97
42. *CAMAC Bulletin* (A journal)*
43. US AEC NIM Committee 1972. *CAMAC — A Modular Instrumentation System for Data Handling-Description and Specifications.* Y3AT7:22, TID-25875. Washington DC: GPO (EUR-4100e)*
44. US AEC NIM Committee. 1972. *CAMAC — Organization of Multicrate Systems.* Y3AT7:22, TID-25876. Washington DC: GPO (EUR-4600)*
45. US NIM Committee. 1975. *CAMAC — The Definition of IML.* Y3AT7:22, TID-26615. Washington DC: GPO
46. US AEC NIM Committee. 1972. *Supplementary Information on CAMAC Instrumentation System.* Y3AT7:22, TID-25877. Washington DC: GPO; also Suppl. to CAMAC Bull. No. 6
47. CAMAC Tutorial Issue. 1973. *IEEE Trans. Nucl. Sci.* NS-20(2):1–71
48. Machen, D. R. 1974. *IEEE Trans. Nucl. Sci.* NS-21(1):876–80
49. Hooton, I. N., Hagen, P. J. 1974.

IEEE Trans. Nucl. Sci. NS-21(1): 903–8
50. Zacharov, B. 1972. *CAMAC Systems: A Pedestrian's Guide.* Daresbury Nucl. Phys. Lab. Rep. DPNL/R23
51. *What is CAMAC?* 1973. CERN-NP CAMAC Note 45-00
52. Oakes, A., Andreae S., Rudden, R. 1973. *IEEE Trans. Nucl. Sci.* NS-20(1):685–89
53. Peatfield, A. C., Spurling, K., Zacharov, B. 1974. *IEEE Trans. Nucl. Sci.* NS-21(1):867–69
54. Zacharov, B. 1974. *IEEE Trans. Nucl. Sci.* NS-21(1):898–902
55. Biswell, L. R. et al 1973. *IEEE Trans. Nucl. Sci.* NS-20(1):675–79
56. Costrell, L. 1974. *IEEE Trans. Nucl. Sci.* 1974. NS-21(1):870–75
57. Eichholz, J. J., Lenkszus, F. R., Strauss, M. G. 1971. *IEEE Trans. Nucl. Sci.* NS-18(1):292–98
58. *CC11 CAMAC Crate-PDP-11 Interface.* 1972. CERN-NP CAMAC Note 43-00
59. Halling, H., Zwoll, K., Muller, K. D. 1972. *IEEE Trans. Nucl. Sci.* NS-19(1):699–703
60. Biswell, L. R., Rajala, R. E. 1972. *A Micro-programmed Branch Driver (MBD) for a PDP-11 Computer.* Los Alamos Sci. Lab. Rep. LA-4916MS
61. Halling, H. 1974. *IEEE Trans. Nucl. Sci.* NS-21(1):886–88
62. Bobbitt, J. 1975. *IEEE Trans. Nucl. Sci.* NS-22(1):508–10

* See footnote, p. 553. Reference 43 has also been issued in the US as IEEE Standard 583.

THE PARTICLE DATA GROUP: GROWTH AND OPERATIONS—Eighteen Years of Particle Physics[1]

✕5569

Arthur H. Rosenfeld

Department of Physics and Lawrence Berkeley Laboratory,
University of California, Berkeley, California 94720

CONTENTS

[1] This review was written under the auspices of the US Energy Research and Development Administration.

555

1 INTRODUCTION

1.1 Manpower, Budgets, and Publications

In 1975, worldwide, there are probably 7000 particle physicists (plus graduate students). Particle physics annual research budgets total to ~ $500 M, and some 300 experiments are underway at 16 major accelerator centers. Some details are given in Table 1.

Communication is remarkably good. Most high-energy physicists speak English, give papers in English at conferences, and publish their experimental results in English. A single international group, the Particle Data Group (PDG), compiles all the data on particle properties, and most data compilations are coordinated by one of two groups, PDG (with headquarters in the US) and the HERA (High Energy Reaction Analysis) group at CERN (Organisation Européenne pour la Recherche Nucléaire) in Geneva. Particle physicists produce about 4000 preprints and reports each year; most are received by two large institutional libraries, one at SLAC (Stanford Linear Accelerator Center) and one at CERN, and each library distributes a list of preprints received. Most of these preprints are theoretical papers; of 4000 preprints received by SLAC in 1974, only 400 contained new experimental data. Not all papers appear as preprints. There are 50–100% more papers than the 4000 quoted, depending on how much one includes of astrophysics, nuclear physics, instrumentation, and other related fields.

Worldwide, about 150 experiments are completed each year, so each experiment seems to generate about 3 papers, in addition to some PhD theses.

In this review we briefly discuss how the data rate grew from a trickle

Table 1 Estimates of 1975 effort in particle physics

	Physicists[a]	Budget[b] (M$/yr)	Major Accelerator Centers
Japan	800	20	1
USA	2000	150	4
USSR	1000	50	3
Europe (excluding USSR)	3000	250	8
	~ 7000	~ 500	16

[a] Excludes graduate students.
[b] Excludes construction.

to a fairly steady flood, and how PDG has grown along with it. We outline how PDG has learned to collect, evaluate, correct, verify, analyse, and distribute the data, and we discuss its plans for the future.

PDG has taken on the responsibility of critically reviewing the results of experiments—in this article we take a critical look at our own record.

2   HISTORICAL OUTLINE

In this section, we make some historical comments on our publications, starting with 1957. This sketch serves as a convenient outline for the rest of the article, shows how particle data collection has grown to match the expanding needs of particle physics, and mirrors how some of the understandings and interests of particle physicists have developed.

1957   The current form of the Particle Properties Tables stems directly from a 1957 article in the *Annual Review of Nuclear Science,* by Gell-Mann & Rosenfeld (1). Data on particles were becoming available at an increasing rate, and even before the *Annual Review* volume was published, Walter H. Barkas and A. H. Rosenfeld decided to make the first update of the table of masses and mean lives. It appeared as Lawrence Radiation Laboratory Report UCRL-8030 (unpublished). Barkas was using nuclear emulsions to measure precise particle masses; Rosenfeld was doing bubble chamber experiments and preparing to automate the processing of bubble chamber events; and both needed current values of particle properties for computer programs to process measurements of events. Thus 1957 saw the first edition of UCRL-8030: *Data for Elementary Particle Physics.*

1958   The first revision of UCRL-8030 appeared, accompanied by a "wallet card," the front side of which is reproduced in Figure 1. Only one resonance was known, so instead of the now familiar Baryon Table, there is a single entry for the "(3/2, 3/2) πp Resonance" at the center left of panel 3. Since it was not then considered to be a particle, its mass (1238 MeV) was not given; instead its energy was characterized by the momentum of the pion beam used to form the resonance.

During the next few years, bubble chamber experiments became very successful, exciting, and time-consuming, and nobody had any spare time for revisions of "8030."

1961   Scintillation counter experiments on πp scattering had uncovered the broad peak now identified with $\Delta(1950, 7/2+)$, and the "2nd" and "3nd" πN "resonances" (now known as the 2nd and 3nd "resonance *regions*" since the regions contain 3 and 4 resonances, respectively) (2, 2a). But the big surprise was the discovery, with bubble chambers, of 7 more striking, narrow resonances in systems other than πN. Specifi-

Masses and mean lives of elementary particles; November, 1957
(The antiparticles are assumed to have the same spins, masses, and mean lives as the particles listed)

Particle	Spin	Mass (Errors represent standard deviation) (Mev)	Mass difference (Mev)	Mean life (sec)	Decay rate (number per second)
Photon γ	1	0		stable	0
ν	$\frac{1}{2}$	0		stable	0
e^-	$\frac{1}{2}$	0.510976 (a)		stable	0
μ^-	$\frac{1}{2}$	105.70 ±0.06 (a)		$(2.22 \pm 0.02) \times 10^{-6}$	0.45×10^6
π^+	0	139.63 ±0.06 (a)	4.6 (a)	$(2.56 \pm 0.05) \times 10^{-8}$ (a)	0.39×10^8
π^0	0	135.04 ±0.16 (a)		$< 4 \times 10^{-16}$ (d)	$> 2.5 \times 10^{15}$
K^+	0	494.0 ±0.2 (g)	0.4 ± 1.8	$(1.224 \pm 0.013) \times 10^{-8}$ (h)	0.815×10^8
K^0	0	494.4 ± 1.8 (i)		$K_1: (0.95 \pm 0.08) \times 10^{-10}$ (e)	1.05×10^{10}
				$K_2: (4 < \tau < 13) \times 10^{-8}$ (c)	$(0.07 < \tau < 0.25) \times 10^8$
p	$\frac{1}{2}$	938.213 ±0.01 (a)		stable	0.0
n	$\frac{1}{2}$	939.506 ±0.01 (a)		$(1.04 \pm 0.13) \times 10^{+3}$ (a)	0.96×10^{-3}
Λ	$\frac{1}{2}$	1115.2 ±0.14 (j)		$(2.77 \pm 0.15) \times 10^{-10}$ (k)	0.36×10^{10}
Σ^+	$\frac{1}{2}$	1189.4 ±0.25 (l)	7.1 ± 0.4	$(0.83 \, ^{+.06}_{-.05}) \times 10^{-10}$ (m)	1.21×10^{10}
Σ^-	$\frac{1}{2}$	1196.5 ±0.5 (n)		$(1.67 \pm 0.17) \times 10^{-10}$ (o)	0.60×10^{10}
Σ^0	$\frac{1}{2}$	1190.5 $^{+0.9}_{-1.4}$ (p)	6.0 $^{+1.4}_{-0.4}$	$(< 0.1) \times 10^{-10}$ (b) theoretically ~10^{-19}	$> 10 \times 10^{10}$ theoretically ~10^{19}
Ξ	?	1320.4 ± 2.2 (q)		$(4.6 < \tau < 200) \times 10^{-10}$ (f)	$(>0.005, <0.2) \times 10^{10}$
Ξ^0	?	?		?	?

Table IV

Atomic and nuclear constants in units of Mev, cm, and sec[a]

GENERAL ATOMIC CONSTANTS

$N = 6.0249 \times 10^{23}$ molecules/gram

$c = 2.99793 \times 10^{10}$ cm/sec

$e = 4.80286 \times 10^{-10}$ esu = 1.6021×10^{-1} coulumb.

1 Mev = 1.6021×10^{-6} erg [1 ev = $e(10^8/c)$]

$\hbar = 6.5817 \times 10^{-22}$ Mev sec = 1.054×10^{-27} erg sec.

$\hbar c = 1.9732 \times 10^{-11}$ Mev cm [= λ for p = 1Mev/c]

$k = 8.6167 \times 10^{-11}$ Mev/°C [Boltzmann constant]

$\alpha = \frac{e^2}{\hbar c} = 1/137.037, \quad e^2 = 1.44 \times 10^{-13}$ Mev cm

QUANTITIES DERIVED FROM THE ELECTRON MASS, m

Mass and Energy

$m = 0.510976$ Mev = $1/1836.12$ m_p = $1/273.26$ m_π

Rydberg, $R_\infty = \frac{me^4}{2\hbar^2} = mc^2 \times \frac{\alpha^2}{2} = 13.605$ ev

Length (1 fermi = 10^{-13} cm; 1 Å = 10^{-8} cm)

$r_e = e^2/mc^2 = 2.81785$ fermi

$\lambda_{Compton} = \frac{\hbar}{mc} = r_e \alpha^{-1} = 3.8612 \times 10^{-11}$ cm

$a = $ Bohr $= \frac{\hbar^2}{me^2} = r_e \alpha^{-2} = 0.52917$ Å

Cross Section

$\sigma_{Thompson} = \frac{8}{3} \pi r_e^2 = 0.6652 \times 10^{-24} cm^2 = 0.6652$ barn

Magnetic Moment and Cyclotron Angular Frequency

$\mu_{Bohr} = \frac{e\hbar}{2mc} = 0.57883 \times 10^{-14}$ Mev/gauss

$\frac{1}{2}\omega_{cyclotron} = \frac{e}{2mc} = 8.7945 \times 10^6$ rad sec^{-1}/gauss

$g_{electron} = 2[1 + \frac{\alpha}{2\pi} + 0.328(\frac{\alpha}{\pi})^2] = 2[1.0011631]^b$

$g_{muon} = 2[1 + \frac{\alpha}{2\pi} + 0.75 (\frac{\alpha}{\pi})^2] = 2[1.0011773]^b$

QUANTITIES DERIVED FROM THE PROTON MASS, m_p

Rest mass = 938.211 Mev/c^2 = 1836.12 m_e = 6.719 m_π

1.007593 m_p (1m_p = 1 amu = $\frac{1}{16}$ O^{16})

Magnetic Moment and Cyclotron Angular Frequency

$\mu_p = \frac{e\hbar}{2m_p c} = 3.1524 \times 10^{-18}$ Mev/gauss

$\frac{1}{2}\omega_{cyclotron} = \frac{e}{2m_p c} = 4.7896 \times 10^3$ rad sec^{-1}/gauss

$g_p = (\frac{\mu}{\mu_p})_{proton} = 2.79275; \quad g_n = (\frac{\mu}{\mu_p})_{neutron} = -1.9128$

Table IV (continued)

QUANTITIES DERIVED FROM THE MASS OF THE CHARGED PION, m_π

Rest mass = 139.63 Mev/c^2 = 273.26 m_e = 0.14882 m_p

Length

$\frac{\hbar}{m_\pi c} = 1.4132$ fermi (~$\sqrt{2}$ fermi)

Natural (= "geometrical") Nucleon Cross Section

$\pi (\frac{\hbar}{m_\pi c})^2 = 62.7344$ mb (1 mb = 10^{-27} cm^2)

(3/2, 3/2)πp Resonance

Center-of-mass momentum: $p_\pi = 230$ Mev/c

Lab-system momentum: $P_\pi = 303$ Mev/c ($T_\pi = 194$ Mev)

RADIOACTIVITY

1 curie = 3.7×10^{10} disintegrations/sec

1 r = 87.8 ergs/g air = 5.49×10^7 Mev/g air

Fluxes (per cm^2) to liberate 1 r in carbon:

3×10^7 minimum ionizing singly charged particles

0.9×10^9 photons of 1 Mev energy.

(These fluxes are actually correct to within a factor of two for all materials.)

Natural background: 100 mr/year

"Tolerance" 100 millirem/week [Note, 1 r may produce up to 10 "rem" (r equivalent for man), depending on type of radiation.]

MISCELLANEOUS

Physical Constants

1 year = 3.1536×10^7 sec (= $\pi \times 10^7$ sec)

Density of air = 1.205 mg/cm^3 at 20°C

Acceleration by gravity = 980.67 cm/sec^2

1 calorie = 4.184 joules

1 atmosphere = 1033.2 g/cm^2

Numerical Constants

1 radian = 57.29578 deg; e = 2.71828

ln 2 = 0.69315; log$_{10}$ e = 0.43429;

ln 10 = 2.30259; log$_{10}$ 2 = 0.30103.

Stirling's approximation

$\sqrt{2\pi m} (\frac{n}{e})^n < n! < \sqrt{2\pi m} (\frac{n}{e})^n (1 + \frac{1}{12n-1})$

Gaussianlike Distributions

For n > -1 but not necessarily integral:

$\int_0^\infty x^{2n+1} \exp{\frac{-x^2}{2\sigma^2}} dx = 2^n \sigma^{2n+2} (\frac{n}{2})! = \sqrt{\pi/2}$

Relation between standard deviation σ and mean deviation a:

$2\sigma^2 = \pi a^2$; $\sigma = 1.4826$ probable error.

Odds against exceeding one standard deviation = 2.15:1; two, 21:1; three, 370:1; four, 16,000:1; five, 1,700,000:1

Figure 1   Front side of the original 1958 Wallet Card.

cally, K*(885) and Y*(1380) were discovered (2) in 1960, and in 1961 there followed two more mesons (ρ and ω) and three more Y*'s (3a). Accordingly the 1961 edition of UCRL-8030 included a Table VI "Possible Resonances of Strongly Interacting Particles" (see Figure 2). It is interesting to look back and note that this Table VI was not considered important enough to add to the wallet card.

1962   The η(548) meson had been predicted by Gell-Mann's Eightfold Way (3) in 1961. It was expected to decay into two photons or four pions. A meson (apparently unrelated because it decayed into three pions (3a) was discovered late in 1961, and properly identified (4) as the predicted pseudoscalar meson, the η, early in 1962. This completed the first meson octet, but by later standards it attracted little attention (no press conference, no flurry of theoretical papers, and no 1962 edition of UCRL-8030). Later in 1962 the ϕ(1020) bump was discovered (5), but it

Possible resonances of strongly interacting particles (as of August 1961)

	Mass (Mev)	Half-width $\Gamma/2$ (Mev)	Spin I	Spin and parity J	Decay properties					
					Orbital wave	Products	Branching fraction	Q^j (Mev)	k (Mev/c)	Ref.
ρ	750	±50	1	1-	p	$\pi+\pi$	100%	480	350	a
ω	790	±<15	0	1-		3π	100%	510	——	b
K*	885	± 8	1/2?	?	?	$K+\pi$	100%	252	282	c
N*	1238	±45	3/2	3/2+	p	$N+\pi$	100%	163	234	d
	1510	±30	1/2	3/2-	d	$N+\pi$ + others	?	435	449	d
	1680	±50	1/2	5/2+	f+?	$N+\pi$ + others	?	605	567	d
	1900	±100	3/2	?	?	?	?	-		e
Y*	1380	±25	1	?	?	$\Lambda+\pi$	96%	130	205	f
						$\Sigma^0+\pi$	4%	54	122	
	1405	±10	0	?	?	$\Sigma^0+\pi^0$ $\Lambda+2\pi$	100%	79 20	153 ——	g
	1525	±20	0	≥ 3/2	?	$\Sigma+\pi$ $\Lambda+2\pi$ $K+p$	4 only 1 this ? ratio known	199 130 89	271 —— 246	h
	1815	±60	0	≥ 3/2	?	many	-	-	-	i

Figure 2   Table of resonances, 1961.

TENTATIVE DATA ON STRONGLY INTERACTING STATES (April 1963, A. H. Rosenfeld)

Particle	Established quantum No. $I(J^{PG})$	Possible assignment Quantum No. $I(J^{PG})$	Possible assignment Regge[1] trajectory	Mass (MeV)	Γ[2] (MeV)	Mass^2 (BeV)^2	Dominant decays Mode	%	Q[4] (MeV)	p or p_{max} (MeV/c)
$K_1 \overline{K}_1$	$0(J^{++}_{even})$	$0(0^{++})$	$^3+\alpha$	$\sim 2m_K$	?		Even number of pions $K\overline{K}(K_1^+K_1^-, K_2K_2,$ not $K_1K_2)$		<0	<0
f = Vacuum ?	$0(\geq 2^{++})$	$0(2^{++})$	$^3+\alpha$	1250	75	1.56	2π 4π $K\overline{K}(K_1K_1, K_2K_2,$ not $K_1K_2)$	large <30 ?	980 710 256	690 550 380
F	$0(0^{-+})$		$^3+\beta$	548	<10	.30	$\pi^+\pi^-\pi^0$ $\pi^0\pi^0\pi^0$[3] $\pi^+\pi^-\gamma$ $\gamma\gamma$	23 39 7 31	134 143 269 548	174 182 235 274
3	$0(1^{--})$		$^3-\gamma$	782	<15	.62	$\pi^+\pi^-\pi^0$[3,5] $\pi^+\gamma$- $\pi^+\pi^-$	84 12±4 4	368 647 503	326 379 364
ϕ	$0(J^{--}_{odd})$	$0(1^{--})$	$^3-\gamma$	1020	< 5	1.04	$K\overline{K}(K_1K_2,$ not $K_1K_1, K_2K_2)$ Odd number of pions		24	111
$\pi^0 \pi^\pm \pi$	$1(0^{--})$		$-\pi\beta$	π^0 135 $\pi^\pm$ 140	0 0	0.018 .02	$\pi^0\to\gamma\gamma$[6] $\pi^\pm\to\mu\nu$	100 58	135 34	67 30
ρ	$1(1^{-+})$		$+\pi\gamma$	750	100	.56	$\pi\pi$[3] [p-wave]	100	471	348
K $\{K^0, K^\pm\}$	$\frac{1}{2}(0^-)$		$\kappa\beta$	K^0 498 $K^\pm$ 494	0 0	.24	$K^0\to\pi^+\pi^-$[6] $K^\pm\to\mu\nu$	2/3 K_1 58	219 388	206 236
$K^*_{1/2}(888)$	$\frac{1}{2}(1^-)$		$\kappa\gamma$	888	50	.78	$K\pi$(p-wave)	100	251($K^0\pi^-$)	283
$K^*_{1/2}(725)$	$\frac{1}{2}(?)$	?	?	725	<15	.53	$K\pi$	?	101($K^-\pi^0$)	161

Particle	I(J^P)	I(J^P)	Symbol	Mass (n940 p938)	0	.88	Decay modes (e⁻ν̄p[6])	100	.78	1.2
$N\begin{cases}n\\p\end{cases}$	$\frac{1}{2}(\frac{1}{2}+)$		N_α	n 940 / p 938	0	.88	$e^-\bar{\nu}p$ [6]	100	.78 / –	1.2 / –
$N^*_{1/2}(1688)="900\ MeV\ \pi p"$	$\frac{1}{2}(\frac{5}{2}+)$		N_α^{II}	1688	100	2.84	$N\pi$(f-wave) / ΛK(f-wave)	80 / <2	610 / 76	572 / 235
$N^*_{1/2}(1512)="600\ MeV\ \pi p"$	$\frac{1}{2}(\frac{3}{2}-)$		N_γ	1512	100	2.28	$N\pi$(d-wave)	80	434(πp)	450
$N^*_{3/2}(1238)="Isobar"$	$\frac{3}{2}(\frac{3}{2}+)$		Δ_δ	1238	100	1.53	$N\pi$(p-wave)	100	160(πp)	233
$N^*_{3/2}(1920)$	$\frac{3}{2}(\frac{7}{2}+)$	$\frac{3}{2}\ \frac{7}{2}$	Δ_δ^{II}	1920	~200	3.69	$N\pi$ / ΣK	30 / <4	842(πp) / 233	722 / 425
Λ	$0(\frac{1}{2}+)$	$0(\frac{1}{2}+)$	Λ_α	1115	0	1.24	π^-p [6]	67	38	100
$Y^*_0(1815)$	$0(\frac{5}{2}+)$	$0(J\geq\frac{5}{2})$	Λ_α	1815	120	3.29	$\bar{K}N$ / $\Sigma\pi$	60 / <33	383 / 490	541 / 504
$Y^*_0(1405)$	$0(\frac{1}{2}-)$	0(?)	Λ_β	1405	50[5]	1.97	$\Sigma\pi$ / {$\Lambda 2\pi$}	100	69($\Sigma^+\pi^-$) / 10($\Lambda\pi^+\pi^-$)	144 / 69
$Y^*_0(1520)$	$0(\frac{3}{2}-)$		Λ_γ	1520	16	2.31	{$\Sigma\pi$(d-wave), $\bar{K}N$(d-wave), $\Lambda 2\pi$}	55 / 30 / 15	194($\Sigma^0\pi^0$) / 88($\bar{K}^0p$) / 125($\Lambda\pi^+\pi^-$)	267 / 244 / 253
$\Sigma\begin{cases}\Sigma^+\\\Sigma^0\\\Sigma^-\end{cases}$	$(\frac{1}{2}+)$		Σ_α	1189 / 1193 / 1197.4	0 / 0 / 0	1.42 / 1.42 / 1.42	$n\pi^+$[6] / $\Lambda\gamma$ / $n\pi\gamma$	50 / 100 / 100	110 / 76 / 117	185 / 74 / 192
$Y^*_1(1385)$	$1(\frac{3}{2}+)$	$1(J\geq\frac{3}{2})$	Σ_δ	1385	50	1.92	$\Lambda\pi$ / $\Sigma\pi$	98 / 4±4	135($\Lambda\pi^0$) / 49($\Sigma^-\pi^+$)	210 / 119
$Y^*_1(1660)$	$1(\frac{3}{2}-)$	$1(\frac{3}{2})$	Σ_γ	1660	40	2.76	$\bar{K}N$ / $\Lambda\pi$ / $\Sigma\pi$ / $\Sigma\pi\pi$ / $\Lambda\pi\pi$	~10 / 25 / 30 / 20 / 15	225 / 335 / 410 / 200 / 275	406 / 386 / 441 / 328 / 394
$\Xi\begin{cases}\Xi^0\\\Xi^-\end{cases}$	$\frac{1}{2}(\frac{1}{2}+)$	$\frac{1}{2}(\frac{1}{2}?)$	Ξ_α	? / 1321	0	1.72	$\Lambda\pi^0$[6] / $\Lambda\pi^-$	– / –	66	138
$\Xi^*(1530)$	$\frac{1}{2}(\frac{3}{2}+)$	$\frac{1}{2}(\frac{3}{2}+)$	Ξ_δ	1530	<7	2.34	$\Xi\pi$	100	74($\Xi^-\pi^0$)	148

Figure 3   Table of resonances, 1963.

was not identified until 1963 (6) as the vector meson which would complete the vector nonet.

1963   "Discoveries," both real and imagined, of resonances were pouring in. We decided to try to limit our tables of resonances to those states that had at least a 90% probability of surviving. For the 1963 edition this entailed critical reading and comparison of about 300 papers and reports, a job of such magnitude that we persuaded Pierre Bastien and Janos Kirz to join us. From the 300 documents, we selected and tabulated 18 resonant states (see Figure 3), all of which have survived to date, except the claim called the "kappa" or $K^*(725)$. Then we also had to establish a best value for the mass, mean life, and typically two partial-decay modes of these 18 resonances plus nine additional stable or metastable particles: e, μ, π, η, K, N, Λ, Σ, and Ξ. We realized that the 1963 edition should be the last attempted without the help of a computer.

We started to prepare data and reference cards, and a program called TABSU to process them. We are still using both the cards and TABSU. Angela Barbaro-Galtieri joined our team. Over the years Galtieri and Rosenfeld have remained permanently. The other three (Barkas, Bastien, and Kirz) have dispersed (Barkas died of a heart attack in 1969) and have been replaced by a present staff of seven physicists (only two of whom serve full time).

In 1963 we printed two different wallet cards to make room for the enlarged Table of Hadrons.

Matts Roos's "Tables of Elementary Particles and Resonant States" appeared in the *Reviews of Modern Physics* (7) and illustrated that it was no longer possible for a single person to compile data critically. Roos's Table IIa listed 19 baryon resonances; his Table IIb listed 30 unstable mesons; (instead of the eight accepted by the Berkeley group, and displayed in Figure 3). Of these 21 additional meson candidates, none has survived, although one wonders how many experimental searches have been made for them, and theories based on them. In December 1963, Roos revised his tables (7a, 7b) and eliminated most of the dubious entries.

Roos had planned to update his tables for the October 1964 issue of the *Reviews of Modern Physics,* but when we sent him a draft of our computerized 1964 edition, he suggested that we combine our efforts and consider publishing, for the first time, in the *Reviews of Modern Physics.*

1964   Our first published edition (8) of "Data on Elementary Particles and Resonant States" was written by five Berkeley authors (A. H. Rosenfeld, A. Barbaro-Galtieri, W. H. Barkas, P. L. Bastien, and J. Kirz) and Matts Roos from Copenhagen. There were 27 pages in all (compared with about 200 today). In addition to 9 pages of text and tables, there were 11 pages of computer printout, 4 pages of ideograms warning of all inconsistent data, and 4 pages of auxiliary tables such as Clebsch-Gordan coefficients.

In terms of physics, the Table of Stable Particles was complete! It had the $J^P = 0^-$ meson octet and the $\frac{1}{2}^+$ baryon octet, as well as the Ω^- (1673 MeV).

If we look in the Data Card Listings we find data for only the first three Ω^- events: two from the Brookhaven bubble chamber (9) and one found in emulsion ten years previously in 1954 (10, 11), although not finally interpreted until 1973 (11).

In 1964 there were 3 wallet cards, including a Meson Table with 12 unstable mesons, and a Baryon Table with 16 unstable baryons. All 28 states have survived to date. (Sociological note: the wallet cards could now be requested in two sizes—small ones to fit American wallets, larger, more readable ones which would still fit in European wallets.)

1965   The resonance population was exploding, as illustrated by Figures 4 and 5. Our 1965 edition (12) added the D, f′, and K*(1420) mesons; and it noted that the $J^P = 0^+$, 1^-, and $2+$ meson *nonets* were now complete. We also added three baryon resonances. This was the last edition carrying the name of Walter Barkas, who left Berkeley in 1965 to head the high-energy physics group at the University of California at Riverside. We were dismayed by news of his death in 1969.

1967   There were now eight authors (13) from LRL, CERN, and Yale, and the publication had grown to 51 pages. We added the $\delta(965)$, $\pi_V(1050)$, g, S, T, U, Q, L to the Meson Table, as well as several baryon resonances.

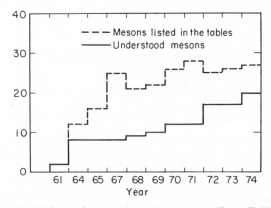

Figure 4   Growth of information on Meson Resonances. From F. Wagner, Ref. 22, pp. II–27.

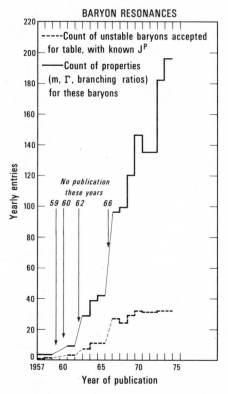

Figure 5   Growth of information on Baryons. The higher curve is a count of those numbers on the Baryon Table (any given year) which were well enough known to be accompanied by an error or an upper limit. The dip in 1971 occurs because we moved some upper limits into footnotes, and forgot to include footnotes in the counts for the figure.

We commenced printing our reports simultaneously at Berkeley and CERN. In an appendix on the "elusive" H and κ mesons, we gained some interesting experience in the hunting and wounding of rumored persistent states. The κ was at that time taken seriously because there were by then five impressive claims for a narrow $K\pi$ mass peak at 725 MeV. We compiled and histogrammed (by computer) 60,000 new $K\pi$ events, found no substantial further evidence, and went on to ask how frequently such striking statistical fluctuations should be expected at some given mass in the $K\pi$ system. (At that time about 2 million bubble chamber events were being measured annually, and about a thousand physicists were hunting through 10,000 to 20,000 mass histograms each year, in search of striking features, real or imagined.) We concluded that the five κ claims were just about what we should expect (14).

At the following year's Philadelphia meson spectroscopy conference, Rosenfeld summarized the Appendix and pointed out that, at the present rate of data-processing, once attention had been drawn to the κ it *should* be observed at the 3σ level about once a year. He concluded,

> So far there have been five impressive bumps, and a number of smaller claims, all near 725 MeV in the $K\pi$ spectrum. But every time that an experiment has been run to confirm a specific claim, it has failed. At the same time (if it is a large statistics experiment) the second experiment often reports the κ in a different channel. The result is that the κ's existence currently depends entirely on three uncorroborated sightings. This strikes me as similar to the problem with flying saucers; they keep appearing, but always in different places, and can never be reproduced. It seems to me that we are going to have to learn not to be fooled by frequent large fluctuations.

In the next talk, Prof. Aihud Pevsner said that he had intended to present new 3σ evidence for the κ, but had changed his mind. Since then the annual rate of reported sightings of the κ has slowly decreased to zero.

September 1967 The wallet cards grew too bulky; we started issuing sheets of paper instead (with print so small we can no longer read it!). These we hopefully called "wallet sheets." In Russia, where there are two words for "sheet," one for a small sheet of paper and one for a bed sheet, physicists chose the latter, suggesting perhaps that we had not sufficiently reduced the bulkiness problem.

1968 On our tenth birthday we printed our first data booklet, and adopted a name for ourselves: "Particle Data Group" (15).

1969 Our annual publication was entitled *Review of Particle Properties* and our eight authors represented four institutions (16).

So far our work had concentrated on particle *properties*, but in 1969 both the CERN-HERA group, led by D.R.O. Morrison, and the PDG started to publish cross sections of particle reactions. Good communication between the two groups has avoided undue duplication of effort. The Berkeley compilations include all information, so that two thirds of our reports are filled with data on angular distributions ($d\sigma/dt$) and on polarization measurements (P). CERN excludes $d\sigma/dt$ and P, i.e. compiles only "channel cross sections." For a given beam particle, e.g. π^+, the CERN reports are then much thinner, but also more up to date. We at Berkeley have considered eliminating our channel cross-section compilations, to avoid any overlap at all, but this turns out to be impractical: Our publication dates never agree with those of HERA, and so our lists of experiments covered are never the same; yet for completeness and ease in

566   ROSENFELD

checking the data it is desirable to have channel cross sections easily related to the other data from an experiment. Hence we accept some minor redundancy, and include channel cross sections.

In addition, PDG started the UCRL-20030 series of partial-wave amplitude compilations. Both the HERA reports and the PDG Cross-Section series are printed and distributed by both CERN and Berkeley.

1970 Our 11 authors represented 7 institutions (3 in Europe), and we specialized into three compiling teams: 1. Stable Particles and Weak Interactions; 2. Mesons; 3. Baryons. We started to alternate in publication between *Physics Letters B* (17a) and the *Reviews of Modern Physics* (17b).

1971 We established for the first time a system for having authors whose data we cite in the *Review* verify our encoding and interpretation of their data. This procedure consists in sending to each author a listing of the subsection of the *Review* in which his entry appears, with his "line" printed in boldface. This is done shortly before the publication of each edition. We have found that we do indeed get responses to our letters, with most of the errors found so far fortunately minor in nature. Our feeling is that this process has contributed both to our accuracy and to good relations with our "constituency."

1972 We routinely present a figure in our annual review, plotting the rate of production of data on particle properties (see Figure 6). In 1972 we noticed that the number of results published had dropped from 600 in 1970 to 415 in 1971. The reader in 1975 may attribute this to the decline of budgets for particle physics; but, as we discuss in the text of our 1972 edition (18), this seemed not to be the answer, since the rate of publication of experimental papers was holding nearly constant at 400 per year. More likely the explanation is that spectroscopy was becoming a less dominant subfield of particle physics. Other more technical reasons are discussed in the 1972 text.

PDG itself went along with the shift in interest to higher energies. In collaboration with Harvard, Stony Brook, Vanderbilt, and Yale, we printed LBL-80, a 757-page compilation of inclusive reactions. We soon regretted the large printing bill, and resolved to make better use of microfiche and magnetic tape, as is discussed in the section on 1973.

By 1972 the CERN-HERA Group had produced 14 compilations of cross sections and PDG had produced 10 (see Figure 14 in Section 5). At Berkeley we were dissatisfied with our original computerized compilation system, so we decided to stop compiling data temporarily and to integrate our document library and data bank into one system, described in Section 6. The challenge of designing, programming, and operating the combined document/data system whetted the interest of Geoffrey Fox of the California Institute of Technology, who set up a Pasadena office of PDG (linked by computer terminal), hired, jointly with LBL, a research associate to head our programming effort. While Berkeley has thus far concentrated on the document portion of the system, Cal Tech has been primarily responsible for data.

1973   The *Review of Particle Properties* had grown to 175 pages (19). On the cross-section side of our activities, we experimented with a new format when we produced Lawrence Berkeley Laboratory Report LBL-53, a compilation of πN interactions. We printed only the tables and figures that would normally have occupied the first half of the book, and reproduced references and backup material on microfiche contained in an

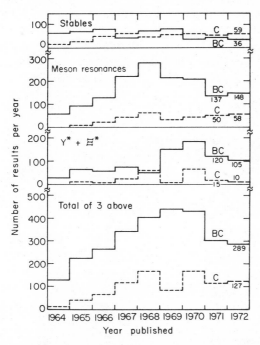

Statistics on the rate of production of data on particle properties. From the top to the bottom, the number of results per year are presented for stable particles, meson resonances, Y^*+ Ξ^*'s, and the total of the three above. The full lines correspond to bubble-chamber techniques (BC) and interrupted lines correspond to counters, spark chambers and spectrometers (C). Note that the figure omits N^* and Z^*, the field where counters have overwhelmed bubble chambers, because we punch mainly results from partial wave analyses instead of primary data.

Figure 6   Rise and fall of rate of publication of data on particle properties. From *Review of Particle Properties,* 1973, Ref. 19.

envelope at the back of the book. This approach has been so well accepted that we intend in future to further reduce the printed fraction of our reports to mere surveys, with the complete data and references on microfiche.

1974   Our 1974 edition of the *Review of Particle Properties* (20) had grown to 200 pages, with 13 authors based at 5 institutions (3 in Europe). In the fall of 1974 the J/Ψ meson was independently discovered at Brookhaven and at Stanford. This welcome surprise changed the whole mood of particle physicists, and almost at once set off a flurry of experimental papers and a flood of theoretical speculations. (See Figure 7)

1975   Except for a special J/Ψ section (21) we decided to postpone the complete *Review,* and will now publish it biennially. This will leave us more manpower to produce indexes and compilations of cross sections. In 1975 we hope to issue LBL-90, a complete, continuously updated index of all documents in the field of experimental particle physics.

Historical Note   We have shown, in Figure 6, a decrease in the rate of production of individual results bearing on particle properties. We must contrast this with the fact that the quality of each result from a large modern experiment is higher than ever, so that our knowledge of hadron spectroscopy continues to advance steadily. To discuss this point further

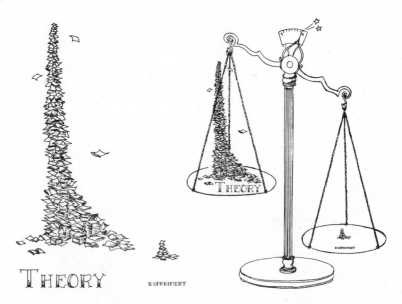

Figure 7   Two cartoons on the discovery of the J/Ψ mesons. (*a*) J. D. Jackson compares the theoretical flood started by the experimental trickle. (*b*) Bob Gould and Roy Schwitters show the importance of a weighted average.

Baryon Table
April 1972

[See notes on N's and Δ's, on possible Z* 's, and on Y* 's at the beginning of those sections in the
Baryon Data Card Listings; also see notes on individual resonances in the Baryon Data Card Listings.]

Particle[a]	I (J^P) ⊢—⊣ estab.	π or K Beam T(GeV) p(GeV/c) σ = $4\pi\lambda^2$ (mb)	Mass M^b (MeV)	Full Width Γ^b (MeV)	M^2 ± ΓM^c (GeV^2)	Partial decay mode Mode	Fraction %	p or p_{max}^d (MeV/c)
p	1/2($1/2^+$)		938.3		0.880	See Stable Particle Table		
n			939.6		0.883			
N' (1470)	1/2($1/2^+$) P'_{11}	T=0.53πp p=0.66 σ=27.8	1435 to 1505	165 to 400	2.16 ±0.34	Nπ Nππ	60 40	420 368
N'(1520)	1/2($3/2^-$) D'_{13}	T=0.61 p=0.74 σ=23.5	1510 to 1540	105 to 150	2.31 ±0.18	Nπ Nππ [Δ(1236)π]^e [dominant]^e Nη	50 50 ~0.6	456 410 224
N'(1535)	1/2($1/2^-$) S'_{11}	T=0.64 p=0.76 σ=22.5	1500 to 1600	50 to 160	2.36 ±0.18	Nπ Nη^n Nππ^n	35 55 ~10	467 182 422
N'(1670)[i]	1/2($5/2^-$) D'_{15}	T=0.87 p=1.00 σ=15.6	1655 to 1680	105 to 175	2.79 ±0.24	Nπ Nππ [Δ(1236)π]^e [44]^e ΛK Nη	40 60 <.3 <1^j	560 525 357 200 368
N'(1688)[i]	1/2($5/2^+$) F'_{15}	T=0.90 p=1.03 σ=14.9	1680 to 1692	105 to 180	2.85 ±0.21	Nπ Nππ [Δ(1236)+π]^e [26]^e ΛK Nη	60 40 <.2 <.5^j	572 538 371 231 388
N''(1700)[i]	1/2($1/2^-$) S''_{11}	T=0.92 p=1.05 σ=14.3	1665 to 1765	100 to 400	2.89 ±0.42	Nπ ΛK Nη	-65 5	580 250 340
N''(1780)[i]	1/2($1/2^+$) P''_{11}	T=1.07 p=1.20 σ=12.2	1650 to 1860	50 to 450	3.17 ±0.51	Nπ ΛK Nη	30 ~7 ~10^j	633 353 476
N(1860)	1/2($3/2^+$) P_{13}	T=1.22 p=1.36 σ=10.4	1770 to 1900	180 to 330	3.46 ±0.57	Nπ Nππ ΛK Nη	25 ~5$_j$ ~4^j	685 657 437 545
N(2190)	1/2($7/2^-$) G_{17}	T=1.94 p=2.07 σ=6.21	2000 to 2260	270 to 325	4.80 ±0.67	Nπ Nππ	25	888 868
N(2220)	1/2($9/2^+$) H_{19}	T=2.00 p=2.14 σ=5.97	2200 to 2245	260 to 330	4.93 ±0.65	Nπ Nππ	15	905 887
N(2650)	1/2($?^-$)	T=3.12 p=3.26 σ=3.67	2650	360	7.02 ±0.95	Nπ Nππ	(J+1/2)x =0.45^f	1154 1140
N(3030)	1/2(?)	T=4.27 p=4.41 σ=2.62	3030	400	9.18 ±1.21	Nπ Nππ	(J+1/2)x =0.05^f	1366 1354
Δ(1236)^m	3/2($3/2^+$) P'_{33}	T=0.195 (++) p=0.304 σ=91.8	1230 to 1236	110 to 122	1.53 ±0.14	Nπ Nπ$^+$π$^-$ Nγ	99.4 0 ~0.6	231 90 262
		Pole Position^m:	1211	± i50				
Δ(1650)	3/2($1/2^-$) S_{31}	T=0.83 p=0.96 σ=16.4	1615 to 1695	130 to 200	2.72 ±0.28	Nπ Nππ	28 72	547 511
Δ(1670)	3/2($3/2^-$) D_{33}	T=0.87 p=1.00 σ=15.6	1650 to 1720	175 to 300	2.79 ±0.40	Nπ Nππ	15	560 525
Δ(1890)	3/2($5/2^+$) F_{35}	T=1.28 p=1.42 σ=9.88	1840 to 1920	135 to 350	3.57 ±0.49	Nπ Nππ	17	704 677
Δ(1910)	3/2($1/2^+$) P_{31}	T=1.33 p=1.46 σ=9.54	1780 to 1935	230 to 420	3.65 ±0.62	Nπ Nπππ	25	716 691
Δ(1950)	3/2($7/2^+$) F_{37}	T=1.41 p=1.54 σ=8.90	1930 to 1980	140 to 220	3.80 ±0.39	Nπ Δ(1236)π ΣK Σ(1385)K	45 ≈50 ~2 1.4	741 571 460 232

Figure 8 Beginning of the 1972 Baryon Table. The complete table with footnotes
occupied four journal pages.

we return to Figure 4, which is taken from Fritz Wagner's survey of mesons at the 1974 *London Conference* (22). The dashed line, now hovering around 25 mesons, represents the count of mesons whose existence is well enough established to admit them to the Meson Table. It would seem to tell us that progress has stopped. However, the monotonically rising solid line is the count of mesons for which all quantum numbers are established. This is the sort of information that can be used to stimulate or test theories; it still has an encouraging trend.

Figure 5 shows similar numbers for baryons. Since many baryon resonances are discovered by partial-wave analysis (so their quantum numbers are known by definition), we must find some other test to be equivalent to the solid curve for mesons. We chose simply to count and plot the total number of properties that have been established for all states. This point is now further explained in the next two paragraphs (23).

Figure 8 shows the first few N* entries in the 1972 Baryon Table. The information came mainly from "elastic" partial-wave analyses, i.e. from studies of the reaction $\pi N \rightarrow \pi N$. Other two-body channels like ΛK or $N\eta$ were known only as upper limits, and the $N\pi\pi$ cross sections were not independent information; they were simply obtained by subtracting the sum of the two-body reactions from the total cross section. Finally, the photon couplings were so poorly known that they were not even tabulated.

We contrast the 1972 Baryon Table with the 1974 Table shown in Figure 9. The flood of new data has two explanations: First, experimental technique continues to improve. The advent of polarized photon beams enables far better photoproduction partial-wave analyses. Very fast measurement of bubble chamber events permits large statistics, and therefore permits partial-wave analyses of the reaction $\pi N \rightarrow N\pi\pi$ via the "isobar model": $N\pi\pi = \Delta\pi + N\rho + N\varepsilon$. Second, a great amount of programming and computer time has gone into the partial-wave programs themselves. The useful increase in the available data is even more impressive than suggested by the increased length of the table. Each partial-wave analysis contributes a complex amplitude **A**, (or coupling constant g), whereas we tabulate only a branching fraction proportional to $|\mathbf{A}|^2$. But any theory must predict correctly the complex amplitude, including its sign. Seen from this point of view the steeply rising upper line of Figure 5 shows that we are certainly still in a population explosion of facts about baryons.

$\longrightarrow$

Figure 9 Beginning of the 1974 Baryon Table. The complete table had grown since 1972 by one journal page.

Baryon Table

April 1974

The following short list gives the status of all the Baryon States in the Data Card Listings. In addition to the status, the name, the nominal mass, and the quantum numbers (where known) are shown. States with three or four star status are included in the main Baryon Table, the others have been omitted because the evidence for the existence of the effect and/or for its interpretation as a resonance is open to considerable question.

N(939) P11 ****	Δ(1232) P33 ****	Λ(1116) P01 ****	Σ(1193) P11 ****	Ξ(1317) P11 ****	
N(1470) P11 ****	Δ(1650) S31 ****	Λ(1330) Dead	Σ(1385) P13 ****	Ξ(1530) P13 ****	
N(1520) D13 ****	Δ(1670) D33 ***	Λ(1405) S01 ****	Σ(1440) Dead	Ξ(1630) **	
N(1535) S11 ****	Δ(1690) P33 *	Λ(1520) D03 ****	Σ(1480) *	Ξ(1820) ***	
N(1670) D15 ****	Δ(1890) F35 ***	Λ(1670) S01 ****	Σ(1620) S11 **	Ξ(1940) ***	
N(1688) F15 ****	Δ(1900) S31 *-	Λ(1690) D03 ****	Σ(1620) P11 **	Ξ(2030) **	
N(1700) S11 ****	Δ(1910) P31 ***	Λ(1750) P01 **	Σ(1670) D13 ****	Ξ(2250) *	
N(1700) D13 **	Δ(1950) F37 ****	Λ(1815) F05 ****	Σ(1670) **	Ξ(2500) **	
N(1780) P11 ***	Δ(1960) D35 **	Λ(1860) P03 **	Σ(1690) **		
N(1810) P13 ***	Δ(2160) **	Λ(1870) S01 **	Σ(1750) S11 ***	Ω(1672) P03 ****	
N(1990) F17 **	Δ(2420) H311 ***	Λ(2010) D03 **	Σ(1765) D15 ****		
N(2000) F15 **	Δ(2850) ***	Λ(2020) F07 **	Σ(1840) P13 *		
N(2040) D13 **	Δ(3230) ***	Λ(2100) G07 ****	Σ(1880) P11 **		
N(2100) S11 *		Λ(2110) ?05 *	Σ(1915) F15 ****		
N(2100) D15 *	Z0(1780) P01 *	Λ(2350) ****	Σ(1940) D13 ***		
N(2190) G17 ***	Z0(1865) D03 *	Λ(2585) ***	Σ(2000) S11 *		
N(2220) H19 ***	Z1(1900) P13 *		Σ(2030) F17 ****		
N(2650) ***	Z1(2150) *		Σ(2070) F15 *		
N(3030) ***	Z1(2500) *		Σ(2080) P13 **		
N(3245) *			Σ(2100) G17 **		
N(3690) *			Σ(2250) ****		
N(3755) *			Σ(2455) ***		
			Σ(2620) ***		
			Σ(3000) **		

**** Good, clear, and unmistakable.   *** Good, but in need of clarification or not absolutely certain.
** Needs confirmation.   * Weak.

[See notes on N's and Δ's, on possible Z*'s, and on Y*'s and Ξ*'s at the beginning of those sections in the Baryon Data Card Listings; also see notes on individual resonances in the Baryon Data Card Listings.]

Particle[a]	$I\,(J^P)$[a] —— estab.	π or K Beam[b] $\frac{}{P_{beam}}$(GeV/c) $\sigma = 4\pi\lambda^2$ (mb)	Mass M[c] (MeV)	Full Width Γ[c] (MeV)	M^2 $\pm\Gamma M$[b] (GeV²)	Partial decay mode		
						Mode	Fraction %	p or p_{max}[d] (MeV/c)
p	$1/2(1/2^+)$		938.3		0.880	See Stable Particle Table		
n			939.6		0.883			
N(1470)	$1/2(1/2^+)$ P'11	p = 0.66 σ = 27.8	1400 to 1470	165 to 300 (250)	2.16 ±0.37	Nπ	~60	420
						Nη	~9	d
						Nππ	~35	368
						[Nε	~6]e	d
						[Δπ	~16]e	177
						[Nρ	<5]e	d
						pγ^f	0.12-0.32	435
						nγ^f	<0.24	435
N(1520)	$1/2(3/2^-)$ D'13	p = 0.74 σ = 23.5	1510 to 1540	105 to 150 (125)	2.31 ±0.19	Nπ	~55	456
						Nππ	~45	410
						[Nε	<5]e	d
						[Nρ	~18]e	d
						[Δπ	~23]e	228
						Nη	0.2-1.4	d
						pγ^f	0.57-0.79	471
						nγ^f	0.36-0.56	471
N(1535)	$1/2(1/2^-)$ S'11	p = 0.76 σ = 22.5	1500 to 1600	50 to 160 (100)	2.36 ±0.15	Nπ	~35	467
						Nη	~55	182
						Nππ	~10	422
						[Nρ	~2]e	d
						[Nε	~2]e	d
						pγ^f	0.04-0.32	481
						nγ	0.04-0.12	481
N(1670)[g]	$1/2(5/2^-)$D'15	p = 1.00 σ = 15.6	1670 to 1685	115 to 175 (140)	2.79 ±0.23	Nπ	~40	560
						Nππ	~60	525
						[Δπ	~50]e	360
						ΛK	<1	200
						Nη	<1^h	368
						pγ^f	<0.04	572
						nγ^f	0.01-0.13	572

3 CRITICAL EVALUATION
AND REVIEW OF DATA

The text of each edition of the *Review of Particle Properties* is considerably longer than this entire article, and it is pointless to repeat that text here—we give only a brief survey of our procedures, successes, and failures. A major revision of our normal text will appear in the April 1976 edition of the *Reviews of Modern Physics,* to which we refer the reader for details.

3.1   The "Listings" of Our Data Base

3.1.1   DESCRIPTION   The lengths of the various sections of our 1974 Review are as follows:

	Pages
Text	16
Tables of Particle Properties	14
Miscellaneous Tables	20
Listings (of data cards, reference cards, minireviews, ideograms, etc)	154

Thus, although our tables are our trademark (particularly because of our annual worldwide distribution of 8500 copies of the *Particle Properties Data Booklet),* the body of our *Review* and of our whole operation is the Listings, which are, literally, a photographed paste-up of the computer listing of the data, our evaluations, and the references in our data base.

All substantial claims for resonances (and all data) go into the Listings, but only half these resonances satisfy our admission standards for the Tables. We discuss the Tables in Section 4.

Figure 10 is a reproduction of the illustrative key to the Listings, for a hypothetical XX meson of mass 1200 MeV, reported in five papers by "Merrill, Lynch, Pierce, Fenner and Beane, and Smith." We use this key as a discussion outline for our data base.

At the top is a large arrow under the name, pointing to the note "Omitted from Table." A similar arrow would appear on the Meson Table, referring the reader to this point in the Listings.

The first block of data cards stores information on mass; hence each card has an M at the left. Five mass measurements have been encoded, but two of them have been suppressed from the weighted average. The measurement of LYNCH 67 has an "L" punched in column 8, which tells our program to exclude the data from averaging and to print it between parentheses. This "L" is keyed to a comment card immediately below stating "questionable background subtraction."

The measurement of FENNER 69 is also unusable for a weighted average because it has no quoted error. The three acceptable masses have been averaged to give a weighted mean mass and error. The value of χ^2 for the three input data was less than 2.0, so no scale factor was introduced (see Section 3.3). Note, however, that when it comes to averaging the *width* in the next block of data, χ^2 was too high. An ideogram has been plotted to show the reader that the two good measurements have a χ^2 of 1.8 where 1.0 was expected, and to warn him that the weighted average is a poor representation of the situation, even after its error has been scaled up by $(1.8)^{1/2}$, as explained in Section 3.3.

3.1.2 CONSTRAINED FITS   A more interesting part of our data fitting is introduced in the Partial Decay Mode block below the ideogram. Here we postulate that XX(1200) has only two partial modes whose branching fractions, P's, must add up to 100%, i.e. are subject to one constraint: $P_1 + P_2 = 1.0$. From two partial modes, various experiments can (and usually do) report the three different branching ratios listed: $R1 = P1$, $R2 = P2$, $R3 = P1/P2$. From three modes they can report 21 ratios; and for n partial modes, there are $(2^n - 1)(2^{n-1} - 1)$ ratios! Thus our constrained-fit program must have versatile input and calculational logic.

The Listings give the values of the fitted R's and P's (and partial widths Γ_i if relevant), together with an error matrix.

Sometimes the constrained fit is a standard minimization process, but generally the input R's are somewhat inconsistent, so we introduce several scale factors. Details of our constrained fits are in our January 1970 Text (17a), Section IX B.

3.1.3 OTHER FEATURES   Not shown in Figure 10 are two important features of the Listings:

1. *Underground Comments*   If we wish to record a private comment on some data, we can punch a "$" in column 7 and the comment will not appear in the publicly printed part of the listings. There are hundreds of these comments. Some are quite technical (e.g. "Error below limit"— explained in Section 3.3.1); many refer to our own doubts; some refer to still unresolved correspondence between us and the experimenters; some are comments that could perfectly well be public, but that are not very important and so are suppressed to reduce the length of the Listings. Each author of the *Review of Particle Properties* has access to the underground section of the Listings. This feature helps us greatly to understand one another's decisions; thus a curious American physicist can

Illustrative Key

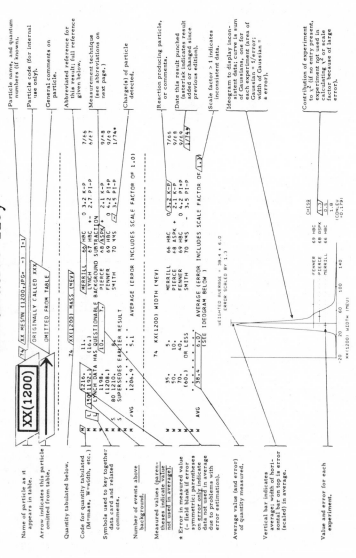

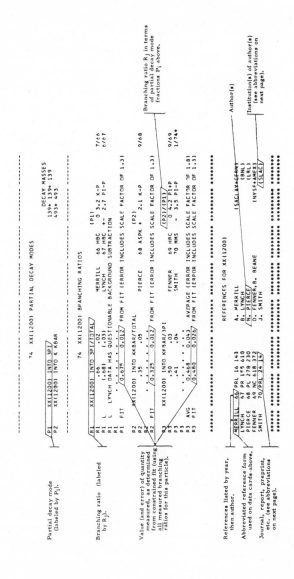

Figure 10 Illustrative key to the Data Card Listings.

phone Berkeley and get a satisfactory explanation of why the Meson Team in Europe excluded certain data.

2. *Minireviews* Of over 154 pages of "Listings," 50 pages are actually not listings, but figures, or *Minireviews*. Figure 11a,b is a typical minireview that we discuss in Section 3.3.1.

3.2 Selection of Documents and Data

3.2.1 PRELIMINARY VS FINAL RESULTS   In high-energy physics, the results

Note on K^* Masses and Mass Differences

This note is divided into two discussions:

I. Basic difficulties in determining the mass difference because of interferences and biases.

II. Impossibly small errors reported by some experiments. We have increased some errors that violate the laws of statistics, and scaled up some errors that are inconsistent; but we warn that most of the errors in our data cards are inconsistent. One cannot then obtain a K^* mass difference by calculating an average mass for K^{*0} and for $K^{*\pm}$ and just subtracting the two.

I. BASIC DIFFICULTIES

There are two difficulties in measuring a mass difference $m(K^{*0}) - m(K^{*\pm})$ of ~ 7 MeV when the half-width $\Gamma/2$ of the K^* is 25 MeV:

1) Interference between the resonant amplitude and background can in general shift the peak in the mass spectrum by some fraction of $\Gamma/2$.

2) The two charges of K^* have different topologies; this introduces differences in the measuring and fitting of the events, which can also produce mass shifts.

Some reactions (symmetric under reflection of I_z) are immune to the first difficulty. Thus compare the mass of K^{*0} produced in

$$\pi^- p \to \Lambda \pi^- K^+$$

with the mass of K^{*+} in the I_z-reflected reaction

$$\pi^+ n \to \Lambda \pi^+ K^0 \ .$$

The final-state amplitudes of each will contain not only the $|K^*\rangle$ with Ispin 1/2, but also an interfering $I = 3/2$ P-wave, which we can call $|K^*_{3/2}\rangle$. But I_z symmetry forces $\langle \pi^- p | \Lambda K^{*0}\rangle$ to equal $\langle \pi^+ n | \Lambda K^{*+}\rangle$; and similarly for the two $K^*_{3/2}$ amplitudes, so that the shifting of the K^* peak is the same in both reactions. Nobody has published a mass difference exploiting this fact.

II. IMPOSSIBLY SMALL ERRORS

Consider a sample of N events, with their invariant masses m distributed as an S-wave Breit-Wigner resonance:

i.e., $$P(\epsilon - \epsilon_R) = \frac{1/\pi}{(\epsilon - \epsilon_R)^2 + 1} \ , \tag{1}$$

where $\epsilon = \dfrac{m}{\Gamma/2}$, $\epsilon_R = \dfrac{m_R}{\Gamma/2}$. One can then show that the minimum possible error on the determination of the central value ϵ_R is

$$\delta_{min}(\epsilon_R) = \pm\sqrt{\frac{2}{N}} \ , \ \text{i.e.,} \ \delta_{min}(m_R) = \pm\sqrt{\frac{2}{N}} \ \frac{\Gamma}{2} \ . \tag{2}$$

This lower limit assumes no background events. In practice, with background, the error will be larger, by another factor $\alpha \approx \sqrt{2}$.

We illustrate errors with small and large backgrounds with a table summarizing the recent experiment ("Unsplit K^*'s") by DAVIS 69.

(a)

Figure 11   Minireview of statistical tests on reported statistical errors and on the K* mass difference. From *Review of Particle Properties,* August 1970, Ref. 17b.

of an important experiment appear many times. The succession may be roughly as follows:

1. Preliminary Results:
   (a) abstracts submitted to conferences and to the *Bulletin of the APS*,
   (b) papers contributed to conferences,
   (c) numbers quoted in conference talks and proceedings.
2. Results, in preprints, "letter" journals, articles, and, finally, theses.

For many years we have kept track of how a quantity q and its error

---

Mass Errors δm of DAVIS 69

● Sample with 5% background/signal at peak.

Events: $K^*(892)$, 10 700 events in resonance, $\frac{\Gamma}{2} \approx 25$ MeV.

Lower limit from Eq. (2), $\delta_{min}(m) = \sqrt{\frac{2}{N}} \frac{\Gamma}{2} = \pm 0.35$ MeV.

Their likelihood fit yields two sorts of errors:

$\underline{\delta_1(m)}$. Ignore correlations, i.e., keep all the parameters (background, width, etc.) fixed, vary m only:

$$\delta_1(m) = \pm 0.41, \quad \delta_1(m)/\delta_{min}(m) = 1.16.$$

$\underline{\delta_2(m)}$. As m is varied, reoptimize other parameters.

$$\delta_2(m) = \pm 0.53, \quad \delta_2(m)/\delta_{min}(m) = 1.5.$$

DAVIS 69 mention $\delta_2 = 0.53$, but to hedge against systematic effects, they quote $\delta_3 = 2$ MeV. We punch 2 MeV.

● Sample with 50% background/signal at peak.

Events: $K^*(1420)$, 2200 events in resonance, $\frac{\Gamma}{2} = 50$ MeV.

$$\delta_{min}(m) = 1.6 \text{ MeV},$$

$$\underline{\delta_1(m)} = \pm 2.2 \text{ MeV}, \quad \delta_1(m)/\delta_{min}(m) = 1.4,$$

$$\underline{\delta_2(m)} = \pm 2.6 \text{ MeV}, \quad \delta_2(m)/\delta_{min}(m) = 1.6.$$

Width Errors $\delta\Gamma$ of DAVIS 69

For width, the equivalent of Eq. (2) is $\delta_{min}(\Gamma) = \pm \sqrt{\frac{8/3}{N}} \frac{\Gamma}{2} = 1.15 \delta_{min}(m)$.

For convenience we neglect the factor 1.15 and use $\delta_{min}(\Gamma) \approx \delta_{min}(m)$.

● 5% background, $K^*(892)$:

$$\delta_2(\Gamma) = \pm 1.6 \text{ MeV}, \quad \delta_2(\Gamma)/\delta_{min}(m) = \frac{1.6}{0.35} = 4.6.$$

● 50% background, $K^*(1420)$:

$$\delta_2(\Gamma) = \pm 10 \text{ MeV}, \quad \delta_2(\Gamma)/\delta_{min}(m) = \frac{10}{1.6} = 6.25.$$

We note that $\delta_2(m)/\delta_{min}(m)$ does not change rapidly with background (1.5 at 5%, 1.6 at 50%) and hence conclude that it is hard to believe an error with $\delta_2/\delta_{min} < 1.4 = \sqrt{2}$. We chose $\sqrt{2}$ because together with Eq. (2) it leads to the simple "realistic" result

$$\delta(m) > \sqrt{2}\sqrt{\frac{2}{N}} \frac{\Gamma}{2} = \frac{\Gamma}{\sqrt{N}} . \tag{3}$$

We conclude that for a sensitive subtraction like $m(K^{*0}) - m(K^{*\pm})$, the experiments as listed are useless, and we must either re-evaluate them all or concentrate on those two experiments that explicitly quote a mass difference. For a detailed discussion of how he have actually treated those experiments, we refer to the January 1970 edition of this note.

The table above also allows us to concoct a criterion for "realistic" errors in width $\delta(\Gamma)$. We average the 5% and 50% background results (to give $\delta(\Gamma)/\delta_{min}(m)$ of 5 to 6) and express the result in terms of Γ, in the style of Eq. (3). We then get the "realistic" test for widths:

$$\delta\Gamma > 4 \frac{\Gamma}{\sqrt{N}}$$

(b)

δq tend to "home in" on its final published value. We find that abstracts may be essential as a way to organize a conference, but as sources of numerical data they are useless. Even the central value q is still several standard deviations (σ) "from home" and δq is too small. Accordingly, we ignore abstracts.

By the time a paper is submitted to a conference, the central value q is typically within σ of "home," but δq is still too small. Accordingly, from major conference contributions we encode the results, but punch them with a "suppress" character in column 8; the result is printed in parentheses and not averaged. As we encode the conference data, we send a copy to the experimenters saying that we will remove the parentheses if they wish to certify that the result is not likely to change. Usually they choose to let the parentheses stand. This is consistent with an informal poll conducted by Geoffrey Manning on behalf of the Organizing Committee for the 1974 *International Conference on High Energy Physics* held in London. It had been suggested that all contributed papers be photographed and made available to preprint libraries on microfiche, but the poll indicated that most of the contributors would prefer not to have their "preliminary" contributions broadly distributed. These remarks on the tentativeness of results that are analyzed at the last minute before a conference deadline date are not intended to detract from the importance of conferences or their printed Proceedings—which constitute the best way to keep up with a field. But we must remember that the results are subject to change. We therefore never accept conference papers unless they are published, or (in very special cases) verified in writing.

As for preprints, we keep the results in parentheses until the authors inform us that they are final and that the preprint is accepted for publication.

When a Letter or paper is published, it replaces the preprint entry. If results in papers are just repeated or slightly reworked numbers from a Letter, we put the earlier entry in parentheses.

In summary then, we see that hours of reading, correspondence with authors, and punching of temporary cards stand behind the single final data card, which represents the end of the chain of talks, preprint, letter, and paper (and perhaps thesis and paper on phenomenology) resulting from one experiment.

3.2.2 SELECTION OF DATA   Roughly 40% of our encoded results are not used for averaging; they are set off in parentheses. Our explanation is then often given in a comment below the data. If no reason is given, it is one of the following:

1. Preliminary data, as from a conference.
2. Error is not stated.
3. Poor quality data, e.g. bad signal-to-noise ratio.
4. The result has been used for some years, but by present experimental standards is of poor quality and may bias the average (see Section 3.4).
5. The result involves an assumption we do not wish to incorporate.
6. The result is inconsistent with others, e.g. because of different methods employed, rendering averaging meaningless. See, for example, the entries listed under the S(1930) meson, which contain both a wide peak formed in $\bar{p}p$ interactions and narrow peaks reported in production experiments.

3.3   Statistical Procedures

Our statistical procedures are set forth in Section IX of our January 1970 text (17a).

3.3.1   CHECKING AND ENLARGING ERRORS ON A SINGLE RESULT   We would have liked at this point to refer to a photograph of our listings for the K*(892) meson, but the listings are already photo-reduced so as to be just readable on an $8\frac{1}{2}'' \times 11''$ page, and further reduction for this book is impractical. However, as a written example, in our 1974 *Review* (20) one can find among ~ 100 K* data cards six with comments like "Mass error enlarged by us to $\Gamma/\sqrt{(N)}$" or "Width error enlarged by us to $4 \Gamma/\sqrt{(N)}$." The reasoning is explained in Figure 11a,b, which is a reproduction of the relevant minireviews. We show there, in the subsection labeled *"II   Impossibly Small Errors,"* that we must check quoted errors (particularly if a background subtraction has been performed) and must sometimes enlarge these quoted errors.

In the case of the mass difference between the neutral and charged K*, we have enlarged the quoted errors by appalling factors (see our January 1970 Minireview (17a)):

	δm(MeV)	
Reference	Paper quotes	We enlarge to
BARASH 71	± 4.1	± 6
FICENEC1 68	± 1.55	± 5
FICENEC2 68	± 0.9	± 5

In the case of Barash, the enlargement was justified purely by statistical considerations. In the case of the two Ficenec experiments, statistical considerations raised the error only to 2.5 MeV each. But then the two results were inconsistent, with a χ^2 of 4.5 where 1.0 was expected, so we introduced a further scale factor $S = \sqrt{(4.5)}$. See the discussion following equation 1 below.

The above discussion illustrates the close scrutiny we give to input errors, and how we sometimes reinterpret them drastically.

We return to the minireview in Figure 11 and note two further historical points:

1. One cannot in general obtain the K* mass difference merely by subtracting the mean mass of K* (charged) from that of K* (neutral). (When necessary we have protected the reader from that mistake.)
2. It suggests a reaction that is immune to the interference difficulty. [In fact, this suggestion was picked up by Hoch (24) and by Taft (25), but neither experiment yielded a result that could compete with the 1971 high statistics, low background experiment of Aguilar (26)].

3.3.2 UNCONSTRAINED AVERAGES—SCALE FACTORS For resonance masses, for example, where there are seldom constraints, we calculate a weighted average of the measurements x_i,

$$\bar{x} \pm \delta\bar{x} = \sum w_i x_i / \sum w_i \pm (\sum w_i)^{-1/2}; \quad w_i = 1/(\delta x_i)^2, \qquad 1.$$

where the sums run over N experiments. We also calculate χ^2 and compare it with its expectation value of $N-1$. If $\chi^2 > N-1$, we increase the error $\delta\bar{x}$ in equation 1 by a factor $S = [\chi^2/(N-1)]^{1/2}$.

It is easy to design statistical tests for determining whether one experiment (or a group of experiments) is consistent with the other experiments. However, statistics do not tell us who is wrong in case of contradictions. When $S \gg 1$, one can conclude either that:

1. some (or all) experiments are wrong, or
2. some (or all) experiments have underestimated their errors, or
3. the experiments do not measure the same quantity (systematic errors).

We do our best to resolve these cases. If we cannot, we assume that all experimentalists underestimated their errors by the same scale factor. If we scale up all input errors by this factor, χ^2 returns to $N-1$, and of course the output error scales up by the same factor.

Empirically, if we sum, say, the χ^2 values for all our weighted averages of meson masses, we find a total χ^2 which is larger than expected by a factor of 1.34^2, i.e. the typical reported error seems to be underestimated and should be raised by 4/3. For more details see Section 3.6.

If all the experiments have errors of about the same size, the above procedure is straightforward. If, however, there are both precise and imprecise (large error) measurements of a particular quantity, one must be very careful not to permit the imprecise ones to "dilute" the scale factor. See our January 1970 edition (17a) for the prescription we use to handle this effect.

We often plot an ideogram to guide the reader in deciding which data he might reject before making his own selected average.

3.3.3 CONSTRAINED FITS   In Section 3.1 near the end of our discussion of the Illustrative Key to the Listings, we gave an example of the need for a constrained fit. Details of our procedure are found in our 1970 text (17a).

3.4   Reliability of Our Averages

We have recently begun a retrospective study of the reliability of our procedures. Figure 12 is a historical plot of the masses that have appeared in our tables. (Our 1976 Review will also include such plots for mean lives and partial decay fractions.) If all errors were estimated correctly, two thirds of the oldest masses should fall within one old standard deviation of the band representing the present mass, or stated graphically, two thirds of the oldest error bars (and somewhat more of the more recent ones) should touch the bands. A study of the plots leads to the conclusions below:

We have always taken the masses of the electron and the proton directly from the current review of the fundamental physical constants starting in 1957 (27–30), so we can first glance at the plots for these particles to see the record of the results of these reviewers. We see in fact that the reliability is poor—apparently the experimenters have consistently underestimated their errors, and the reviewers have not found a mechanism such as our Scale Factor to take this into account.

Our record, by contrast, seems a bit too good, at least as far as masses are concerned. We are well aware that there has just been a change in the mean life of the K meson, and there is a chronic problem with the Λ lifetime, called the "East-West Effect," so we ourselves are eager to see the upcoming mean-life plots. Also note that our averages would fluctuate more if the input data had not already been adjusted by us—40% of the data have been excluded from the averages, errors have been enlarged if they were impossibly small, old experiments have been excluded, etc. In summary, systematic errors in compilations are bound to cause error underestimates; nevertheless our crude but thorough methods, and our Scale Factor, seem to increase our error estimates by just the right amount.

3.5   A Case Study: The Mass of ω(783)

Periodically, we attempt to sift through and "clean up" our data cards: We throw out old experiments with very large errors, which may not contribute much to the average, and may possibly introduce biases; we check, even more carefully than was done originally, for problems with

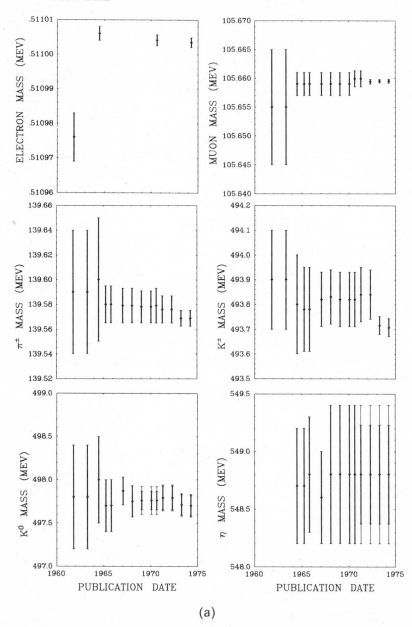

Figure 12 History of masses in Particle Properties Tables. The error bars indicate ±1 standard deviation; the outer bars include the Scale Factor (if any), while the

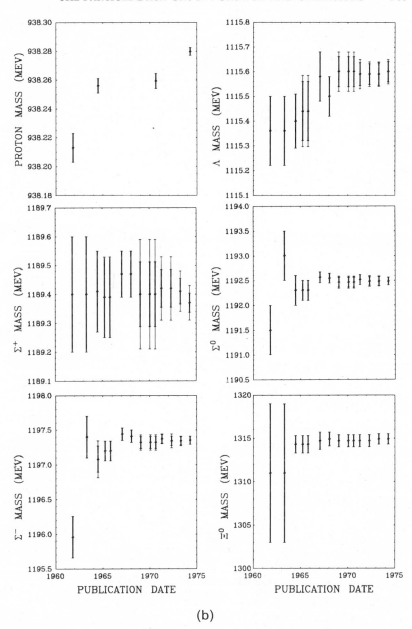

(b)

(heavy) inner bars do not. In the case where the error is from an "educated guess," rather than from a calculation, the inner bar is absent.

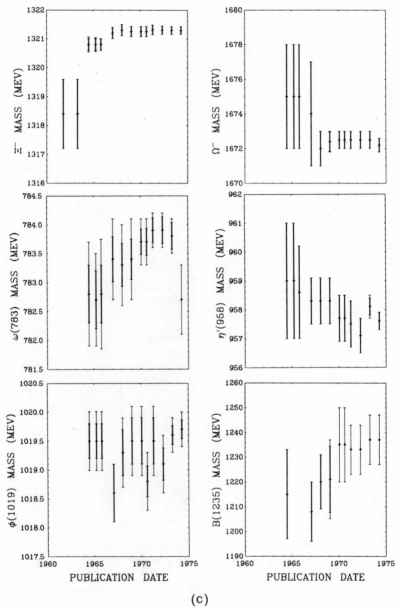

(c)

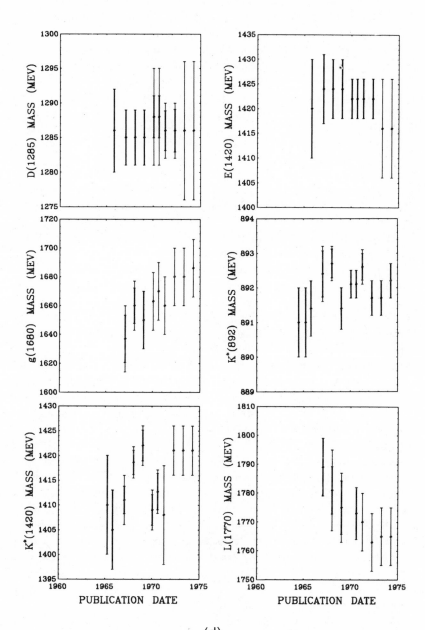

(d)

individual experiments; we look for possible mistakes we may have made, etc.

For the 1974 edition, this procedure was applied to the $\omega(783)$ mass, and as a result the mass shifted downward by ~ 4 old standard deviations. (Actually one new measurement was also added in 1974, but its effect was small.) Part of this shift was due to reincluding in the average an experiment which studied ω's in the reaction $\bar{p}p \rightarrow \omega K_1^0 K_1^0$; three $\bar{p}p$ experiments had all reported values of the ω mass significantly lower than the average of all other experiments. Thus, although nothing was ever found wrong with these studies, they had previously been averaged separately and not used in obtaining the value quoted in the Tables. In the cleanup process, it was decided that this was unfounded. Two of the $\bar{p}p$ experiments were excluded because they did not properly account for the end of phase space, but the best experiment was reintroduced.

Most of the remainder of the post-cleanup mass *shift* resulted from leaving out six measurements [by Danburg (31)] resulting from one experiment (at six different beam momenta). Each of these six values had been quoted only to the nearest MeV, and all such "rough" measurements (rough when compared to the error in the 1974 mean of ~ 0.3 MeV) were removed from the average. In this case, all six values were on the high side, probably because of correlations, and leaving them out produced a noticeable effect. It should be noted that our averaging procedures do not take correlations into account, even in the moderately rare cases when they are given. If we did account for correlations, the six values undoubtedly would not have originally pulled the average as far as they did. Put more bluntly, we should never have averaged all of the six Danburg measurements without folding in a common systematic error.

Assuming, as we do, that this cleanup has resulted in a more accurate value of the ω mass, an interesting picture arises as to the variation with time of our reported value of this quantity. Examining Figure 12, we see that the tabulated ω mass rose gradually from its 1964 value of 782.8 ± 0.9 MeV to its 1973 value of 783.8 ± 0.3 MeV, and seemed to be leveling out. But after the cleanup, the value dropped back to 782.7 ± 0.3 MeV. The conclusion to be drawn from this is that even when reported values seem to be homing in on some particular figure, it may not be the correct value, because mistakes (in experiment or in compilation) may simply lie uncorrected for years.

There is a second conclusion to the 4σ change in the ω mass resulting from this cleanup. It took Vladimir Chaloupka 16 hours to reread and reorganize the data cards. In this relatively brief time he detected and corrected two misjudgments on our part, and excluded 11 other experiments which he felt were doing more harm than good. Chaloupka was so alarmed by the magnitude of the mass change that he has undertaken a major cleanup of all the meson data, and has decided to report many errors as conservative "educated guesses" until the cleanup is complete.

3.6 Empirical Likelihood Distributions of Particle Data

In Section 3.3, equation 1, we stated that we calculate our weighted means $\bar{x}$ and their errors $\delta\bar{x}$ using the method of least squares, i.e. using weights $\omega_i = 1/\delta\chi_i^2$. This method is predicated on the assumption that the likelihood distribution for the individual input quantities x_i is normally distributed, with standard deviation δx_i.

For years we have discussed studying the *empirical* distribution of our residuals and using it to replace our assumed *normal* distributions. We always postponed the project on the grounds that (a) we probably did not have enough data, (b) the gains to be achieved were small, and (c) we would do far better to use the same manpower in regularly cleaning out input decks of old or unreliable data (which are the main contributions to the excess tails) and in studying the experiments that fall in the tails.

Despite this experimentalist-oriented skepticism, our statistical expert, Matts Roos, and his students have now shown (31a) that the distributions of our residuals h_i all seem to have a similar shape, which is well described by a Student distribution $S_{10}(h/1.11)$.

A Student distribution of N degrees of freedom, $S_N(t)$, has the property that $S_1(t)$ is a Breit-Wigner function, and S_∞ is a standard Normal distribution. Thus S_{10} has a slightly wider tail than our postulated Normal, S_∞. Specifically S_∞ and S_{10} are very similar out to 2σ, but at 3σ, S_{10} is larger by a factor of 3. Roos et al even show that S_{10} is a theoretically plausible distribution. The remaining broadening factor of 1.11 in the argument of S_{10} is purely empirical.

Roos et al can now determine by what new scale factor, S_{new}, the least-squares errors should really be extended to cover a 68% confidence interval. They find that even for averages with a good $\chi^2(<N-1)$, S_{new} should be 1.11 (where we have so far used 1.0). But for large $\chi^2/(N-1)$, e.g. 4.0, they find that S_{new} should be only about 4/3, where we have used 2.0; i.e. they find that our large scale factors are slightly too conservative. Doubtless we shall incorporate many of these ideas of Roos et al into our 1976 Review.

Meson Table

April 1974

In addition to the entries in the Meson Table, the Meson Data Card Listings contain all
substantial claims for meson resonances. See Contents of Meson Data Card Listings[1]

Quantities in italics have changed by more than one (old) standard deviation since April 1973.

Name $\frac{-\overline{}\ \begin{array}{c|c}I & 0 \\ \hline \omega/\phi & \pi \\ \hline \eta & \rho\end{array}}{+\ \ }$ $I^G(J^P)C_n$ estab.	Mass M (MeV)	Full Width Γ (MeV)	M^2 $\pm\Gamma M^{(a)}$ $(GeV)^2$	Partial decay mode		
				Mode	Fraction (%) [Upper limits are 1σ (%)]	p or $P_{max}^{(b)}$ (MeV/c)
$\pi^\pm(140)$ $1^-(0^-)+$ $\pi^0(135)$	139.57 134.96	0.0 7.8 eV $\pm.9$ eV	0.019483 0.018217	See Stable Particle Table		
$\eta(549)$ $0^+(0^-)+$	548.8 ±0.6	2.63 keV $\pm.58$ keV	0.301 $\pm.000$	All neutral $\pi^+\pi^-\pi^0 + \pi^+\pi^-\gamma$	71 29	See Stable Particle Table
ε $0^+(0^+)+$ Existence of pole not established. See note on $\pi\pi$ S wave¶.	$\lesssim 700^{(c)}$	$\gtrsim 600^{(c)}$		$\pi\pi$		
$\rho(770)$ $1^+(1^-)-$	$770_\S$ $\pm10^\S$	$150_\S$ $\pm10^\S$	0.593 $\pm.116$	$\pi\pi$ e^+e^- $\mu^+\mu^-$ For upper limits, see footnote (e)	≈ 100 $0.0043\pm.0005$ (d) $0.0067\pm.0012$ (d)	359 385 370
$\omega(783)$ $0^-(1^-)-$	$782.7_\S$ $\pm0.6^\S$	10.0 $\pm.4$	0.613 $\pm.008$	$\pi^+\pi^-\pi^0$ $\pi^+\pi^-$ $\pi^0\gamma$ e^+e^- For upper limits, see footnote (g)	90.0 ± 0.6 S=1.2* 1.3 ± 0.3 S=1.5* 8.7 ± 0.5 $0.0076\pm.0017$ S=1.9*	327 366 380 391
$\eta'(958)$ or X^0 $0^+(\ ^-)+$ J=0 or 2	957.6 ±0.3	< 1	0.917 $<.001$	$\eta\pi\pi$ $\rho^0\gamma$ $\gamma\gamma$ For upper limits, see footnote (h)	70.6 ± 2.5 S=1.4* 27.4 ± 2.2 S=1.6* 1.9 ± 0.3	234 458 479
$\delta(970)$ $1^-(0^+)+$ Possibly a virtual bound state of the I = 1 $K\bar{K}$ system¶.	$976_\S$ $\pm10^\S$	$50_\S$ $\pm20^\S$	0.953 $\pm.049$	$\eta\pi$ $\rho\pi$	seen < 25	315 139
$S^*(993)$ $0^+(0^+)+$ See notes on $\pi\pi$ and $K\bar{K}$ S wave¶.	$\sim 993^{(c)}$ ±5	$40^{(c)}$ ±8	0.986 $\pm.040$	$K\bar{K}$ $\pi\pi$	near threshold	479
$\phi(1019)$ $0^-(1^-)-$	1019.7 ±0.3 S=1.9*	4.2 $\pm.2$	1.040 $\pm.004$	K^+K^- $K_L K_S$ $\pi^+\pi^-\pi^0$ (incl. $\rho\pi$) $\eta\gamma$ e^+e^- $\mu^+\mu^-$ For upper limits, see footnote (i)	46.6 ± 2.5 S=1.6* 34.6 ± 2.2 S=1.6* 15.8 ± 1.5 S=1.2* 3.0 ± 1.1 S=1.6* $.032\pm.002$ S=1.4* $.025\pm.003$	127 111 462 362 510 499
$A_1(1100)$ $1^-(1^+)+$ Broad enhancement in the $J^P=1^+$ $\rho\pi$ partial wave; not a Breit-Wigner resonance¶.	~ 1100	~ 300	1.21 $\pm.33$	$\rho\pi$	~ 100	253
$B(1235)$ $1^+(1^+)-$	$1237_\S$ $\pm10^\S$	$120_\S$ $\pm20^\S$	1.53 $\pm.12$	$\omega\pi$ [D/S amplitude ratio = .24±.06] For upper limits, see footnote (j)	only mode seen	352
$f(1270)$ $0^+(2^+)+$	$1270_\S$ $\pm10^\S$	$170_\S$ $\pm30^\S$	1.61 $\pm.22$	$\pi\pi$ $2\pi^+2\pi^-$ $K\bar{K}$ For upper limits, see footnote (f)	$83\pm5\S$ $4\pm1\S$ 4 ± 3 S=1.5*	619 556 394

4   THE TABLES—HOW TO CHOOSE A RESONANCE

The need for careful standards for admission of Particles and their Properties to the Tables is indicated by the following counts:

	On Table	In Listings
Mesons (Figure 13)	28	54
Baryons (Figure 9)	48	90

Thus we see that a newly discovered resonance has only a 50% probability of being admitted to the Tables and surviving there. Although at times we have been unpopular because we did not admit the experimenter's latest discovery, we are aware that every newly accepted resonance provokes many theoretical papers, and we hate to see theorists trying to bring order into a poorly evaluated mixture of real and imagined states. Hence we try to tabulate only those peaks or resonances that have a large chance of survival, and specifically we strive for criteria which, in retrospect, will work 90% of the time.

4.1   Selection Criteria

In our 1974 text, page ix, we list five "Criteria for Resonances," mainly based on partial-wave analyses. If such analyses are not available we have to be more tolerant of accepting a peak in a cross section, provided that it is experimentally reliable (because of its high statistical significance or because it is seen in several different reactions).

In the past, most of the convincing peaks have later turned out to be associated with one *or more* resonances. Thus, before 1966 we tabulated the πp peaks at 1520 and 1688 MeV as single resonances (with a warning "Assignment still not final"). Later polarization data and partial-wave analyses uncovered seven resonances in these two peaks.

We must also be careful to distinguish between "explaining" a peak, and explaining it away. Thus in 1965 we tabulated the N* (1480, $J^P = 1/2^+$), but added the warning "existence not yet definitely established" for two independent reasons:
1. The peak was seen as a missing-mass or $p\pi\pi$ peak produced peri-

---

←

Figure 13   Beginning of 1974 Meson Table. The complete table with footnotes occupies four journal pages.

pherally in pp collision (Cocconi 32) but could be partially "explained" by the "Deck" Effect (or in more modern usage, by "double-Reggepole exchange").

2. Some partial-wave analyses, particularly by Roper, called for a resonance, but two papers quoted among our 1965 references warn that the onset of a rapid rise in the inelastic reaction $\pi N \rightarrow N\pi\pi$ could simulate resonant behavior.

By 1966, partial-wave analyses clearly supported the Roper resonance, and it was evident that the feared alternative mechanisms should be taken as compatible, not competing, explanations. This problem of dual explanation remains today in the question of how to interpret the A_1, Q, and A_3 mesons.

Thus, we enter experimentally convincing peaks into the Tables of Particle Properties unless there is contradictory information, and we can expect that most of these peaks will eventually be confirmed as one or more resonances. But we can easily give examples of experimentally convincing peaks which in all likelihood have nothing to do with resonances: such are the K^+p and pp total cross-section bumps near 1.2 and 3 GeV/c, respectively, and the low-energy ABC and DEF bumps in the S-wave $\pi\pi$ system.

In conclusion, our criteria for including a resonance in the Tables are based much more on experience than on theoretical insight.

4.2 Reliability of Our Selections

In this section we ask how often we have been fooled into too hastily admitting a peak, and later having to banish it back to the Listings.

4.2.1 MESONS There are a total of six mesons which have appeared on the Meson Table, but were later removed. Five of these were narrow bumps: $\delta(962)$, $S(1930)$, $T(2200)$, $U(2360)$, and $\kappa(725)$. One was a broad bump: $H(990)$.

The narrow bumps all had widths comparable with experimental resolution, and were each "confirmed" subsequently. We discussed one of them, $\kappa(725)$, in Section 2 when we recalled our 1967 Appendix on the κ and H mesons. In that Appendix we stated that once attention was called to a narrow mass peak, it was bound to be confirmed frequently at the 3σ level. We should have realized this earlier. There have also been claims for the opposite effect: a narrow valley splitting the ρ meson, and the A2 meson. Fortunately we had learned enough to remain skeptical.

The $H(990)$ is a different and interesting case. It had been included as a possible bump in the Listings for some time. We moved it onto the

Table for one year, primarily because of a single unpublished paper (prepared hurriedly for a conference), which confirmed the peak clearly and which, alas, included two authors from the *Review of Particle Properties* (33). One of these same authors then proceeded to kill the H less than a year later at the 1968 *Philadelphia Conference on Experimental Meson Spectroscopy,* by explaining the signal as $\eta'(958) \to \pi^+ \pi^- \gamma$ misinterpreted as $H(990) \to \pi^+ \pi^- \pi^0$ (34).

4.2.2   BARYONS   With Baryons, our record is better and less interesting. Only two ever got admitted to the Table, complete with J^P assignments, but were later removed: these were an $N^*(1990, 7/2^+)$ and an $N^*(2040, 3/2^-)$, both removed in 1971.

5   INTERNATIONAL COLLABORATION

In our experience, transatlantic collaboration works surprisingly well, but only after people have worked together and grown to know one another well. Fortunately, Europeans frequently spend leaves and sabbaticals in Berkeley, and Americans like to work in Europe, particularly at CERN.

Our *Review of Particle Properties* is produced by three teams: one concerned with the Weak Interaction and the Stable Particles Table, one with Mesons, and one with Baryons. The Meson team is centered at CERN, and our main data-averaging program runs on the CERN computer. Also, two CERN physicists, J. Engler and F. Mönnig, are responsible for our Table of Atomic and Nuclear Properties of Materials.

Our reports and data booklet (and reports by CERN-HERA) are printed simultaneously at Berkeley and at CERN. Each laboratory prints about 3000 copies of the *Review of Particle Properties* and 2000 copies of other reports. We alternate in printing 8500 copies of the data booklet.

For purposes of distribution, we have divided the world into two parts in such a way as to minimize postage costs. Berkeley mails to the Americas, Australasia, and Japan. CERN supplies the rest of the world. Both Berkeley and CERN use self-liquidating mailing lists: in order to receive future mailings, each physicist must return an IBM card (or mailing label) enclosed with each report mailed.

Heartened by the success of the international particle properties collaboration, we are now undertaking two collaborations to compile cross-section data: one with Durham/Rutherford Laboratory in England, one with KEK in Japan. Both of these will involve exporting our computer system to overseas computers.

6   COMBINED DOCUMENT AND DATA SYSTEM

During 1971 we began to study how to organize the data and experimental documents for the whole of particle physics in somewhat the same comprehensive way that we had organized the subfield of particle properties.

At that time the functions of libraries and of abstracting services were not at all integrated with those of data centers, at least not in high energy physics. There were several current indexes available on magnetic tape (*Nuclear Science Abstracts, the DESY Index of High-Energy Physics,* and—for preprints—*Preprints in Particles and Fields*) but they were indexed only by subject, author, or institution. Thus they were not very useful to the physicist seeking, say, experimental papers, new or old, on a reaction like $\pi^+ p \rightarrow \rho^+ p$ between 5 and 10 GeV/c. In fact, in making our $\pi^+ p$ compilations, we neither read these index tapes nor took advantage of the bibliographic information available from them.

Not only did we ignore these indexes, but we neglected to produce good indexes ourselves; i.e. even after we assembled all the papers on, say, $\pi^+ p$ interactions ordered by beam momentum, it did not occur to us to publish a $\pi^+ p$ index, promptly and separately from our intended data compilation, which would typically take three years to produce, print, and distribute. Hence our indexes were chronically out of date. Finally there was no index of experiments (approved, scheduled, running, completed), and no system for linking these experiments with the publications of their results.

By 1972 we decided to ask *Nuclear Science Abstracts,* DESY, and *Preprints in Particles and Fields,* to add a few useful attributes (beam and momentum, accelerator, etc) to every paper or preprint indexed; for our part, we undertook to collect information on approved proposals. We further decided to link Document and Data information in one computerized system. Thus, we began development of a new complex of computerized data storage and retrieval programs, the Particle Physics Data System (PPDS).

At the heart of this complex is a general-purpose, data-base management system (BKY-DBMS), designed by us to be sufficiently flexible to handle the many kinds of information we need to encode, and to be transportable to machines other than the one on which it was written. (It is, in fact, general enough that several data compilation groups at the Laboratory in fields other than physics are using it, and it has been adopted by the Laboratory's Math and Computing Department as their main data-base management system.) PPDS makes use of BKY-DBMS for three files of information we maintain: the Document, Reaction-Data, and Particle Properties files.

The Document file contains, for all documents in experimental particle physics (journal articles, preprints, reports, theses, accelerator proposals, etc), both bibliographic information and experiment description (accelerator, detector, beam momentum, reactions and particles studied, and properties measured). Status information regarding the processing of each document is also kept in this file. The bibliographic information is input automatically at the preprint stage from tapes produced by the Stanford Linear Accelerator Center Library in conjunction with the weekly publication, *Preprints in Particles and Fields*. Documents not preprinted will have their bibliographic information extracted from *Nuclear Science Abstract* tapes. As soon as a bibliographic document entry is made in our system, we append the experiment description from a copy of the document. To this end we have been placed on the appropriate preprint mailing lists; we think that, as a result, we will be able to provide a detailed and up-to-the-minute literature-searching facility. For instance, it is within the scope of the system to search for all papers reporting density matrix elements for ρ mesons produced in the reaction $\pi^- p \rightarrow \rho^- p$, or reporting the mass of the A_2. In addition, we will periodically produce and distribute an *Index of Particle Physics Data*.

The other two files which comprise PPDS, the Reaction-Data file and the Particle Properties file, will contain the actual physics data extracted from the documents. Various collaborations around the world have agreed, and hopefully more will agree, to help us encode the data. We also plan a pilot project to see if authors themselves will assist us. It should be noted that data encoding is bound to lag behind the experiment description encoding done for the Document file, but the latter file will at least alert the user to the *existence* of the unencoded data. In fact, since the Document file contains information extracted from approved experiment proposals (as well as papers produced by experiments), a user of the system can even find out about yet-to-be-produced data. Then, once the physics data have been encoded, the user can, of course, make more detailed literature searches and, more importantly, can extract the data for plotting, fitting, etc. We ourselves will use the data files to provide printed "trend-of-the-date" guides, reviews, listings of selected data, tapes for distribution, and other services.

7 MANPOWER AND ECONOMICS

Authors The 1974 edition of *Review of Particle Properties* had 13 authors, mainly experimental physicists, from Berkeley, Brandeis, CERN,

For a list of older compilations, see any of the LBL reports below, through LBL-63.

CERN/HERA 69-1   (1969)
G. GIACOMELLI, P. PINI, S. STAGNI
A compilation of Pion-Nucleon
Scattering Data.

CERN/HERA 69-2   (April 1969)
B. SADOULET
Data Compilation of Antiproton-Proton
Reactions into Antihyperon-Hyperon.

CERN/HERA 69-3   (Dec. 1969)
G. GIACOMELLI
A Compilation of Total and Total
Elastic Cross Sections.

CERN/HERA 70-1   (June 1970)
P. SPILLANTINI, V. VALENTE
A Collection of Pion Photoproduction
Data.

CERN/HERA 71-1   (Sept. 1971)
L. D. JACOBS, M. ROOS, S. SANTIAGO
Selective Compilation of $\pi^- p \to \pi\pi N$
Events from Hydrogen Bubble Chambers.

CERN/HERA 72-1   (May 1972)
E. BRACCI, J. P. DROULEZ, E. FLAMINIO,
J. D. HANSEN, D. R. O. MORRISON
Compilation of Cross Sections.
I - π^- and π^+ Induced Reactions.

CERN/HERA 72-2   (Oct. 1972)
E. BRACCI, J. P. DROULEZ, E. FLAMINIO,
J. D. HANSEN, D. R. O. MORRISON
Compilation of Cross Sections.
II - K^- and K^+ Induced Reactions.

CERN/HERA 73-1   (June 1973)
E. BRACCI, J. P. DROULEZ, E. FLAMINIO,
J. D. HANSEN, D. R. O. MORRISON
Compilation of Cross Sections.
III - p and $\bar{p}$ Induced Reactions.

CERN/HERA 75-1   (March 1975)
U. CASADEI, G. GIACOMELLI,
P. LUGARESI-SERRA, G. MANDRIOLI,
A. M. ROSSI, F. VIAGGI
A Compilation of K^+N Cross Sections
Below 2 GeV/c.

CERN/HERA 75-2   (March 1975)
E. BRACCI, C. BURICHETTI, J. P. DROULEZ,
E. FLAMINIO, C. PRETI
Compilation of Differential Cross
Sections; π-induced Reactions.

*UCRL-20000 K^+N (Sept. 1969)
L. R. PRICE, N. BARASH-SCHMIDT,
O. BENARY, R. W. BLAND,
A. H. ROSENFELD, C. G. WOHL
A Compilation of K^+N Reactions.

UCRL-20000 YN   (Jan. 1970)
O. BENARY, N. BARASH-SCHMIDT,
L. R. PRICE, A. H. ROSENFELD
A Compilation of YN Reactions.

*UCRL-20001   (Jan. 1970)
G. C. FOX, C. QUIGG
Compilation of Elastic Data.

UCRL-20030 πN (Feb. 1970)
D. J. HERNDON, A . BARBARO-GALTIERI,
A. H. ROSENFELD
πN Partial-Wave Amplitudes.

UCRL-20000 NN   (Aug. 1970)
O. BENARY, L. R. PRICE, G. ALEXANDER
NN and ND Interactions
(Above 0.5 GeV/c) - A Compilation.

LBL-55   (March 1972)
F. UCHIYAMA, J. S. LOOS
$K_L^0 N$ Interactions - A Compilation.

LBL-58   (May 1972)
J. E. ENSTROM, T. FERBEL,
P. F. SLATTERY, B. L. WERNER,
Z. G. T GUIRAGOSSIAN, Y. SUMI,
T. YOSHIDA
$\bar{N}N$ and $\bar{N}D$ Interactions -
A Compilation.

LBL-80   (Aug. 1972)
M. E. LAW, J. KASMAN, R. S. PANVINI,
W. H. SIMS, T. LUDLAM
A Compilation of Data on Inclusive
Reactions.

LBL-63   (April 1973)
C. LOVELACE, S. ALMEHED, F. UCHIYAMA,
R. L. KELLY, V. P. HENRI
πN Two-Body Scattering Data:
I. A User's Guide to the
Lovelace-Almehed Data Tape.

LBL-53   (May 1973)
D. M. CHEW, V. P. HENRI, T. A. LASINSKI,
T. G. TRIPPE, F. UCHIYAMA,
F. C. WINKELMANN
$\pi^+ p$, $\pi^+ n$, and $\pi^+ d$ Interactions -
A Compilation: Part I and Part II.

------

* Out of print.

Availability of Copies

For:

North and South America, Australasia,
and the Far East,

write to:

Technical Information Divsion
Lawrence Berkeley Laboratory
Berkeley, California 94720
U.S.A.

For:

All other areas,

write to:

CERN Scientific Information Service
CH-1211 Geneva 23
Switzerland

DESY, and Helsinki. Each contributes typically a few month's work each year as a member of one of our three teams (Stable Particles, Mesons, Baryons).

Some 20 more names appear on our list of cross-section compilations (Figure 14), of whom perhaps 10 are still prepared to produce new compilations as soon as we commission our new Particle Physics Data System. Half a dozen more physicists are now working on reports in progress. Figure 14 also shows about 15 more HERA authors. So we can count about 40 authors currently active, worldwide.

Budgets   The Berkeley headquarters, which provides the liaison, meeting place, computer system, and facilites and staff for producing the reports, has a budget of $250,000. It is jointly supported by ERDA (US Energy Research and Development Administration, formerly AEC), by NSF (US National Science Foundation), and by NBS (US National Bureau of Standards). In addition, ERDA supports Prof. Geoffrey Fox and Paul Stevens at Cal Tech, who are just as active as our Berkeley staff. As mentioned earlier, CERN supports HERA, hosts our meson team (through Vladimir Chaloupka), and pays roughly one half of our annual printing and worldwide distribution costs of about $100,000. The British Science Research Council supports two positions at Durham under the leadership of Fred Gault. A small particle data group (a few full-time equivalents) is being formed at KEK in Japan. Thus the worldwide budget for compilation staff and expenses is about $400,000.

Comparison with Other Centers. Most fields of science have their data centers, although usually they are national rather than international. Thus in nuclear physics there are several centers in the US, whose budgets add to about 5% of the US support for research in that field. By contrast we are really tiny—our $250,000 budget represents one part in 600 of the US program in high energy physics. The worldwide compilation effort of $400,000 represents about one part in 1000 of the worldwide program.

7.1 Conclusions

We undertake three different sorts of compilation:

1. Our biannual critical *Review of Particle Properties*. This requires mainly the volunteer time of physicists who can be dispersed at many laboratories, plus a moderate centralized computer system that has not changed much in ten years. We are quite satisfied with our record and our level of support.

← *Figure 14*   Compilations since 1969. For a list of older compilations, see any of the reports listed.

2. Current indexes of all documents (including approved experimental proposals) in particle physics. This requires a new computer system, which is nearly ready. We could have done it in two years instead of four with financial support for a few more physicists and programmers.

3. Compilation of the data in every document. Our new data language is designed to make it possible for the authors themselves to help us compile their data. Even so, about one man-day per article will be needed for correspondence with authors or for encoding directly. With 500–800 experimental papers per year, we do not have enough manpower (even divided among 3 or 4 international centers) to do this steady-state job; in addition we must catch up with the backlog and finish some programming and full documentation.

This work must initially be done at Berkeley; then after a year or so the working system can be exported to centers in many countries. To accomplish this task we need contributions from a visiting physicist and programmer from each of several of those countries.

As we write this review we wonder if we have not been too modest in our requests for support, particularly from institutions other than the US government. An average experiment in high energy physics costs $3 million (and about 150 are completed each year); we feel that PDG is doing an effective job, but if it could spend, each year, one fifth of the typical experiment, it could provide broader and more timely services.

ACKNOWLEDGMENTS

Janos Kirz and Pierre Bastien were the major designers and producers of our system of data and reference cards, and of programs for producing our *Review of Particle Properties*. LeRoy Price pioneered the first computer system for our cross-section compilations, with the help of Odette Benary and Naomi Schmidt. Marge Hutchinson has maintained and improved it. David Richards has produced DBMS and the second generation of programs.

It has been agreeable and instructive to work with Alan Rittenberg as he has progressed from graduate student and η' expert to manager of PDG. He has contributed to both the text and figures of this article. The Stable Table could not appear without Tom Trippe and Naomi Schmidt; nor the Meson Table without Vladimir Chaloupka, Matts Roos, and Paul Söding. Robert Kelly manages to keep up with both the Baryon Table and with πN and KN scattering data, and still finds time to do partial-wave analyses. All of our publications depend heavily on Fumiyo Uchiyama and George Yost and, until recently, on Denyse Chew and Tom Lasinski.

We would be lost without Betty Armstrong, our combined editor, production manager, librarian, and secretary. It has always been a pleasure to cooperate with Dr. A. Günther, Director of Scientific Information Services at CERN. Prof. J. David Jackson has served for years as our dependable advisor and critic and star proofreader. It is a pleasure to acknowledge the continuous interest and many suggestions of Luis W. Alvarez and Frank T. Solmitz, and to thank Alvarez for his careful reading of this article.

Most impressive is the remarkable contribution of Angela Barbaro-Galtieri, who for ten years simultaneously led two of our three teams (Stable Particles and Baryons). She herself must have written or checked half the cards in our decks—who else could accomplish that in her "spare time"?

I have really enjoyed all the hours of work (even the long hours after midnight) with every one of our past and present authors. May our future associations continue to be as pleasant and productive as in the past.

Literature Cited

1. Gell-Mann, M., Rosenfeld, A. H. 1957. Hyperons and heavy mesons, *Ann. Rev. Nucl. Sci.* 7:407
2. Alston, M. H. et al 1961. *Phys. Rev. Lett.* 5:520
2a. Anderson, J. A. et al 1961. *Phys. Rev. Lett.* 6:365; Stonehill, D. et al 6:624; Erwin, A. R. et al 6:628; Maglic, B. C. et al 7:178; Alston, M. H. et al 6:698; Bastien, P. et al 6:705; Ferro-Luzzi, M. et al 1962. 8:25; Chamberlain, O. et al 1962. *Phys. Rev.* 125:1696
3. Gell-Mann, M. 1961. *The eightfold way.* Synchrontron Lab., Rep. CTSL-20 Calif. Inst. Technol., Pasadena, Calif. Unpublished
3a. Pevsner, A. et al 1961. *Phys. Rev. Lett.* 7:421
4. Bastien, P. L. et al 1962. *Phys. Rev. Lett.* 8:114
5. Bertanza, L. et al 1962. *Phys. Rev. Lett.* 9:180
6. Schlein, P., Slater, W. E., Smith, L. T., Stork, D. H., Ticho, H. K. 1963. *Phys. Rev. Lett.* 10:368; Connolly, P. L. et al 1963. *Phys. Rev. Lett.* 10:371
7. Roos, M. 1963. *Rev. Mod. Phys.* 35:314

7a. Roos, M. 1964. *Nucl. Phys.* 52:1–24
7b. Roos, M. 1963. *Phys. Lett.* 8:1
8. Rosenfeld, A. H. et al 1964. *Rev. Mod. Phys.* 36:977
9. Barnes, V. E. 1964. *Phys. Rev. Lett.* 12:204
10. Eisenberg, Y. 1954. *Phys. Rev.* 96:541
11. Alvarez, L. W. 1973. *Phys. Rev. D* 8:702
12. Rosenfeld, A. H. et al 1965. *Rev. Mod. Phys.* 37:633–51
13. Rosenfeld, A. H. et al 1967. *Rev. Mod. Phys.* 39:1–51
14. Rosenfeld, A. H. 1968. *Meson Spectroscopy,* p. 467. New York: Benjamin
15. Rosenfeld, A. H. et al 1968. *Rev. Mod. Phys.* 40:77–128
16. Barash-Schmidt, N. et al 1969. *Rev. Mod. Phys.* 41:109–92
17a. Roos, M. et al 1970. *Phys. Lett. B* 33:1
17b. Barbaro-Galtieri, A. 1970. *Rev. Mod. Phys.* 42:87
18. Söding, P. et al 1972. *Phys. Lett. B* 39:1
19. Lasinski, T. A. et al 1973. *Rev. Mod. Phys.* 45:1
20. Chaloupka, V. et al 1974. *Phys. Lett.*

B 50:1
21. Chaloupka, V., Bricman, C., Barbaro-Galtieri, A., Chew, D. M., Kelly, R. L., Lasinski, T. A., Rittenberg, A., Rosenfeld, A. H., Trippe, T. G., Uchiyama, F., Yost, G. P. 1974. *Phys. Lett. B* 50:535
22. Wagner, F. 1974. *Proc. Int. Conf. High Energy Phys., 17th, London, England, July 1–10, 1974*
23. Rosenfeld, A. H. 1973. *Proc. Sch. Subnucl. Phys., Erice, Italy, July 10–24, 1973*; also Almost Everything About Baryon Resonances. LBL-2098. Lawrence Berkeley Lab.
24. Hoch, P. L. 1972. A Measurement of the K* (890) Mass Difference in the Reactions $\pi^+ d \to (p)\Lambda K^{*+}$ and $\pi^- p \to \Lambda K^{*\circ \prime\prime}$. LBL-1053. Lawrence Berkeley Lab.
25. Taft, H. D. 1971. Private communication.
26. Aguilar-Benitez, M., Chung, S. V., Eisner, R. L., Samios, N. P. 1972. *Phys. Rev. D* 6:29
27. Cohen, E. R., Crowe, K. M., DuMond, J. W. M. 1957. *Nuovo Cimento* 5:541. Used in our 1958–1963 editions for masses of electron and proton
28. Cohen, E. R., DuMond, J. W. M. 1965. *Rev. Mod. Phys.* 37:537. Used for 1964–1971
29. Taylor, B. N., Parker, W. H., Langenberg, D. N. 1969. *Rev. Mod. Phys.* 41:375. Used for Aug. 1971 through 1973
30. Cohen, E. R., Taylor, B. N. 1973. *J. Phys. Chem. Ref. Data* 2:663. Used since 1974
31. Danburg, G. et al 1970. *Phys. Rev. D* 5:2564
31a. Roos, M., Hietanen, M., Luoma, J. 1975. *Phys. Fenn.* 10:21
32. Cocconi, G. et al 1964. *Phys. Lett.* 8:134
33. Chadwick, G. B., Guiragossian, Z. G. T., Pickup, E., Barbaro-Galtieri, A., Matison, M. J., Rittenberg, A. 1967. *Stanford Linear Accelerator Center Report*. SLAC-PUB-347
34. Barbaro-Galtieri, A., Söding, P. 1968. *Meson Spectroscopy*, 137. New York: Benjamin

AUTHOR INDEX

CUMULATIVE INDEXES

CONTRIBUTING AUTHORS VOLUMES 16-25

CHAPTER TITLES VOLUMES 16-25